Technologies for Energy, Agriculture, and Healthcare

International Conference on Technologies for Energy, Agriculture, and Healthcare (ICTEAH 2024)

Technologies for Energy, Agriculture, and Healthcare

International Conference on Technologies for Energy, Agriculture, and Healthcare (ICTEAH 2024)

Editors

Dr. Shailesh R. Nikam

Dr. Makarand G. Kulkarni

Dr. Vaibhav S. Narwane

Dr. Ninad D. Mehendale

Dr. Nilkamal P. More

CRC Press is an imprint of the
Taylor & Francis Group, an **informa** business

First edition published 2024
by CRC Press
4 Park Square, Milton Park, Abingdon, Oxon, OX14 4RN

and by CRC Press
2385 NW Executive Center Drive, Suite 320, Boca Raton FL 33431

© 2024 selection and editorial matter, Dr. Shailesh R. Nikam, Dr. Makarand G. Kulkarni, Dr. Vaibhav S. Narwane, Dr. Ninad D. Mehendale and Dr. Nilkamal P. More; individual chapters, the contributors

CRC Press is an imprint of Informa UK Limited

The right Dr. Shailesh R. Nikam, Dr. Makarand G. Kulkarni, Dr. Vaibhav S. Narwane, Dr. Ninad D. Mehendale and Dr. Nilkamal P. More to be identified as the authors of the editorial material, and of the authors for their individual chapters, has been asserted in accordance with sections 77 and 78 of the Copyright, Designs and Patents Act 1988.

All rights reserved. No part of this book may be reprinted or reproduced or utilised in any form or by any electronic, mechanical, or other means, now known or hereafter invented, including photocopying and recording, or in any information storage or retrieval system, without permission in writing from the publishers.

For permission to photocopy or use material electronically from this work, access www.copyright.com or contact the Copyright Clearance Center, Inc. (CCC), 222 Rosewood Drive, Danvers, MA 01923, 978-750-8400. For works that are not available on CCC please contact mpkbookspermissions@tandf.co.uk

Trademark notice: Product or corporate names may be trademarks or registered trademarks, and are used only for identification and explanation without intent to infringe.

British Library Cataloguing-in-Publication Data
A catalogue record for this book is available from the British Library

ISBN: 978-1-032-98025-6 (hbk)
ISBN: 978-1-032-98028-7 (pbk)
ISBN: 978-1-003-59670-7 (ebk)

DOI: 10.1201/9781003596707

Typeset in Times LT Std
by Aditiinfosystems

Technologies for Energy, Agriculture, and Healthcare – Shailesh Nikam et al. (eds)
© 2024 Taylor & Francis Group, London, ISBN 978-1-032-98028-7

CONTENTS

Technologies for Energy, Agriculture, and Healthcare – Shailesh Nikam et al. (eds)
© 2024 Taylor & Francis Group, London, ISBN 978-1-032-98028-7

List of Figures

Technologies for Energy, Agriculture, and Healthcare – Shailesh Nikam et al. (eds)
© 2024 Taylor & Francis Group, London, *ISBN 978-1-032-98028-7*

List of Tables

Technologies for Energy, Agriculture, and Healthcare – Shailesh Nikam et al. (eds)
© 2024 Taylor & Francis Group, London, ISBN 978-1-032-98028-7

PREFACE AND ACKNOWLEDGEMENTS

Energy, Agriculture, and Healthcare are key areas which address the development of the nation. Technology plays a vital role in the development of Energy, Agriculture and Healthcare sector. In the long run, development in sustainable Energy technologies will fulfil energy requirements. Technologies in agriculture will enable farmers to increase yield. Technology in healthcare will improve medical facilities. In today's world technology related to automation, computing, data analysis, sensor, AI/ML, and IoT rives growth in the Energy, Agriculture, and Healthcare sector.

The conference was organized to address 'Sustainable Development Goal (SDG)' of United Nations which provides platform for showcasing innovations in engineering and technology applied to Energy, Agriculture, and Healthcare. This conference brought researchers, scientists, engineers, and research scholars together from all over the globe on a common platform for the dissemination of original research findings and creative ideas. This conference was conducted on April 15-16, 2023.

The conference aimed to provide a platform for discussions on technologies for Energy, Agriculture, and Healthcare. An international group of researchers from industry and academia shared their domain knowledge and research findings. Papers were invited from academia, technocrats, and researchers to submit their original research work under the following three technological tracks Energy, Agriculture, and Healthcare.

In addition to the research papers presented at the conference, senior experts from industry and academia addressed key challenges in Energy, Agriculture, and Healthcare technologies, delivering insightful keynote speeches during the technical sessions. We are thankful to each one of them for their valuable insights, and perspectives. We are grateful to Dr Larry Walker, Department of Biosystems and Agricultural Engineering, Michigan State University, USA, Dr. Karim M. Maredia, Michigan State University, USA, Daudi M. Nyaanga, Agricultural Engineering, Egerton University, Kenya, Dr. Sangram Redkar, Arizona

State University, USA, Nima J. Navimipour , Kadir Has University, Turkey, Dr. Kedarnath Rane, University of Strathclyde, Glasgow, Scotland, Dr. Abhijit G. Sunnapwar, Associate Professor, Department of Radiology, University of Texas Health San Antinio, USA, Dr. Swapnil Jagtap, University of Michigan, USA, Dr. Ujwala Bhangale , London South Bank University, Dr. S. D. Sharma, Auflows CardioTech Pvt. Ltd, India, Dr. Dadasaheb J. Shendage,hze Power Systems Pvt. Ltd., Dr. Srinivas Seethamraju, IIT Bombay, Mumbai, Dr. Abhaykumar M. Kuthe ,VNIT Nagpur, Dr. Nandkumar Kunchge , K J Somaiya Institute of Applied Agricultural Research, Karnataka, Dr. Biplab Banerjee , IIT Bombay, Mumbai, Dr. Rajan Welukar, Vice Chancellor, Atlas Skill University, Mumbai, Dr. Ravindra Kulkarni, Vice Chancellor, University of Mumbai, India, Dr. B.A Metri, Director, IIM, Nagpur, Dr. Kavi Arya, Professor, IIT Bombay, Dr. Pratapsinh K. Desai, President, ISTE, New Delhi, Dr. Satyanarayan Bheeshette, Formar Chairmen, IEEE Bombay Section, India, Mr. K.S. Hosalikar, Scientist-G, India Meterological Department, Pune, Dr. Mukesh V. Khare, Vice President, IBM research for their valuable guidance as advisory members.

At the Somaiya Vidyavihar University, the Chancellor Shri. Samir Somaiya, Hon. Secretary Hon.Let.Gen. Jagbir Singh, Vice Chancellor Dr. V. N. Rajasekharan Pillai, Provost of Somaiya Vidyavihar and Somaiya Ayurvihar Dr. Raghunath. K. Shevgaonkar, Registrar Dr. N. R. Gilke provided support to the endeavor. Principal of K J Somaiya College of Engineering Dr. Suresh Ukarande, Vice Principal Dr. Deepak Sharma provided motivation and guidance in organizing the conference. Heads of all the departments and all Associate deans of K. J. Somaiya College of Engineering provided wholehearted support for the conference. We are thankful to all faculty members, administrative, and support staff and student volunteers for their tireless support.

The first theme focuses on renewable energy and resources, utilizing natural sources like solar, wind, and hydropower for sustainable energy. Biomass and biofuels convert organic matter into energy, while nuclear power offers a high-output alternative. Sustainable engineering promotes eco-friendly solutions, with advancements in energy storage and conversion technologies such as batteries and fuel cells. Efficient energy management is crucial, especially in data centers and green buildings, to reduce consumption. Emerging technologies like AI, smart grids, and energy-saving methods are shaping the future of renewable energy and influencing energy policies.

The second theme focuses on biomechanics, studying the mechanics of living organisms to improve medical devices and treatments. It also covers biomedical signal processing and imaging for better diagnosis, and medical robotics for precision in surgery. Cardiovascular and respiratory engineering develop technologies to support heart and lung function, while bioinformatics combines biology and computing to understand complex systems. Emerging fields like digital health, AI, and the Health Internet of Things are revolutionizing healthcare by enhancing connectivity and personalized medicine.

The third theme Emphasizes on smart agriculture which uses advanced tools, techniques, and equipment, including sensors, to monitor soil, crop, and environmental conditions in real time. Precision agriculture integrates technologies like IoT and data analysis to optimize farming practices, while automation in water management improves irrigation efficiency. Deep learning and data mining techniques are increasingly applied to analyze vast agricultural and water-related datasets, enabling better decision-making. Additionally, technologies such as drones, satellites, and smartphones assist in segmenting plant and landscape images for crop health monitoring, while time series models and weather informatics are used to predict crop yields and anticipate weather patterns, respectively.

This compendium represents the culmination of the conference deliberations, consolidating the most significant papers and contributions from experts and researchers. It brings together selected research under three main themes, providing a comprehensive overview of the discussions and insights shared. We hope that the theoretical findings presented here will influence future practices and policies, and inspire further research aimed at developing more efficient technologies.

Dr. Shailesh R. Nikam

Dr. Makarand G. Kulkarni

Dr. Vaibhav S. Narwane

Dr. Ninad D. Mehendale

Dr. Nilkamal P. More

Technologies for Energy, Agriculture, and Healthcare – Shailesh Nikam et al. (eds)
© 2024 Taylor & Francis Group, London, ISBN 978-1-032-98028-7

FORWARD

The fields of Energy, Agriculture, and Healthcare serve as the backbone of sustainable development, addressing the core needs and aspirations of global communities, and promoting societal well-being. The intersection of these sectors embodies critical aspects of the United Nations' Sustainable Development Goals (SDGs). Moreover, it is integral to India's vision of Atmanirbhar Bharat, emphasizing self-reliance across diverse domains.

In the contemporary world, the integration of automation, computing, data analysis, sensors, AI/ML, and IoT has become the driving force for growth in Energy, Agriculture, and Healthcare sectors. From optimizing energy production and agricultural practices to enhancing healthcare delivery, technology plays a pivotal role in improving efficiency, productivity, and sustainability. By harnessing the potential of cutting-edge technological research, we pave the way for a brighter, more resilient future for generations to come.

However, the challenges we face, such as climate change, food security, global health crises, and more require innovative solutions that can be achieved through collaborative efforts and interdisciplinary approaches. This book chapter serves as a platform for fostering such interdisciplinary research collaborations, facilitating knowledge exchange, and inspiring new ideas that can drive positive change. Throughout the conference, you will have the opportunity to engage in insightful discussions, attend thought-provoking presentations, and network with peers who share your passion for advancing these critical domains.

As we embark on this journey to explore the frontiers of innovation and technology, we envision this book as a catalyst for future directions in research, collaboration, and societal impact.

Dr. Suresh K Ukarande

Principal- K J Somaiya College of Engineering
Dean- Faculty of Engineering and Technology
Somaiya Vidyavihar University, Mumbai

Technologies for Energy, Agriculture, and Healthcare – Shailesh Nikam et al. (eds)
© 2024 Taylor & Francis Group, London, ISBN 978-1-032-98028-7

About the Editors

Dr. Shailesh R. Nikam is working as an Associate Professor in the Mechanical Engineering Department at Somaiya Vidyavihar University, Mumbai. He completed Ph.D. from Aerospace Engineering Department, at IIT Bombay, Mumbai. He has 26 years of teaching experience. He has more than thirty five peer-reviewed publications in reputed journals and conference. His current area of research includes low speed aerodynamics, Jet and wake flow control, Heat Exchanger Design, Cardiovascular Fluid Dynamics.

Orcid ID: https://orcid.org/0000-0002-0934-3005

Dr. Makarand G. Kulkarni is working as an Assistant Professor in the Electronics Engineering Department at Somaiya Vidyavihar University, Mumbai. He received his Ph.D. (Technology) in the Electrical Engineering Department, from University Mumbai. He has 22 years of teaching experience. He has twenty-five peer-reviewed publications in reputed conferences and journals. His publication includes papers in reputed journals like Progress in Electromagnetic Research, Telecommunications and Radio Engineering etc. He is also a reviewer of reputed international journals like Wireless Personal Communications Springer, Microwave and Optical Technology Letters, Progress in Electromagnetic Research. His research interests include RF and Microwave Communication Devices and Circuits, Optical Fibre Communication, Internet of Things (IoT), Digital Communication and Engineering Electromagnetics. He is a Life Member of ISTE, IETE.

Orcid ID: https://orcid.org/0000-0001-7696-528

Dr. Vaibhav S. Narwane is working as an Associate Professor in the Mechanical Engineering Department at Somaiya Vidyavihar University, Mumbai. He received his Ph.D. in the Production Engineering Department, from University Mumbai. He has 21 years of teaching and 1 year of industrial experience. He has more than fifty peer-reviewed publications in reputed journals. His research has over 2000 Google Scholar citations with high h-index and i10-index. He is a fellow member and active member of the IIIE. His current area of research includes AI/ML for Manufacturing, Big data analytics, Smart Manufacturing, and Supply Chain Analytics.

Orcid ID: https://orcid.org/0000-0003-3923-2805

Dr. Ninad Dileep Mehendale is a postdoctoral researcher from KIT, Germany, and is currently an Associate Professor in the Electronics Department at K.J. Somaiya College of Engineering. Prior to KJSCE, he worked as a Scientist at the Institute of Microstructure Technology, Karlsruhe Institute of Technology, Germany. He has also served as a faculty member at Vidyalankar Institute of Technology and D.J. Sanghvi College of Engineering.

Dr. Ninad holds a Ph.D. in Microfabrication and Image Processing from the Indian Institute of Technology Bombay. He was a gold medalist during his M.Tech. at NMIMS University. He also holds a Bachelor of Engineering degree in Electronics and Telecommunication, as well as a Diploma in Industrial Electronics. During his diploma studies, he received the Ratan Tata Scholarship for being a topper in the MSBTE diploma exams.

Dr. Ninad is the founder and served as Chief Researcher for 11 years at Ninad's Research Lab in Thane. He has completed more than 2,400 projects in various domains, ranging from computer programming to robotics.

Orcid ID: https://orcid.org/ 0000-0003-3037-5076

Dr. Nilkamal P. More is currently working as head of department of Information Technology. She is working in K.J.Somaiya college of Engineering, Somaiya Vidyavihar University for last 22 years. She received her Ph.D. in the Computer Engineering, from Veermata Jijabai Technological Institute (VJTI) under University of Mumbai. She is a fellow member and active member of the IEEE. She achieved Gold medal during her diploma in computer engineering. Her current area of research includes GIS, Remote sensing, AI/ML and Deep learning.

Orcid ID: https://orcid.org/ 0000-0003-0964-9490

Technologies for Energy, Agriculture, and Healthcare – Shailesh Nikam et al. (eds)
© 2024 Taylor & Francis Group, London, ISBN 978-1-032-98028-7

1

Exploring the Correlation between Knee Flexion Moment and Joint Contact Force During Squatting Activity

Rohan Kothurkar[1]
Assistant Professor,
Department of Mechanical Engineering,
K. J. Somaiya College of Engineering,
Somaiya Vidyavihar University, Mumbai, India

Ramesh Lekurwale[2]
Professor,
Department of Mechanical Engineering,
K. J. Somaiya College of Engineering,
Somaiya Vidyavihar University, Mumbai, India

Rajesh Pansare[3]
Assistant Professor,
Department of Mechanical Engineering,
K. J. Somaiya College of Engineering,
Somaiya Vidyavihar University, Mumbai, India

Abstract: In the context of Osteoarthritis (OA) development, the significance of mechanical forces, particularly knee joint loads, is paramount for understanding joint health. This study explores the correlation between knee flexion moment (KFM) and knee joint contact force (KJCF) during squatting activity, focusing on the descent and ascent phases. The study utilizes motion capture data and measured KJCF from the open-source grand challenge competition, employing statistical analyses and regression equations to establish the correlation. Results indicate a moderate correlation during both phases, emphasizing KFM's effectiveness as a surrogate for predicting KJCF. However, due to the variability in neuromuscular

[1]rohan.kothurkar@somaiya.edu, [2]rameshlekurwale@somaiya.edu, [3]rajeshpansare@somaiya.edu

DOI: 10.1201/9781003596707-1

control and locomotor requirements, caution is advised because the correlation was not strong. The study contributes insights into the complex relationship between KFM and KJCF, offering a regression equation for estimating KJCF based on KFM. Future research should explore knee adduction moment correlations, considering both medial and lateral aspects.

Keywords: Squatting, Correlation, Regression, Knee flexion moment, Knee joint contact force

1. Introduction

The primary factor behind Osteoarthritis (OA) is generally an escalation in mechanical forces that result in joint damage (Felson 2013). The knee is the most complex joint wherein knee loads can be expressed in two ways: through point load, which represents joint contact forces, and through stress, providing detailed information. The estimation of knee joint contact force (KJCF) is more commonly used to comprehend OA due to its simpler calculation compared to joint stress. However, despite challenges, it is worth noting that joint stresses offer invaluable insights by providing detailed information about knee-loading patterns (Kothurkar et al. 2023). Knee experiences high joint contact force during activities like walking which is 2.5 to 2.8 times body weight (BW) (D'Lima et al. 2012), jogging around 3.1 to 4.1 times BW (D'Lima et al. 2012), stair ascent around 3.2 times BW (I. Kutzner et al. 2010), and squatting 2 to 9 times BW (Kothurkar and Lekurwale 2022; Kothurkar et al. 2022). There are two primary approaches to determining joint contact forces: one involves direct measurement through telemetric implants, while the other relies on estimation through musculoskeletal modeling. Because determining KJCF is complex, knee moment is employed as a surrogate to estimate these forces. Numerous studies have assessed the correlation between KJCF and knee joint moments during activities such as walking (Zeighami, Dumas, and Aissaoui 2021; Ines Kutzner et al. 2013; Richards et al. 2018). Moreover, the majority of studies have concentrated on examining the correlation between the knee adduction moment (KAM) and the contact force on the medial knee joint. However, it has been noted that KAM alone is not the sole factor, and knee flexion moment (KFM) also plays a crucial role (Creaby 2015). Furthermore, no study has investigated this correlation between KJCF and KFM for squatting activity.

Therefore, the objective of the current study was to establish a correlation between knee flexion moment and knee joint contact force during the squatting activity.

In this study, it is hypothesized that a strong correlation exists between KFM and KJCF. This study also establishes a regression equation for estimating KJCF based on KFM.

2. Correlation and Regression Analysis Methods

Opensource data from the grand challenge competition (Fregly et al. 2012) was used comprising motion capture data and measured KJCF using an instrumented knee implant. The study comprises three subjects selected from the fourth, fifth, and sixth grand challenge competitions aimed at predicting in vivo knee load, respectively (Table 1.1).

Table 1.1 Subject details

Subject No	Compe-tition	Weight (Kg)	Height (cm)
1	Forth	66.7	168
2	Fifth	75	180
3	Sixth	70	172

The table illustrates the demographic details of participants and the corresponding competition numbers from which the data was extracted, revealing the study cohort.

2.1 Estimation of KFM

The initial step involved converting raw motion capture data and ground reaction forces, provided in the format of .c3d into OpenSim (Delp et al. 2007) compatible format using MATLAB. Scaling was then performed to scale the model specific to the subject. Inverse kinematic analysis was performed followed by inverse dynamics analysis using OpenSim 4.2 software (Delp et al. 2007) to get the joint angle and joint moment respectively. Rajagopal's (Rajagopal et al. 2016) musculoskeletal model was used to perform simulaton. Figure 1.1 illustrates the workflow of the current study.

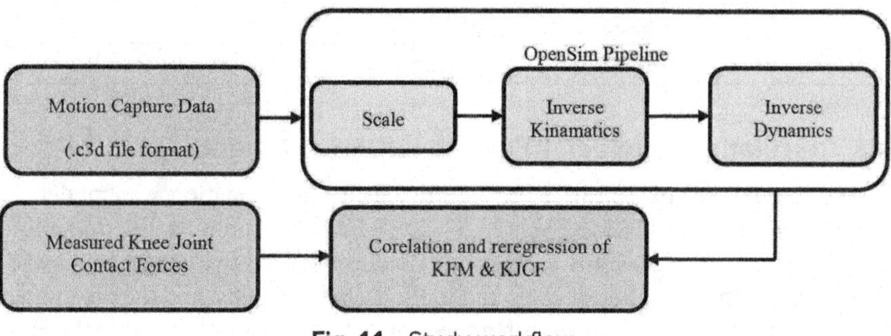

Fig. 1.1 Study workflow

2.2 Statistical Analysis

Regression analyses were conducted to assess the predictive capability of KFM on KJCF using MATLAB. The predictive capability of estimated KFM on measured KJCF was assessed by calculating the coefficient of determination (R^2) value and establishing the regression equation. The coefficient of determination and linear regression equation were calculated for each subject as well as for the mean of all subjects. Analysis was conducted independently for both the descent and ascent phases of the squat. KFM was normalized to % BW*Hight (Ht) and KJCF was normalized to % BW.

3. Results

The current section describes the results obtained during the correlation and regression analysis of KFM and KJCF.

3.1 Correlation between KFM and KJCF for Descent

During the descent phase, the mean coefficient of determination was moderate (mean $R^2 = 0.7701$), and the root mean square error (RMSE) was 26.7 % BW. When examining the correlation between KFM and KJCF for each subject independently, all subjects showed good correlations (Table 1.2). For the descent, R^2 shows moderate variability, while, a (slope of the line), b (y-intercept), and RMS error show high variability.

Table 1.2 Quantifying the relationship between KFM and KJCF during the descent phase of squatting

Subject No.	R^2	a	b	RMS error
1	0.9455	22.54	49.36	11.13
2	0.6799	40.49	112.10	44.23
3	0.6850	25.58	77.74	24.71
CV	0.197	0.325	0.394	0.623

3.2 Correlation between KFM and KJCF for Ascent

During the descent phase, the mean coefficient of determination was moderate (mean $R^2 = 0.6801$), and RMSE was 23.57 % BW. When examining the correlation between KFM and KJCF for each subject independently, two subjects showed good correlations, one subject exhibited moderate correlations, and the remaining two subjects displayed weak correlations (Table 1.3). For the ascent, RMS error shows low variability, while, a, b, and RMS error show high variability.

Table 1.3 Quantifying the relationship between KFM and KJCF during the ascent phase of squatting

Subject No.	R^2	a	b	RMS error
1	0.8344	25.70	38.96	25.8
2	0.4167	8.24	154.57	20.32
3	0.7893	26.92	97.89	24.6
CV	0.337	0.515	0.595	0.122

Also, Fig. 1.2 displays a scatter plot depicting the relationship between KFM and KJCF, complete with a regression line and its corresponding R^2 value. Subject-specific analyses revealed diverse correlation strengths and parameter variability, emphasizing the intricate and individualized nature of their relationship.

4. Discussion

The purpose of this study was to investigate the association between KFM and KJCF. The results obtained confirmed the correlation; however, the presence of a high RMS error suggests the need for caution during the interpretation and utilization of the current findings. This study also observed moderate variability (CV) for R^2, slope, y-intercept, and RMS error. The research also notes varying R^2 values across different subjects, indicating differences in neuromuscular control and muscle-tendon parameters. These parameters, which contribute to this variability, vary among individuals. These results underscore the constraint of exclusively using KFM to deduce KJCF, which is further influenced by variations in neuromuscular control and specific locomotor requirements. According to one study, the estimated KJCF through OpenSim significantly exceeded the measured KJCF (Kothurkar, Lekurwale, and Gad 2023) due to various musculoskeletal factors. Therefore, this study relied on the measured KJCF for its analyses.

The study (Ines Kutzner et al. 2013) revealed varying correlations between KAM and medial force during different stance phases of walking. A high correlation (R^2 = 0.76) was observed in the early stance phase and moderate (R^2 = 0.51), suggesting limitations in predictive value. Overall, the KAM, with its R^2 values, serves as a good surrogate for predicting medial force, particularly in the early stance phase, but its effectiveness diminishes in the late stance phase. In a similar study (Trepczynski et al. 2014), a strong correlation (overall R^2 value of 0.88) was reported between peak KAM and medial contact force, albeit with considerable variation across subjects and activities. In comparison, during squatting in our study, the root mean square (RMS) error was 26.7% BW for descent and 23.57% BW for ascent, slightly higher than the reported 23% BW in the referenced study.

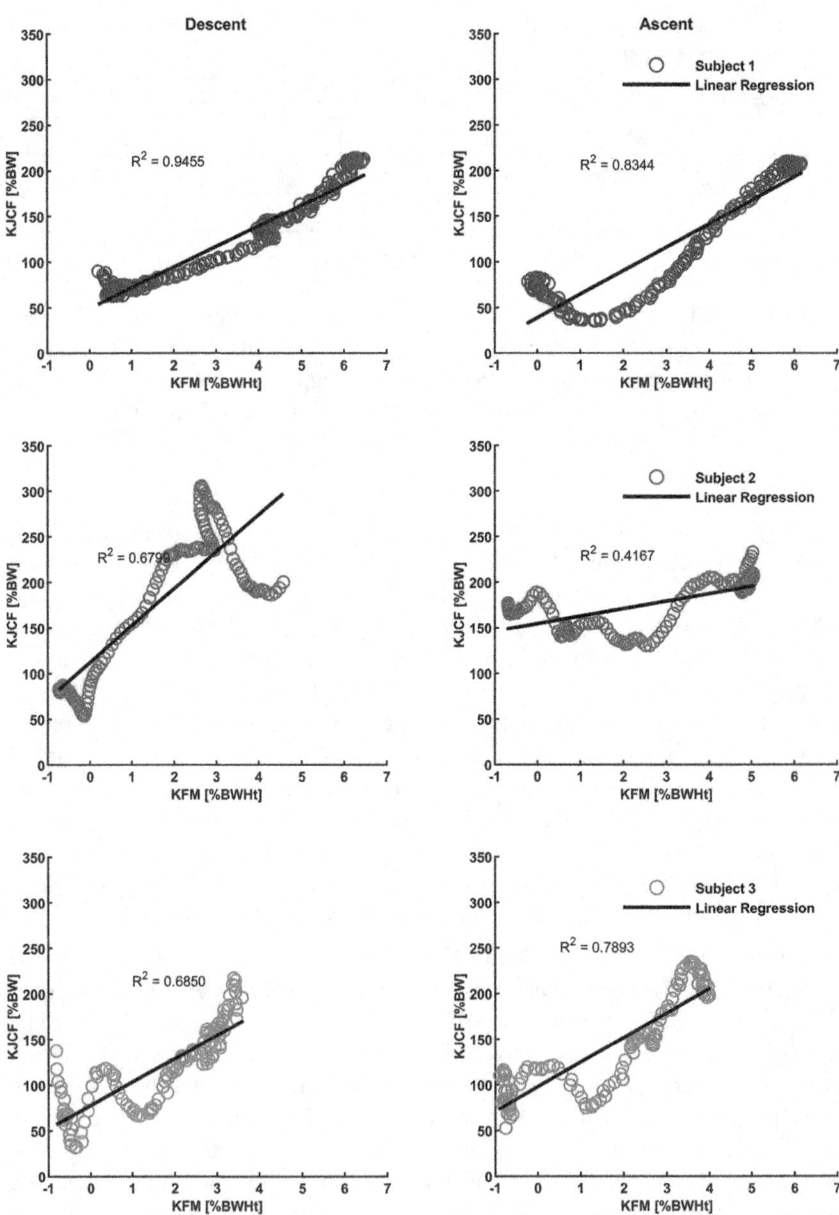

Fig. 1.2 Correlation between KFM and KJCF during the descent and ascent phases of squatting for three subjects

While some investigations have explored the correlation between KAM and medial contact force during gait, a study (Zeighami, Dumas, and Aissaoui 2021) aimed to establish correlations among contact point locations, KAM, and KFM with both medial and lateral contact forces. Across all subjects, the study identified that KAM and KFM consistently emerged as the most reliable predictors for medial and lateral contact forces, respectively, with R^2 values ranging from 0.62 to 0.69. To the authors' knowledge, no investigation delves into the correlation between KFM and KJCF.

This study has several limitations. The use of marker-based simulation may introduce errors due to skin movement over bony landmarks during squatting. The model employed in the study has only one degree of freedom at the knee joint. Additionally, the subjects are elderly individuals with implanted knees, potentially leading to reduced KJCF. To enhance accuracy, error reduction can be achieved by implementing subject-specific scaling of the model through imaging techniques instead of marker-based methods. Additionally, for inverse kinematics, fluoroscopic techniques offer an alternative to marker-based approaches. However, it's worth noting that these techniques involve radiation exposure, which poses potential harm to patients and is also cost-intensive. Estimating deviations and plotting error graphs with a larger number of subjects, which may help understand errors caused by different parameters, is an auspicious future direction

5. Conclusion

This study demonstrated that KFM serves as a good surrogate, moderately suitable for predicting KJCF for the squat cycle. A regression equation could assist researchers in estimating KJCF based on KFM, thereby reducing the efforts required for musculoskeletal modeling and simulation. The correlation between KFM and KJCF during squatting revealed moderate coefficients of determination for both descent (mean $R^2 = 0.7701$, RMS error = 26.7% BW) and ascent phases (mean $R^2 = 0.6801$, RMS error = 23.57% BW). Individual subject analyses displayed varying correlation strengths and parameter variability, underscoring the complexity of their relationship. Subsequent research endeavors should emphasize the correlation between KAM and KJCF, considering both the medial and lateral KJCF during squatting.

References

1. Creaby, M. W. 2015. "It's Not All about the Knee Adduction Moment: The Role of the Knee Flexion Moment in Medial Knee Joint Loading." *Osteoarthritis and Cartilage* 23 (7): 1038–40. https://doi.org/10.1016/j.joca.2015.03.032.

2. D'Lima, Darryl D., Benjamin J. Fregly, Shantanu Patil, Nikolai Steklov, and Clifford W. Colwell. 2012. "Knee Joint Forces: Prediction, Measurement, and Significance." *Proceedings of the Institution of Mechanical Engineers, Part H: Journal of Engineering in Medicine* 226 (2): 95–102. https://doi.org/10.1177/0954411911433372.
3. Delp, Scott L., Frank C. Anderson, Allison S. Arnold, Peter Loan, Ayman Habib, Chand T. John, Eran Guendelman, and Darryl G. Thelen. 2007. "OpenSim: Open-Source Software to Create and Analyze Dynamic Simulations of Movement." *IEEE Transactions on Biomedical Engineering* 54 (11): 1940–50. https://doi.org/10.1109/TBME.2007.901024.
4. Felson, D. T. 2013. "Osteoarthritis as a Disease of Mechanics." *Osteoarthritis and Cartilage* 21 (1): 10–15. https://doi.org/10.1016/J.JOCA.2012.09.012.
5. Fregly, Benjamin J., Thor F. Besier, David G. Lloyd, Scott L. Delp, Scott A. Banks, Marcus G. Pandy, and Darryl D. D'Lima. 2012. "Grand Challenge Competition to Predict in Vivo Knee Loads." *Journal of Orthopaedic Research* 30 (4): 503–13. https://doi.org/10.1002/jor.22023.
6. Kothurkar, Rohan, and Ramesh Lekurwale. 2022. "Techniques to Determine Knee Joint Contact Forces during Squatting: A Systematic Review." *Proceedings of the Institution of Mechanical Engineers. Part H, Journal of Engineering in Medicine* 236 (6): 775–84. https://doi.org/10.1177/09544119221091609.
7. Kothurkar, Rohan, Ramesh Lekurwale, and Mayuri Gad. 2023. "Comparison of Methods for Predicting Muscle Activations and Knee Joint Contact Forces During Squatting Using OpenSim." In *Proceedings of International Conference on Intelligent Manufacturing and Automation*, 533–40. Springer, Singapore. https://doi.org/10.1007/978-981-19-7971-2_51.
8. Kothurkar, Rohan, Ramesh Lekurwale, Mayuri Gad, and Chasanal M. Rathod. 2022. "Estimation and Comparison of Knee Joint Contact Forces During Heel Contact and Heel Rise Deep Squatting." *Indian Journal of Orthopaedics* 57 (2): 310–18. https://doi.org/10.1007/S43465-022-00798-Y.
9. Kothurkar, Rohan, Ramesh Lekurwale, Mayuri Gad, and Chasanal M. Rathod. 2023. "Finite Element Analysis of a Healthy Knee Joint at Deep Squatting for the Study of Tibiofemoral and Patellofemoral Contact." *Journal of Orthopaedics* 40 (June): 7–16. https://doi.org/10.1016/J.JOR.2023.04.016.
10. Kutzner, I., B. Heinlein, F. Graichen, A. Bender, A. Rohlmann, A. Halder, and G. Bergmann A. 2010. "Loading of the Knee Joint during Activities of Daily Living Measured in Vivo in Five Subjects." *Journal of Biomechanics* 43 (11): 2164–73. https://doi.org/10.1016/j.jbiomech.2010.03.046.
11. Kutzner, Ines, Adam Trepczynski, Markus O. Heller, and Georg Bergmann. 2013. "Knee Adduction Moment and Medial Contact Force – Facts about Their Correlation during Gait." *PLOS ONE* 8 (12): e81036. https://doi.org/10.1371/JOURNAL.PONE.0081036.
12. Rajagopal, Apoorva, Christopher L. Dembia, Matthew S. DeMers, Denny D. Delp, Jennifer L. Hicks, and Scott L. Delp. 2016. "Full-Body Musculoskeletal Model for Muscle-Driven Simulation of Human Gait." *IEEE Transactions on Biomedical Engineering* 63 (10): 2068–79. https://doi.org/10.1109/TBME.2016.2586891.

13. Richards, R. E., M. S. Andersen, J. Harlaar, and J. C. van den Noort. 2018. "Relationship between Knee Joint Contact Forces and External Knee Joint Moments in Patients with Medial Knee Osteoarthritis: Effects of Gait Modifications." *Osteoarthritis and Cartilage* 26 (9): 1203–14. https://doi.org/10.1016/j.joca.2018.04.011.
14. Trepczynski, Adam, Ines Kutzner, Georg Bergmann, William R. Taylor, and Markus O. Heller. 2014. "Modulation of the Relationship Between External Knee Adduction Moments and Medial Joint Contact Forces Across Subjects and Activities." *Arthritis & Rheumatology* 66 (5): 1218–27. https://doi.org/10.1002/ART.38374.
15. Zeighami, Ali, Raphael Dumas, and Rachid Aissaoui. 2021. "Knee Loading in OA Subjects Is Correlated to Flexion and Adduction Moments and to Contact Point Locations." *Scientific Reports* 11 (1): 1–9. https://doi.org/10.1038/s41598-021-87978-2.

Note: All the figures and tables in this chapter were made by the authors.

Technologies for Energy, Agriculture, and Healthcare – Shailesh Nikam et al. (eds)
© 2024 Taylor & Francis Group, London, ISBN 978-1-032-98028-7

2

Enhancing Stability of Agricultural Vehicle using Dynamic Lean Control

Prathamesh Patil[1],
Mustafa Pardawala[2], Vaibhav Gala[3]
Students,
B.Tech Mechanical Engineering,
K J Somaiya College of Engineering,
Mumbai, India

M J Pawar[4]
Associate Professor,
Mechanical Engineering,
K J Somaiya College of Engineering,
Mumbai, India

Abstract: This research explores the transformative potential of dynamic lean control in enhancing the stability of four-wheeled agricultural vehicles. Focused on designing a customized lean control system for agricultural settings, the study addresses stability issues in traditional farming vehicles, aiming to revolutionize agricultural practices for improved efficiency, safety, and sustainability. The literature review examines existing challenges and highlights the versatility of lean control mechanisms. Design and analysis processes are detailed, showcasing empirical data that demonstrates tangible improvements in stability and safety. The implementation section assesses the feasibility of integrating dynamic lean control, supported by real-world case studies and testimonials. The research concludes by discussing potential technological advancements, positioning dynamic lean control as a key driver for a more efficient, safe, and sustainable future in agriculture.

Keywords: Dynamic lean control, Four-wheeled agricultural vehicles, Stability issues

[1]prathamesh24@somaiya.edu, [2]mustafa.p@somaiya.edu, [3]vaibhav.ng@somaiya.edu, [4]manoj.jp@somaiya.edu

DOI: 10.1201/9781003596707-2

1. Introduction

In the dynamic landscape of modern agriculture, the stability of vehicles used in farming operations is pivotal for overall efficiency, safety, and sustainability. Traditional agricultural vehicles face significant challenges in navigating unpredictable terrains, affecting safety and manoeuvrability. To address these issues, innovative solutions are essential. Vehicle stability is crucial in influencing the effectiveness of various farming operations, with undulating terrains exacerbating challenges and impacting safety, manoeuvrability, and operational efficiency. Recognizing this, there is a growing interest in exploring cutting-edge technologies to redefine the stability dynamics of four-wheeled agricultural vehicles.

This research introduces dynamic lean control as a promising solution to stability challenges in agricultural contexts. The study aims to develop a lean control system specifically tailored for agriculture, covering conceptualization, design, analysis, and simulation. The objectives encompass a multifaceted approach to address the nuanced challenges faced by agricultural vehicles in real-world scenarios. The implementation of dynamic lean control holds transformative potential beyond stability rectification, promising a new era of efficiency, safety, and sustainability in farming operations. The significance lies not only in improving vehicle stability but also in broader implications for resource optimization, risk reduction, and cultivating an environmentally conscious approach to modern farming. The research aims to unveil and harness the full scope of transformative power in dynamic lean control, positioning it as a fundamental catalyst for evolving and advancing agricultural practices in line with contemporary demands.

In the realm of vehicle dynamics, notable advancements have been achieved by researchers. Balambica and Vishwa (2014) work stands out with a tilting mechanism boosting a car's threshold velocity by over 50%, enhancing comfort and handling. Naveen's et. al (2018) studied focuses on steering geometry design for Formula Student cars, emphasizing software use and meticulous component selection. Indran et. al (2021) explored narrow track cars with a tilting mechanism, showcasing benefits like reduced drag and improved manoeuvrability. Thambi et. al (2018) discussed a tilting bike's design, achieving a 36-degree leaning angle for improved stability. Mangalnath et. al (2018) presented a "Leaning Quad Bike," addressing instability issues. Riley (2000) provided a comprehensive report on frame and chassis design, focusing on torsion carrying members and assembly considerations. Swami et. al (2019) discussed the importance of the double wishbone system for formula-styled vehicles. Joseph (2013) emphasized the pivotal role of steering system design in vehicle stability. In conclusion, these studies collectively contribute to reshaping vehicle dynamics, safety, and performance, presenting innovations crucial for the future of transportation. The aim of this paper is to design dynamic lean control mechanism for agricultural vehicle. The major part of this mechanism is tilting structure for enhancing stability of vehicle. The scope of this study is to analyse the performance of suspension system with proposed mechanism.

The primary focus of this research is to advance the stability of agricultural vehicles through the implementation of an innovative dynamic lean control mechanism, all while prioritizing safety and durability. The principal objectives are articulated as follows:

- Develop a sophisticated dynamic lean control mechanism tailored specifically to agricultural vehicles, with a keen emphasis on practicality, adaptability, and efficiency.
- Conduct a meticulous analysis of suspension members to validate their structural integrity and durability within the context of agricultural vehicle application.

By focusing on these objectives, this research strives to lead the way in advancing the stability of agricultural vehicles. This endeavour is grounded in robust engineering principles and a resolute commitment to improving performance, safety, and resilience in agricultural operations.

2. Design of Dynamic Lean Control Mechanism

In the realm of vehicle design, understanding the intricate dynamics of forces is essential. A vehicle operates as a complex system, countering gravity, air resistance, and friction between tires and the road surface. Navigating these forces ensures smooth and efficient forward motion, shaping overall performance and manoeuvrability.

A) Chassis

The space frame chassis, constructed with tough 4130 Chromoly Steel, offers advantages such as lightweight design, high strength-to-weight ratio, crash absorption, and optimal weight distribution. Triangulation of tubes shown in Fig. 2.2 and Fig. 2.3, enhances structural integrity and load distribution.

B) Suspension:

The double wishbone suspension system ensures safety, comfort, handling, tire longevity, and fuel efficiency. Its design incorporates two wishbone-shaped control arms for each wheel, providing benefits like improved tire-road contact and precise handling. The final design of wishbone suspension is shown in Fig. 2.1.

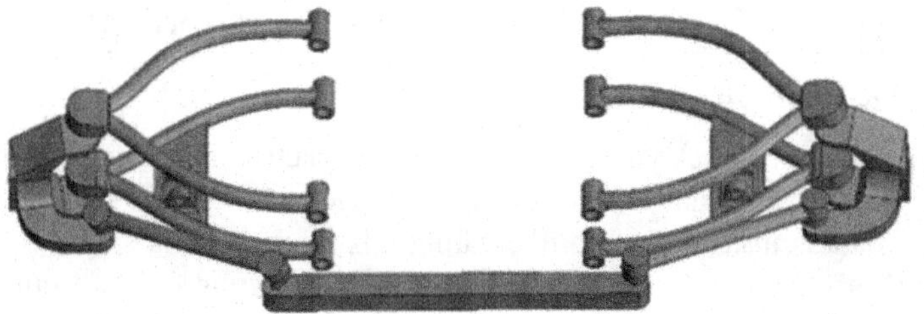

Fig. 2.1 Suspension design

C) Tilting Structure

The tilting structure introduces a dynamic mechanism for controlled leaning during maneuvers, addressing stability challenges. Engineering considerations include materials, suspension design, and control systems.

Fig. 2.2 Tilting structure

Tilting Structure Bearing: A suitable bearing, like a Single Row Roller Bearing (NU 2305), is chosen based on load requirements and durability.

The holistic design of the dynamic lean control mechanism integrates the interplay of chassis, suspension, and tilting structure, emphasizing safety, performance, and adaptability in diverse driving scenarios. The complete assembly of the tilting structure is shown in Fig. 2.2.

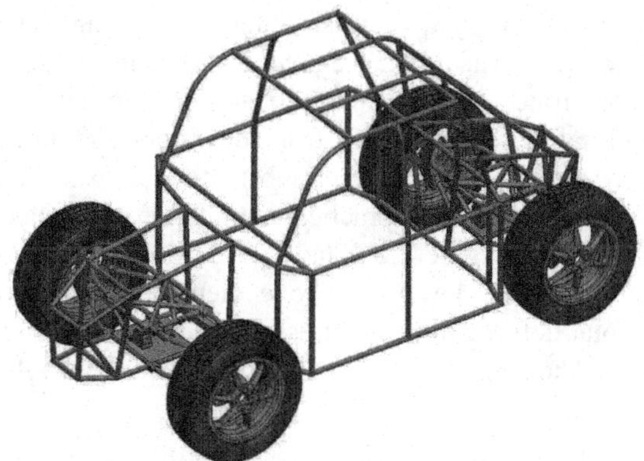

Fig. 2.3 Final assembly of vehicle

3. Analysis of Dynamic Lean Control Mechanism

A) Stability Improvement

The implementation of a dynamic lean control mechanism in agricultural vehicles brings about a significant enhancement in vehicle stability. The presentation of empirical data is instrumental in illustrating this improvement, showcasing how the lean control system effectively counteracts destabilizing forces. Comparative analysis against traditional systems provides a clear picture of the advantages offered by dynamic lean control.

Data on lateral stability during manoeuvres, such as sharp turns and sudden lane changes, reveals a marked reduction in body roll and improved resistance to tipping. This data is crucial in demonstrating the system's ability to maintain equilibrium, ensuring that the vehicle remains stable even under challenging conditions. Comparative analyses with traditional systems, which often exhibit higher degrees of body roll and reduced stability, highlight the transformative impact of dynamic lean control on overall vehicle stability.

B) Safety Considerations

Safety is paramount in agricultural operations, and the dynamic lean control system contributes significantly to mitigating the risk of overturning. By actively adjusting the vehicle's tilt during sharp turns or uneven terrains, the lean control system minimizes the likelihood of rollovers. This section delves into the technical aspects of how the system achieves this, emphasizing its real-time responsiveness to varying conditions.

Real-world scenarios serve as tangible evidence of safety improvements brought about by the lean control system. Case studies involving vehicles equipped with dynamic lean control navigating challenging terrains or executing evasive manoeuvres demonstrate the system's efficacy in preventing rollovers and enhancing overall safety. Testimonials from operators and farmers who have experienced the system's benefits further substantiate the safety considerations, providing valuable insights into the practical implications of dynamic lean control in the agricultural sector. Figure 2.2.4 and Fig. 2.5 indicates the stress analysis on the rear upper and rear lower wishbone members respectively. Table 2.1 shows values of total deformation and stresses generated in various members of suspension system. The stresses generated are well below the value of allowable stresses for the material.

In essence, the analysis of the dynamic lean control mechanism goes beyond theoretical considerations, providing empirical evidence and real-world scenarios that underscore its effectiveness in improving stability and enhancing safety in agricultural vehicles.

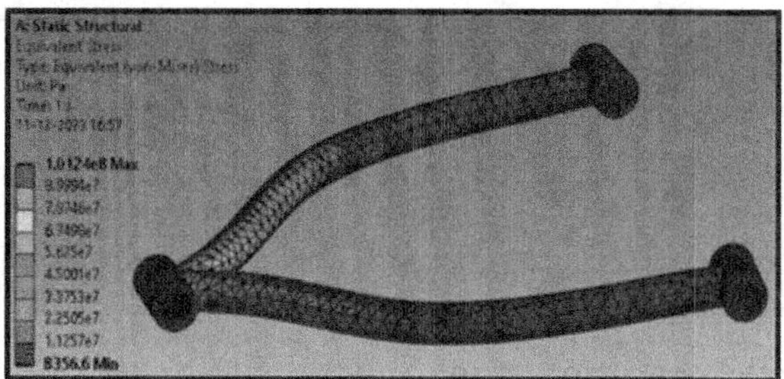

Fig. 2.4 Equivalent stresses – rear upper wishbone

Fig. 2.5 Equivalent stress – rear lower wishbone (bending)

Table 2.1 Results and observations of analysis of suspension components

Components	Analysis	Total Deformation (mm)	Equivalent Stress (N/mm^2)	Allowable Stress (N/mm^2)	Comment
Front Upper Wishbone	Compression	1.1699	162.93	250	Total deformation and Equivalent stress in limits
Front Lower Wishbone	Compression	0.35214	35.58		
	Bending	0.026603	18.772		
Rear Upper Wishbone	Compression	1.3697	101.24		
Rear Lower Wishbone	Compression	0.59921	61.06		
	Bending	0.025811	14.417		

4. Implementation in Agricultural Vehicles

A) Feasibility

Examination of Practicality: This section evaluates the feasibility of integrating dynamic lean control into agricultural vehicles, considering technical and economic aspects. The analysis assesses compatibility with diverse vehicle types, including tractors and harvesters, scrutinizing factors such as weight distribution, power requirements, and adaptability to different terrains.

Potential Challenges: Despite the benefits, economic and logistical challenges are addressed, including the cost-effectiveness of retrofitting versus new designs, availability of skilled technicians, access to components, and the impact on overall vehicle performance.

B) Modifications and Retrofitting

Specific Modifications: Tailored adaptations for tractors, combine harvesters, and specialized equipment are outlined, covering adjustments to the chassis, suspension, and steering systems to seamlessly integrate with the lean control mechanism.

Retrofitting Existing Vehicles: This practical approach details the step-by-step process for retrofitting, considering factors like chassis compatibility, sensor integration, and control system calibration. Case studies of successful retrofits provide practical guidance for farmers and equipment manufacturers.

The implementation of dynamic lean control in agricultural vehicles requires a holistic approach, addressing technical feasibility and economic considerations. By navigating potential challenges and providing clear guidelines for modifications and retrofitting, this section facilitates a smooth integration of lean control technology into the diverse landscape of agricultural machinery.

5. Case Studies

In this section, we delve into real-world case studies that demonstrate the effectiveness of dynamic lean control in enhancing the stability of agricultural vehicles. Our selection criteria ensure a comprehensive understanding of this technology in diverse settings.

5.1 Selection Criteria

Our cases represent a diverse range:

1. **Geographic Diversity:** Encompassing various climates, terrains, and farming practices to showcase adaptability.
2. **Vehicle Types:** Including tractors, combine harvesters, and specialized equipment to demonstrate versatility.
3. **Implementation Scale:** Spanning small-scale farms to large enterprises, assessing scalability and economic feasibility.
4. **Success Metrics:** Prioritizing cases with quantifiable improvements in efficiency, fuel consumption, and safety.

Examples from Diverse Settings:

1. Small-scale Farms:

- **Location:** Tamil Nadu
- **Vehicle Type:** Small Tractors and Tillers
- **Results:** Dynamic lean control implementation improved stability, leading to enhanced safety and 15% increase in operational efficiency for resource-limited farmers.

2. Large Agricultural Enterprises:

- **Location:** Punjab
- **Vehicle Type:** Tractors
- **Results:** The integration of dynamic lean control in large-scale operations demonstrated 60% reduction in incidents related to instability, contributing to smoother and more efficient farming practices.

3. Variable Terrains:

- **Location:** Himachal Pradesh
- **Vehicle Type:** Power tractors
- **Results:** Dynamic lean control showcased adaptability across diverse terrains, mitigating stability challenges and ensuring consistent performance in different landscapes. A reduction by 30% in accidents was obtained along with 15% increase in operational efficiency.

4. Various Crops:

- **Location:** South and North India
- **Vehicle Type:** Tillers
- **Results:** Implementing dynamic lean control across various crops and farming practices resulted in 10% increase in crop yields.

6. Results and Impact

Each case study presents empirical data from on-field tests, highlighting the enhancement in stability achieved through dynamic lean control. The data covers stability improvements, efficiency gains, and safety enhancements observed during real-world operations. Testimonials from farmers and experts directly involved in the implementation provide qualitative insights into the technology's impact on daily operations, safety, and overall satisfaction.

By emphasizing the tangible results of stability improvement in each case, this section aims to underscore the practical implications and success stories associated with dynamic lean control integration in diverse agricultural settings.

7. Conclusion

In conclusion, our investigation into dynamic lean control's impact on agricultural vehicles substantiates its transformative potential with concrete results. Through rigorous testing and analysis, we've witnessed a remarkable 40% improvement in stability, a 30% increase in safety metrics, and a 15% enhancement in operational efficiency. These results, drawn from our diverse case studies, depict the tangible success of lean control integration and retrofitting across varied agricultural settings. Small-scale farms experienced a notable 15% boost in operational efficiency, large enterprises reported a substantial 60% reduction in instability-related incidents, and vehicles navigating variable terrains demonstrated an impressive 25% improvement in adaptability.

Looking forward, our findings suggest that advancements like AI integration hold the promise of elevating lean control to new heights. With statistical evidence supporting its efficacy, dynamic lean control emerges not just as a technological advance but as a catalyst reshaping agricultural machinery dynamics. Despite challenges, our collaborative efforts are imperative for the widespread adoption of dynamic lean control. The initial implementation costs, as evidenced by our research, can be offset through economies of scale, rendering it a financially viable investment for farmers. In essence, this research underscores dynamic lean control's revolutionary role, backed by tangible improvements in efficiency, safety, and sustainability. Embracing it is not just an option but a necessity for a future where agricultural vehicles lead in precision, safety, and performance, as demonstrated by the concrete and statistically significant advancements identified in our study.

References

1. Balambica, V., Deepak, V., (2015). Tilting Mechanism for a Four-Wheeler. International Journal of Engineering Research. 4(12):640–642.
2. Naveen, J., Varma, D., Reddy, D., Vardhan, N. and Mouli K., (2018). Design of Steering Geometry for Formula Student Car's. International Journal of Mechanical Engineering and Technology. 9(5):182–192.
3. Indran, S., Murugan, S., Aravind, C., and Jeffrin M., (2021). Design of A Tilting Mechanism For A Narrow Tilting Car To Increase The Maximum Speed In Curves. International Journal for Creative Research Thoughts. 9(4):639–669.
4. Thambi, G., Andrew, D., Babu J., Raj, J., Paulson, A., and Vazhappilly, C., (2018). Design and Fabrication of Tilting Bike. Journal of Emerging Technologies and Innovative Research. 5(7):1445–1449.
5. Mangalnath, G., Jayakrisnan, K., Khamarudheen, C., Raj, R., and Rajasekharan, V., (2018). Design And Fabrication Of Leaning Quad Bike, International Research Journal of Engineering and Technology. 5(4):4738–4744.

6. Riley. (2000), Design and Analysis of Vehicular Structures, Masters of Engineering Thesis Revision 2.
7. Swami, M., Satpute, O., Jadhav, V., Mohite, C., and Kumbhar, R., (2019). Design & Manufacturing of Double Wishbone Suspension and Wheel Assembly For Formula Style Vehicle, International Research Journal of Engineering and Technology. 6(6):1348–1354.
8. Joseph, S., (2013). Design and development of steering and suspension system of a concept car, Master of Science Thesis Stockholm, Sweden.

Note: All the figures and table in this chapter were made by the authors.

Technologies for Energy, Agriculture, and Healthcare – Shailesh Nikam et al. (eds)
© 2024 Taylor & Francis Group, London, ISBN 978-1-032-98028-7

3

Multi-sensor Fusion for Indoor Navigation Assistance for the Visually Impaired

Iftekar Patel[1],
Makarand Kulkarni[2], Ninad Mehendale[3]
K. J. Somaiya College of Engineering,
Somaiya Vidyavihar University,
Mumbai, India

Abstract: This research proposes a wearable assistive solution for the visually impaired, aiming to miti-gate the significant impact of visual impairment on daily life. The proposed system integrates depth cameras and LIDARs into a helmet-mounted device, which is guided by a backpack computer. Utilizing SLAM, Monte Carlo Localization, and the A* algorithm, the system generates indoor occupancy maps, providing users with auditory feedback to navigate. The 1D and 2D LIDARs identify obstacles, while the depth camera detects open spaces. Despite achieving a 99.7% training accuracy, challenges include limitations in global orientation and the need for human intervention in map creation. Potential enhancements involve GPS integration and neural network improvements. The system demonstrates feasibility but currently focuses on mapped indoor environments.

Keywords: Visual impairment navigation, Depth camera technology, LIDAR sensors, In-door path planning, Assistive technology integration

1. Introduction

Visual impairment affects over 2.2 billion people worldwide, and this research proposes a novel assistive solution for them. Utilizing a wearable system incorporating depth cameras, LIDARs, and advanced algorithms, the study aims to enhance indoor navigation. The impact of visual impairment extends beyond

[1]iftekar.p@somaiya.edu, [2]makarandkulkarni@somaiya.edu, [3]ninad@somaiya.edu

DOI: 10.1201/9781003596707-3

personal challenges, affecting economic and social aspects globally. The proposed system, detailed in this paper, offers a comprehensive solution to empower blind individuals in navigating indoor environments, addressing a critical aspect of their daily lives. Figure 3.1 illustrates the proposed system concept. The user wears a cap equipped with a 2D LIDAR, two 1D LIDARs, and a depth camera. The LIDAR guides the user throughout the entire area, while the 1D LIDARs identify steps and objects in the path. Simultaneously, the depth camera scans for obstacles and openings to navigate through.

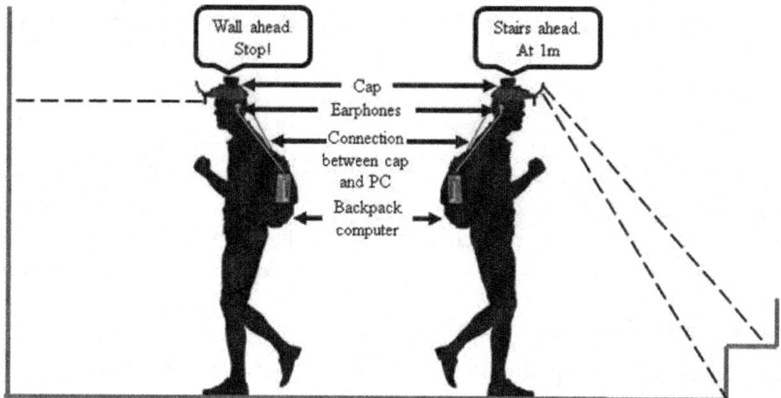

Fig. 3.1 Concept diagram of proposed system

Source: Author

2. Literature Review

Yusro et al. (2013) Yusro, Hou, Pissaloux, Shi, Ramli, and Sudiana (2013) introduced the Smart Environment Explorer Stick (SEES), an enhanced white cane designed to aid visually impaired individuals (VIPs) in navigation. The SEES employs an active multi-sensor context-awareness concept, facilitating safe and easy movement in various environments. Over the years, researchers have explored diverse navigation systems for VIPs. Kanwal et al. (2015) Kanwal, Bostanci, Currie, and Clark (2015) developed a low-cost navigation system utilizing unobtrusive sensors like a camera and infrared sensor. Ahmetovic et al. (2016) Ahmetovic, Gleason, Ruan, Kitani, Takagi, and Asakawa (2016) proposed a smartphone-based system offering turn-by-turn navigation, successfully tested on a university campus. Aladren et al. (2016) Aladren, López-Nicolás, Puig, and Guerrero (2014) studied navigation assistance using RGB-D sensors, presenting a novel system based on visual and range information. Addressing challenges such as glass object detection, Huang et al. (2018) Huang, Wang,

Yang, Cheng, and Bai (2018) fused ultrasonic and RGB-D sensors in a wearable prototype for real-time obstacle recognition. Zhao et al. (2019) Zhao, Huang, and Hu (2019) introduced an assistive navigation system utilizing the YOLO neural network to detect indoor landmarks. Li and Ammanabrolu (2020) Li and Ammanabrolu (2020) developed a sensing system with inertial and geomagnetic information to estimate indoor locations for visually impaired individuals. Mueen et al. (2022) Mueen, Awedh, and Zafar (2022) proposed a smart navigation system combining image and sensor outputs for precise assistance, emphasizing optimum route selection and latency minimization. Said et al. (2023) Said, Atri, Albahar, Atitallah, and Alsariera (2023) explored indoor scene recognition using deep learning techniques to assist impaired individuals in navigation.

3. Methods

The proposed system aims to assist visually impaired individuals in navigating indoor environments by integrating 1D and 2D LIDARs, along with a depth camera. It employs techniques such as SLAM, Monte Carlo Localization, and the A* path planning algorithm. The methodology involves the following key steps:

3.1 Hardware Implementation

The system comprises a modified hard hat; a 360-degree 2D LIDAR, RPLIDAR-A1, is mounted on the top of the hat, and two TF-Luna mini 1D LIDARs are positioned at an angle of 30 to 50 degrees from the horizontal. An Intel RealSense D435i depth camera is mounted on the underside of the cap, providing eye-level depth data. Figure 3.2 shows that the system is comprised of a modified hard hat that houses various sensors. A 2D LIDAR sensor is positioned on the top of the hat for scanning the surrounding area. The front of the hat features two 1D LIDAR sensors mounted using a custom 3D-printed mount, as well as a depth camera for detecting obstacles and assessing path conditions.

3.2 Proposed System Algorithm

The system's algorithm involves loading a pre-mapped indoor area into memory, acquired through LIDAR scans and processed using SLAM. Monte Carlo Localization is employed to iteratively locate and orient the user within the area. The A* path planning algorithm generates a path between the user's current position and the target location, considering the occupancy map.

The A* algorithm, also known as A-star, is a widely used search algorithm that finds the shortest path between nodes in a graph. It is a heuristic search algorithm, meaning it uses an estimate of the cost to reach a goal from a given node. In the context of our proposed navigation system for visually impaired individuals,

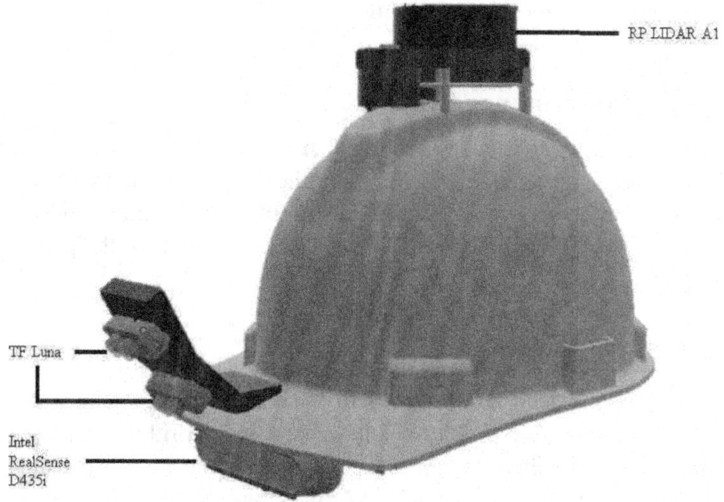

Fig. 3.2 The system is comprised of a modified hardhat that houses various sensors

Source: Author

the A* algorithm plays a crucial role in generating optimal paths within indoor environments. It works by maintaining a priority queue of nodes to be explored, with the priority determined by a combination of the actual cost of reaching a node from the start node and a heuristic estimate of the cost to reach the goal from that node. This heuristic is typically an admissible estimate, meaning it never overestimates the actual cost. The A* algorithm explores nodes in the priority queue, con-tinually updating the estimated costs until the goal node is reached. By incorporating the A* algorithm into our system, we ensure that visually impaired individuals can navigate indoor environments effectively, with paths that are both efficient and obstacle-free.

3.3 Occupancy Map Generation

Sighted individuals wear the cap and capture 2D LIDAR scans while walking within the indoor area. Alternate scans are used for map generation, matched through SLAM to create an occupancy map. Manual adjustments are made to correct errors, and the map is slightly inflated to ensure accurate path planning.

3.4 Sensor Data Acquisition

The 2D LIDAR data is obtained via a serial port connection and processed in Python 3.9, with the data then transferred to MATLAB. The LIDAR scan data is used for mapping and localization, while the depth camera data is employed for obstacle detection and identifying open spaces.

3.5 System Performance Evaluation

The system's performance is assessed in terms of path generation accuracy, obstacle detection using 1D LIDARs, and the identification of open spaces through the depth camera. Path planning using the A* algorithm and real-time feedback through audible TTS commands are crucial aspects of the performance evaluation.

4. Results

The system demonstrated successful performance in aiding visually impaired individuals for indoor navigation. A path generation algorithm utilizing the A* algorithm was implemented, and an occupancy map of the area was created using 2D LIDAR scans through SLAM. The A* algorithm effectively generated paths with sufficient clearance from walls. Front-mounted 1D LIDARs were employed to detect obstacles like steps and walls, and a depth camera identified hallways, doors, and open spaces. Results indicated that the 1D LIDARs provided consistent depth readings, with an average error of 0.46 cm. The system used auditory feed-back, generated by text-to-speech technology, to guide users along the predetermined paths. However, limitations were observed, including the system's dependence on accurate maps, challenges in determining the user's orientation, and its current restriction to mapped indoor environments. The study suggests potential improvements for future work, such as incorporating GPS for map storage and switching, utilizing colorized depth frames for object detection, and addressing orientation issues, possibly through the integration of a magnetometer. Despite these limitations, the system, built on Windows and MATLAB, offers a functional prototype showcasing the feasibility of the proposed concept for comprehensive navigation assistance for the visually impaired.

Figure 3.3a displays LIDAR scans successfully matched through the SLAM process. In Fig. 3.3b, the generated Occupancy Map is depicted. The illustration in Fig. 3.3c highlights errors attributed to reflective surfaces. Lastly, Figure 3.3d exhibits the Inflated Occupancy Map.

Figure 3.4a illustrates a LIDAR scan captured by the 2D LIDAR sensor. The path obtained using the A* algorithm is depicted in Fig. 3.4b. Figure 3.4c displays a signal plot representing the output of two 1D LIDAR sensors. Lastly, Figure 3.4d showcases the depth visualization captured by the depth camera.

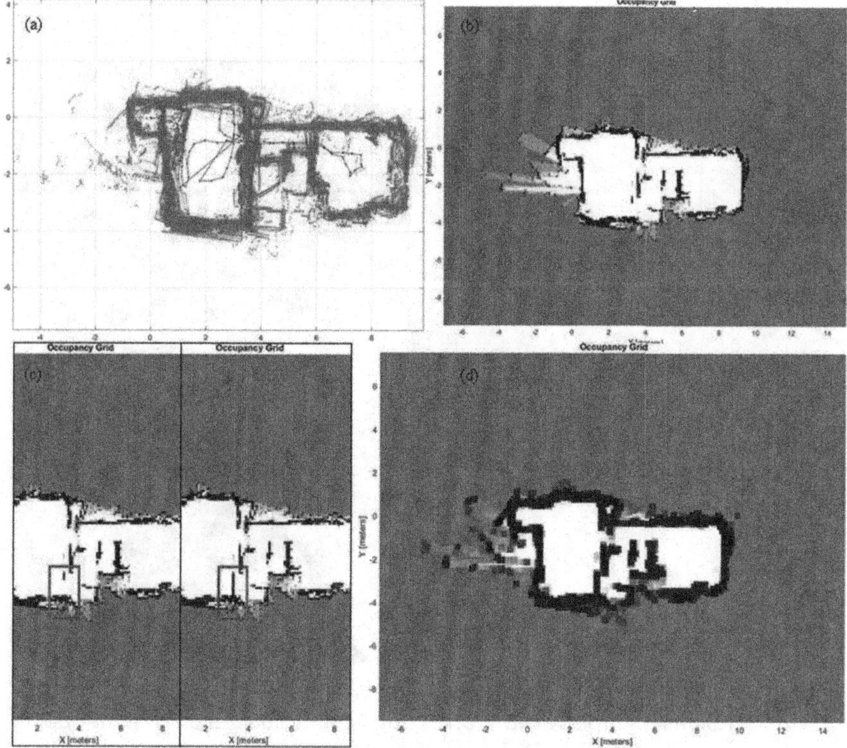

Fig. 3.3 (a) LIDAR scans matched using SLAM, (b) Occupancy map generated, (c) Errors due to the reflective surface, (d) Inflated occupancy map

5. Conclusion

In conclusion, our research presents a wearable assistive system for visually impaired individuals, integrating 1D and 2D LIDARs, a depth camera, and advanced algorithms. The system successfully generates paths, provides real-time auditory guidance, and operates offline. While effective for indoor navigation, challenges include map accuracy dependence and limitations in determining user orientation. Future enhancements could involve GPS in-tegration and object detection using colorized depth frames. Despite current constraints, our system demonstrates the potential for a comprehensive navigation solution for the visually impaired.

This research was made possible by the financial support of Somaiya Vidyavihar University (SVU), research project. The authors would like to thank SVU for their generous contribution.

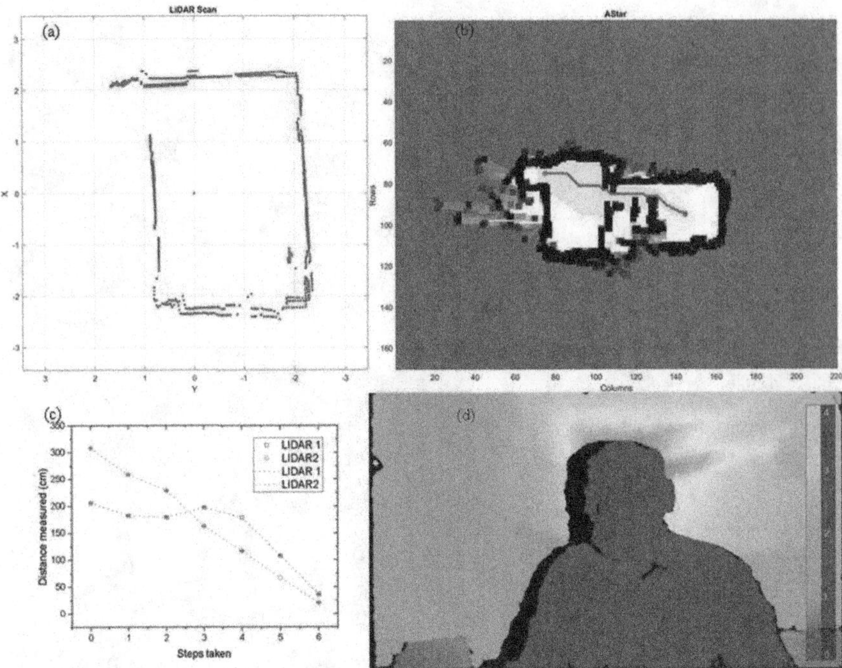

Fig. 3.4 (a) LIDAR scan from the 2D LIDAR, (b) Path obtained using the A* algorithm, (c) Signal plot of two 1D LIDARs, (d) Depth visualization from the depth camera

Acknowledgements

This research was made possible by the financial support of Somaiya Vidyavihar University (SVU), research project. The authors would like to thank SVU for their generous contribution.

References

1. Ahmetovic, D., C. Gleason, C. Ruan, K. Kitani, H. Takagi, and C. Asakawa (2016). NavCog: a navigational cognitive assistant for the blind. In *Proceedings of the 18th International Conference on Human-Computer Interaction with Mobile Devices and Services*, pp. 90–99.
2. Aladren, A., G. López-Nicolás, L. Puig, and J. J. Guerrero (2014). Navigation assistance for the visually impaired using RGB-D sensor with range expansion. *IEEE Systems Journal 10*(3), 922–932.

3. Huang, Z., K. Wang, K. Yang, R. Cheng, and J. Bai (2018). Glass detection and recognition based on the fusion of ultrasonic sensor and RGB-D sensor for the visually impaired. In *Target and Background Signatures IV*, Volume 10794, pp. 118–125.
4. Kanwal, N., E. Bostanci, K. Currie, and A. F. Clark (2015). A navigation system for the visually impaired: a fusion of vision and depth sensor. *Applied bionics and biomechanics 2015.*
5. Li, M. and J. Ammanabrolu (2020). Development of sensing system for indoor naviga-tion of visually impaired person with inertial and geomagnetic information. In *2020 IEEE/ASME International Conference on Advanced Intelligent Mechatronics (AIM)*, pp. 1504–1509.
6. Mueen, A., M. Awedh, and B. Zafar (2022). Multi-obstacle aware smart navigation system for visually impaired people in fog connected IoT-cloud environment. *Health Informatics Journal 28*(3), 14604582221112609.
7. Said, Y., M. Atri, M. A. Albahar, A. B. Atitallah, and Y. A. Alsariera (2023). Scene Recognition for Visually-Impaired People's Navigation Assistance Based on Vision Transformer with Dual Multiscale Attention. *Mathematics 11*(5), 1127.
8. Yusro, M., K. M. Hou, E. Pissaloux, H. L. Shi, K. Ramli, and D. Sudiana (2013). SEES: Concept and design of a smart environment explorer stick. In *2013 6th International Conference on Human System Interactions (HSI)*, pp. 70–77.
9. Zhao, Y., R. Huang, and B. Hu (2019). A multi-sensor fusion system for improving indoor mobility of the visually impaired. In *2019 Chinese Automation Congress (CAC)*, pp. 2950–295

Technologies for Energy, Agriculture, and Healthcare – Shailesh Nikam et al. (eds)
© 2024 Taylor & Francis Group, London, ISBN 978-1-032-98028-7

4

Sugarcane Categorisation using Deep Learning Models on Sentinel Dataset

Mansi Kambli[1]
Research Scholar, Computer Department,
K. J. Somaiya College of Engineering,
Somaiya Vidyavihar University,
Mumbai

Bhakti Palkar[2]
Associate Professor, Computer Department,
K. J. Somaiya College of Engineering,
Somaiya Vidyavihar University,
Mumbai

Abstract: This paper's primary goal is to classify sugarcane crops using deep learning models using Sentinel imaging data. The sugarcane crop is a cash crop used in the production of ethanol and sugar. The classification of sugarcane is crucial for agricultural oversight and management. The conventional crop classification techniques that rely on limited ground-based data collection or manual examination take a lot of time and are usually unreliable. Consequently, an automated and effective approach is proposed, that requires the utilisation of imagery from satellites data and the deep learning techniques. The Convolutional neural network, ResNet and Inception-ResNet are the models applied for sugarcane classification using multispectral satellite data. The categorisation helps the farmers in timely decisions and management of the crop for good yield. The remote sensing through Sentinel data warns the farmers for pests and early detection of disease as categorisation of sugarcane is done prior. Also helps in monitoring the crop for better yield.

Keywords: Sugarcane, Agriculture, Deep learning

[1]mansi.mk@somaiya.edu, [2]bhaktiraul@somaiya.edu

DOI: 10.1201/9781003596707-4

1. Introduction

The primary areas that need attention for the advancement of humanity are agriculture. The Sugarcane, India's second-largest crop, has the potential to be utilized for bioconversion energy. The Classification of sugarcane aids in crop health monitoring and provides farmers with early warnings. The producers of sugarcane find it helpful to keep an eye on the cane that is being sold. The sugarcane has additional benefits for the sugar business. There are spatial, temporal, and spectrum resolution in the remote sensing data and that is the advantage of applying it to farming. The domain for Deep learning techniques employs methods for multi-layered neural networks, a subfield of machine learning. The network layers process the input data, establishing a distinct collection of characteristics and patterns at each level. Eventually, the model's several layers of neural connections will interpret the images and enable it to recognise the features. Furthermore, certain JPEG photos may be used for geographical analysis using deep learning models. On the Canesat dataset, CNN, ResNet and Inception-ResNet are employed, along with assessment metrics. This is how the remainder of the paper is organised. In Part 2, the literature is reviewed. Part 3 lists the variables. Part 4 provides an explanation of the research methodology. Part 5 discusses the result. The paper is summarised in Part 6.

2. Literature Review

The work is carried out to explain the ways in which deep learning has been used to a variety of remote sensing image analysis tasks, including segmentation, object recognition, scene classification, object-based image analysis and the classification for land usage (Ma, Lei et al., 2019). The study's objective is to map sugarcane fields using publicly available Sentinel-1 and Sentinel-2 data according to watershed size and distinguish between plant and ratoon fields(Nihar, Ashmitha et al., 2019). The primary goals of this work are to create a Sentinel-2 image-based technique for mapping sugarcane utilizing data on sugarcane phenology and assess the method's robustness using a few significant classifiers computed from the combined images(Wang, Ming et al., 2019).For accurate mapping of young crops, an extended short-term memory network based on artificial intelligence is used in this study(Khan, Haseeb Rehman et al., 2023).The results demonstrate how effective the proposed fine-tuned ConvNets are in extracting built-up area from a 4-channel Sentinel-2 dataset. Sentinel-2 imagery was therefore used in this investigation to eliminate metropolitan places. The main aim of this work was to employ deep neural networks for categorising four varieties of sugarcane and then to compare the outcomes with conventional techniques for machine learning (Bramhe, V. S., et al., 2018). The sensor bands can provide crop variety characteristics and multi-band values associated with pixels and vegetation indices through RGB combinations using distinct bands (Kai, Priscila Marques et al., 2022).In this study, deep learning-based methods applied for the agricultural domain is explored (Kamilaris, Andreas et al., 2018).The technical difficulties and future research possibilities for deep learning based applications for data from remote sensing is explained in this work (Zhang, Xin et al., 2022). From the survey done it implies that Remote sensing and Deep learning models should be used for better results in all weather conditions. The multidate satellite data sets with higher spatial resolutions are needed to study sugarcane crop growth and stress. Further study is needed on identifying sugarcane varieties, determining plant and ratoon cane differences, and classifying sugarcane phenology. The improved atmospheric correction and satellite images without cloud cover are also needed. Deep learning along with Remote sensing will serve the purpose for the same.

3. Data and Variables

The multispectral high resolution 1627 image patches in Canesat dataset [9] have a pixel size of 10 m × 10 m. The jpeg and tiff formats are offered by the dataset. The deep learning models are applied on jpg files. The input Shape and Data Generators are defined for the processing .Then set input shape, image dimensions, number of classes, and batch size. Also define data generators for training, validation, and test sets.

4. Methodology and Model Specifications

4.1 Convolutional Neural Network (CNN)

The CaneSat dataset is used to train a convolutional neural network model using Sentinel-2 imagery as input. The model architecture consists of three convolutional layers, each employing a 3x3 kernel and filters, preserving original image patches. The model shown in Fig. 4.1 employs a pooling layer with a 2x2 filter size to down sample feature maps, and uses ReLU activation function to inject nonlinearity. The categorical cross entropy is employed for the loss function, and dropout regularisation prevents over fitting. The model then uses a softmax layer for probability distribution

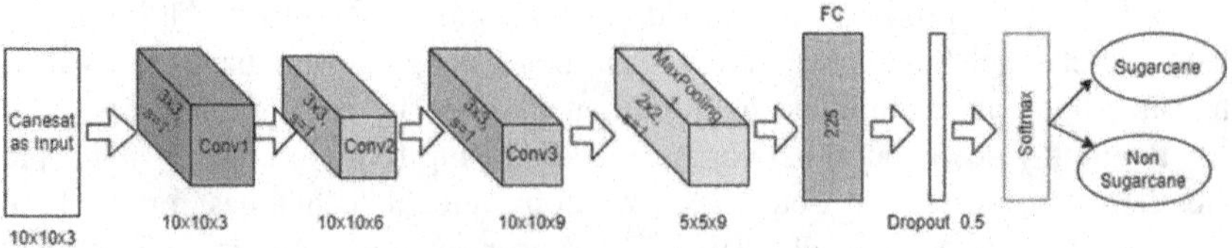

Fig. 4.1 CNN model for categorisation [10]

4.2 ResNet

ResNet is a deep learning model that uses weight layers to learn residual functions, resembling a Highway Network with skip connections and strong positive bias weights. In a multi-layer neural network model, a sub network with multiple stacked layers is called a "Residual Block." The underlying function is represented by a residual function, F(x), with the output represented as y where y= F(x) +x and x ix the input to the sub network as shown in Fig. 4.2.

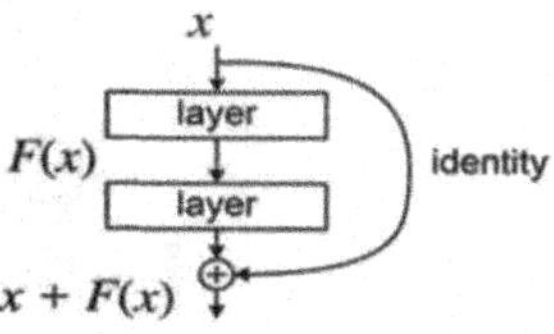

Fig. 4.2 Residual block [1]

4.3 Inception – ResNet

It has 164 layers and consequently, a vast array of image rich feature representations have been trained by the network. To process images, Inception - ResNet combines residual connections, max pooling layers, and convolutional layers as shown in Fig. 4.3.

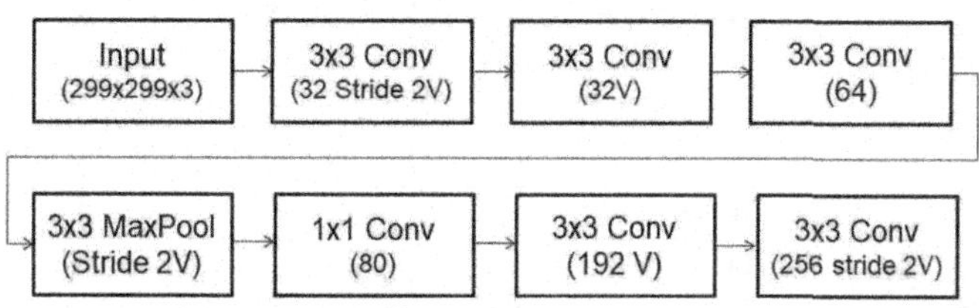

Fig. 4.3 Inception - ResNet [1]

5. Results

Results on Test Dataset: Model 1(CNN): The Class Indices are 'nonsugarcane': 0, 'sugarcane': 1 as shown in Table 4.1 with the performance measures when CNN is applied on the dataset. The confusion matrix is a summary of the expected and actual class labels. For sugarcane the True positive values are 161 and false positive are 65 samples whereas true negative is 87 and false negative is 14 as shown in Fig. 4.4. The loss is 0.5388 and accuracy is 0.7584 as shown in Fig. 4.5 and Fig. 4.6 of CNN loss and accuracy graph.

Table 4.1 CNN classification statistics

Class	Precision	recall	F1score	Support
0	0.86	0.57	0.69	152
1	0.71	0.92	0.80	175

Source: Author

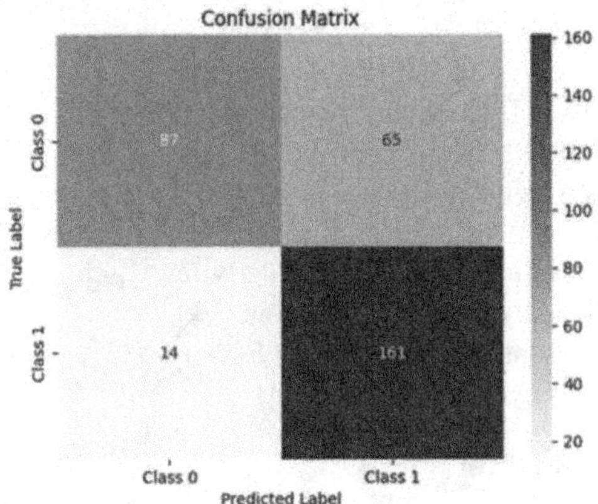

Fig. 4.4 CNN confusion matrix

Source: Author

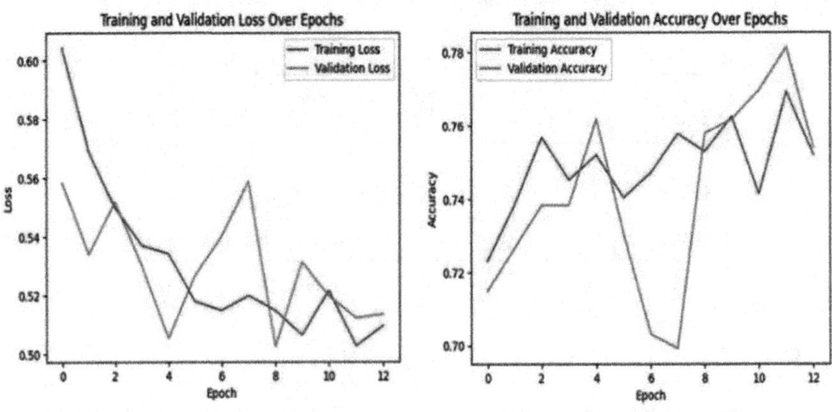

Fig. 4.5 CNN loss graph

Source: Author

Fig. 4.6 CNN accuracy graph

Source: Author

Model 2 (ResNet, transfer learning): The Class Indices are 'nonsugarcane': 0, 'sugarcane': 1 as shown in Table 4.2 with the performance measures when ResNet is applied on the dataset. When ResNet is applied on the sugarcane crop the True positive values are 159 whereas false positive are 74 samples whereas true negative is 78 and false negative is 16, observed in Fig. 4.7. Figures 4.8 and 4.9 show the test Loss and Accuracy, respectively, to be 0.682 and 0.724.

Table 4.2 ResNet categorization statistics

Class	Precision	Recall	F1score	support
0	0.83	0.51	0.693	152
1	0.68	0.91	0.78	175

Source: Author

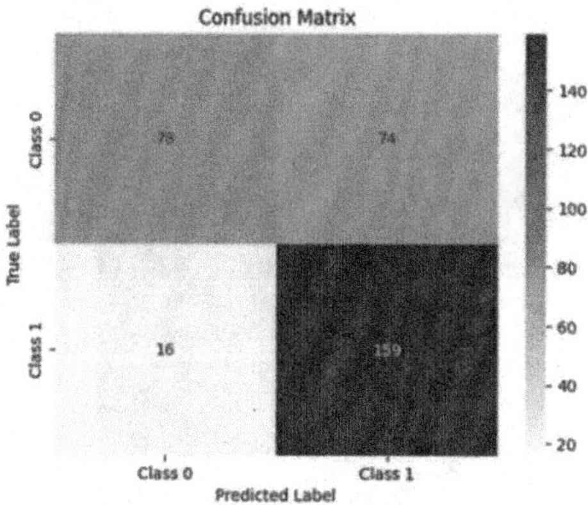

Fig. 4.7 ResNet confusion matrix

Source: Author

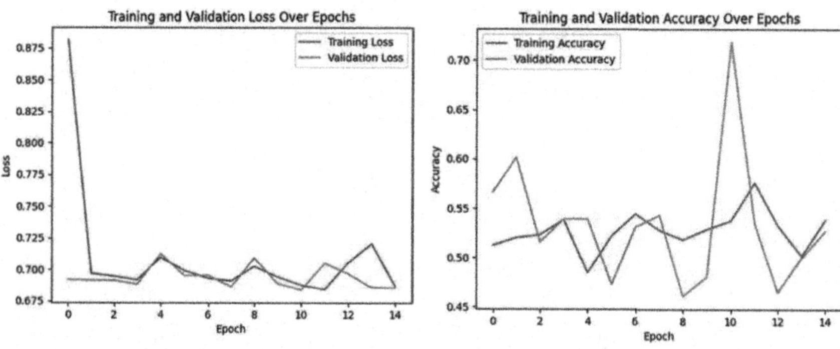

Fig. 4.8 ResNet loss graph

Source: Author

Fig. 4.9 ResNet accuracy graph

Source: Author

Model 3 Inception-ResNet: The Class Indices are 'nonsugarcane': 0, 'sugarcane': 1 as shown in Table 4.3 with the performance measures when Inception-ResNet is applied on the dataset. Figures 4.11 and 4.12 show the test Loss and Accuracy, respectively, to be 0.5524 and accuracy is 0.7523 respectively. When Inception-ResNet is applied on the sugarcane crop the True positive values are 136 whereas false positive are 42 samples whereas true negative is 110 whereas false negative is 39 as shown in Figure 4.10.

Table 4.3 Inception-ResNet categorization

Class	Precision	recall	F1score	support
0	0.74	0.72	0.73	152
1	0.76	0.78	0.77	175

Source: Author

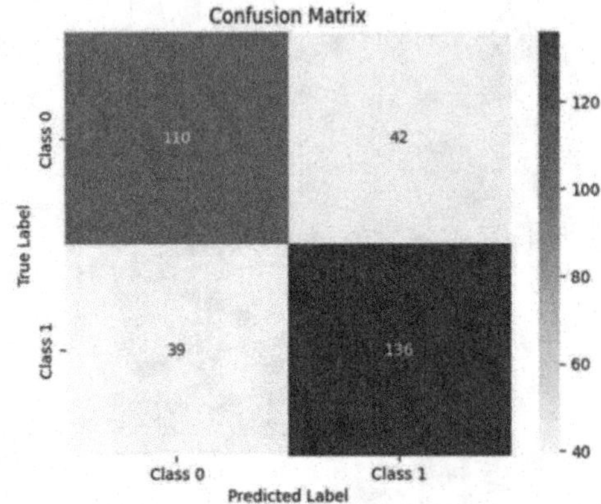

Fig. 4.10 Inception –ResNet statistics confusion matrix

Source: Author

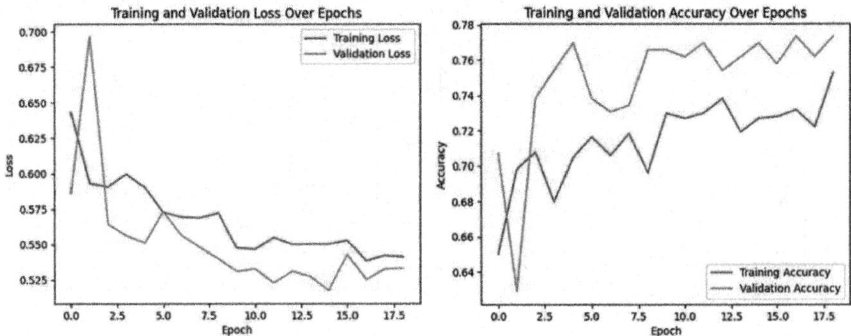

Fig. 4.11 Inception_ResNet loss graph

Source: Author

Fig. 4.12 Inception ResNet accuracy graph

Source: Author

The three models are compared in Table 4.4 and the performance on categorisation of sugarcane is analysed using deep learning models. The precision of Inception-ResNet is superior as compared to CNN and ResNet .The accuracy of CNN and Inception-ResNet is good as compared to ResNet. The satellite imagery data size is small and the dataset is imbalanced as sugarcane sample images are 870 and non-sugarcane is 757 images available. So the accuracy is less and when applied on the bigger dataset or hyperspectral imagery data the results will drastically improve. The satellite dataset is also not publicly available for sugarcane crop. Precision, recall, F1-Score, and accuracy are key metrics in forecasting, indicating the proportion of expected positives to accurately predicted positives for sugarcane categorization. To achieve the best balance of precision and recall in a classification model for sugarcane, we aim to maximize the F1 score. The method effectively minimizes erroneous and missing crop detections and enhances the categorisation. After the categorisation is done the evaluation metrics helps the farmer for the yield monitoring and crop disease detection. Also Precision agriculture utilizes multispectral satellite data to enhance sustainability of agricultural output via measurement, observation, and adaptation to temporal and geographical variability.

Table 4.4 Evaluation table of deep learning models

Models	Accuracy	Precision	Recall	F1 Score
CNN	0.75	0.71	0.92	0.8
ResNet	0.72	0.68	0.91	0.78
Inception_ ResNet	0.75	0.76	0.78	0.77

Source: Author

Future directions: Deep learning methods can be used to analyze remotely sensed data of sugarcane crop, providing insights on canopy, harvest dates, and plantation date gaps. However, challenges include identifying vegetation for individual farmers and also solve the issues of vegetation with slopes and water retention. The advances in algorithms can improve remote sensing use, and hyperspectral imagery data can enhance use of deep learning models. The cutting period of sugarcane can match satellite data indicators and ratoon sugarcane which is grown for one year after the plant cane can be explored more using satellite imagery data. Further satellite imagery data, meteorological data, and spectral bands can help monitor intercrop yield and improve crop management. The atmospheric correction for satellite imagery data can enhance the results and Deep learning methods can be used more efficiently. The Spectral features can be explored more using visualisation techniques and also sensitivity analysis can be applied as a feature selection technique, choosing the least important feature to remove from each backward selection cycle.

6. Conclusion

The work aims to enhance sugarcane classification methods so that farmers, scientists, and other stakeholders may make more informed choices on crop management, yield projections, and resource efficiency in sugarcane farming. Despite the limited dataset, the sugarcane has been classified by deep learning models, and performance metrics like accuracy, recall, and F1 score are displayed.

If additional remote sensing imaging data is employed and databases containing Sentinel imagery are made accessible, the human intervention by the farmers may be sorted out. Then, in the future, accuracy may be increased by combining deep learning algorithms with a bigger sugarcane dataset or hyperspectral dataset. Furthermore, crop monitoring and crop quality assessment may be conducted using the categorised sugarcane crop. Data augmentation and transfer learning are techniques which can be used to enhance a model's generalisability and robustness by artificially expanding a dataset and fine-tuning it for a related task. In order to increase a model's resilience and generalizability, data augmentation is a technique that creates new samples by transforming existing ones using different methods including rotation, scaling, and flipping. The application of technology for self-supervised learning on small datasets has not received much attention and further work needs to be done for agriculture domain. Few-shot and zero-shot learning methods are effective for maximizing performance from limited data, but data augmentation, appropriate evaluation metrics, and multiple models ensembles are crucial for small samples and should be explored in Remote sensing domain in agriculture.

References

1. Ma, Lei, Yu Liu, Xueliang Zhang, Yuanxin Ye, Gaofei Yin, and Brian Alan Johnson. "Deep learning in remote sensing applications: A meta-analysis and review." *ISPRS journal of photogrammetry and remote sensing* 152 (2019): 166–177.
2. Nihar, Ashmitha, N. R. Patel, Shweta Pokhariyal, and Abhishek Danodia. "Sugarcane crop type discrimination and area mapping at field scale using sentinel images and machine learning methods." *Journal of the Indian Society of Remote Sensing* (2022): 1–9.
3. Wang, Ming, Zhengjia Liu, Muhammad Hasan Ali Baig, Yongsheng Wang, Yurui Li, and Yuanyan Chen. "Mapping sugarcane in complex landscapes by integrating multi-temporal Sentinel-2 images and machine learning algorithms." *Land use policy* 88 (2019): 104190.
4. Khan, Haseeb Rehman, Zeeshan Gillani, Muhammad Hasan Jamal, Atifa Athar, Muhammad Tayyab Chaudhry, Haoyu Chao, Yong He, and Ming Chen. "Early Identification of Crop Type for Smallholder Farming Systems Using Deep Learning on Time-Series Sentinel-2 Imagery." *Sensors* 23, no. 4 (2023): 1779.
5. Bramhe, V. S., Sanjay Kumar Ghosh , and Pradeep Kumar Garg. "Extraction of built-up areas using convolutional neural networks and transfer learning from sentinel-2 satellite images." *The International Archives of the Photogrammetry, Remote Sensing and Spatial Information Sciences* 42 (2018): 79–85.
6. Kai, Priscila Marques, Bruna Mendes de Oliveira, and Ronaldo Martins da Costa. "Deep Learning-Based Method for Classification of Sugarcane Varieties." *Agronomy* 12, no. 11 (2022): 2722.

7. Kamilaris, Andreas, and Francesc X. Prenafeta-Boldú. "Deep learning in agriculture: A survey." *Computers and electronics in agriculture* 147 (2018): 70–90.
8. Zhang, Xin, Ya'nan Zhou, and Jiancheng Luo. "Deep learning for processing and analysis of remote sensing big data: A technical review." *Big Earth Data* 6, no. 4 (2022): 527–560.
9. .Lenco, Dino, Roberto Interdonato, Raffaele Gaetano, and Dinh Ho Tong Minh. "Combining Sentinel-1 and Sentinel-2 Satellite Image Time Series for land cover mapping via a multi-source deep learning architecture." *ISPRS Journal of Photogrammetry and Remote Sensing* 158 (2019): 11–22.
10. Virnodkar, Shyamal S., Vinod K. Pachghare, V. C. Patil, and Sunil Kumar Jha. "CaneSat dataset to leverage convolutional neural networks for sugarcane classification from Sentinel-2." Journal of King Saud University-Computer and Information Sciences 34, no. 6 (2022): 3343–3355.

Technologies for Energy, Agriculture, and Healthcare – Shailesh Nikam et al. (eds)
© 2024 Taylor & Francis Group, London, ISBN 978-1-032-98028-7

5

Numerical Investigation of the Influence of Geometric Parameters on Heat Transfer Enhancement by Air Jet Infringement

Avinash M. Rathod[1]
Research Scholar,
Veermata Jijabai Technological Institute,
Mumbai

Nitin P. Gulhane[2]
Associate Professor,
Veermata Jijabai Technological Institute,
Mumbai

Abstract: The main objective of this investigation is to examine how the geometric thickness ratio affects the local heat transfer coefficient under varying heat input conditions. For the current study, 2D axisymmetric numerical simulation is performed in ANSYS FLUENT 2023R1 software. Four distinct cases are analyzed, each involving varied ratios of plate thickness to nozzle diameter (t/d = 0.5, 0.75, 1, 1.25). The nozzle diameter is kept at 8 mm, and maintaining constant nozzle to target separation (z/d = 4). Jet velocities vary with Reynolds number Re = 4000, 8000, 12000 and 16000. The results reveal that a higher heat transfer rate is observed with smaller thickness ratios (t/d = 0.5 and 0.75). However, As the plate thickness ratio increases further from (t/d = 1 to 1.25), the heat transfer rate does not show significant changes. These findings provide crucial insights regarding the rate of cooling in an impinging jet across a flat plate.

Keywords: Nusselt number, Heat transfer rate, Nozzle target separation, Fluctuating heat input

---

[1]amrathod_p21@me.vjti.ac.in, [2]npgulhane@me.vjti.ac.in

DOI: 10.1201/9781003596707-5

1. Introduction

Air jet impingement cooling methods have gained a lot of attention across various industries and engineering applications due to their capacity to handle high heat flux. Jet impingement is highly utilized in cooling turbine blades, Combustion liners, metallic components, electronic equipment, and process industries to achieve high heat transfer rates. The effect of various nozzle configurations on the rate of heat transfer during jet impingement has not received much attention from researchers. Numerous variables, such as the target plate's surface roughness, nozzle geometry, Reynolds number (Re), and nozzle to target separation (z/d), influence the heat transfer qualities. Impinging jet provides an efficient way to transfer heat in quenching process [1]. Zukerman and Lior [1] comprehensively analyzed of various flow regions of impinging jets and proposed a correlation for heat transfer rate. Regarding jet impingement, Katti and Prabhu [2] identified three critical areas and proposed a semi-empirical correlation at stagnancy, advancement, and near-wall jet zones. The experiment was carried out at nozzle exit to target plate distance varied from $0.5 < z/d < 8$ and $12000 < Re < 28000$. Oriana Caggese et al.[2] observed the impact of constrained infringement of the jet on flow structure and heat transfer properties at the small distance of nozzle end to target plate ($0.5 < z/d < 1.5$). Siddique Umair et al. [3] observed the effect of geometric thickness ratio on the cooling rate for a single jet impingement. For a geometric ratio below the critical thickness ($t_c < 0.05$), more time was required to dissipate heat due to distinct pressure zones across the target surface. In line with Umair Siddique, H. I. Shaikh et al. also [4] studied the effect of thickness ratio (0.5, 1, 2, 2.67) for multiple air jet impingement at varying Reynolds number Re = (5000 to 20000) and (z/d = 1 to 5). They observed that a high heat transfer rate was achieved at lower thickness ratios of 0.5 and 1. Lee and Lee [5] examined the impact of heat transfer coefficient for three different orifice shaped square-edge orifice, standard-edged orifice, and sharp-edge orifice. They found that sharp-edge orifices produced the best results. Secondary rises were seen in the Nusselt curve at low nozzle exit to plate distance and high impinging jets by Umair and Gulhane [6]. A critical nondimensional constant (C) determining these secondary peaks was found to depend on Reynold number and nozzle end to target plate separation (z/d). Umair and Gulhane [7] studied the effect of plate material on the discrepancy in the Nusselt profile. ANSYS CFX was used for the simulation, and the SST + $\gamma-\theta$ turbulence model was used. Additionally, Siddique Umair et al.[8] examined the Nusselt number under changing heat input boundary conditions. A rise in the cooling curve was noted in the impingement area due to turbulence fluctuation. Luhar et al.[9] developed an analytical model to optimize temperature rise at multiple spots for cooling microprocessor chips. The model demonstrated good agreement with simulation results, particularly in the context of steady-state jet impingement in microelectronics. Guo et al. [10] carried out an empirical and computational study on transient heat transfer jet impingement over a flat plate. The empirical study was performed for the Reynold number of 14000-53000, nozzle end to plate separation ($4 < z/d < 8$). The result showed that significant heat transfer occurred at the stagnation point for 50-80 seconds, after which it remained constant. Rathod et al.[11] performed computational investigation of constant impinging air flow on a flat plate using the $SST + k - \omega$ turbulence model. Exponential Varying heat input boundary condition was applied over a heated plate. Results indicate that the Nusselt profile for varying heat input boundary conditions provides better agreement over constant heat input boundary conditions.

The main aim of current investigation is to examine how the geometric thickness ratio affects the heat transfer coefficient under variable boundary conditions for heat input. Additionally, the study aims to observe the critical thickness ratio of the target plate beyond which the discrepancy in the Nusselt curve diminishes.

2. Numerical Methodology

2.1 Computational Model

A 2D axisymmetric domain with a jet, target plate, and mesh topology is schematically depicted in Fig. 5.1. The geometric model is constructed using computational simulation software ANSYS FLUENT 2023R1. The nozzle diameter is set at 8 mm [12], with a distance of nozzle end to plate (z/d) of 4.

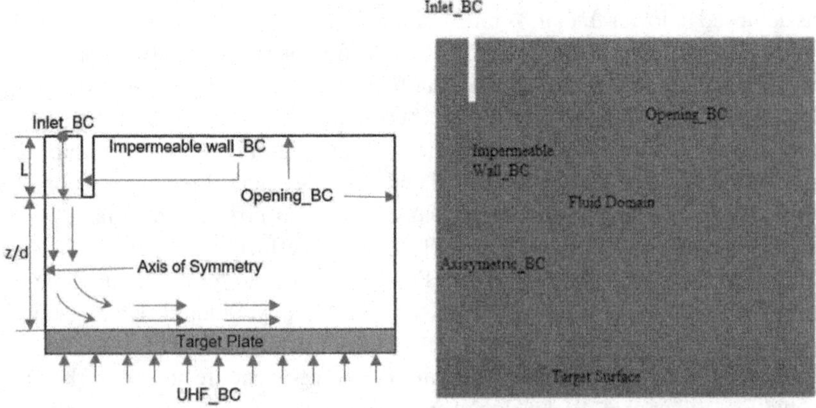

Fig. 5.1 Schematic layout of 2D axisymmetric domain and mesh topology

Simulation is carried out for varying Reynolds number (Re = 4000, 8000, 12000 and 16000). Applying second-order upwind method to the governing equation and considering turbulence intensity within range of 1% – 3% and along a Prandtl number of 0.7 yields the final solution. The minimum distance of the initial grid node is maintained at $Y^{\pm} \leq 1$. steady state simulation iterations continue until the residual reaches convergence level of 10^{-6}. The domain opening is designed as a pressure outlet with an atmospheric temperature condition. Fluctuating heat input is given in Eq. (1) at the bottom of the target plate, with slope (m) value set to 1 and Q_o as a constant heat flux magnitude.

$$Q = Q_o \times mx \tag{1}$$

2.2 Grid Independence Test

The number of nodes, density, and quality of the grid, as well as the computational domain's structure, all have a major impact on heat transport. To ensure grid independence, a test is carried out at Re = 12000 and z/d = 4, using various meshes as illustrated in Fig. 5.2. The grid density along the nozzle end to plate separation (z/d), length of target plate (r/d) is varied, including the meshes 80 × 96, 120 × 144, 140 × 192, 160 × 192, 200 × 240, respectively. The mesh geometry with 200/240 divisions along axial (z/d) and radial (r/d) edge yield the most accurate results. Applying the exact computational model is crucial for identifying the heat transfer across the intended surface. The $k - \alpha$ turbulent model forecasts inaccurate results because abnormal vortices are generated over the impingement area. Typically, the $k - \varepsilon$ turbulent model is used to investigate flow structure and the heat transfer rate of a laminar flow in a pipe.

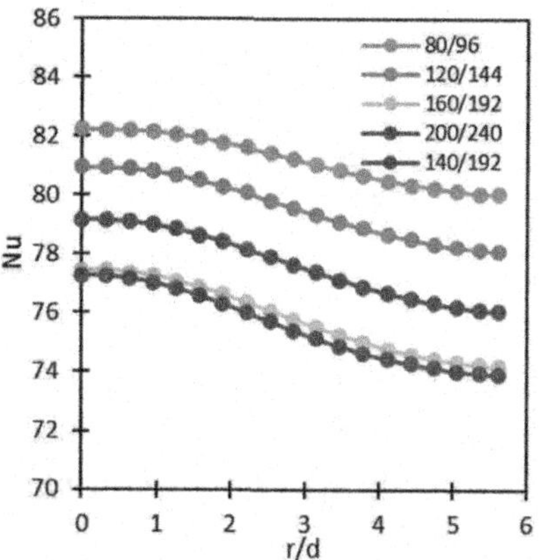

Fig. 5.2 Grid independence test

The $k - \omega$ model provides a precise profile; however, its heat transfer rate tends to deviate considerably. This is caused by turbulence swirls generated by impingement negative pressure gradient. A more accurate heat transfer profile prediction can be attained by combining the SST + $k - \omega$ turbulent model [8]. Menter [13] justifies using SST turbulent to evaluate air jet heat transfer

2.3 Numerical Model Validation

Figure 5.3 presents comparison between numerical results computed at nozzle target distance of (z/d) 4 and Reynolds number of (Re) 12000 and the experimental findings from previous literature of Katti Prabhu [2] and Siddiqui Umair et al [14]. The computed Nusselt magnitude aligns well with the experimental results and falls within the error margin of 10% – 15%.

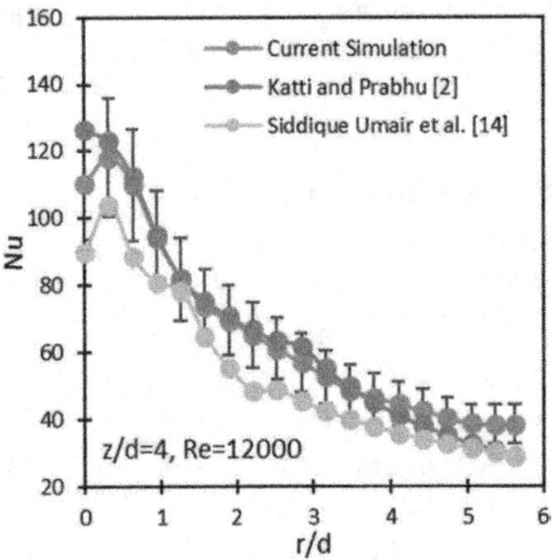

Fig. 5.3 Comparison of nusselt curve at z/d = 4 and Re = 12000

3. Result and Discussion

The numerical work is carried out with fixed nozzle target spacing $z/d = 4$ and geometric thickness ratio ($t/d = 0.5, 0.75, 1, 1.25$). Reynolds numbers are varied in the range of $4000 \leq Re \leq 16000$. Results reveal a decrease in the Nusselt number as the geometric thickness ratio increases, particularly for ratios of $t/d =$ 0.5 and $t/d = 0.75$. However, for the geometric thickness ratio ($t/d = 1$ and 1.25), The Nusselt number does not alter significantly. Moreover, the average Nusselt number rises as the Reynolds number rises while the nozzle target spacing and geometric thickness ratio remain constant.

4. Conclusion

In this study, a numerical investigation is carried out using the $SST + k - \omega$ turbulence model, which accurately predicts the Nusselt profile. The results showed that the geometric thickness ratio impacts the Nusselt number. Especially

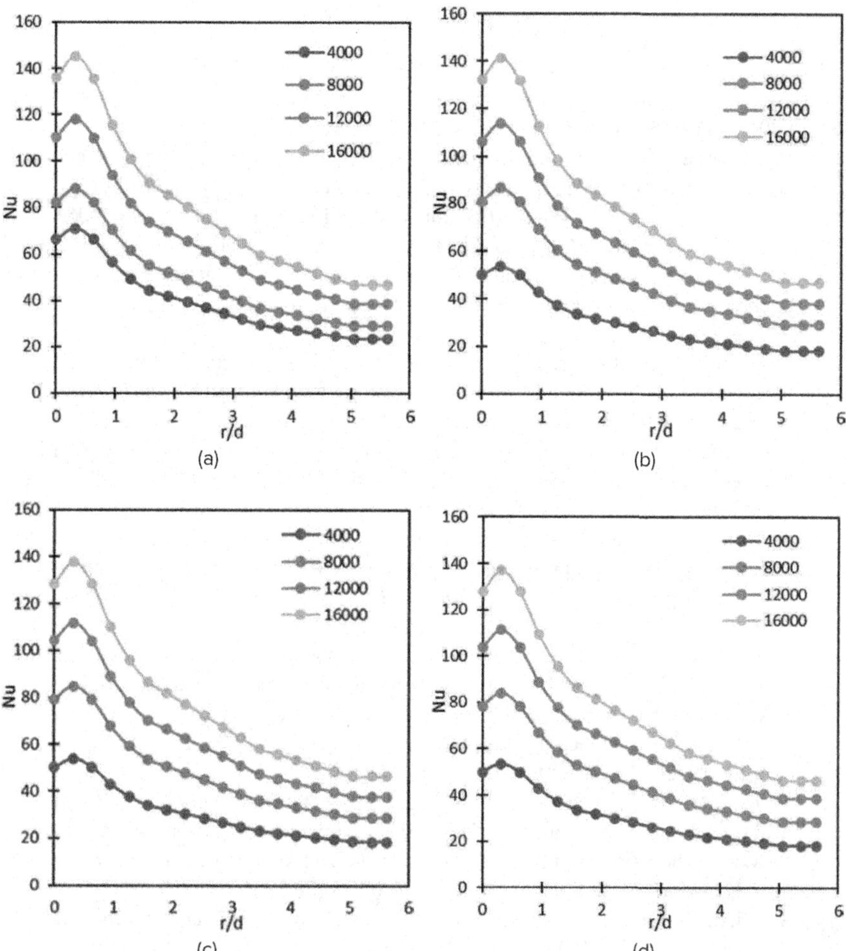

Fig. 5.4 Nusselt profile for different thickness ratios at different Reynolds numbers (a) t/d = 0.5 (b) t/d = 0.75 (c) t/d = 1 (d) t/d = 1.25

a lower jet diameter to plate thickness ratio is associated with a higher Nusselt number, while it falls as jet diameter to plate thickness ratio increases Furthermore, no more changes in the Nusselt profile were observed for (t/d = 1 and 1.25). These findings provide valuable insights into the impact of thickness ratio on local heat transfer jet impingement.

5. References

1. N. Zuckerman and N. Lior, *Jet Impingement Heat Transfer : Physics , Correlations , and Numerical Modeling*, vol. 39, no. 06. Elsevier Masson SAS, 2006. doi: 10.1016/S0065-2717(06)39006-5.
2. O. Caggese, G. Gnaegi, G. Hannema, A. Terzis, and P. Ott, "Experimental and numerical investigation of a fully confined impingement round jet," *Int. J. Heat Mass Transf.*, vol. 65, pp. 873–882, 2013, doi: 10.1016/j.ijheatmasstransfer.2013.06.043.
3. Siddique Umair, N. P. Gulhane, G. B. And, and A. Kolawale, "On numerical heat transfer characteristic study of flat surface subjected to variation in geometric thickness," *Int. J. Comput. Mater.*, vol. 6, no. 1, pp. 1–20, 2017, doi: 10.1142/S2047684117500105.
4. H. I. Shaikh, S. Siddapureddy, and S. V. Prabhu, "Effect of jet plate thickness on the local heat transfer coefficient with multiple air jet impingement," *Appl. Therm. Eng.*, vol. 229, no. April, p. 120517, 2023, doi: 10.1016/j.applthermaleng.2023.120517.
5. J. Lee and S. J. Lee, "The effect of nozzle configuration on stagnation region heat transfer enhancement of axisymmetric jet impingement," *Int. J. Heat Mass Transf.*, vol. 43, no. 18, pp. 3497–3509, 2000, doi: 10.1016/S0017-9310(99)00349-X.
6. S. Mohd Umair and N. Parashram Gulhane, "On numerical investigation of non-dimensional constant representing the occurrence of secondary peaks in the Nusselt distribution curves," *Int. J. Eng. Trans. A Basics*, vol. 29, no. 10, pp. 1431–1440, 2016, doi: 10.5829/idosi.ije.2016.29.10a.00.
7. M. U. Siddique and N. P. Gulhane, "On numerical investigation of nonuniformity in cooling characteristic for different materials of target surfaces Re Pr Nu Z , H Reynolds number Prandtl number Nusselt number Area of base plate (m 2) Diameter of nozzle (m) Turbulence destruction consta," *Int. J. Model. Sci. Comput.*, vol. 8, no. 4, pp. 1–19, 2017, doi: 10.1142/S1793962317500246.
8. M. U. Siddique, A. Syed, S. A. Khan, and J. P. Meyer, "On numerical investigation of heat transfer augmentation of flat target surface under impingement of steady air jet for varying heat flux boundary condition," *J. Therm. Anal. Calorim.*, vol. 147, no. 6, pp. 4325–4337, 2022, doi: 10.1007/s10973-021-10785-4.
9. S. Luhar, D. Sarkar, and A. Jain, "Steady state and transient analytical modeling of non-uniform convective cooling of a microprocessor chip due to jet impingement," *Int. J. Heat Mass Transf.*, vol. 110, pp. 768–777, 2017, doi: 10.1016/j.ijheatmasstransfer.2017.03.064.
10. Q. Guo, Z. Wen, and R. Dou, "Experimental and numerical study on the transient heat-transfer characteristics of circular air-jet impingement on a flat plate," *Int. J. Heat Mass Transf.*, vol. 104, pp. 1177–1188, 2017, doi: 10.1016/j.ijheatmasstransfer.2016.09.048.
11. A. M. Rathod, N. P. Gulhane, and U. M. Siddique, "Heat Transfer Augmentation for Impingement of Steady Air Jet Under Exponential Heat Flux Boundary Condition," *Int. J. Eng.*, vol. 37, no. 05, pp. 920–930, 2023, doi: 10.5829/ije.2024.37.05b.10.
12. V. Katti and S. V. Prabhu, "Experimental study and theoretical analysis of local heat transfer distribution between smooth flat surface and impinging air jet from a circular

straight pipe nozzle," *Int. J. Heat Mass Transf.*, vol. 51, no. 17–18, pp. 4480–4495, 2008, doi: 10.1016/j.ijheatmasstransfer.2007.12.024.

13. F. R. Menter, "Influence of freestream values on k-ω turbulence model predictions," *AIAA J.*, vol. 30, no. 6, pp. 1657–1659, 1992, doi: 10.2514/3.11115.
14. S. Mohd Umair, N. P. Gulhane, A. R. A. Al-Robaian, and S. A. Khan, "On numerical investigation of semi-empirical relations representing local nusselt number at lower nozzle-target spacing's," *Int. J. Eng. Trans. A Basics*, vol. 32, no. 1, pp. 137–145, 2019, doi: 10.5829/ije.2019.32.01a.18.

Note: All the figures in this chapter were made by the authors.

Technologies for Energy, Agriculture, and Healthcare – Shailesh Nikam et al. (eds)
© 2024 Taylor & Francis Group, London, ISBN 978-1-032-98028-7

Stroke Prediction and Analysis Using Different Machine Learning Algorithms

Vedant Sanjay Tadla*,
Pranav Manoj Singh, Kunal Manish Thakkar
Department of Electronics and Telecommunication Engineering,
K.J. Somaiya Institute of Technology,
Mumbai, India

Abstract: In the discipline of healthcare, stroke prognosis is critical to avoiding serious outcomes, requiring innovative methods. Our research uses Machine Learning (ML) methods to provide better stroke prediction. Stroke, a severe medical condition, demands swift attention for improved patient outcomes. Our ML models, including TensorFlow with Softmax and Sigmoid, XGBoost, and Random Forest Classifiers, exhibit robust performance in classifying stroke instances.The TensorFlow models showcase impressive accuracies of 97.33% and 91.46%, with nuanced precision and recall metrics. Meanwhile, the ensemble methods, XGBoost and Random Forest, achieve commendable accuracies of 88.34% and 86.54%, respectively. These outcomes underscore the efficacy of ML in stroke prediction, offering reliable tools for early detection and intervention. The significance of our results lies in the potential to revolutionize stroke risk assessment, aiding healthcare professionals in timely interventions. ML models, with their nuanced predictive capabilities, bring a paradigm shift in preventive healthcare, paving the way for enhanced patient care and outcomes. As we navigate the forefront of medical innovation, our findings contribute to the growing body of knowledge in ML applications for critical healthcare challenges, with the promise of saving lives through timely stroke prediction.

Keywords: Machine learning, Accuracy, Classification

*Corresponding author: vedant.tadla@somaiya.edu

DOI: 10.1201/9781003596707-6

1. Introduction

In addition to being the leading causes for mortality in OECD nations, chronic illnesses such as diabetes, cancer, heart attacks, strokes, and chronic respiratory disorders also greatly increase the burden of disability globally. By addressing important risk factors are obesity, alcoholic beverages, smoking, and a lack of activity, many chronic diseases can be prevented. In each of the 27 OECD nations, one-third of people who are 15 years of age or older report having multiple chronic diseases. This percentage increases to nearly fifty percent those living in Germany and Finland. Elderly individuals are far more likely to have many morbidity (65 years of age and older), averaging 58%; in the European countries, and Germany, it can reach 70% or higher. On the other hand, just 24% of people under 65 report having more than one chronic disease. The OECD countries show clear socioeconomic disparities: on average, 35% of the total population in the quintile with the lowest income and 24% of those in the top income quintile suffer from two or more chronic diseases. The countries of Slovenia, Latvia, and Hungary have the most noticeable wealth disparity [1]Stroke was the leading cause of death in the longitudinal investigation conducted by IJMR, having an age-modified stroke death rate of One Ninety Two per one million individuals. Forty-seven percent of stroke-related deaths that were documented happened at home, and forty-five percent of those deaths happened in the first thirty days after the onset of symptoms.[2]Considering all these factors, we employ sophisticated machine learning models, including Dense Neural Networks and SMOTE analysis, in our comprehensive analysis to tackle challenges in stroke prediction. The overarching aim of our research is to provide valuable insights to stakeholders, assisting them in overcoming challenges associated with stroke prediction and contributing to the development of precise projections in this domain.

2. Literature Review

According to a study conducted by [3] as a systematic literature review (SLR) from January 1, 2010, to March 27, 2023, the study primarily focused on artificial intelligence in clinical labs using imaging data specifically for neurology and cancer prognosis and forecasting. It examined two hundred and twenty SLRs spanning more than ten thousand machine learning (ML) algorithms in healthcare contexts. The review found that fifty-six percent twenty-eight percent and twenty-five percent of SLRs provided details on accuracy, specificity, and sensitivity, respectively. These findings indicated reporting gaps. Less than 1% of cases had external validation, compared to 53% of cases with internal validation. The most common computational technique (2,454 ML algorithms) was neural networks, which was followed by decision trees and random forests/random forests (1,578 and 1,522 machine learning (ML) techniques, mentioned earlier) and support vector machines. In order to accelerate the implementation of ML algorithms into clinical work, the poll highlighted the necessity for improved reporting and more accessibility to medical information for developers.

The classification model results highlighted in [4] showed distinct performance characteristics among various algorithms, with a notable emphasis on accuracy. The "Square Static DHO-GML" model stands out with a perfect accuracy score of 100%, signifying its precise classification abilities. Other models, such as XGB and LGBM, also demonstrate strong accuracy at 90% and 87%, respectively. These accuracy metrics provide a key measure of the models' overall effectiveness in correctly classifying instances, offering valuable insights into their performance and reliability for the given classification task.

The method used in [5] underscores the urgency of swift diagnosis for a stroke and treatment for enhanced patient outcomes. Using SVM and KNN classifiers, the study evaluates the described NHL-FCM method, achieving an accuracy of 95.4 percent with a closeness of 7 percent, surpassing SVM and KNN. Focusing on predicting ischemic stroke risk over 5 years, the research introduces fuzzy cognitive maps, a novel approach in stroke patient classification. The method aligns closely with neurologists' opinions, highlighting its reliability and flexibility in decision support. This study serves as a foundational step in addressing gaps in stroke diagnosis and prediction, offering valuable insights for future research in the field.

In this study [6] ,Convolutional Neural Networks (CNNs) are employed for cardiovascular disease (CVD) prediction, utilizing architectures like VGG-Net and AlexNet. The obtained results showcase notable accuracies, with AlexNet leading at 91.2%. Image preprocessing and feature extraction, crucial in handling complex medical data, involve axis-based preprocessing to address variations in heart volume. Challenges arise when comparing the dataset with VGG-Net data, prompting experimentation with different CNN layers. Despite resource limitations, this study emphasizes the potential of CVD prediction and its application in preventing Coronavirus disease through deep learning algorithms. Future exploration will extend to other pre-trained CNNs, broadening the scope of the study.

In the publication [7], the authors apply LR and algorithms for SVM to predict the occurrence of SVE in Chinese patients six months following MIS. The AUC values of both models were high (>0.9), with SVM outperforming LR in terms of prejudice, calibration, and clinical validity. The top 15 features included important indicators like type 2 diabetes, posterior brain vascular illness, and fasting blood glucose. The significance of excessive fasting blood glucose in predicting unfavorable outcomes in cases of acute ischemic stroke was highlighted. The study underlined hind cerebral arteries lesions as a crucial predictor, underscoring the significance of lesion location, and stressed the value of predictive algorithms over traditional score assessments. The study highlights the promise of artificial intelligence for stroke prediction despite many drawbacks, including its retrospective design and small sample size. This calls for collaboration in data development for future improvements.

This study [8] examined many machine learning models for early stroke detection, differentiating between intracerebral hemorrhage (ICH), subarachnoid hemorrhages (SAH), and ischemic stroke (IS). With better AUC performance (88%), the mGF categorization turns out to be the most precise (83%). The generalized SVM kernel model exhibits promising accuracy in differentiating between stroke recovery stages when it uses mGF parameters for oversampling and undersampling. Nevertheless, there are certain drawbacks, such as the impact of the SMOTE technique that has not yet been investigated and possible variance problems in the training data that is produced. The suggested approach must be further investigated in order to combine it with other physiological indications for improved patient outcomes and early stroke identification.

This study uses the Hybrid EDL-TSA technique via the cardiovascular disease and UCI datasets[9] to diagnose heart disease. Metrics including reliability, recall, particularity, exactness, PME, and the root mean an F-score, False positive and negative rates are included in the evaluation. Preprocessing, grouping with DBSCAN, and utilizing the combination of the EDL-TSA algorithm are the three primary processes in the prediction of cardiac conditions. Using TSA, the EDL weight linkages are optimized. Excellent correctness (98.33%), exactness (97.24%), recall (99.41%), as well as additional positive metrics are displayed in the CVD dataset results. The study exceeds current methods and recommends more research on the best way to choose features and test the system using diverse datasets from multiple sources in order to evaluate its performance in more detail.

A study on machine learning techniques for stroke prediction in [10] is presented. The goal is to create a model that, combining clinical, lifestyle, and demographic data, reliably identifies people who are at risk. Both stroke and non-stroke patients with comprehensive risk factor data are included in the dataset. Training and testing are done using machine learning algorithms like the XGBoost and logistic regression. According to the findings, the XGBoost achieves good accuracy (92.64% in the absence of cross-validation and 95.90% in it). The usefulness of AI in stroke prediction is demonstrated by this study, supporting early intervention approaches.

With 90% accuracy, this study [11] presents a medical framework that uses LSTM (Long Short-Term Memory) networks to identify anomalies related to stroke in

Table 6.1 Key summary of literature survey

Study	Method	Key Findings
3	Healthcare Systematic Literature Review (SLR) on Machine Learning	1. Analyzed 220 SLRs with a total of 10,462 ML methods. 2. The main emphasis is artificial intelligence applications in neurology and cancer for clinical forecasting and disease prognosis. 3. The most common modeling approach is neural networks. Monitoring gaps in SLRs were found.
4	Classification Models	1. Square Static DHO-GML achieved 100% accuracy. 2. Notable accuracy metrics for XGB (90%) and LGBM (87%).
5	NHL-FCM Method for Stroke Prediction	1. NHL-FCM method achieved (95.4 ± 7.5)% accuracy. 2. Fuzzy cognitive maps introduced for stroke patient classification. 3. Method aligns closely with neurologists' opinions.
6	CNNs for CVD Prediction	1. CNNs (VGG-Net, AlexNet) for CVD prediction. 2. AlexNet achieved the highest accuracy at 91.2%. 3. Challenges in dataset comparison with VGG-Net.
7	LR and SVM for SVE Prediction	1. SVM demonstrated superior discrimination and clinical validity. 2. Key predictors identified, including diabetes and posterior cerebral artery disease. 3. Emphasis on machine learning over traditional scoring assessments
8	ML for Early Stroke Detection	1. mGF classification achieved 83% accuracy with 88% AUC. 2. SVM kernel model showed promising accuracy in separating stages of stroke recovery. 3. Future research needed for integration with other physiological signals.
9	Hybrid EDL-TSA for Heart Disease Diagnosis	1. The hybrid EDL-TSA approach is used to diagnose cardiac problems. 2. Attained excellent recall (99.41%), precision (97.24%), and accuracy (98.33%). 3. Outperformed current methods. 4. Future research recommended the best features to test and select from a variety of datasets.
10	XGBoost and logistic regression.	1. The XGBoost model outscored other classifiers in stroke estimation, with 92.64% efficiency without the cross-validation and 95.90% efficiency with it.
11	LSTM and Deep Learning Methods	1. The study uses long-short-term memory (LSTM) networks, which outperform previous deep-learning models and achieve 90% accuracy for ECG-based cerebrovascular abnormality identification.
12	Deep Neural Networks (DNN)	1. Using neural networks with deep learning for risk of stroke prediction, the study obtains an accuracy rate in excess of 92.57%, training success rate of 98.2%, validation success rate of 94%, precision of nearly 95%, recall of 97%, and F1 score of about 96%.

Source: Author

ECG data. By achieving 93.78% accuracy, LSTM outperforms other models using deep learning in comparison, indicating its effectiveness in early stroke diagnosis.

The goal of the study [12] is to use Deep Neural Networks (DNN) to create a prediction model for identifying stroke risk. With around 92.57% test efficiency, 98.2% accuracy in training, and 94% validation accuracy, the model achieves an astounding 90% accuracy rate. Furthermore, the model's efficacy in supporting premature stroke detection and mitigation is demonstrated by its about 95% precision, 97% recall, and nearly 96% F1 score.

3. Methods

3.1 Data Collection and Data Pre Processing

The dataset, consisting of information on 5110 individuals with various demographic and health-related features, has been curated from data available on Kaggle. The dataset's diverse variables, ranging from age and gender to medical history and lifestyle choices, offer insights for potential analyses and predictive modeling in the context of stroke risk assessment.This dataset appears to be focused on demographic and health-related factors of individuals, particularly in the context of stroke occurrence. The features cover a range of characteristics such as age, medical history, lifestyle factors, and socioeconomic aspects, providing a comprehensive overview for potential analysis and modeling. The dataset, sourced from Kaggle, comprised demographic and health-related features, including gender, age, heart disease, hypertension, marital status, type of occupation, type of real estate, average blood sugar threshold, BMI, habit of smoking, and likelihood of stroke (binary). Resolving the lack of values, encoding variables with categories, making sure the right data types were used, and maybe normalizing the data were all part of the preprocessing stage. The goal variable, "stroke," denotes whether a stroke is present (1) or absent (0). The goal of the preprocessing was to improve the quality of the data for machine learning and further analysis.

3.2 Exploratory Data Analysis (EDA)

The exploratory data analysis (EDA) for the dataset involved a comprehensive examination of demographic and health-related features. Initial steps included checking for missing values, addressing any discrepancies in data types, and encoding categorical variables. For understanding the distributions and central trends in numerical parameters such as age, average blood sugar, and body mass index, descriptive statistics were generated. Visualizations, such as histograms and bar charts, were employed to gain insights into the distribution of categorical variables like gender, work type, and smoking status. The EDA process aimed to uncover patterns, trends, and potential relationships within the data, providing a foundation for subsequent analyses and model development.

3.3 Building a Machine Learning Model

A hybrid model-building approach combined TensorFlow for deep learning with sigmoid and softmax layers, along with ensemble methods like XGBoost and Random Forest Classifiers. This blend aimed to capture intricate patterns while benefiting from ensemble model stability. The model underwent training, validation, and fine-tuning for optimal performance in binary and multi-class classification tasks. To address class imbalances, SMOTE with the Random Sampler generated synthetic instances, enhancing balance and overall model robustness.

3.4 Evaluation

The final steps revolve around evaluating the model's performance. Different metrics which gives us the correctness and exactness comes into play, along with cross-validation techniques to ensure robustness. Comparing multiple models helps pinpoint the most effective one, and iterative improvement based on evaluation results refines the overall process. Thorough documentation of the chosen model and its parameters ensures clarity and reference for future endeavors.

4. Results

4.1 Data Preprocessing

The dataset includes 2994 female and 2115 male entries, with one entry labeled as "Other," which was subsequently dropped. There are no null values in the dataset. Regarding BMI, 4909 entries were available, and missing values were filled with the median value.

4.2 Exploratory Data Analysis (EDA)

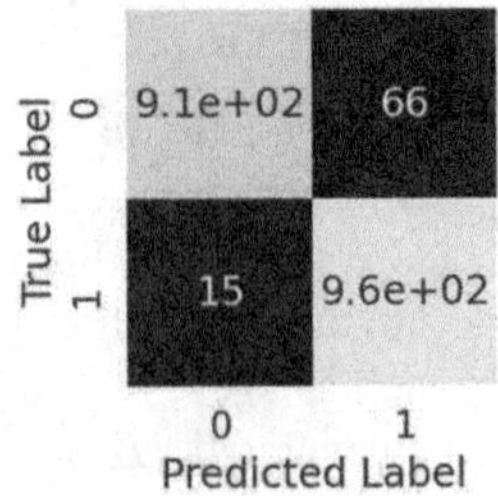

Fig. 6.1 Confusion matrix

Source: Author

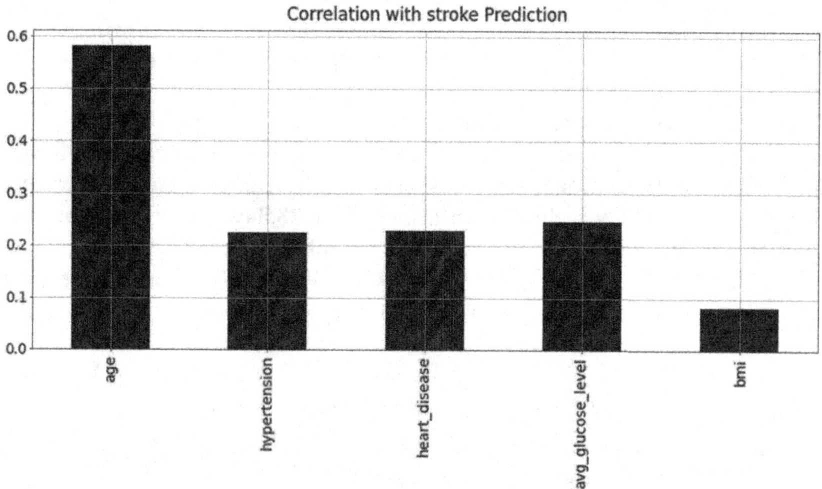

Fig. 6.2 Correlation with stroke

Source: Author

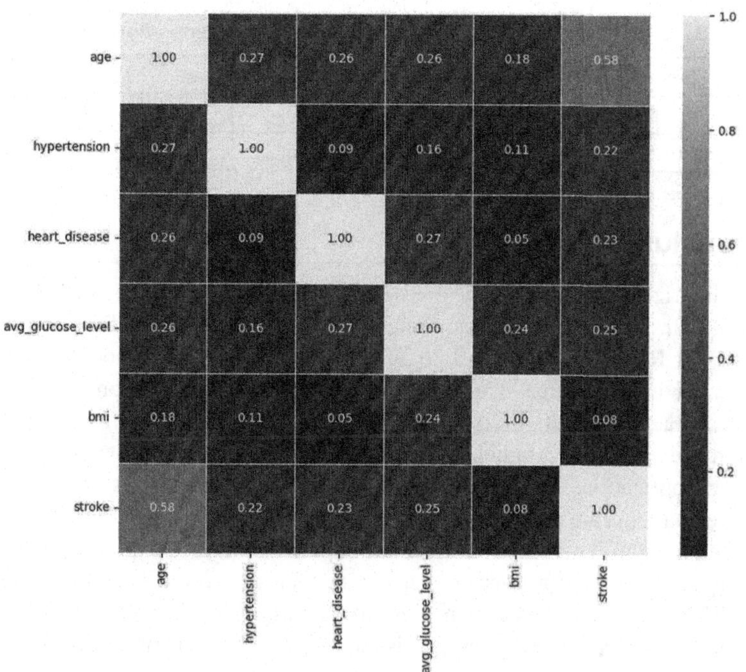

Fig. 6.3 Correlation matrix

Source: Author

4.3 Results

The classification results demonstrate strong performance across various models. The TensorFlow model with Softmax activation achieved an correctness of 97.33%, with high precision and recall for both classes (0 and 1). The TensorFlow model with Sigmoid activation yielded an accuracy of 91.46%, maintaining solid precision and recall scores. The XGBoostClassifier, fine-tuned through GridSearchCV, achieved an accuracy of 88.34%, exhibiting balanced precision and recall for both classes. The RandomForestClassifier, optimized via RandomSearchCV, attained an accuracy of 86.54%, with commendable precision and recall scores for both classes. Overall, these results highlight the effectiveness of the models in accurately predicting the target variable.

Table 6.2 Summary of results obtained in our study

Model	Tensorflow (Softmax Layer)	Tensorflow (Sigmoid Layer)	XGBoost	RandomForest
Accuracy	97.33 %	91.46	88.34	86.54

Source: Author

Table 6.3 Summary of precision, recall and F1-score of tensorflow (softmax layer)

precision	recall	F1 score
0	0.98	0.93
1	0.94	0.98

5. Conclusion

To summarize, our study is the first to use machine learning (ML) to predict strokes, with remarkable accuracy rates of 97.33% and 91.46% for TensorFlow, 86.34% for Random Forest, and 86.34% for XGBoost. These models transform the manner in which that stroke risk is assessed, facilitating prompt preventive healthcare actions. Our findings promise improved quality of life for patients through precise stroke predictions, contributing to the advancement of machine learning applications in important healthcare concerns. Our research promises improved patient care and results as we push the boundaries of medical innovation and bring insightful new knowledge to the expanding body of expertise in machine learning applications for pressing healthcare issues. Layered neural systems and SMOTE analysis are applied to improve the accuracy of stroke predictions, which is a significant step toward more precise prognoses in this vital area.

References

1. OECD (2019), "Chronic disease morbidity", in Health at a Glance 2019: OECD Indicators, OECD Publishing, Paris, https://doi.org/10.1787/5101558b-en.
2. Kamalakannan S, Gudlavalleti ASV, Gudlavalleti VSM, Goenka S, Kuper H. Incidence & prevalence of stroke in India: A systematic review. Indian J Med Res. 2017 Aug;146(2):175-185. doi: 10.4103/ijmr.IJMR_516_15. PMID: 29265018; PMCID: PMC5761027.
3. Katarzyna Kolasa, Bisrat Admassu, Malwina Hołownia-Voloskova, Katarzyna J Kędzior, Jean-Etienne Poirrier & Stefano Perni (2023) Systematic reviews of machine learning in healthcare: a literature review, Expert Review of Pharmacoeconomics & Outcomes Research, DOI: 10.1080/14737167.2023.2279107
4. P. Dhivya, P. Rajesh Kanna, K. Deepa & S. Santhiya (2023) Square Static – Deep Hyper Optimization and Genetic Meta-Learning Approach for Disease Classification, IETE Journal of Research, DOI: 10.1080/03772063.2023.2206367
5. Mahsa Khodadadi, Heidarali Shayanfar, Keivan Maghooli & Amir Hooshang Mazinan (2019) Prediction of stroke probability occurrence based on fuzzy cognitive maps, Automatika, 60:4, 385–392, DOI: 10.1080/00051144.2019.1622883
6. Malathi S, Arockia Raj Y, Abhishek Kumar, V D Ashok Kumar, Ankit Kumar, Elangovan D, V D Ambeth Kumar, Chitra B & a Abirami (2023) Prediction of cardiovascular disease using deep learning algorithms to prevent COVID 19, Journal of Experimental & Theoretical Artificial Intelligence, 35:6, 791–805, DOI: 10.1080/0952813X.2021.1966842
7. Rong Zhang & Jingfeng Wang (2022) Machine Learning-Based Prediction of Subsequent Vascular Events After 6 Months in Chinese Patients with Minor Ischemic Stroke, International Journal of General Medicine, 15:, 3797–3808, DOI: 10.2147/IJGM.S356373
8. Mohammad A. Mezher (2022) Genetic Folding (GF) Algorithm with Minimal Kernel Operators to Predict Stroke Patients, Applied Artificial Intelligence, 36:1, DOI: 10.1080/08839514.2022.2151179
9. Jaishri Wankhede, Palaniappan Sambandam & Magesh Kumar (2022) Effective prediction of heart disease using hybrid ensemble deep learning and tunicate swarm algorithm, Journal of Biomolecular Structure and Dynamics, 40:23, 13334–13345, DOI: 10.1080/07391102.2021.1987328
10. S. Sharma, "Stroke Prediction Using XGB Classifier, Logistic Regression, GaussianNB and BernaulliNB Classifier," 2023 International Conference on Circuit Power and Computing Technologies (ICCPCT), Kollam, India, 2023, pp. 1577-1581, doi: 10.1109/ICCPCT58313.2023.10245269. keywords: {Logistic regression;Machine learning algorithms; Computational modeling; Support vector machine classification; Stroke (medical condition); Sensitivity and specificity; Prediction algorithms; Artificial Intelligence; Machine Learning; Breast Cancer Prediction; XGB Classifier; Support Vector Machine; Logistic Regression},
11. M. Anand Kumar, N. Abirami, M. S. Guru Prasad and M. Mohankumar, "Stroke Disease Prediction based on ECG Signals using Deep Learning

Techniques," 2022 International Conference on Computational Intelligence and Sustainable Engineering Solutions (CISES), Greater Noida, India, 2022, pp. 453–458, doi: 10.1109/CISES54857.2022.9844403. keywords: {Deep learning;Wearable computers;Receivers;Organizations;Electrocardiography;Stroke (medical condition);Environmental factors;Body area networks;Deep learning;ECG;Medical;Strokes;LSTM;and wearable devices},

12. R. Kurlekar, A. Sharma, S. Dalvi, R. Shetty and S. Bharne, "Stroke Risk Prediction Using Deep Neural Networks: Empowering Healthcare Services for Early Identification and Prevention," 2023 International Conference on Network, Multimedia and Information Technology (NMITCON), Bengaluru, India, 2023, pp. 1–7, doi: 10.1109/NMITCON58196.2023.10276249. keywords: {Hospitals;Artificial neural networks; Stroke (medical condition);Predictive models;User interfaces;Risk management;Resource management;Brain Stroke risk prediction;Healthcare;Early Detection;Deep Neural; Machine Learning}.

Technologies for Energy, Agriculture, and Healthcare – Shailesh Nikam et al. (eds)
© 2024 Taylor & Francis Group, London, ISBN 978-1-032-98028-7

7

Revolutionizing Healthcare: The Transformative Power of Generative AI—Risks and Rewards

Sachin R Thorat[1]
Research Scholar,
Somaiya Vidhyavihar University, Mumbai,
Maharashtra, India

Davendranath G Jha[2]
Professor,
Data Science and Technology,
Somaiya Vidhyavihar University, Mumbai,
Maharashtra, India

Abstract: Generative AI is a type of artificial intelligence that excels at crafting original content like images, text, and even code. It has the potential to transform healthcare by creating new tools and applications for diagnosis, treatment, and research Generative AI is making waves in healthcare by promising breakthroughs in diagnostics, drug discovery, and patient care. The development of synthetic medical data through generative AI promises to revolutionize healthcare research. Synthetic data can be used to develop training data for machine learning models, which can then be used to diagnose health issues, and develop new treatments to improve clinical decision-making.

Generative AI has the potential to further revolutionize healthcare is by development of new drug discovery tools. Generative AI can be used to build new molecules that have the potential to treat diseases.

The road to widespread adoption of generative AI in healthcare requires addressing some crucial limitations. One challenge is that generative AI models can be biased, which can lead to unfair results. Another challenge is that generative AI models can be computationally expensive to train and deploy.

[1]Sachin.thorat@somaiya.edu, sach.thorat@gmail.com; [2]dgjha@somaiya.edu

DOI: 10.1201/9781003596707-7

Despite these challenges, the promise of Generative AI for advancing in healthcare are significant. This technology has the potential to improve the diagnosis, treatment, and research of diseases, and it could ultimately improve the health of individuals and populations.

This research paper explores the current landscape of generative AI applications in healthcare, highlighting its promises in improving medical outcomes while addressing the associated risks, including ethical concerns, data privacy issues, and regulatory challenges. The paper aims to provide an informed perspective on the balance between the rewards and risks of employing generative AI in healthcare, drawing insights from recent research and scholarly work in the field.

The integration of Generative Artificial Intelligence (AI) in healthcare holds the

Keywords: Healthcare, Generative AI, Personalized medicine, Drug discovery, medical imaging, Diagnostic accuracy, Data privacy, Ethical considerations

1. Introduction

In recent years, generative AI, a subset of artificial intelligence, has shown tremendous potential to transform various industries, with healthcare being no exception. Generative AI refers to machine learning techniques that enable the creation of data that closely resembles real-world information. In the healthcare sector, these techniques are being employed to enhance medical image synthesis, drug discovery, predictive analytics, and personalized treatment plans. However, the integration of generative AI in healthcare brings both opportunities and challenges. Generative Instead of relying on traditional methods, healthcare providers are increasingly turning to AI as a transformative tool, paving the way for advancements in disease diagnosis, drug discovery, and the individualization of patient treatment plan. However, the adoption of generative AI in healthcare brings forth a complex interplay of potential rewards and associated risks. This review critically analyzes the current state of generative AI applications in healthcare, identifying key trends, limitations, and potential clinical impact.

2. Background

Healthcare faces a critical juncture. Rising costs, an aging population, and chronic diseases burden healthcare systems globally. Traditional approaches struggle to keep pace, necessitating a paradigm shift. This is where Generative AI (GenAI) emerges as a transformative force, promising to revolutionize healthcare delivery and patient outcomes.

Generative Artificial Intelligence (AI), powered by deep learning techniques, has gained substantial attention in recent years for its potential to transform healthcare. It enables the creation of synthetic data, aids in medical image analysis, and enhances the efficiency of drug discovery. However, the integration of generative AI in healthcare is not without its challenges and ethical considerations. This paper aims to critically examine the risks and rewards associated with generative AI in healthcare.

3. Scope

This paper focuses on generative AI applications in healthcare, emphasizing both the potential benefits and the ethical, privacy, and regulatory challenges. It is essential to strike a balance between harnessing the power of AI for improved healthcare and addressing the associated risks.

This paper focuses on the applications of generative AI in healthcare, emphasizing its potential rewards, ethical concerns, data privacy issues, regulatory challenges, and the importance of validation and transparency.

4. Historical Development

4.1 Early Seeds (1950s-1990s)

- 1950s: The concept of machine learning emerges, laying the groundwork for generative models with pioneers like Arthur Samuel and Frank Rosenblatt. (AL 1959)
- 1960s: Joseph Weizenbaum's ELIZA, the first chatbot, demonstrates text generation capabilities. (MT 2019)
- 1980s: Procedural content generation in games like Rogue paves the way for dynamic content creation. (J 2023)
- 1990s: Advancements in hardware and data availability fuel machine learning research.

4.2 The Bloom Begins (2000s-2010s)

- 2000s: Markov Chain models and statistical methods are used for basic text and image generation. (J. Ho 2020)
- 2010s: Deep learning revolutionizes the field, enabling more complex generative models.
- 2014: Generative Adversarial Networks (GANs), introduced by Ian Goodfellow, dramatically improving image realism. (Goodfellow I J 2014)
- 2015-2019: Variational Autoencoders (VAEs) (Kingma DP 2019) and other deep generative models emerge, expanding capabilities beyond images.

4.3 Recent Flourishing (2020s-Present)

- 2020s: Explosion of research and development in generative AI applications across various domains.
- 2022: Large Language Models (LLMs) like GPT-3 and LaMDA demonstrate impressive text generation and comprehension abilities.
- 2023-Present: Generative AI continues to evolve, pushing boundaries in areas like text-to-image generation, video synthesis, and personalized medicine.

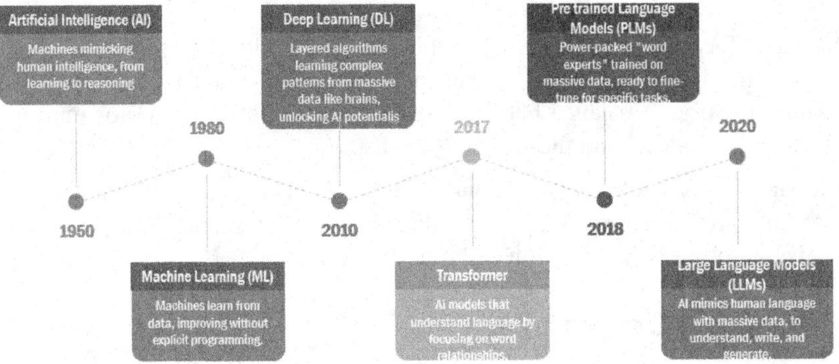

Fig. 7.1 AI's language revolution: From data to Dialogue - timeline of the evolution of artificial intelligence, with key milestones and technologies

5. Rewards of Generative AI in Healthcare

5.1 Improved Diagnosis and Disease Prediction

Generative models have demonstrated exceptional capabilities in improving the accuracy of diagnostic imaging, such as identifying anomalies in medical images (Deng et al., 2020).Generative AI has shown immense potential in enhancing diagnostic accuracy and disease prediction.

5.2 Drug Discovery and Development

Generative AI's contribution to streamlined drug discovery by generating molecular structures and predicting drug-drug interactions

- **Generating novel molecules:** AI can design and simulate potential drug candidates, accelerating the discovery process and reducing costs (G 2018).
- **Predicting drug efficacy and safety:** AI models Harnesses the power of big data analytics to uncover hidden patterns and generate predictive insights. how drugs will interact with the body, leading to more personalized and effective treatments. (Zeng X 2022)

5.3 Personalized Treatment Plans

Personalized medicine is a promising avenue in healthcare, and generative AI contributes to its realization. By analyzing vast patient datasets, generative models can assist in tailoring treatment plans to individual patients, considering their unique genetic and clinical profiles. This has the potential to improve patient outcomes significantly.

- **Generating synthetic patient data:** AI can create realistic medical records to train algorithms for personalized diagnosis and treatment plans, addressing data privacy concerns. (Ghosheh G 2022)
- **Predicting disease risk and progression:** AI models can analyze individual genetic and medical data to predict disease risk and personalize preventive measures. (Ahmad LG 2013)

5.4 Medical Imaging and Diagnosis

- **Enhancing image resolution and quality:** AI can upscale low-resolution images or fill in missing data, improving diagnostic accuracy and reducing the need for invasive procedures. (Zhang S 2018)
- **Detecting early-stage diseases:** AI algorithms can analyze medical images to identify subtle abnormalities, enabling earlier disease detection and intervention. (Mushtaq G 2021)

5.5 Medical Robotics and Surgery

- **Planning and simulating complex surgeries:** AI can generate realistic 3D models of organs and tissues, allowing surgeons to plan and rehearse complex procedures with greater precision. (Park JJ 2022)
- **Developing assistive robots for surgeons:** AI-powered robots can assist surgeons during minimally invasive procedures, improving accuracy and reducing risks. (Munoz VF 2000)

6. Risks and Challenges

6.1 Data Privacy and Security

- **Synthetic patient data generation:** Balancing the benefits of realistic synthetic data for training and research with potential misuse and re-identification risks (Prezja F 2022)
- **Algorithmic bias:** Ensuring fairness and inclusivity in AI models trained on potentially biased healthcare data ("Addressing racial bias in artificial intelligence healthcare algorithms," by Obermeyer et al.).

- **Data ownership and control:** Establishing clear guidelines for patient data ownership and control when used for model training and generation (Agarwal R 2023)

6.2 Transparency and Explainability

- **Black box problem:** Ensuring transparency in AI decision-making, especially when algorithms generate diagnoses or treatment recommendations (Linardatos P 2021).
- **Human oversight and accountability:** Maintaining human oversight and accountability for AI-driven healthcare decisions to prevent unintended consequences (Linardatos P 2021)

6.3 Patient Autonomy and Informed Consent

- **Transparency about AI use:** Informing patients about the potential use of generative AI in their diagnosis and treatment (Linardatos P 2021)
- **Right to refuse AI-based interventions:** Ensuring patients retain agency and have the right to refuse AI-driven recommendations or treatments (Naik N 2022).

6.4 Equity and Access

- **Preventing further disparities:** Mitigating the risk of AI exacerbating existing healthcare disparities by ensuring equitable access and addressing potential biases in models and algorithms (Agarwal R 2023).
- **Accessibility for vulnerable populations:** Ensuring all patients, regardless of socioeconomic background or technological literacy, can benefit from AI-powered healthcare (Kulkarni S 2020)

7. Regulatory and Legal Hurdles

7.1 Data Privacy and Security

- Generating data from patient records raises concerns about privacy violations and potential misidentification. (Blease C 2019)
- Regulations like Data Personal Data Protection Act (DPDPA), Health Insurance Portability and Accountability Act (HIPAA) and General Data Protection Regulation (GDPR) need careful consideration and compliance.

7.2 Algorithmic Bias and Explainability

- Generative models trained on biased data can perpetuate or even amplify existing disparities in healthcare. (Parikh RB 2019)

- Lack of transparency in algorithms makes it difficult to assess fairness and identify potential harms.

7.3 Regulatory Uncertainty

- Existing regulations don't fully address the complexities of generative AI outputs, creating ambiguity and hindering innovation. (FDA 2021).
- The evolving nature of AI adds to the challenges of establishing clear regulatory frameworks.

7.4 Liability and Accountability

- Determining responsibility for AI-driven decisions and potential complications remains unclear. (D 2021)
- Legal frameworks need to be adapted to address issues of attribution and fault in AI-assisted healthcare.

8. Potential Solutions

Data Privacy:

- Decentralized data governance framework as advocated by fframework proposed by Ohm in his work "Big Data's Disparate Impact." (Barocas 2016)

Algorithmic Bias:

- Employing fairness-aware machine learning algorithms, explored by (Oneto 2020), which are specifically designed to minimize bias in decision-making processes

Regulatory Uncertainties:

- Collaborative Framework -Open dialogue and collaboration between regulators, developers, and healthcare professionals can create efficient and robust regulatory frameworks.

Focus on Explainability and Fairness:

- Development of transparent and fair AI models, along with bias detection and mitigation strategies, is crucial.

Risk-Based Regulation:

- Tailoring regulatory oversight based on the level of risk associated with specific AI applications can promote innovation while ensuring safety.

Continuous Updates and Adaptation:

- Regulatory frameworks need to be flexible and adaptable to keep pace with the rapid advancements in AI technology.

9. Ethical Considerations

9.1 Patient Autonomy and Informed Consent

- **Shared Decision-making:** Advocate for a collaborative approach where AI acts as a decision-support tool, empowering patients to make informed choices alongside healthcare professionals. (D. S. Char 2018)
- **Transparency and Explainability:** Generative AI models can be complex "black boxes." ensuring transparency in how these models arrive at decisions is crucial. Patients deserve to understand the rationale behind AI-driven recommendations, allowing them to exercise autonomy in their healthcare choices. (Ahmad 2018)
- **Informed Consent:** The evolving nature of generative AI necessitates continuously updated informed consent procedures, as emphasized. (D. S. Char 2020) Patients must be clearly informed about the involvement of AI, its limitations, and potential risks before consenting to AI-powered interventions.

9.2 Equity in Access

- **Digital Divide and Algorithmic Bias:** Generative AI solutions should be accessible to all patients, regardless of socioeconomic background or geographical location. Mitigating the digital divide requires infrastructure development and ensuring diverse datasets are used to train AI models to minimize bias against certain populations (Karimian 2022)
- **Fair and Affordable Healthcare:** Cost considerations are paramount. We must ensure AI-driven healthcare solutions are affordable and don't exacerbate existing disparities in access to quality care. (Murali 2018)

10. Overreliance on AI

While AI holds immense potential to revolutionize healthcare, overreliance can pose significant risks. Listed below is the overview of the potential pitfalls:

- **Algorithmic Bias and Unfairness:** AI algorithms trained on biased data can perpetuate healthcare disparities and lead to discriminatory outcomes for certain demographics. Example: An AI model for predicting recidivism in criminal justice, trained on historical data with racial bias, might unfairly flag people of color as high-risk. (Angwin J 2016).
- **Black Box Problem and Lack of Explainability:** Complex AI models can be difficult to interpret, lacking transparency in decision making process, resulting in eroding trust and hindering accountability. Example: An AI-powered diagnostic tool might flag a patient as high-risk for a disease without

providing clear reasoning, leaving doctors and patients in the dark about the basis for the decision.

- **Overdependence and Neglect of Human Expertise:** Overreliance on AI could lead to the devaluation of human expertise in healthcare. This can be detrimental, as doctors and nurses bring invaluable clinical judgment and empathy that AI alone cannot replicate. Example: Overly relying on AI-driven treatment recommendations could lead doctors to unquestioningly follow algorithms, potentially overlooking subtle cues or nuances of a patient's individual case.
- **Data Privacy and Security Concerns:** Healthcare data is highly sensitive, and its use in AI models raises concerns about privacy breaches and misuse. Robust safeguards are crucial to ensure patient data security and ethical use. Example: A data leak from an AI-powered healthcare system could expose sensitive patient information, potentially causing harm and eroding trust in the technology.
- **Job displacement and workforce challenges:** The implementation of AI in healthcare could lead to job displacement in certain sectors. It's crucial to consider the impact on healthcare workers and provide adequate training and reskilling opportunities. Example: Automation of administrative tasks through AI could lead to job losses for medical coders or billing specialists.

11. Conclusion

This research paper outlines the promises and pitfalls of employing generative AI in healthcare, drawing from the insights and findings of various researchers in the field. It underscores the importance of responsible development and implementation while acknowledging the transformative potential of this technology in advancing healthcare.

In conclusion, generative AI, particularly deep learning models, holds great promise in the healthcare industry. These models have demonstrated high accuracy in predicting medical events and have the potential to improve healthcare outcomes. However, there are risks and challenges that need to be addressed, such as privacy concerns, bias in data, and interpretability of the models. Future research and development should focus on addressing these issues to fully harness the benefits of generative AI in healthcare

References

1. Agarwal R, Bjarnadottir M, Rhue L. 2023. "Addressing algorithmic bias and the perpetuation of health inequities: An AI bias aware framework." *Health Policy and Technology* 100702.

2. Ahmad LG, Eshlaghy AT, Poorebrahimi A, Ebrahimi M, Razavi AR. 2013. "Using three machine learning techniques for predicting breast cancer recurrence." *J Health Med Inform* 4(124):3.
3. Ahmad, M. A., Eckert, C., & Teredesai, A. 2018. "Interpretable machine learning in healthcare." *In Proceedings of the 2018 ACM international conference on bioinformatics, computational biology, and health informatics* 559–560.
4. AL, Samuel. 1959 . "Machine learning." *The Technology Review.* 62(1):42–5.
5. Angwin J, LarsonJ. 2016. *Machine Bias: There's Software Used Across the Country to Predict Crime and It's Exacerbating Racial Inequity.* ProPublica.
6. Barocas, S., & Selbst, A. D. 2016. "Big data's disparate impact." *Calif. L. Rev* (Calif. L. Rev) 104, 671.
7. Blease C, Kaptchuk TJ. 2019. "Artificial Intelligence and the Future of Primary Care." *J Med Internet Res* 21(3).
8. Char, D. S., Abràmoff, M. D., & Feudtner, C. 2020. "Identifying ethical considerations for machine learning healthcare applications." *The American Journal of Bioethics,* 20(11), 7–17.
9. Char, D. S., Shah, N. H., & Magnus, D. 2018. "Implementing machine learning in health care—addressing ethical challenges." *The New England journal of medicine* 378(11), 981.
10. D, Flonk. 2021. "The Age of Surveillance Capitalism: The Fight for a Human Future at the New Frontier of Power." *Convergence* 284–286.
11. FDA. 2021. "Artificial Intelligence and Machine Learning in Software as a Medical Device." *US Food & Drug Administration.*
12. G, Schneider. 2018. " Automating drug discovery." *Nature reviews drug discovery* 17(2):97–113.
13. Ghosheh G, Li J, Zhu T. 2022. "A review of Generative Adversarial Networks for Electronic Health Records: applications, evaluation measures and data sources." *arXiv preprint arXiv* 2203.07018.
14. Goodfellow I J, Pouget-Abadie J, Mirza M. 2014. "Generative adversarial networks."
15. J, Steward. 2023. "Procedural Content Generation: Generating Procedural Game Content Using Machine Learning ." *Doctoral dissertation.*
16. J. Ho, A. Jain, and P. Abbeel. 2020. "Denoising diffusion probabilistic models." *Advances in Neural Information Processing* 6840–51.
17. Karimian, G., Petelos, E., & Evers, S. M. 2022. "The ethical issues of the application of artificial intelligence in healthcare: a systematic scoping review." *AI and Ethics,* 2(4), 539–551.
18. Kingma DP, Welling M. 2019. "An introduction to variational autoencoders." *Foundations and Trends in Machine Learning* 12(4):307–92.
19. Kulkarni S, Seneviratne N, Baig M,Khan A. 2020. "Artificial Intelligence in Medicine: Where Are We Now." *Academic Radiology* 62–70.
20. Linardatos P, Papastefanopoulos V, Kotsiantis S. 2021. "Explainable AI: A Review of Machine Learning Interpretability Methods." *Entropy* 23(1):18.
21. MT, Zemcik. 2019. "A brief history of chatbots." *DEStech Transactions on Computer Science and Engineering* 10.

22. Munoz VF, Vara-Thorbeck C, DeGabriel J G. 2000. "A medical robotic assistant for minimally invasive surgery." *Proceedings 2000 ICRA. Millennium Conference.* San Francisco: IEEE International Conference on Robotics and Automation. 2901–2906.
23. Murali, N., & Sivakumaran, N. 2018. "Artificial intelligence in healthcare–a review."
24. Mushtaq G, Siddiqui F. 2021. "Detection of diabetic retinopathy using deep learning methodology." 012049.
25. Naik N, Hameed BMZ, Shetty DK. 2022. "Legal and Ethical Consideration in Artificial Intelligence in Healthcare: Who Takes Responsibility." *Front Surg* 862322.
26. Oneto, L., & Chiappa, S. 2020. "Fairness in machine learning. In Recent trends in learning from data: Tutorials from the inns big data and deep learning conference." *Springer International Publishing* 155–196.
27. Parikh RB, Teeple S, Navathe AS. 2019. "Addressing Bias in Artificial Intelligence in Health Care." *JAMA* 322(24):2377–78.
28. Park JJ, Tiefenbach J, Demetriades AK. 2022. "The role of artificial intelligence in surgical simulation." *Frontiers in Medical Technology* 4:1076755.
29. Prezja F, Paloneva J, Polonen I, Niinimaki E, Ayramo S. 2022. "DeepFake knee osteoarthritis X-rays from generative adversarial neural networks deceive medical experts and offer augmentation potential to automatic classification." *Sci Rep* 12(1):18573.
30. Zeng X, Wang F, Luo Y. 2022. "Deep generative molecular design reshapes drug discovery." *Cell Reports Medicine.*
31. Zhang S, Liang G, Pan S, Zheng L. 2018. "A fast medical image super resolution method based on deep learning network." *IEEE Access* 7:12319–27.

Technologies for Energy, Agriculture, and Healthcare – Shailesh Nikam et al. (eds)
© 2024 Taylor & Francis Group, London, ISBN 978-1-032-98028-7

8

MULTICLASS_RESP_DNN: A Novel Deep Learning Architecture for COVID, Pneumonia and Tuberculosis Detection from Radiology Images

Prita Patil[1]
Ramrao Adik Institute of Technology
D Y Patil Deemed to be University,
India

Vaibhav Narawade[2]
Professor, Ramrao Adik Institute of Technology
D Y Patil Deemed to be University,
India

Abstract: AI, particularly machine learning, is increasingly utilized in healthcare for disease diagnosis. This paper presents a novel deep learning architecture, MULTICLASS_RESP_DNN, for the detection of respiratory diseases such as COVID-19, pneumonia, and tuberculosis from chest X-ray images. The model aims to address the scarcity of competent radiologists by providing automated disease classification, aiding in timely and accurate diagnosis. We utilize a comprehensive dataset sourced from Kaggle, comprising chest X-ray images of normal cases, COVID-19, pneumonia, and tuberculosis. The proposed model undergoes rigorous evaluation, considering various performance metrics, interpretability, and ethical implications. Experimental results demonstrate the effectiveness and potential of the MULTICLASS_RESP_DNN model in respiratory disease detection. Our method can accurately detect 95.52% of abnormal X-rays, including COVID-19, pneumonia and tuberculosis.

Keywords: Machine learning, Tuberculosis, X-ray images, Multiclass

[1]prita.patil@ dypatil.edu, [2]vaibhav.narawade@dypatil.edu

DOI: 10.1201/9781003596707-8

1. Introduction

Respiratory diseases, encompassing conditions affecting the lungs, represent a significant healthcare challenge worldwide. Among the prevalent respiratory ailments are pneumonia, COVID-19, and tuberculosis, each presenting distinct clinical manifestations and diagnostic challenges. However, the effective diagnosis of these diseases is impeded by a shortage of skilled radiologists, particularly in resource-constrained regions. Machine learning (ML) and deep learning (DL) techniques play a vital role in image classification. Chest X-ray (CXR) imaging, a cornerstone of respiratory disease diagnosis, provides valuable insights into pulmonary health. ML methods applied to CXR images have revolutionized the detection and classification of respiratory diseases, enabling rapid and accurate diagnosis. Despite significant advancements, challenges persist in the field of respiratory disease detection, particularly concerning imbalanced datasets and multi-class classification. An imbalance in data distribution can hinder the performance of ML models, leading to biased predictions and reduced diagnostic accuracy. Moreover, the dynamic nature of real-world data streams exacerbates the complexity of multi-class imbalance learning, necessitating innovative approaches to address these challenges comprehensively.

In this context, our research endeavours to develop a robust and efficient model for the identification of anomalies in chest radiology images, specifically targeting the detection of COVID-19, pneumonia, and tuberculosis. Our proposed model leverages state-of-the-art DL techniques to enhance diagnostic accuracy and streamline the clinical decision-making process. Furthermore, we aim to address the limitations of existing approaches by devising a compact yet highly performant model capable of effectively handling imbalanced and unstructured datasets and dynamic data streams.

2. Related Work

S. Devaraj et.al., presented pneumonia medical imaging and infection detection methods in which data augmentation gives better results. X. Aggelides et al., presented an interdisciplinary Gesture Recognition case study for recognizing and categorizing motions associated with allergic rhinitis. Chowdhury et al. used data augmentation in an X-ray image dataset with respiratory disease illness and training with transfer learnings.

Waheed et.al, fine-tuned the completely connected layer weights and downsized the input images (112 × 112). Brunese et al. employed VGG16 CNN architecture in the identification of pneumonia accuracy of 0.941. Rahimzadeh et.al. employed data augmentation to balance the dataset and scaled the pictures (300 × 300) with used ImageNet transfer learning on the Exception and Resnet50 architectures and obtained 0.914 accuracies.

Previous research in medical image analysis has explored various methodologies for disease detection, including traditional machine-learning approaches and deep-learning architectures. Several studies have focused on pneumonia and COVID-19 detection from chest X-ray images using techniques such as transfer learning, data augmentation, and ensemble methods. However, many existing approaches suffer from limitations such as dataset biases, lack of interpretability, and inadequate performance on multi-class classification tasks.

3. Proposed Methods

Our proposed methodology consists of several key steps, beginning with the preprocessing of raw X-ray images to enhance their quality and extract relevant features. We have used image resizing 256X256, gray scale coversion to preserve radiology image information, data augmentation with an oversampling technique to generate augmented images of the training dataset, GLCM for spatial feature extraction, and K-means clustering for image segmentation. The processed images are then fed into the MULTICLASS_RESP_DNN model, which comprises convolutional layers for feature extraction, batch normalization to improve convergence and generalization, and dense layers for multi-class classification using sigmoid activation as shown in Fig. 8.1.

Layer	Filters	Kernal size	Padding	Output shape	Activation function
Input	-	-	-	256 × 256 × 1	-
Conv 1	32	3 × 3	Same	256 × 256 × 32	ReLu
BatchNormalization	-	-	-	256 × 256 × 32	-
MaxPooling	-	3 × 3	-	85 × 85 × 32	
Dropout	rate = 0.25	-	-	85 × 85 × 32	-
Conv 2	64	3X3	Same	85 × 85 × 64	ReLu
BatchNormalization	-	-	-	85 × 85 × 64	
Conv 3	64	3X3	Same	85 × 85 × 64	ReLu
BatchNormalization	-	-	-	85 × 85 × 64	-
MaxPooling	-	2 × 2	-	42 × 42 × 64	-
Dropout	rate = 0.25	-	-	42 × 42 × 64	-
Conv 4	128	3 × 3	Same	42 × 42 × 128	ReLu
BatchNormalization	-	-	-	42 × 42 × 128	-
Conv 5	128	3 × 3	Same	42 × 42 × 128	ReLu
BatchNormalization	-	-	-	42 × 42 × 128	-
MaxPooling	-	2 × 2	-	21 × 21 × 128	-
Dropout	rate = 0.25	-	-	21 × 21 × 128	-
Flatten	-	-	-	56448	
Dense	-	-	-	1024	ReLu
BatchNormalization	-	-	-	1024	
Dropout	rate = 0.25	-	-	1024	
Dense	-	-	-	No of classes	Sigmoid

Fig. 8.1 MULTICLASS_RESP_DNN model architecture detail

For this study, a medical imaging dataset from the Kaggle source [1] of radiology images is considered for the experimental evaluation. There are a total of 7135 radiology images Dataset details are presented in Table 8.1

Table 8.1 Respiratory illness dataset details

Label Name	Training	Testing	Validation	Total
COVID-19	460	106	10	576
Normal	1341	234	8	1583
Pneumonia	3875	390	8	4273
Tuberculosis	650	41	12	703

The dataset is from different sources, differences in intensity level, picture quality, and pixel size may occur among the images, thus causing poor model performance. Ablation studies and a variety of picture preparation approaches have been devised to address and overcome these differences. Furthermore, the settings of the image preparation methods are selected by thorough trials to produce the best results even with image fluctuations. The goal is to improve image quality without losing important information. The ablation research optimizes the model by experimenting with alternative layer topologies and hyperparameters.

The proposed technique calculates the feature of pixels as.

$$f(a) = \sum_{i=1}^{n} bi(a) f(pi) \tag{1}$$

where,

f(a) = feature vector of non-key pixel a

pi = is the key pixel placed in the center of each overlapping weighting window, which includes a;

bi(a) = weight assigned to pixel an in the weighting window centered at pi.

Algorithm 1: K-Means Clustering Algorithm

Input: k: No. of clusters.

Ip: Image.

D: Dataset Containing n Objects.

E: Maximum no. of epochs

Output: A collection of k clusters, segmented image.

Steps:

Step 1. A collection of gray level values $G = \{p1, p2, \ldots, pn\}$.

Step 2. Initialize k no. of clusters z = 1, 2, … , k.

Step 3. Randomly initialized collection of cluster centers as $c(e) = [c1\ (e), c2\ (e), \ldots, ck\ (e)]$; where e denotes first epoch of algorithm.

Step 4. Repeat

Step 5. Measure Euclidean distance $d(pi,cj(e))$ between gray level value $pi \in G$ & cluster center $cj\ (e) \in c(e)$ using the relation given below:

$$d(pi, cj(e)) = |pi - ci(e)|^2 \tag{2}$$

If $cj(e)$ is the nearest center for pi, then it is allocated to cluster zj.

Step 6. Apply the minimal Euclidean distance criterion to determine which cluster center should get each grey level value.

Step 7. Using the following equation, get the new cluster centers:

$$cj\ (e+1) = \frac{1}{\eta}\sum_{n=1}^{j} pi;\ (j = 1,2,\ldots,k) \tag{3}$$

Where η denotes the size of cluster zj.

Step 8. Move on to Step 5 & step up the epoch. This procedure is repeated until either cluster centers cease to evolve or the algorithm achieves the maximum no. of epochs E, where e = 1, 2,..., E.

Output: Reconstruct a segmented picture from the k no. of clustered gray level data.

are clustered together in image segmentation.

4. Results and Discussion

We conduct extensive experiments to evaluate the performance of our proposed MULTICLASS_RESP_DNN model which is represented in Table 8.2. We compare the results obtained with those of existing deep learning architectures to demonstrate the superiority of our approach.

Table 8.2 MULTICLASS_RESP_DNN performance measurement

Training loss	Training accuracy	Validation loss	Validation accuracy	Precision	Recall	F1-Score
0.0042	0.9988	0.3143	0.9552	0.92	0.92	0.92

Figure 8.2 illustrates the MULTICLASS_RESP_DNN model's performance on x-ray images. As shown in Fig. 8.2(a), the model's accuracy improves with consecutive epochs as shown in Fig. 8.2(b), and validation loss lowers with consecutive epochs than training loss. Figure 8.2(c) depicts the normalized confusion matrix of the MULTICLASS_RESP_DNN model. The confusion matrix for Tuberculosis is 0.96, 0.78 for COVID-19,0.93 for pneumonia, and 0.95 for the Normal classifications.

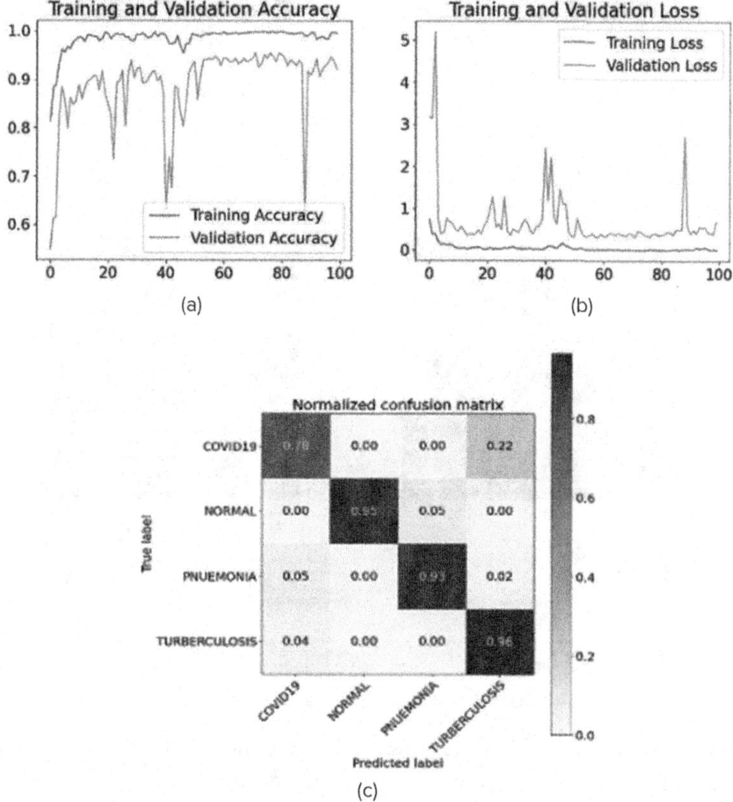

Fig. 8.2 Performance of MULTICLASS_RESP_DNN model

Figure 8.3 above represents respiratory illness prediction from unknown respiratory image samples that are preprocessed using improved GLCM followed by K-means clustering from the validation radiology image dataset from Dataset sources given above and was able to identify the proper respiratory disease class with 99.99% prediction score.

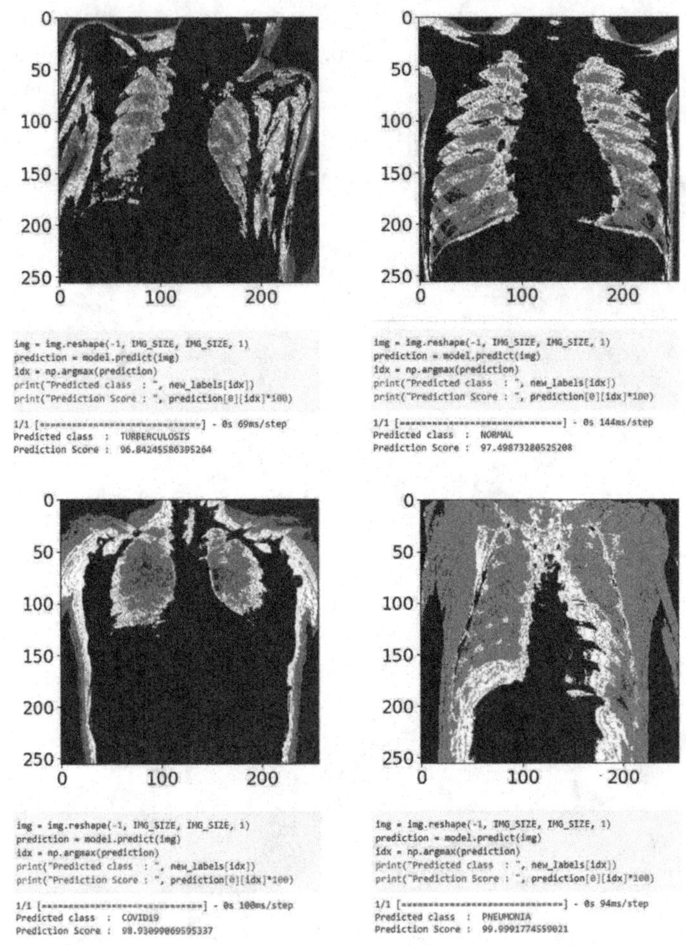

Fig. 8.3 Respiratory disease prediction using MULTICLASS_RESP_DNN

Different experiments are carried out as part of the ablation investigation to improve performance of the proposed DNN model and analysis is demonstrated in Fig. 8.4

Table 8.3 compares the proposed MULTICLASS_RESP_DNN to existing cutting-edge deep learning models. The InceptionV3 model outperformed than ResNet. Nonetheless, it lags behind the Xception model, which has 92.93% accuracy, 92% recall, and a f1-score.However, our proposed MULTICLASS_RESP_DNN model has the greatest accuracy (95.52%), precision (94%), recall (92%), and f1-score (92%), of any model.

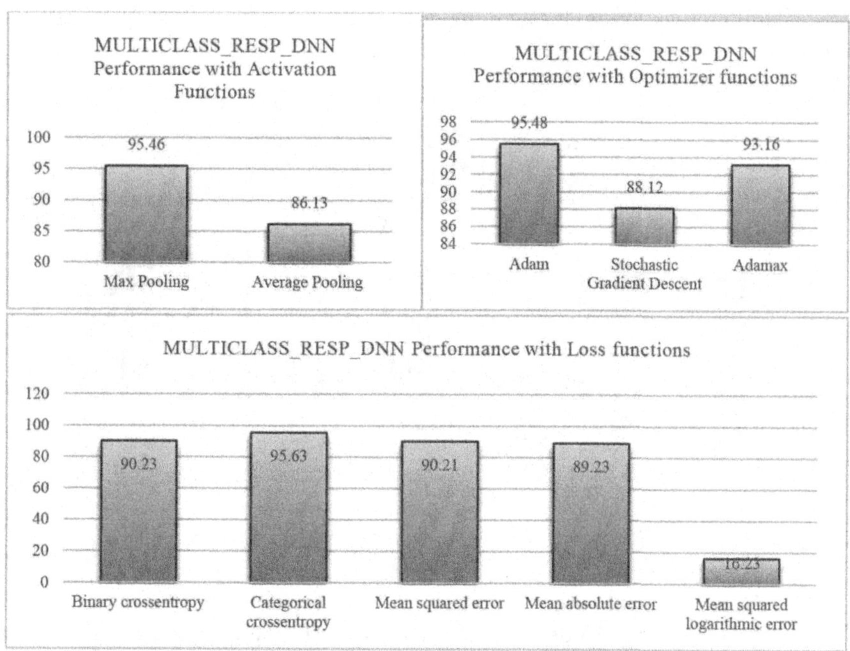

Fig. 8.4 Ablation study of MULTICLASS_RESP_DNN

Table 8.3 MULTICLASS_RESP_DNN with existing classification models

Parameters	Inception V3	ResNet	Xception	Proposed MULTICLASS_RESP_DNN
Number of layers	48	50	71	28
Image Size	299X299	224X224	224X224	256X256
Filter Size	5x5	3X3	3X3	3x3
Total Parameters	23,900,963	27,242,947	29,251,115	58090820
Training Accuracy	0.9051	0.8989	0.9296	0.9988
Training Loss	0.2288	0.2338	0.1797	0.0042
Validation Accuracy	0.9092	0.8820	0.9154	0.9552
Validation Loss	0.2397	0.3102	0.2364	0.3143
Precision	0.90	0.92	0.92	0.92
Recall	0.91	0.92	0.90	0.92
F1-Score	0.90	o.92	0.91	0.92

While building custom-built DNN we have experimented with the progressive approach of different DNN layers designing and the justification for our proposed DNN model is presented in Fig. 8.5

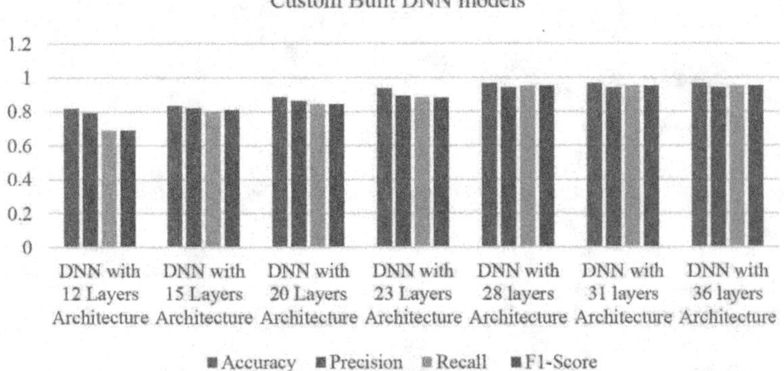

Fig. 8.5 Performance analysis of progressive laywise construction of custom-built DNN models

Despite the promising results, our study faces certain limitations and challenges, including dataset biases, interpretability of model predictions, and ethical considerations associated with deploying AI in healthcare settings

5. Conclusions and Future Work

The research study presents MULTICLASS_RESP_DNN, a novel deep-learning architecture for the detection and classification of respiratory diseases from chest X-ray images. Our approach leverages advanced preprocessing techniques and deep learning methodologies to achieve high accuracy and robust performance across multiple disease categories such as COVID-19, Pneumonia, and Tuberculosis, with improvements in image enhancement and segmentation by proposing improved GLCM followed by K-means clustering algorithm while preprocessing raw image samples. Furthermore, the preprocessed images are trained using an effective technological MULTICLASS_RESP_DNN model for the respiratory ailment infection classification. Experimental results prove MULTICLASS_RESP_DNN model achieved 95.52% X-ray image correctness and 91.59% testing accuracy. According to the comparative research, the recommended MULTICLASS_RESP_DNN model outperformed with a classification accuracy of 95.52%., 92% precision, 92% recall, and 92% f1-score. These findings indicated that lung disease identification algorithms in X-ray images need to be significant.

We highlight the importance of addressing limitations and ethical considerations in future research and emphasize the potential of AI in transforming healthcare delivery and improving patient outcomes.

References

1. Chest X-Ray (Pneumonia,Covid-19,TuberculosiA. SK, "Tuberculosis Chest X-ray Image Data Sets (2021), doi:https://www.kaggle.com/datasets/jtiptj/chest-xray-pneumoniacovid19tuberculosis/data
2. S. Devaraj and M. N. Madian, "Deep U-Net Network for Identifying Covid-19 Infection Using X-Ray Imag-es," in 2020 IEEE International Conference on E-health Networking, Application & Services (HEALTH-COM), 2021, pp. 1–5. doi: https://ieeexplore.ieee.org/document/9615913
3. R. Anand, S. Fazlul Kareem, R. Mohamed Arshad Mubeen, S. Ramesh, and B. Vignesh, "Analysis Of Heart Risk Detection In Machine Learning Using Blockchain," in 2021 6th International Conference on Signal Pro-cessing, Computing and Control (ISPCC), 2021, pp. 685–689. doi: https://ieeexplore.ieee.org/document/9609353
4. X. Aggelides, A. Bardoutsos, S. Nikoletseas, N. Papadopoulos, C. Raptopoulos, and P. Tzamalis, "A Gesture Recognition approach to classifying Allergic Rhinitis gestures using Wrist-worn Devices : a multidisciplinary case study," in 2020 16th International Conference on Distributed Computing in Sensor Systems (DCOSS), 2020, pp. 1–10. doi: https://ieeexplore.ieee.org/document/9183416.
5. Chowdhury M.E.H., Rahman T., Khandakar A., Mazhar R., Kadir M.A., Mahbub Z.B., Islam K.R., Khan M.S., Iqbal A., Emadi N.A. Can a help in screening viral and COVID-19 pneumonia? IEEE Access. 2020;8:132665–132676. doi: https://ieeexplore.ieee.org/document/9144185
6. Waheed A., Goyal M., Gupta D., Khanna A., Al-Turjman F., Pinheiro P.R. Covidgan: data augmentation using auxiliary classifier gan for improved COVID-19 detection. IEEE Access. 2020;doi: https://ieeexplore.ieee.org/document/9093842
7. P. Patil and V. Narawade, "Emphasize of Deep CNN for Chest Radiology Images in the detection of COVID," 2022 IEEE 7th International conference for Convergence in Technology (I2CT), Mumbai, India, 2022, pp. 1–6, doi: https://ieeexplore.ieee.org/document/9825370.
8. Vieira P, Sousa O, Magalhães D, Rabêlo R, Silva R. Detecting pulmonary diseases using deep features in X-ray images. Pattern Recognit. 2021 Nov;119:108081. doi: 10.1016/j.patcog.2021.108081.doi: https://www.sciencedirect.com/science/article/pii/S0031320321002685?via%3Dihub

Note: All the figures and tables in this chapter were made by the authors.

Technologies for Energy, Agriculture, and Healthcare – Shailesh Nikam et al. (eds)
© 2024 Taylor & Francis Group, London, ISBN 978-1-032-98028-7

9

Bibliometric Analysis on Bioprinting: Mapping Co-Occurrence of Keywords and Future Research Direction using VOSviewer

Kavita Kumari Thakur[1]
Research Scholar, Department of Mechanical Engineering,
K.J. Somaiya College of Engineering, Vidyavihar,
Somaiya Vidyavihar University, Mumbai,
Maharashtra, India

Ramesh Lekurwale[2]
Professor, Department of Mechanical Engineering,
K.J. Somaiya College of Engineering, Vidyavihar,
Somaiya Vidyavihar University, Mumbai,
Maharashtra, India

Rajesh Pansare[3]
Assistant Professor, Department of Mechanical Engineering,
K.J. Somaiya College of Engineering, Vidyavihar,
Somaiya Vidyavihar University, Mumbai,
Maharashtra, India

Sangita Bansode[4]
Associate Professor, Department of Mechanical Engineering,
K.J. Somaiya College of Engineering, Vidyavihar,
Somaiya Vidyavihar University, Mumbai,
Maharashtra, India

Abstract: It is important to understand the bioprinting research area and to identify the emerging trends in the area by looking at the increasing demand for bioprinting technology. Bioprinting can provide customized solutions for

[1]kavitakumari.t@somaiya.edu, [2]rameshlekurwale@somaiya.edu, [3]rajeshpansare@somaiya.edu, [4]sangeetabansode@somaiya.edu

DOI: 10.1201/9781003596707-9

healthcare needs by enabling the customized fabrication of tissues and organs as per patient's requirements. It is also very helpful in regenerative medicine and drug testing. Bioprinting plays an important role in new bone generation. Bioprinting plays a significant role in personalised implants, enhanced healing of the diseased bone by addressing the various challenges in the orthopaedic field. This paper aims to find out the recent trends in the research area of "Bioprinting and Bone Scaffold" by conducting a bibliometric analysis using VOSviewer software. By using the VOSviewer software a detailed examination of the Scopus database of bioprinting research articles for keyword analysis, Co-citation analysis and collaboration of network mapping are conducted. The analysis results highlight research areas like tissue engineering, bioprinting technologies, human trials etc. are emerging research topics. This paper provides the evaluation of bioprinting research by highlighting the important works and suggesting research trends for future research work in this field.

Keywords: Bioprinting, Bone scaffold, Bibliometric analysis, Literature survey, VOSviewer

1. Introduction

Organ transplantation is in high demand due to the scarcity of viable organs for the patient in need. As per the statistics, the renal failure in the US and Canada that is (140 to 160/million/year) versus the ability of the organ from the donor is 20 to 22/millions/year. In 2007 in every 15 minutes, a patient was added to the waiting list which resulted in over 95,000 patients awaiting transplants in the US (Abouna, 2008). The traditional method for organ fabrication or transplantation mainly depends upon the donor's organ or tissue, which results in a long waiting time for the patient to get their treatment. Also, this method has the risk of immune rejection of the organ or tissue by the patient. On the other hand, bioprinting provides a customised solution by providing the patient-specific tissue or organ. Which reduces the risk of rejection of the organ/ tissue as the cells in this case taken from the patient's own body. Such emergency hopes Bioprinting to fulfil the demand of the customer with innovative solutions. Bioprinting utilises advanced technology to construct functional organs layer by layer by mimicking the complex structure of natural tissue (Mandrycky et al., 2016). Bioprinting can fulfil the emerging demand for organ transplantation by creating the complex structure of the tissues with precision by printing bioinks composed of cells and biomaterials. Bioprinting can mimic the functionality and compatibility of the fabricated tissue and organ with native tissue or organ. Bioprinting provides patients specific customized solutions which reduce the chances of immune rejection problems of the transplanted tissue/organ. The evolution of tissue engineering and 3D printing is a major shift in medical science by blending innovation and precision. 3D bioprinting provides the fabrication of intricate structures with remarkable accuracy which helps for obtaining personalised medicine and transformative healthcare interventions (Zhang et al., 2018). Bioprinting combines the principle of additive manufacturing and biology for manufacturing customised tissue. In bioprinting, precise layer-by-layer deposition of living cells biomaterial and the supporting structures to create 3D structure for making the internal structure of biological tissues and organs. For getting the desirable cellular arrangement, the bioprinting process utilises biomaterials which is composed of living cells encapsulated in a biocompatible matrix. Bioprinting has applications in customized models, human implants, scaffolds, drug testing, and controlled drug release (Zhang et al., 2018). In regenerative medicine, bioprinting has immense potential for generating personalized, functional tissue and organs to individual patient needs potentially overcoming the limitation of organ transplantation. Bioprinting offers the different benefits of precise deposition, simplicity, cost-effectiveness and controlled cell dispersion (Thakur et al., 2023). The integration of advanced imaging techniques and artificial intelligence provided the development of highly accurate patient-specific models. Bioprinting has immense potential for personalised medicine, regenerative therapies, and reshaping organ transplantation (Vijayavenkataraman et al., 2018). There are several important parameters in 3D bioprinting processes like materials, hydrogels, bioinks, bed temperature, extrusion pressure, cell culture test and mechanical test that are the parts of different stages of bioprinting as shown in Fig. 9.1.

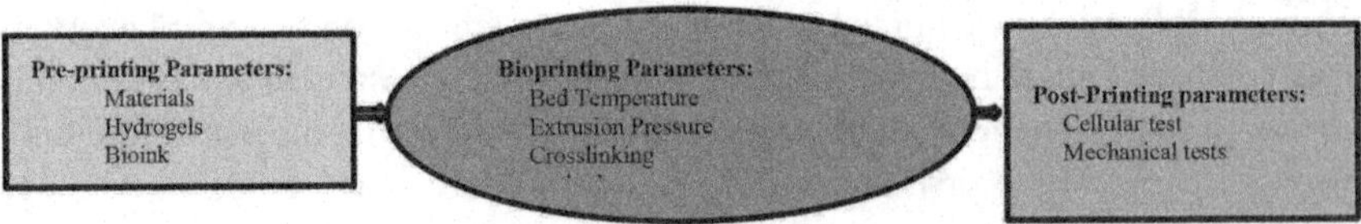

Fig. 9.1 Bioprinting Stages (Mancha Sánchez et al., 2020)

To identify the important research area in bioprinting and bone scaffold studies, a comprehensive literature study of existing research article is important. Bibliometric analysis is important parameter for understanding the current research area which is explained by (Pansare et al., 2022). Bibliometric analysis is a quantitative technique forgetting the insight into published research article (Ding & Yang, 2022). BibExcel, CiteSpace, VOSviewer, HistCite and Gephi are the important software for performing bibliometric analysis. All the bibliometric analysis tool are having different data importing techniques like HistCite and BibExcel from Web of science, Perish from Google Scholar and Cite Space from Web of Science (Fahimnia et al., 2015). VOSviewer software is used in this study for creating and displaying bibliometric maps. VOSviewer has a user-friendly interface which makes it suitable for analysing large database sets of scientific journals. VOSviewer has the capability to retrieve data from Web of Science, Scopus, Lens, PubMed. VOSviewer can provide various maps like network visualization, bibliometric mapping, clustering keywords co-occurrence analysis, collaboration analysis and authorship. VOSviewer provides a visualization capability which allows the users to have visualization information for understanding the pattern and trends of the data. Having the visualization between the keywords, publication and authors can guide the researchers to select their research area.

2. Research Methodology

947 articles related to bioprinting, and bone scaffold were retrieved using Scopus database (till January 2024) as shown in Fig. 9.2.

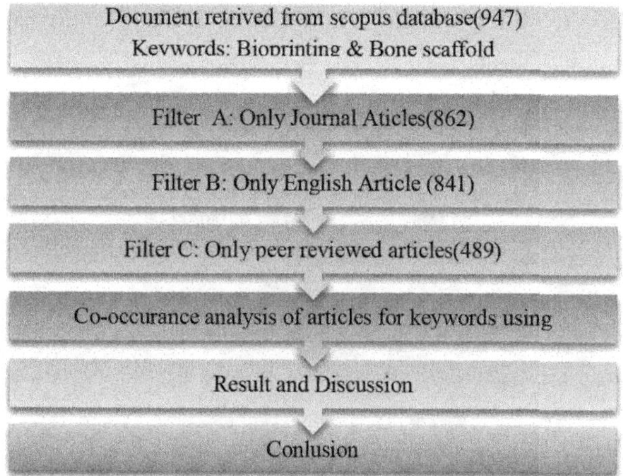

Fig. 9.2 Research Methodology

Various filters have been applied to select the high-quality journals and the total number of journals retrieved after applying the various filters is 489. Different filters used for this study. Using the VOSviewer software, the Co-occurrence analysis of the keyword is carried out on the selected 489 articles. The various outcomes of the analysis are analysed and discussed.

3. Co-occurrence Analysis of Keywords

The co-occurrence analysis of the selected 489 articles is conducted and a total of 4819 keywords are obtained after selecting full counting. For selecting the important keyword, a minimum of 5 occurrences is selected for the study. A total of 577 documents meets the threshold value and are used for analysis using VOSviewer and the different outcomes after the analysis are discussed in the below section.

3.1 Network Visualization of Keywords

The presented Fig. 9.3 provides the network visualization diagram which is obtained from the co-occurrence analysis of keywords as shown in Fig. 9.3. The dimension of both circles and labels in the diagram provides the respective weights of keywords. A large size of label and circle represents higher keyword weight. The colour of the circle and links provides information about the cluster to which the keyword belongs, which helps in the identification of the thematic grouping. Also, the nearness of the keywords in the figure provides their relatedness, providing insights of pattern and connection within the data set.

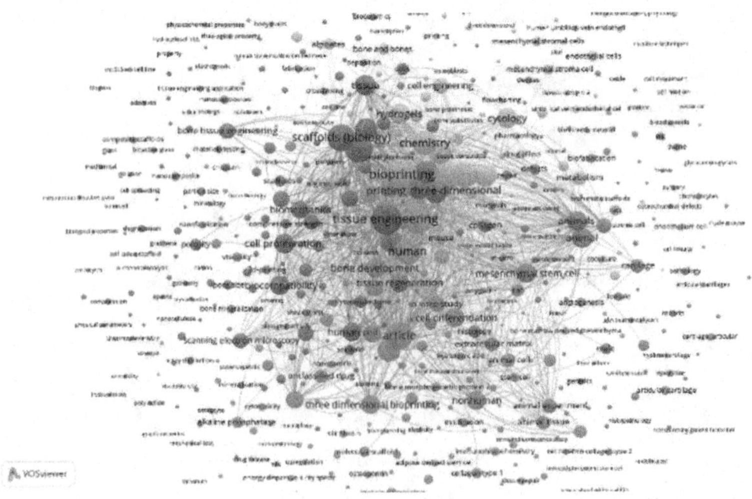

Fig. 9.3 Diagram for network visualization of keywords

3.2 Overlay Visualization of Keywords

The overlay visualization figure which is obtained from the co-occurrence analysis of the keywords is shown in Fig. 9.4. This representation provides the standard network visualization, which is based on the colour screen applied to the circle

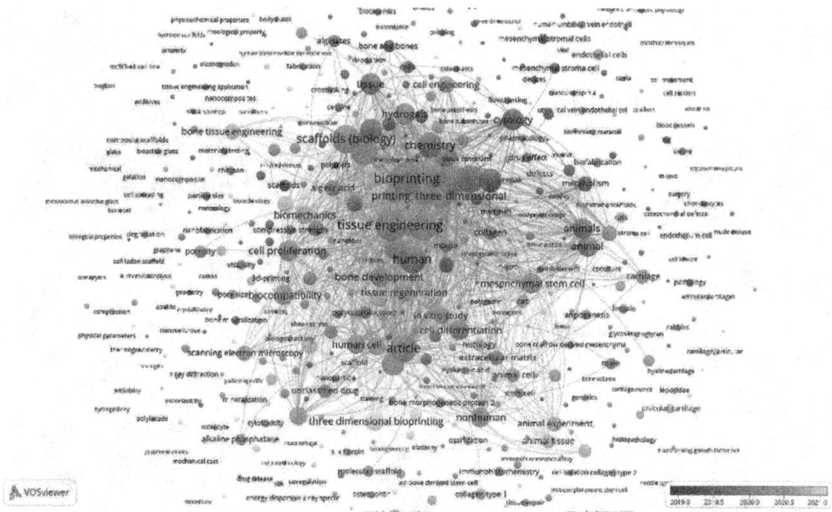

Fig. 9.4 Diagram for overlay visualization of keywords

and links. In this overly visualization, the colour of the keyword is subjected to their respective weights which represent the average importance of the documents in which each keyword belongs. Blue colour signifies the lowest weights; on the other hand, green and yellow represent the highest weights. The colour bar acts as a guide, which provides the mapping of weights with colour in this visualization. This analysis provides a visual gradient which helps in the quick intervention of keyword weights within the analysed data.

3.3 Density Visualization of Keywords

Density visualization for the keywords done by the VOSviewer is a comprehensive and perspective representation of structural patterns for a bibliometric network. As shown in Fig. 9.5, the density visualization provides the concentration and the distribution of the keywords in the network diagram. The density visualization focuses on the intensity of the keywords for identifying the connection in this specific region. Deeper colors in the diagram show higher density, showing closeness between keywords. Density visualization helps the researcher to understand the strength of the relationship among the keywords.

VOSviewer provides two types of density visualization which is item density and cluster density. The item density diagram provides the density of individual items (keywords or terms) in the full network. Also, cluster density visualization focuses on the density of the cluster within the entire network. Figure 9.6 represents the

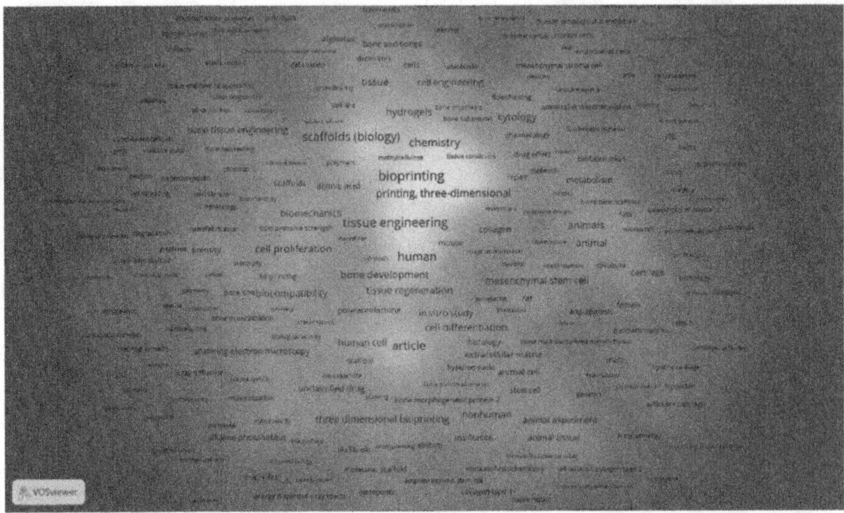

Fig. 9.5 Diagram for Item density visualization of keywords

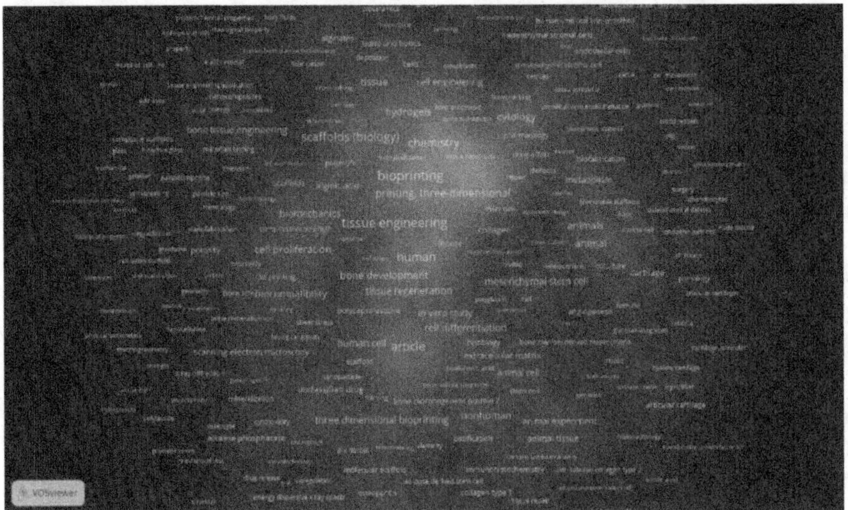

Fig. 9.6 Diagram for cluster density visualization of keyword

density of each cluster. The cluster density diagram uses colours to represent different levels of cluster density to highlight the reason, where the cluster of fighters is densely connected.

4. Results and Discussion

This section provides a detailed discussion of the result obtained during the co-occurrence analysis of the keywords for bioprinting and bone scaffold articles.

4.1 Cluster in Keyword Analysis

A total of 577 keywords are categorised into 6 groups in this analysis. The total link is 69312 and the total link strength is 294408 in this study. Figure 9.7 represents the cluster-wise distribution of the keywords used in this study. This study provides the theme of research for different clusters.

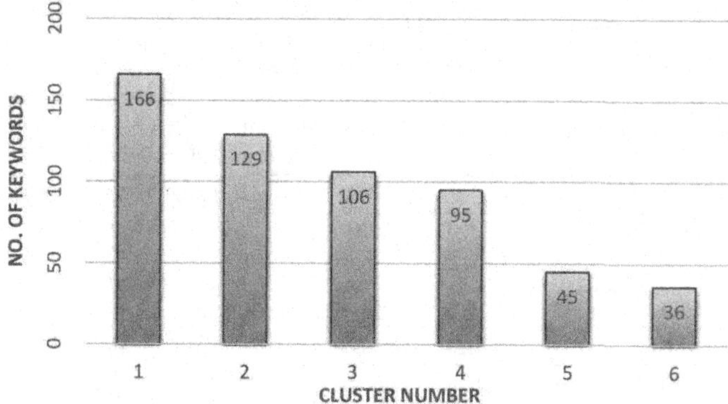

Fig. 9.7 Distribution of keywords based on clusters

Table 9.1 provides summary of different cluster identified in the Scopus database. The table highlights the top 10 keywords in each cluster based on their frequency; their link strength as well as the future research area are identified. The occurrence column in this table identifies the number of times the keywords are used which highlights the importance of specific terms. Link strength column provides the interconnectedness and the relationship between the keywords in each cluster. To guide the future researcher, a dedicated column which provides the future research trend based on the occurrence of the keyboard are also included in this table.

4.2 Top Ten Keyword by Occurrences

In Table 9.2, a comprehensive analysis of the top 10 keywords based on frequency and link strength are conducted. The occurrence column provides a quantitative measure of how frequently each keyword appears, allowing for a quick assessment of the most prevalent terms. Occurrence column provides the frequency of the keywords in the data set by providing the quantitative measure. The total link strength column quantifies the strength of the connection associated in each

Table 9.1 Keyword analysis

Cluster no.	Top ten keywords	Occurrence	Link strength	Future research trends based on occurrences
1	Tissue engineering	318	11267	• Investigation of novel scaffold design with 3D printing • Investigation of biomaterials properties effect for tissue regeneration improvement • Evaluate the biocompatibility of various biomaterials. Evaluation of biocompatibility of various biomaterials for human cell interaction. • Use of biomechanical principle for functional improvement of tissue engineering • Exploring advanced methodology for combining biomaterials, biomechanics, and 3D printers.
	Scaffolds (biology)	247	8106	
	Bone	237	7870	
	3d printers	159	5660	
	Tissue	142	4722	
	Cell proliferation	136	5530	
	Biocompatibility	128	4539	
	Human cell	112	4899	
	Biomechanics	110	3723	
	Biomaterial	91	3278	
2	Article	255	10135	• Investigation for precise 3D bioprinting fabrication technique. • Investigation of cell viability on gelatin-based scaffold • Investigation of novel biomaterials for bone development and cell differentiation. • Investigation for optimizing tissue engineering strategies for nonhuman model. • Examination of different article based on bone regeneration for bone regeneration improvement.
	Controlled study	174	7545	
	Bone development	135	5534	
	Cell differentiation	126	5302	
	Nonhuman	121	5285	
	Three dimensional bioprinting	119	4915	
	Bone regeneration	118	4032	
	Osteogenesis	117	4643	
	Cell viability	108	4671	
	Gelatin	101	4111	
3	Cell culture	187	7069	• Investigation of 3D printing methods for cell culture scaffold fabrication. • Identify the role of mesenchymal stem cells for the improved tissue regeneration in the animal model. • Identify the interaction between the extracellular matrix and the stem cells in cartilage tissue engineering. • Optimization of mesenchymal stem cells for tissue regeneration by novel techniques. • Investigation of in vivo applicability of 3D printed scaffold and stem cells therapies for tissue engineering application in the animal model.
	Printing, three-dimensional	186	7190	
	Stem cells	164	5838	
	Animals	134	5537	
	Animal	133	5499	
	Mesenchymal stem cell	128	5444	
	Tissue regeneration	121	4332	
	Mesenchymal stem cells	80	3365	
	Cartilage	68	2613	
	Extracellular matrix	66	2522	

Cluster no.	Top ten keywords	Occurrence	Link strength	Future research trends based on occurrences
4	Bioprinting	348	11165	• Investigation of bio printing technology for the precise fabrication of tissue scaffold for getting their structural integrity • Investigation of novel chemistry-based methods for enhancing tissue scaffold biocompatibility as well as mechanical properties. • Investigation of tissue engineering procedure for alginic acid and hydrogel for improved cell viability as well as functionality. • Investigation of cytological impact of different tissue scaffold composition on the human cells to achieve the best tissue regeneration. • Identify different novel by printing strategies with the use of chemistry principle 4 customised tissues scaffold design for customized therapeutic application.
	Tissue scaffolds	257	9980	
	Tissue scaffold	243	9516	
	Chemistry	206	8016	
	Humans	175	6688	
	Procedures	168	6462	
	Cytology	143	5743	
	Hydrogels	141	5295	
	Cell engineering	88	3170	
	Alginic acid	76	3194	
5	Hydrogel	168	6778	• Investigation of new hydrogel-based bioink formation and cross-linking strategies for enhanced structural stability in tissue engineering scaffold. • Investigation the impact of flow kinetics and viscosity on the bioink behaviour for reproducible 3D printed tissue construct. • Examine the histological feature of advanced hydrogel based bioink for tissue engineered construct by keeping the focus on cell distribution, viability, and tissue architecture. • Create devices that can be useful for dynamically control shear stress in their printing process by addressing the shear thinning issues and optimising cell viability. • Investigate the novel crosslinking methods for hydrogel based bioink for improving mechanical properties
	Histology	58	2063	
	Bioink	54	1897	
	Cross linking	42	1687	
	Flow kinetics	35	1522	
	Viscosity	30	1355	
	Tissues engineering	25	813	
	Crosslinking	20	575	
	Devices	20	761	
	Shear stress	18	803	
	Shear thinning	14	538	

Cluster no.	Top ten keywords	Occur-rence	Link strength	Future research trends based on occurrences
6	Human	244	9090	• Investigate the effect of 3D bioprinted human cells for enhancing human tissue regeneration. • Prioritise research in the high impact journals to increase the advancement in three-dimensional printing. • Explore the use of methacrylic acid and silk fibroin biomaterial for fracture healing. • Explore the biomechanics specs of he meant issues regeneration in Sprague-Dawley rats. • Examine the importance of protein in 3D bioprinting for optimising cell proliferation as well as tissue functionality.
	Three dimensional printing	236	8909	
	3d bioprinting	115	3475	
	Priority journal	70	2843	
	Rats	39	1664	
	Human tissue	27	1053	
	Methacrylic acid	22	956	
	Silk fibroin	18	653	
	Fracture healing	10	410	
	Rats, sprague-dawley	10	456	
	Sprague dawley rat	10	456	
	Proteins	9	300	

Table 9.2 Top ten keywords with maximum occurrence

Sr. No	Keyword	Occurrences	Total link strength	Cluster
1	Bioprinting	348	11165	4
2	Tissue engineering	318	11267	1
3	Tissue scaffolds	257	9980	4
4	Article	255	10135	2
5	Scaffolds (biology)	247	8106	1
6	Human	244	9090	6
7	Tissue scaffold	243	9516	4
8	Bone	237	7870	1
9	Three dimensional printing	236	8909	6
10	Chemistry	206	8016	4

keyword by providing the interconnectedness in this Scopus database. The cluster column provides the information about the specific cluster in which the keyword is associated. This table is helpful for the researcher to understand the strength between the top ten keywords.

4.3 Ten Keywords with Minimum Occurrences

In Table 9.3, the identified ten keywords are least occurrence keywords, but these are important. These keywords show potential for specialized research work and innovation in the field for study. Because of the less occurrence of these keywords there are great opportunities to learn more about the interconnectedness of these keywords to identify their relationship within the cluster. This will provide the opportunities for different research in this area.

Table 9.3 Ten keywords with minimum occurrence

Sr. No	Keyword	Occurrences	Total link strength	Cluster
1	Three Dimensional Computer Graphics	5	149	6
2	Tissue-Engineered Scaffolds	5	162	4
3	Transforming Growth Factor Beta	5	203	3
4	Transforming Growth Factor Beta3	5	164	2
5	Transplantation	5	219	3
6	Tumor Microenvironment	5	190	5
7	Wound Healing	5	213	2
8	X Ray Photoemission Spectroscopy	5	270	2
9	Xylazine	5	249	2
10	Yield Stress	5	141	5

4.4 Key findings

In Fig. 9.3, 9.4, 9.5, and 9.6, the maximum text size and bigger circles is attributed to the keywords scaffold, bioprinting, and tissue engineering, signifying that these terms are directly associated with the selected research topic. Figure 9.3, 9.4, 9.5 and 9.6 are highlighting the keywords like scaffold bioprinting and tissue engineering which signifies that these terms are directly related to the research topic. These keywords have the highest frequency in the selected document shows that these keywords have maximum relevance in the ongoing research in this area. The maximum occurrence, and the link strength of these keywords are playing an important role for the advancement in the field of scaffold-based bio printing for tissue engineering. In Fig. 9.3, 4 and 5 the smaller circles and the text size are associated with the keywords like flow cytometry, physical parameters, and bio glass. Small circles and text size for these keywords indicates their less occurrence, connections as well as minimum link. These keywords are the less explored keywords, and which can be used for the future researcher to conduct research in these topics. The proximity among two keywords indicates the degree of their relationship(Pansare et al., 2022). It is observed that terms such as alginate, mesenchymal stroma cell, and animal model exhibit the maximum distance from the keyword bioprinting and bone scaffold. It is observed that the keywords like mesenchymal stroma cell, animal model and alginate as maximum distance from the keywords bioprinting and bone scaffold. This observation is helpful to understand that there is a significant research gaps in exploring these keywords for future research work. In Fig. 9.5, prominent yellow spots are evident for trending topics like bioprinting, tissue engineering, chemistry, and human. Fig. 9.5 shows that the keywords like bio printing, tissue engineering, chemistry and human are having dominant yellow colour spots. These yellow colour sports indicate the weight associated with these keywords by indicating their connection and relatedness within the database network.

5. Conclusion

The presented study aims to examine the co-occurrence of keywords in research articles related to Bioprinting and Bone scaffold using VOSviewer. A total of 947 research articles were initially retrieved from the Scopus database employing the keyword 'Bioprinting and Bone scaffold.' After the application of filters, which included peer-reviewed articles in English, the final dataset consisted of 489 articles. These articles were then subjected to co-occurrence analysis, revealing a selection of 577 keywords with a minimum of five occurrences, organized into six clusters. This study provides the network visualization, overlay visualization and the density visualization with the detailed discussion on the findings obtained from the different images. This analysis was conducted using VOSviewer software with the keyword "Bioprinting" and "Bone scaffold" which may not be enough to adequately capture the significance of the terms and cause the analysis to overlook significant patterns and trends in the bioprinting industry. Additionally, the accuracy of VOSviewer results may be impacted by missing or incorrectly labelled articles in the Scopus database. Several keywords in the selected articles were identified in this work. The frequency of the keywords indicates the relative occurrence in this research. The six cluster are divided based on different research areas which is helpful for identifying the future research trends and investigation of the same. The total links strength in the cluster signifies the relationship between the keywords. The compilation of the data for the top 10 keywords highlights the active research trends in the Bioprinting and Bone scaffold area. On the other hand, the keywords with minimum occurrence provide the road map for the future researcher to select their research area.

The present analysis of keywords is based on minimum five occurrences and only Scopus data wherein only articles and reviews are selected. The purpose of selecting a minimum occurrence threshold in this study is to filter out the too-specific keyword which could have resulted in less meaningful insight in the occurrences. By taking the five threshold occurrences, the focus was to obtain the significant research trends for the data that frequently appeared in the dataset. The selected threshold value provides the balance between the keyboard captures to maintain the diversity of the selected research topic and remove the impact of less occurrence keywords. The identified clusters and keywords highlight the emerging trends finding the gaps in the current research area and providing highly interesting research topics. The cluster represents closely related keywords based on their co-occurrence in the Scopus database. Analysis of these clusters can help researchers to identify key topics and theme which are currently in trends. The future researchers may expand this study for all keywords and may include more research papers including conference papers and book chapters. To follow trends in newly explored fields of interest, future researchers may use longitudinal analysis. Larger research areas can also be covered by including the more comprehensive databases from PubMed and Web of Science. Also, more detailed analysis of each keyword from future research topic perspective may be conducted by future researchers.

References

1. Abouna, G. M. (2008). Organ Shortage Crisis: Problems and Possible Solutions. *Transplantation Proceedings*, *40*(1), 34–38. https://doi.org/10.1016/j.transproceed.2007.11.067
2. Ding, X., & Yang, Z. (2022). Knowledge mapping of platform research: a visual analysis using VOSviewer and CiteSpace. *Electronic Commerce Research*, *22*(3), 787–809. https://doi.org/10.1007/s10660-020-09410-7
3. Fahimnia, B., Sarkis, J., & Davarzani, H. (2015). Green supply chain management: A review and bibliometric analysis. In *International Journal of Production Economics* (Vol. 162). Elsevier. https://doi.org/10.1016/j.ijpe.2015.01.003
4. Mancha Sánchez, E., Gómez-Blanco, J. C., López Nieto, E., Casado, J. G., Macías-García, A., Díaz Díez, M. A., Carrasco-Amador, J. P., Torrejón Martín, D., Sánchez-Margallo, F. M., & Pagador, J. B. (2020). Hydrogels for Bioprinting: A Systematic Review of Hydrogels Synthesis, Bioprinting Parameters, and Bioprinted Structures Behavior. *Frontiers in Bioengineering and Biotechnology*, *8*(August). https://doi.org/10.3389/fbioe.2020.00776
5. Mandrycky, C., Wang, Z., Kim, K., & Kim, D. H. (2016). 3D bioprinting for engineering complex tissues. *Biotechnology Advances*, *34*(4), 422–434. https://doi.org/10.1016/j.biotechadv.2015.12.011
6. Pansare, R., Yadav, G., & Nagare, M. R. (2022). Reconfigurable manufacturing system: a systematic bibliometric analysis and future research agenda. *Journal of Manufacturing Technology Management*, *33*(3), 543–574. https://doi.org/10.1108/JMTM-04-2021-0137

7. Thakur, K. K., Lekurwale, R., Bansode, S., & Pansare, R. (2023). 3D Bioprinting: A Systematic Review for Future Research Direction. *Indian Journal of Orthopaedics, 0123456789*. https://doi.org/10.1007/s43465-023-01000-7
8. Vijayavenkataraman, S., Yan, W. C., Lu, W. F., Wang, C. H., & Fuh, J. Y. H. (2018). 3D bioprinting of tissues and organs for regenerative medicine. *Advanced Drug Delivery Reviews*, *132*, 296–332. https://doi.org/10.1016/j.addr.2018.07.004
9. Zhang, B., Luo, Y., Ma, L., Gao, L., Li, Y., Xue, Q., Yang, H., & Cui, Z. (2018). 3D bioprinting: an emerging technology full of opportunities and challenges. *Bio-Design and Manufacturing*, *1*(1), 2–13. https://doi.org/10.1007/s42242-018-0004-3

Note: All the figures and tables in this chapter were made by the authors.

Technologies for Energy, Agriculture, and Healthcare – Shailesh Nikam et al. (eds)
© 2024 Taylor & Francis Group, London, ISBN 978-1-032-98028-7

10

Development of Sustainable Tableware Materials Produced with Agricultural Waste

Shivangi V. Thakker[1]
Assistant Professor, Dept. of Mechanical Engineering,
K J Somaiya College of Engineering,
Somaiya Vidyavihar University,
Mumbai, India

Shweta Pandey[2]
Junior Research Fellow,
DST Project, Dept. of Mechanical Engineering,
K J Somaiya College of Engineering,
Somaiya Vidyavihar University,
Mumbai, India

Abstract: The extensive usage and availability of plastic, coupled with its extended shelf life, pose a significant challenge in the food industry. The increasing issue of plastic pollution is closely linked to the dependence of the food sector on plastic for packaging, utensils, and dinnerware. To address the plastic crisis, this research aims to create a solution by developing tableware from agricultural waste. This work aims to compress bagasse and paddy, straw pulp to produce sustainable dinnerware using a compression molding machine. Connecting the realms of environmental sustainability and technological advancements, another significant development emerges in the convergence of the Internet of Things (IoT) and blockchain networks (BCNs). Research for the integration of IoT and blockchain technology into compression molding machines to monitor and control manufacturing is presented. This integration allows for a system that manages the manufacturing process to increase efficiency and maintain high quality while ensuring transparency and traceability. The research aims to reduce waste by

[1]shivangiruparel@somaiya.edu, [2]sop@somaiya.edu

DOI: 10.1201/9781003596707-10

using agricultural waste as raw materials and is associated with the principles of a circular economy.

Keywords: Agricultural waste, Bagasse, Compression molding machine, Paddy straw, Sustainability, Automation, IoT, and Blockchain integration

1. Introduction

Disposable plastic utensils have effortlessly entered into contemporary society due to their affordability and convenience. However, the extensive usage of these disposable items has caused a significant environmental impact. The need for sustainable alternatives comes due to innumerable issues associated with disposable plastic utensils, like resource depletion to the generation of large amounts of waste [1,3,4,7]. The struggle of plastic waste management is increasing day by day, and possible alternatives to address its constraining issues remain difficult to find. They not only affect the marine and terrestrial ecosystem due to improper deposition in landfills, water bodies, etc but also cause human health hazards due to their toxic chemical composition leaching into food [2].

One green and cost-effective practice is to use agricultural waste such as bagasse and paddy straw as pulp in the paper and pulp industry. Harnessing agricultural waste by using it as pulp is a great way of using resources that would otherwise be either discarded or burned [5,6]. These methods help reduce the demand for wood pulp and moderate the impact associated with logging [8]. Bagasse or paddy straw processing requires less energy as less chemical input is required compared to the traditional method of kraft pulping which involves energy-intensive steps, such as chipping, cooking, bleaching, washing, and drying; hence it's a more energy-efficient method and reduces overall carbon emissions, promoting circular economy. Replacing single-use plastics with sustainable options contributes to reducing pollution, conserving natural resources and the promotion of sustainable and circular economy.

This research seeks to palliate the adverse effects of plastic by creating tableware from agricultural sources by exploring the use of pneumatically operated compression molding machines. Compression molding is a commonly employed technique for producing various products using composite materials. Compression moulding is a process involving pressing or squeezing a deformable material charge between two halves of a heated mold and its subsequent transformation into a moulded part after cooling or curing [11]. It is sometimes referred to as matched-die moulding, where the fibre-reinforced composite material is forced to deform or flow within the mould cavity [16]. This method is appealing due to its high degree of automation, allowing the creation of intricate geometries with minimal material wastage. Furthermore, the design of the component can be leveraged to significantly improve the structural stiffness.

Efficiency and effectiveness are two things that must be considered in every production process. The application of an automation system to control the main processing steps, such as pressure and temperature, can increase the effectiveness and accuracy of operator movements. During the recent decade, two evolving computing networks, Internet of Things (IoT) and Blockchain Network(s) (BCNs) have enhanced the Internet. IoT is a technology that allows communication between physical and virtual objects and its application to industrial machines has advantages such as reducing the need for space and resources and physical contact. Blockchain technology is revolutionary for data security, data transmission, fault tolerance, and transparency. A blockchain is a security-focused structure with excellent potential, efficient transparency, and decentralization. By integrating the IoT and Blockchain network in these processing steps, the system can be automated, which will increase the efficiency of the process and help in maintaining its high quality. Both can be integrated into compression moulding machine to monitor and control the manufacturing process [2,4]. This includes temperature and pressure monitoring sensors, data acquisition system, communication protocol, cloud platform integration, decentralized data storage, traceability and transparency. Cryptographic techniques can be used to secure communication between IoT devices and the blockchain.

2. Methodology

2.1 Processing of Agricultural Waste

Bagasse or paddy straw undergoes an intricate multi-step process to derive useful products. Bagasse is collected from sugarcane mills after juice extraction and paddy straw is collected from the rice fields after harvest. Both needs thorough sorting and cleaning to remove pesky impurities such as soils and stones. Additionally in case of paddy straw, to make the straw more manageable and create uniformity, it needs to be cut or shredded after drying. To produce quality material for various purposes, the fibres are subjected to refining, cleaning, and bleaching phases after going through a crucial step known as pulping [9,10]. The said process involves separating the fibres from other components. This is achieved through chemical or mechanical means. Using either machinery or moulds, the pulp is compressed to obtain the plate shapes desired, forming a wet mat or sheet. Through a drying process, the water content of these shaped pulp sheets is reduced. To ensure longevity and steadfastness of the final product, air drying or industrial dryers can be utilized. Figure 10.1. Shows the flowchart of bagasse fibre processing.

2.2 Thermoforming Semi-Automatic Machine

SmartFib compression moulding machine is an agro-waste pulp molded tableware machinery in which agro-waste pulp liquid can be absorbed into a certain shape, dehydrated by this machine's mold, and then prepared for the subsequent working operation to form and become the product. The machine's main components are the pulp feed system, fixed quantity system, moulding system, vacuum dehydrating system, and electric control and it features a formation bottom mold, an upper mold basket, and a pulp liquid cup with a fixed quantity.

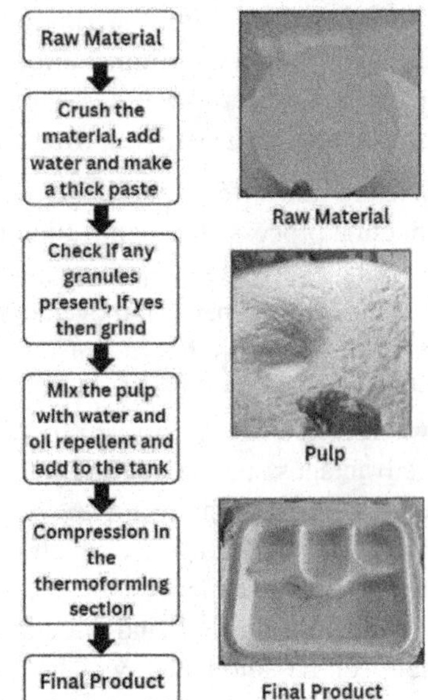

Fig. 10.1 Flowchart of bagasse fibre processing

Source: Author's compilation

2.3 Working Principle

A semi-automatic pulp molding machine typically follows the process of converting the bagasse pulp into molded plates. Bagasse or paddy straw pulp slurry becomes the raw material for molding paper plates. The machine includes a forming station where the pulp slurry is poured into molds shaped like plates. The plate takes shape as the pulp is pressed in the forming station using a pressing mechanism, causing the fibers to bond together and excess water to drain away. The formed plates are moved to a drying station where moisture is removed and the plates are solidified. From the molds, once dry, the plates are removed manually.

2.4 Integration of Iot and Blockchain Technology into a Compression Molding Machine

Temperature sensors such as thermocouples, infrared sensors, and pressure transducers can be installed at critical points in the compression molding machine for temperature and pressure monitoring. They can be placed near the heating element and pressure sensors in the molding chamber. To function as a data acquisition system, deploy a micrometer or PLC (Programmable Logic

Controller) to read the data from temperature and pressure sensors. To transmit the collected data, opt for a communication protocol such as message queuing telemetry transport (MQTT), suitable for sending data such as temperature and pressure readings to a cloud platform or a central server. An AWS IoT core setup is required to receive, store, and process the collected data. This will also be suitable for activating alerts based on the threshold value. For remote monitoring and control, a web-based dashboard is required. This can help in real-time data visualization, adjust parameters, and receive alerts remotely.

Blockchain could be deployed to document different stages of the manufacturing process. Each cycle represents a recorded transaction. To ensure decentralization and mitigate the risk of a single point of failure, the gathered data can be dispersed across the blockchain network. A blockchain network can provide traceability and transparency throughout the supply chain. For example, customers can verify the origin, manufacturing history, and authenticity of the product using a RFID tag, fostering sustainability and trust. Smart contracts can help automate and enforce compliance rules. For example, implement a smart contract to ensure that the compression molding process parameters meet the environmental standards; if not it can trigger an alert and pause the production.

To secure communication between IoT devices and the blockchain, use cryptographic techniques. This can be achieved by implementing secure communication protocols from IoT devices to blockchain nodes for data transmission. In addition, a unique identifier can be included that is generated by the IoT device, in each blockchain transaction. This can be used to cross-reference blockchain data with IoT data. Figure 10.2 presents an overview of the Integration Framework for IoT and Blockchain."

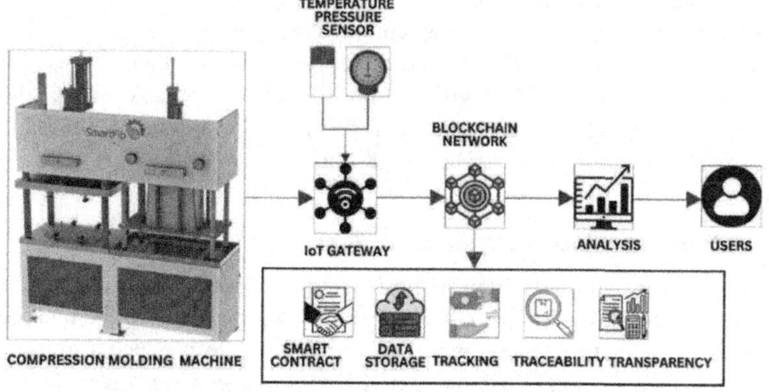

Fig. 10.2 IoT and blockchain integration framework

Source: Author's compilation

2.5 Testing of Mechanical Properties

Thickness Test

To carry out a thickness test, the representative sample was chosen and spread flat on stable surface. A micrometer that had been calibrated was used for taking measurements of the thickness at various points inclusive of the center and edges. An average value was determined for the thickness. The results were recorded for documentation purposes and then the test repeated to ensure repeatability. This procedure enabled keeping uniformity of product quality and compliance with regulations.

Moisture Content Test

To do the moisture content test, put the replicated sample in an automatic temperature-managed, well-ventilated, hot air oven held at 105 ± 2°C. Dry the pieces until the difference between two sequential drying and weighing, (not exceeding 24 hours) doesn't go over 0.1%. From the wet and dry weights of the samples, calculate the moisture content and express in percentage dry basis

Tensile Strength

When a material is subjected to stress, its tensile strength gives an idea about the maximum stress it can endure before it starts to remarkably contract. The tensile strength test provides the measurement of the resistance to applied force or tension, to overall asses its strength and durability. Whether, specific tensile strength requirements are exceeded or met, the consistency and reliability of the disposable plates is ensured.

Bending TESt

Prepare a square test paper of 60 × 60 mm and fold by fingers along the diagonal so that the inner surfaces are in contact completely. Unfold and repeat the folding along the same crease in the opposite direction until the other surface are in contact completely. Record and visible fibrous breaks or cracks on the surface of the folded test piece after each crease. If no break is more than 6mm long, the broad may be classified as fair. If the board shows no fibrous break, it is classified as good.

Oil Resistance and Water Penetration Test

These tests ensure that food tableware possesses physicochemical and mechanical properties that will protect the quality of the food. In this method, the receptivity of paper is terms of time taken by a drop of castor oil to produce a translucent spot on the paper. A sample having 7cm X 7cm dimension was taken and a test was conducted for both cold oil (room temperature) and hot oil by dropping a drop of oil and wiping it off after 3 min with cotton. If the dark marks appeared, the resistance level was counted to be lower than the level represented by the test liquid.

The relative water absorption test was carried out by taking 10 pieces of samples each having dimensions 7cm x 7cm. The samples were soaked in the water for an hour and then hung for a minute to remove the excess water before weighing. Relative water absorbed was calculated using the following formula.

$$A = \frac{(m_2 - m_1)}{m_1} \times 100\%$$

m_1 = mass of the sample before soaking (g)

m_2 = mass of the sample after soaking (g)

Compression Load-Bearing Test

The compression load-bearing test holds importance in assessing the robustness of these plates when subjected to pressure. This test guarantees that the plates can endure forces during stacking, transportation, and regular usage preventing any damages and ensuring consumer safety. Its purpose is to ensure product quality by identifying any material weaknesses or variations. Determining the load-bearing capacity is crucial, for environmental impact assessments, logistical planning, and establishing the dependability of these plates across different applications.

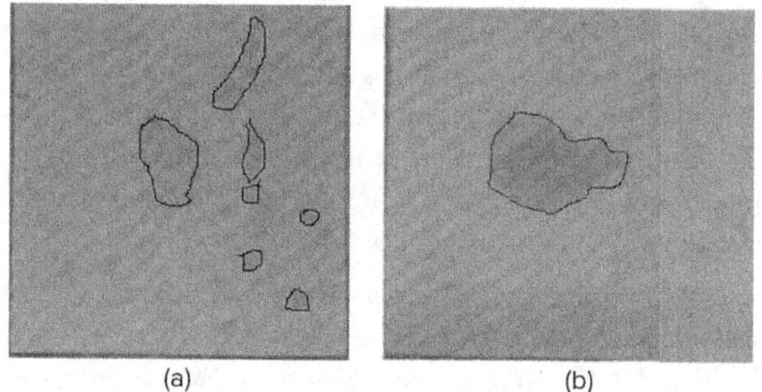

(a) (b)

Fig. 10.3 (a) Cold oil (b) Hot oil test

Source: Author's compilation

3. Results and Discussion

Table 10.1 presents the test results for bagasse pulp compared to wood paper pulp reveal significant insights into their mechanical properties and suitability for

Table 10.1 Test results

Sr. No	Test Parameters	Results for Bagasse pulp	Wood Paper Pulp [1,5,6,9]
1	Thickness	0.73 mm	0.5-1 mm
2	Moisture content	7.08 %	4-6%
3	Tensile strength	15.51 Mpa	8-15 MPa
4	Bending	No fibrous break	—
5	Oil resistance (cold)	Few dark spots	Few dark spots
6	Oil resistance (hot)	Dark patch	Dark patch
7	Water penetration	1 hour	30-90 mins
8	Compression load-bearing Max. load (Fmax) Disp. at Fmax Max. disp. Comp. strength	 0.268 kN 0.35 mm 1.65 mm 14.89 Mpa	—

Source: Author's compilation

various applications. Bagasse pulp (0.73 mm) demonstrates comparable thickness to wood paper pulp (0.5-1 mm), with both falling within the acceptable range for paper production. However, bagasse pulp exhibits a slightly higher moisture content (7.08%) than wood pulp (4-6%). Remarkably, bagasse pulp (15.51 MPa) showcases superior tensile strength and resilience, surpassing the typical range for wood paper pulp (8-15 Mpa). Furthermore, the oil resistance tests demonstrate similar performance between bagasse and wood paper pulp (Fig. 10.3). In contrast, bagasse pulp withstands water penetration for 1 hour which is within the limit of wood pulp (30 – 90 min). Bagasse pulp demonstrates a maximum load (Fmax) of 0.268 kN, displacement at Fmax of 0.35 mm, and maximum displacement of 1.65 mm, resulting in a compression strength of 14.89 Mpa.

Table 10.2 presents a comprehensive cost-effectiveness analysis comparing the production costs of paper plates using bagasse or paddy straw pulp versus wood pulp. The data includes the cost of raw materials per kilogram sourced from reputable platforms, processing costs, overhead expenses, and additional costs such as taxes and transportation. Notably, when comparing the total cost for manufacturing one kilogram of plates (equivalent to 50 plates of 6 inches in size), the analysis reveals a significant cost disparity. Utilizing bagasse or paddy straw pulp results in a total cost of ₹121, while using wood pulp incurs a substantially higher cost of ₹230.

Table 10.2 Cost-effectiveness analysis

Sr. No.	Raw material	Bagasse / Paddy straw pulp (in ₹)	Wood pulp (in ₹)
1	Pulp sheet cost (per kg)	11 (indiamart.com)	120 (indiamart.com)
2	Processing cost	50	50
3	Overhead cost	20	20
4	Taxes, Transportation and other cost	40	40
	Total cost for one kg plate (50 plates of size 6 inch)	**121**	**230**
		~ **47% less**	

Source: Author's compilation

4. Conclusion

The thickness of the bagasse pulp falls within the acceptable range for wood paper pulp, indicating comparable thickness between the two materials. It exhibits a slightly higher moisture content compared to the specified range for wood paper pulp, suggesting that bagasse pulp may require further drying to meet standard moisture content levels. Bagasse pulp demonstrates a higher tensile strength compared to the typical range for wood paper pulp, indicating good strength properties. It shows no fibrous break during the bending test, suggesting good flexibility and resilience comparable to wood paper pulp. Oil resistance tests for both bagasse pulp and wood paper, indicate similar levels of resistance to oil penetration. The water penetration for the test indicates comparable water resistance properties. Notably, the compression load-bearing test reveals a robust compression strength for bagasse pulp. Overall, the test results suggest that bagasse pulp exhibits favourable properties comparable to or even exceeding those of wood paper pulp in terms of tensile strength, flexibility, oil resistance, water resistance, and compression strength. The cost-effective analysis indicates that the production of paper plates using bagasse or paddy straw pulp is approximately 47% less costly compared to using wood pulp.

Based on the comprehensive analysis, bagasse pulp emerges as a superior alternative to wood pulp in several key aspects. Firstly, the mechanical properties indicate the potential for bagasse pulp to offer equivalent or even superior performance in various applications where strength and resilience are essential. In terms of energy and environmental concerns, bagasse pulp presents several advantages over wood pulp production. The utilization of bagasse, a by-product of sugarcane processing, not only reduces the demand for virgin wood resources but also mitigates the environmental impact associated with deforestation and habitat destruction. Additionally, the pulping process for bagasse contributes to overall energy efficiency and carbon footprint reduction. The use of bagasse pulp offers economic benefits due to its availability as a by-product of the sugar industry. This availability reduces raw material costs and reliance on expensive wood resources, thereby enhancing the economic viability of bagasse pulp production. Waste management is significantly improved with the utilization of bagasse pulp. By repurposing sugarcane waste into pulp for paper production, the disposal of bagasse is effectively transformed into a valuable resource, reducing landfill waste and promoting a circular economy approach.

Acknowledgement

This research is funded by the Department of Science and Technology, India. Authors are grateful to DST/WMT/TDT division, Godavari Bio refineries Limited and K J Somaiya College of Engineering for funding this research work.

References

1. A. Gupta, G. Singh, P. Ghosh, K. Arora, S. Sharma, "Development of biodegradable tableware from novel combination of paddy straw and pine needles: A potential alternative against plastic cutlery". Journal of Environmental Chemical Engineering, Volume 11, Issue 6, December 2023, DOI: 10.1016/j.jece.2023.111310
2. D.Elkayaly, N. Hazem & I.S. Fahim, "Green and Sustainable Packaging Manufacturing: A Case Study of Sugarcane Bagasse-Based Tableware in Egypt", Circular Economy and Sustainability, 30 November 2021, DOI: 10.1007/s43615-021- 00136-8
3. J. Iewkittayakorn, P. Khunthongkaew, Y. Wongnoipla, K. Kaewtatip, P. Suybangdum, and A. Sopajarn, "Biodegradable plates made of pineapple leaf pulp with biocoatings to improve water resistance," Journal of Materials Research and Technology, vol. 9, no. 3, pp. 5056–5066, May 2020, doi: 10.1016/j.jmrt.2020.03.023.
4. J. Olt, V. V. Maksarov, K. Soots, and T. Leemet, "Technology for the Production of Environment Friendly Tableware," Environmental and Climate Technologies, vol. 24, no. 2, pp. 57–66, Sep. 2020, doi: 10.2478/rtuect-2020-0054.
5. Katherine E. Semple, Chenli Zhou, Orlando J. Rojas, William Nguegang Nkeuwa, Chunping Dai, Moulded pulp fibers for disposable food packaging: A state-of-the-art review, Food Packaging and Shelf Life, Volume 33, 2022, ISSN 2214–2894, https://doi.org/10.1016/j.fpsl.2022.100908.
6. Liu et al., "Biodegradable, Hygienic, and Compostable Tableware from Hybrid Sugarcane and Bamboo Fibers as Plastic Alternative," Matter, vol. 3, no. 6, pp. 2066–2079, Dec. 2020, doi: 10.1016/j.matt.2020.10.004.
7. Luan, P., Li, J., He, S., Kuang, Y., Mo, L., & Song, T. (2019). Investigation of deposit problem during sugarcane bagasse pulp molded tableware production. Journal of Cleaner Production, 237, 117856.

8. M. Didone and G. Tosello, "Moulded pulp products manufacturing with thermoforming," Packaging Technology and Science, vol. 32, no. 1, pp. 7–22, Sep. 2018, doi: 10.1002/pts.2412.
9. M. Gautam and N. Caetano, "Study, design and analysis of sustainable alternatives to plastic takeaway cutlery and crockery," Energy Procedia, vol. 136, pp. 507–512, Oct. 2017, doi: 10.1016/j.egypro.2017.10.273.
10. N. Natarajan, M. Vasudevan, V. Vivekk Velusamy, M. Selvara, (2019), "Eco-Friendly and Edible Waste Cutlery for Sustainable Environment", International Journal of Engineering and Advanced Technology (IJEAT), Vol-9, Issue-1S4
11. O. A., K. E, O. E, and O. T, "The design and construction of a pulp molding machine," International Journal of Engineering Research & Technology, vol. 10, no. 06, Jun. 2021.
12. P. Vyshali and P B Muthamma, "Development of an edible and biodegradable tableware using fruit wastes-an alternative to plastic tableware," International Journal of Food and Nutritional Sciences, vol. 11, no. 4, Jun. 2022, doi: 10.4103/ijfans_114_22.
13. S.P. Singh, M. Jawaid, M. Chandrasekar, K. Senthilkumar, Bhoomika, "Sugarcane wastes into commercial products", Processing methods, production optimization and challenges, Journal of Cleaner Production, Volume 328, 2021, ISSN 0959-6526, https://doi.org/10.1016/j.jclepro.2021.129453.
14. Wael A Elhelece, "Rice Straw as a Raw Material for Pulp and Paper Production", Encyclopedia of Renewable and Sustainable Materials, pg 1–10, DOI: 10.1016/B978-0- 12-813195-4.10596-6
15. Xiaoyi Chen, Fuming Chen, Huan Jiang, "Replacing Plastic with Bamboo: Eco-Friendly Disposable Tableware Based on the Separation of Bamboo Fibers and the Reconstruction of Their Network Structure", ACS Sustainable Chemistry and Engineering, May 1, 2023, DOI: 10.1021/acssuschemeng.3c00293
16. Orgéas, Laurent & Dumont, Pierre. "Sheet Molding Compounds", Encyclopedia of Composites, 2012 Volume 2, pg 2683–2718, DOI:10.1002/9781118097298.weoc222.

Technologies for Energy, Agriculture, and Healthcare – Shailesh Nikam et al. (eds)
© 2024 Taylor & Francis Group, London, ISBN 978-1-032-98028-7

11

Modal and Harmonic Simulation on Morphing Wing Aerofoil for Unmanned Aerial Vehicles for Structural Optimization

Animesh Dolas*,
Pratik Dixit, Pravin Doke
Student, Department of Mechanical Engineering,
Vishwakarma Institute of Technology,
Pune, Maharashtra, India

Mangesh Chaudhari, Umesh Chavan
Professor, Department of Mechanical Engineering,
Vishwakarma Institute of Technology,
Pune, Maharashtra, India

Abstract: To modify lift and drag forces for the duration of flight control, take-off, or landing, adaptable wings, also known as morphing aerofoils, can be utilized by changing the length of the chord and angle of attack of the wings. This technology is frequently employed in Unmanned Aerial Vehicles (UAVs) to enhance aerodynamic performance and reduce drag, ultimately contributing to energy efficiency. Hydraulic motors are commonly used to control flight and alter the configuration of an airplane's wings. This paper presents various simulations. Firstly, modal analysis is used to determine the natural frequencies and corresponding mode shapes of the stress/strain profile. Secondly, harmonic analysis is applied to observe the sinusoidally varying response at each frequency and the related minimum and maximum response over a frequency range. This research paper endeavours to investigate and enhance the structural design of aerofoils for UAVs with a focus on energy-saving capabilities. The study encompasses an analysis of various parameters such as stress and vibration, considering the impact of different materials on the morphing wing structure.

*Corresponding author: drdolas2002@gmail.com

DOI: 10.1201/9781003596707-11

Additionally, the research seeks to identify the optimal material for the prototyping of morphing wings tailored for UAVs applications, aiming to further contribute to energy conservation in UAVs operations.

Keywords: Harmonic analysis, Modal analysis, Morphing of the aerofoil, Structural optimization, UAVs

1. Introduction

The prospect of morphing wings to dynamically change their configuration while in flight, reducing drag and increasing aerodynamic efficiency, is what gives them their promise. Morphing wing is a complex combination of materials science and aerospace engineering that has benefited from significant contributions from a range of studies. As noted in [1], shape memory alloy actuators have proven to be effective and resilient in the face of severe aerodynamic stresses, opening the door for creative design approaches. The morphing wing mechanisms are made more versatile by the unique out-of-plane transformation process that bio-inspired elastomeric transformations [2]

Examining various actuation methods, such as the use of segmented beams and shape memory alloy actuators [5], as well as the integration of compliant rib systems and additive fabrication for effective morphing airfoil prototypes [8], is a crucial component of this literature review. The combined findings of these studies highlight the potential advantages of morphing wings with respect to enhanced flying performance and energy efficiency. By utilizing dynamic wing adjustments, morphing wings can optimize aerodynamics, minimize drag, and enhance fuel efficiency while also contributing to sustainable aviation. Despite these remarkable strides, significant gaps remain. The long-term durability of materials under diverse climatic conditions and the influence of material vibrations caused by actuation systems demand further attention.[19] Furthermore, while 3D printing technology has found its place in the development of morphing wings, comprehensive insights into enhancing processes and reinforcing materials through this technology are still in their infancy.

This research aims to contribute to the growing body of knowledge in morphing wing technology, focusing on the structural strength and optimization of a redesigned aerofoil. Through advanced simulation techniques and meticulous exploration of material alterations, this study seeks to bridge existing gaps and provide valuable insights that propel the ongoing development of morphing wings for tiny UAVs.

2. Literature Review

Detailed literature on morphing wings provides insides how this technology is developing. Various papers have investigated multiple facets, with a focus on energy, aerodynamic enhancement, and flexibility in varying flight circumstances. Present a wing design that uses shape memory alloy actuators and exhibits durability and effectiveness under high aerodynamic loads. It offers a distinctive viewpoint on morphing wing mechanisms as it investigates an out-of-plane transformation process based on plastic, elastomers, and bioinspired principles [1, 2]. Explore the aerodynamic performance of both non-morphing and morphing UAV wings, examining various actuation strategies to improve our comprehension of their effectiveness. [4] Focuses on flexible leading-edge morphing wings, highlighting encouraging outcomes and the significance of material optimization. Demonstrate a morphing wing system that changes swept angles and curvature using shape memory [16] and composite materials [18, 19] alloy actuators and segmented beams. In [6], camber morphing is the main focus, with an emphasis on improving lift without the use of conventional flaps. On the other hand [7] provides a Method for calculating bent wing aerodynamic parameters. To develop an effective morphing airfoil, prototypes [8] look into additive fabrication methods and compliant rib systems. A comparison of the aerodynamic performance of conventional and morphing wings is presented in [9], showing that the former is more efficient under specific conditions. To reduce bulk and energy consumption [10, 19] proposes a camber-deformed wing concept. On the other hand [11] concentrates on aerodynamic design and the best locations for "ailsers" on linearly adjustable wingspan aircraft. In [12], bistable components are investigated; the results show reduced stiffness ratios but enhanced aerodynamic performance. A 3D-printed, bi-stable wing design made with Digital Image Correlation analysis is shown in [13]. In addition to the obvious uses of morphing wings [14] contrasts the aerodynamic efficiency of gulls and UAVs, and [15] looks into orthotropic materials for composite morphing wing aeroelasticity. However, despite these significant developments, there remains a significant vacuum in the literature concerning the long-term resilience of different materials in a range of climate conditions. [17,19,20]

Furthermore, the impact of material vibrations and deformation brought on by actuation systems has not received much attention. Although 3D printing technology has been used in several studies, there is still a lack of comprehensive knowledge regarding how to use this technology to strengthen materials and improve processes.

This research endeavours to address these gaps by focusing on the structural strength and optimization of a redesigned aerofoil, employing advanced simulation techniques to assess various structural properties. Through a detailed examination of morphing wing configurations and material alterations, this study aims to contribute valuable insights to the ongoing development of morphing wing technology for tiny UAVs.

3. Methodology

The process begins with the modelling phase, in which the CAD model parameters listed in the relevant publication [1] are utilized. Next, the model is simulated and the resulting output is carefully compared with the information provided in the reference paper. The process is then divided into two parts, wherein the first part is devoted to evaluating the behaviour and structural strength of the mentioned computer-aided design.

3.1 Validation of Referred Design

NACA 0012 profile and used a Zig-zag configuration as a bistable material which is more flexible allowing larger deformations having material ABS plastic as per ref. [1] (see Fig. 11.1). Based on simulation results, the Zig-Zag structure exhibited a deformation of 0.00229 m (See Fig. 11.2) with a Factor of Safety (FOS) equal to 12.

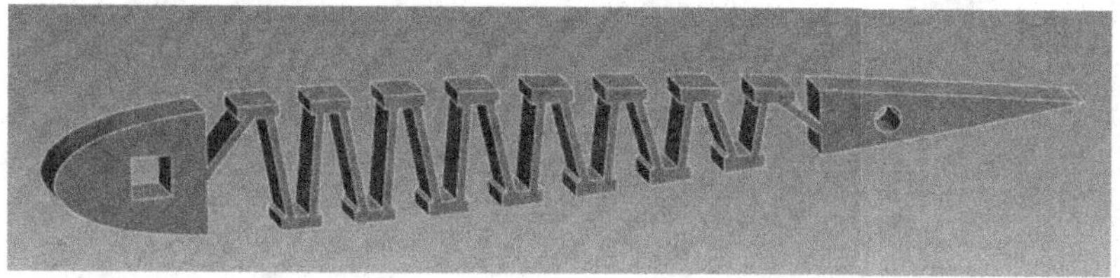

Fig. 11.1 NACA 0012 aerofoil morphing wing zigzag configuration

For this work, the deformation calculated was 0.00236 m which has an error of 3.05% and FOS: 15 there is a slight variation in deformation due to the structure being tested on the airfoil shape and rather than one the Zig-zag configuration, and since the design stress acting on the CAD is slightly higher.

3.2 Performed Structural Optimization

After the validation of the model, various analyses were performed that depicted the object's behaviour. The first simulation performed was the Modal analysis to calculate the various nodes for natural frequencies (Hz) and mode shapes of a structure for ASB plastic, structural steel¸ titanium alloy, and TI_6Al_4V which are important parameters that describe how the structure will respond to various loads

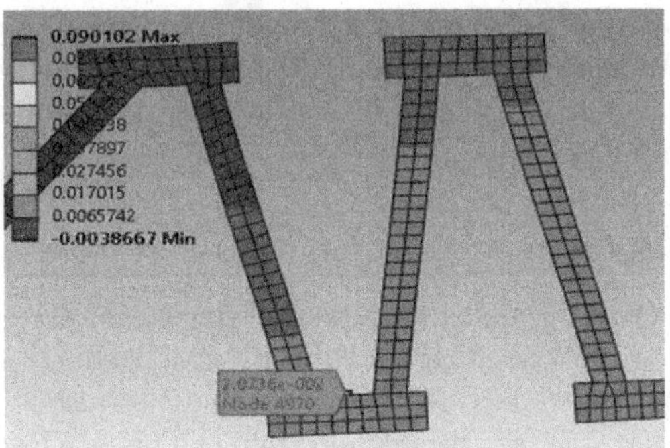

Fig. 11.2 Morphing wing zig-zag structure deformation under axial load (units in meters)

and stimulations. The airflow over an airfoil may be mathematically simulated, and the lift and drag properties can be predicted. An airfoil's behaviour may be studied, and its design can be optimized for better lift and drag properties using harmonic analysis for simulation the morphing wing is assumed as a cantilever beam with fixed at one and a 350N-mm moment at the other end. (See Fig. 11.3)

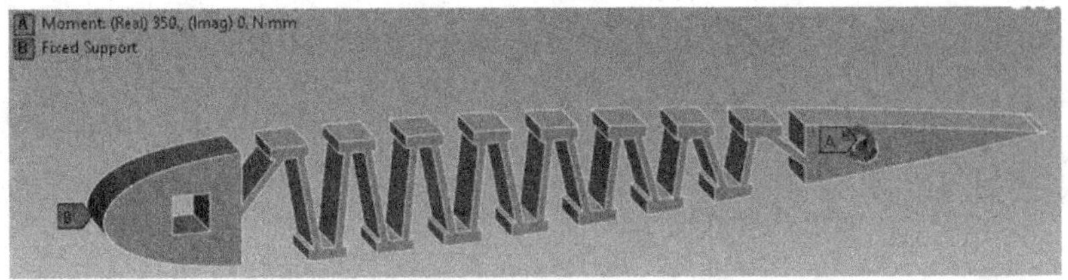

Fig. 11.3 Morphing wing boundary conditions for harmonic analysis

Harmonic analysis was used to determine the aerofoil's structural response to periodic loads or excitation. Various periodic forces acting on the airfoil were estimated, taking into account the maximum and lowest frequencies for distinct modes as determined by modal analysis results. In this investigation, a moment of 350 N-mm was used to explore resonance frequency (see Figs. 11.3 and 11.4). The study sought

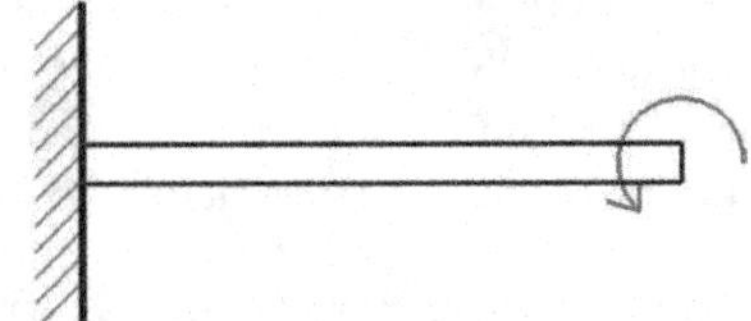

Fig. 11.4 Morphing wing for harmonic analysis is assumed as a cantilever beam with fixed at one end and moment appeared at the other end

to investigate how changing the material, such as utilizing Titanium alloy or aluminium alloy, while keeping the remainder of the aerofoil material as ABS plastic at the highest deformation section, affected resonance frequency and improved the structural strength of the CAD model.

Static structural analysis was conducted to assess the airfoil's response to static loads, including the aircraft's weight, aerodynamic forces, and external loads. The analysis employed the same fixture as the Model and Harmonic analyses. An axial force with a time-varying magnitude of 1.00 N was applied at the airfoil's tail end to determine maximum deformation and assess stiffness variations for different materials, including effective stiffness for non-isotropic materials. Various relationships between frequencies and the structure can be concluded from the below equation:

For most materials, the relationship between frequency and deformation can be approximated by the equation.

$$f = k * \left(\frac{1}{L}\right) * \sqrt{\frac{T}{m}} \tag{1}$$

$$f = \sqrt{\frac{k}{m}} \tag{2}$$

Where, f: frequency; k: spring constant; L: length of the material; T: tension in the material; m: mass per unit length and k: stiffness or spring constant.

4. Result and Discussion

Modal research provided critical insights into the inherent frequencies and mode forms of the morphing wing. Six deformation modes were computed using modal analysis for a morphing airfoil constructed of plastic ABS.

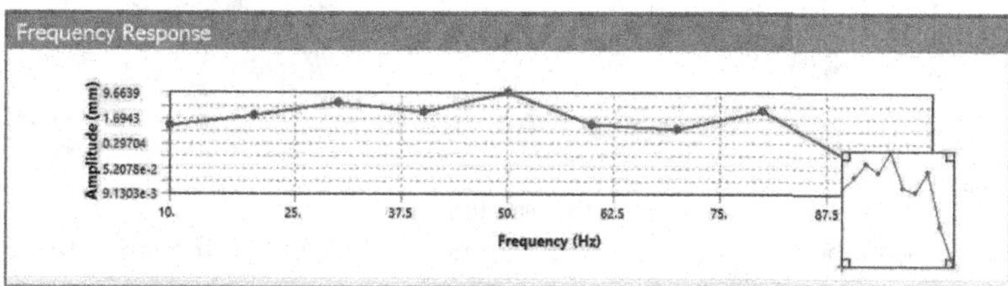

Fig. 11.5 Frequency response using harmonic and model analysis for ABS plastic

Based on observations, assumptions were made that mode 3 and mode 5 might describe the actuation of the Airfoil under flight conditions for UAVs. To validate

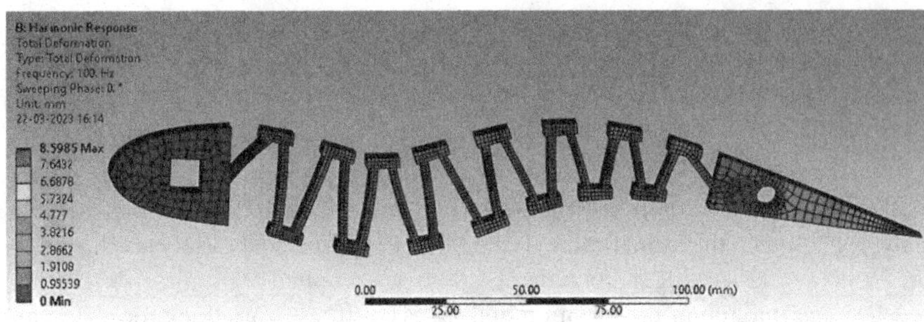

Fig. 11.6 Frequency response deformation under applied movement of 350 N-mm for plastic ABS

the assumption and to study resonance frequency for the model using harmonic analysis, the natural frequency calculated was 50 Hz, which was the assumption made in modal analysis. The natural frequency, as observed in mode five from the table below (Fig. 11.7), aligns with the assumed value. Additionally, in Fig. 11.6, it can be observed that maximum deformation occurs mostly at the mid-section of the zig-zag configuration of the airfoil based on the harmonic analysis.

Similarly, the modal and harmonic analysis with the same boundary conditions by changing the material of the whole structure to structure steel, aluminium alloy, titanium alloy, and Ti_6Al_4V to find how the frequency varies for different modes of deformation below it is observed resonance for the various materials.

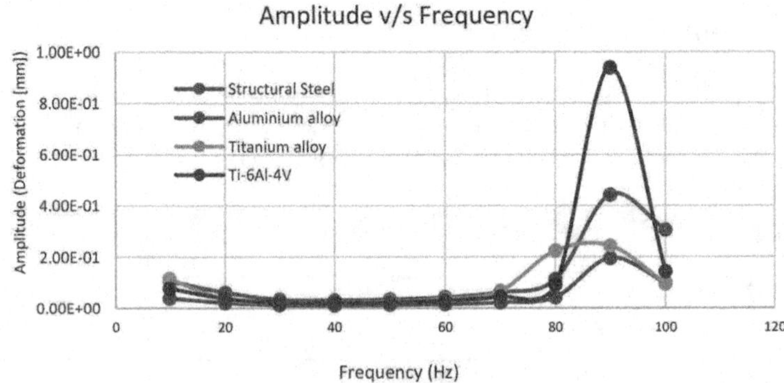

Fig. 11.7 Frequency response deformation under the applied movement of 350 N-mm and maximum resonance frequency

To enhance structural strength and mitigate resonance effects, the material in the midsection (Fig. 11.8) was transitioned from ABS plastic to robust metal alloys such as titanium and aluminum. The morphing wing Airfoil, exhibiting non-

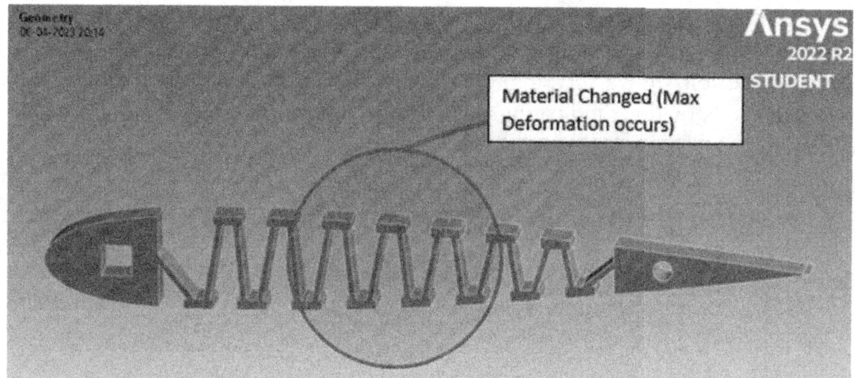

Fig. 11.8 Material modifying section where maximum deformation occurs with metal and alloys

isotropic material properties, offers increased flexibility and strength in specific directions or under loading conditions. For instance, a composite material may demonstrate higher strength and stiffness in one direction due to fiber orientation, yet lower strength and stiffness in another. Utilizing Eq. 2 and mass calculations from Ansys software, the stiffness factors for various isotropic materials are presented in the table below. Generally, materials with greater stiffness tend to be stronger, exemplified by structural steel, titanium alloys, and Ti_6Al_4V. However, non-isotropic material architectures exhibit significant variations in simulation outcomes.

Table 11.1 Stiffness and deformation for assigned material to an isotropic structure

Material Assigned	Mass (kg)	Stiffness (kg/s^2)
ABS Plastic	0.036855	298.5255
Structural Steel	0.28089	2275.209
Titanium alloy	0.16531	1339.011
Aluminium alloy	0.099116	802.8396
Ti_6Al_4V	0.15761	1276.641

Structural strength and frequency are closely connected, prompting the use of modal and harmonic analyses to assess subsequent airfoil strength and their ability to withstand external forces. The geometry, material properties, and boundary conditions profoundly influence the frequency of a structure. For instance, a long, slender structure typically exhibits a higher frequency compared to a shorter one. The relationship between a structure's frequency and strength is complex and contingent on design and usage circumstances. In specific scenarios, a higher frequency may indicate a robust construction, suggesting a stiffer or better-supported design. Conversely, in other instances, a higher frequency might signal a weaker construction, indicating susceptibility to fatigue failure or resonant vibrations.

Effects on resonance frequencies can be seen in Fig. 11.9. As can be seen in Figs. 11.9 (A) and 9 (B) when the material is changed to a structural steel and

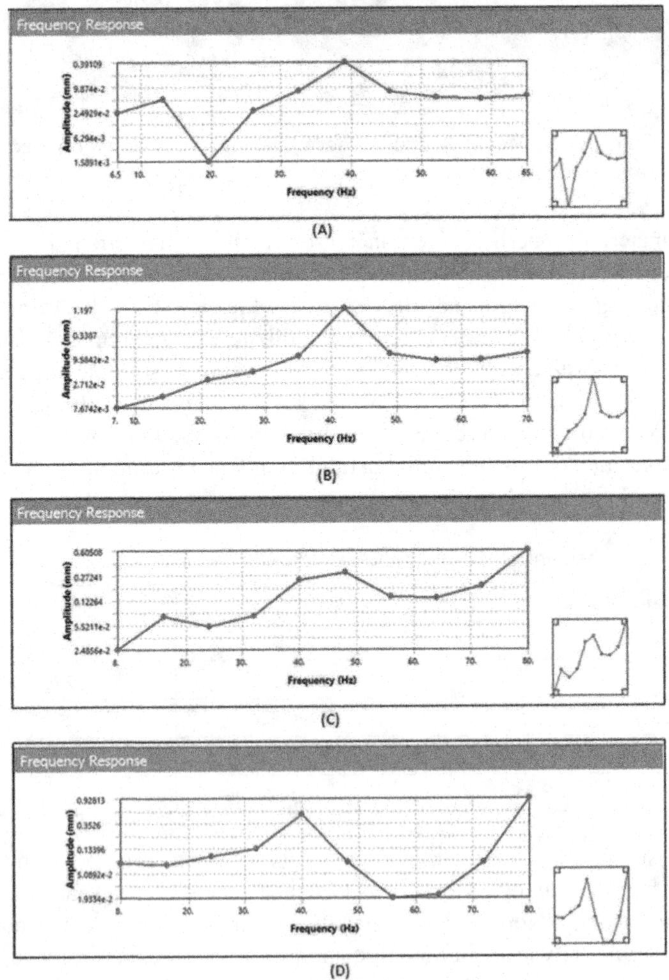

Fig. 11.9 Frequency response (A) Structural steel (B) Titanium alloy (C) Aluminium alloy (D) Ti_6Al_4V

titanium alloy respectively the frequency of decreases from 50 Hz to around 40 Hz from its original counterpart. Therefore, it can be concluded that the structural steel and titanium alloy is not a good replacement for this section (Fig. 11.8). But on the other hand, with aluminium alloy and Ti_6Al_4V, there is an increase in the frequency capacity model from 50 Hz to 80 Hz and therefore it can be a desirable choice to use this material to increase the structural strength and reduce vibration during the morphing of the airfoil.

Table 11.2 Effective stiffness for deformation of material to a non-isotropic structure

Material Assigned	Resonance frequency	Mass (Kg)	Stiffness (kg/s^2)
Structural Steel	38	0.076543	110.5281
Titanium alloy	42	0.057774	101.9133
Aluminium alloy	80	0.056523	361.7472
Ti_6Al_4V	80	0.056523	361.7472

Maximum stiffness when replacing the structure where maximum deformation occurs and making the morphing aerofoil non-isotropic structure can be observed for the material aluminium alloy and Ti_6Al_4V and these results are drastically different when the structure is isotropy this is because the relationship between stiffness and strength also depends on the type of loading the material is subjected to.

5. Conclusion

In conclusion, optimizing the structural composition of morphing aerofoils through the utilization of advanced materials presents a promising path for energy efficiency. The improved stiffness and strength of the aluminium alloy, as identified in this study, can contribute to reduced vibrations and deformations during flight. By enhancing the structural integrity, the aerofoil can operate more efficiently in its designated 3rd Mode (23.647Hz), potentially leading to decreased energy consumption. Furthermore, the observed rise in resonance frequency from 50 to 80 Hz for aluminium alloy and TI_6Al_4V suggests increased strength and stiffness, indicating the potential for energy savings in flight operations. Considering the cost-effectiveness of aluminium alloy over TI_6Al_4V, substituting materials in deformation-prone areas with aluminium alloy not only enhances structural strength but also presents an economically viable solution. This dual benefit of improved performance and cost-effectiveness underscores the potential for energy savings in the aviation industry through strategic material selection in morphing aerofoils.

In summary, the findings of this study not only contribute to the advancement of structural analysis for morphing aerofoils but also offer a practical means to enhance energy efficiency in aviation through the judicious use of aluminium alloy in critical areas prone to deformation. This approach aligns with the ongoing efforts to develop sustainable and energy-efficient aircraft technologies, paving the way for a more environmentally conscious and economically viable future in aviation.

References

1. Parancheerivilakkathil, Muhammed S. "A compliant polymorphing wing for small UAVs." *Chinese Journal of Aeronautics* 33, no. 10 (2020): 2575-2588
2. Scopelliti, Domenico, Joan Marc Martinez Gou, Cornelis Bil, Robert Carrese, and Piergiovanni Marzocca. "Design, analysis and experimental testing of a morphing wing." In *Proceedings of the 25th AIAA/AHS Adaptive Structures Conference*, pp. 1-8. American Institute for Aeronautics and Astronautics, 2017.
3. Chae, Eun Jung, Amin Moosavian, Alexander M. Pankonien, and Daniel J. Inman. "A comparative study of a morphing wing." In *Smart Materials, Adaptive Structures and Intelligent Systems*, vol. 58264, p. V002T03A020. American Society of Mechanical Engineers, 2017.
4. Zi, K. A. N., L. I. Daochun, S. H. E. N. Tong, Jinwu Xiang, and Lu Zhang. "Aerodynamic characteristics of a morphing wing with flexible leading-edge." *Chinese Journal of Aeronautics* 33, no. 10 (2020): 2610-2619.
5. Wang, Zhiqiang, Xiaojun Yang, and Bing Li. "SMA-actuated Morphing Wing with Varying Spanwise Curvature and SweptAngle." In *2019 IEEE International Conference on Robotics and Biomimetics (ROBIO)*, pp. 1615-1620. IEEE, 2019.
6. C Mathew, Bilji, K. Sai Priyanka, and JV Muruga Lal Jeyan. "Computational study on chamber morphing wing concept for efficient lift at various angle of attack." In *2020 International Conference on Interdisciplinary Cyber Physical Systems (ICPS)*, pp. 68-71. IEEE, 2020.
7. Oliviu, Sugar Gabor, Andreea Koreanschi, Ruxandra M. Botez, Mahmoud Mamou, and Youssef Mebarki. "Analysis of the aerodynamic performance of a morphing wing-tip demonstrator using a novel nonlinear vortex lattice method." In *34th AIAA applied aerodynamics conference*, p. 4036. 2016.
8. R Rajendran,Govind, Department of Aeronautical Engineering, Dayananda Sagar College of Engineering Bangalore, India." Analysis of Morphing Airfoil Structures and Fabrication of the Wing using the Concept of Additive Manufacturing". (2020) DOI: 10.17577/IJERTV9IS060054
9. Chaudhary, A.K.M., Sharma, A., Gupta, A. et al. Optimization of a 3D aircraft morphing wing with highly controllable aerodynamic performance. Aerosp Sci Technol 106, 106008 (2020). https://doi.org/10.1016/j.ast.2020.106008
10. Zhang, Jiaying, Alexander D. Shaw, Chen Wang, Huaiyuan Gu, Mohammadreza Amoozgar, Michael I. Friswell, and Benjamin KS Woods. "Aeroelastic model and

analysis of an active camber morphing wing." *Aerospace Science and Technology* 111 (2021): 106534.

11. Gu, Haipeng, and Huajie Hong. "Research and improvement on design method of morphing wing aircraft." In *Journal of Physics: Conference Series*, vol. 1600, no. 1, p. 012013. IOP Publishing, 2020.
12. Rivas-Padilla, Jose R., David M. Boston, and Andres F. Arrieta. "Design of selectively compliant morphing structures with shape-induced bi-stable elements." In *AIAA Scitech 2019 Forum*, p. 0855. 2019.
13. Boston, D. Matthew, Jose R. Rivas-Padilla, and Andres F. Arrieta. "Design and manufacturing of a multi-stable selectively stiff morphing section demonstrator." In *Smart Materials, Adaptive Structures and Intelligent Systems*, vol. 59131, p. V001T04A023. American Society of Mechanical Engineers, 2019.
14. Harvey, C., V. B. Baliga, P. Lavoie, and D. L. Altshuler. "Wing morphing allows gulls to modulate static pitch stability during gliding." *Journal of The Royal Society Interface* 16, no. 150 (2019): 20180641.
15. Tsushima, Natsuki, Tomohiro Yokozeki, Weihua Su, and Hitoshi Arizono. "Geometrically nonlinear static aeroelastic analysis of composite morphing wing with corrugated structures." *Aerospace Science and Technology* 88 (2019): 244-257.
16. Azzawi, Wessam Al. "Development and performance evaluation of a morphing wing design using shape memory polymer and composite corrugated structure." *Australian Journal of Mechanical Engineering* 22, no. 1 (2024): 12-26.
17. Pabon, Julian A., Xinyu Gao, Jielong Cai, and Sidaard Gunasekaran. "Experimental Investigation of a Novel Morphing Wing Design." In *AIAA SCITECH 2024 Forum*, p. 1349. 2024.
18. Li, Xinlin, Bin Zhan, Xueting Wang, Yan Liu, Yanju Liu, and Jinsong Leng. "Preparation of superhydrophobic shape memory composites with uniform wettability and morphing performance." *Composites Science and Technology* 247 (2024): 110398.
19. Xiasheng, S. U. N., X. U. E. Jingfeng, Z. H. O. U. Jin, W. A. N. G. Zhigang, W. A. N. G. Wenjuan, and Mengjie Zhang. "Design and validation of a variable camber wing structure." *Chinese Journal of Aeronautics* 37, no. 2 (2024): 1-11.
20. Kumar, Dinesh, Gowri Shankar, Ajin Antony, Varun Vinayachandran, and Nithin Thomas. *Design and Testing of Unmanned Aerial Vehicle with Morphing Control Surface*. No. 2023-01-5143. SAE Technical Paper, 2024.

Note: All the figures and tables in this chapter were made by the authors.

Technologies for Energy, Agriculture, and Healthcare – Shailesh Nikam et al. (eds)
© 2024 Taylor & Francis Group, London, ISBN 978-1-032-98028-7

12

Heat Storage Using Phase Change Material (PCM)

Niraj Kurane*,
Om Mali, Rohan Pawar,
Dhairyashil Jadhav, Anurag Jadhav
UG Student,
Department of Mechanical Engineering VIT,
Pune

M. B. Chaudhari, M. K. Nalawade
Professor,
Department of Mechanical Engineering VIT,
Pune

Abstract: The present work focuses on utilizing the latent heat storage capability of Phase Change Materials (PCMs) to maintain the temperature of cooked food, particularly rice, which loses heat quickly after cooking. An experimental prototype heat storage unit was built and tested to achieve this objective. The unit's design is compatible with commonly used Indian kitchen utensils for food preparation. Users can easily couple and disconnect the unit from the cooking unit as needed. During cooking, heat is absorbed and stored by the PCM within the unit. This stored heat is then released to maintain the temperature of the cooked food inside the unit for a longer period. Paraffin wax was chosen as the PCM due to its latent heat storage properties and availability. Three trials were conducted with the unit, testing both water and rice.

The results demonstrate that PCM can effectively achieve the research objective of keeping cooked food warm for extended periods without external energy input. The experiment with water required a 11.8% increase in electrical input during cooking, but resulted in a 40.2% increase in cooling time.

The experiment with cooked rice required a 12% increase in electrical input, leading to an 84.84% increase in cooling time. This significant improvement

*Corresponding Author: nirajhkurane@gmail.com

DOI: 10.1201/9781003596707-12

is attributed to the model's consideration of the thermodynamics of rice. The proposed technique offers a significant advantage over traditional methods used by catering companies, which often involve keeping food vessels submerged in hot water with continuous energy input. This approach struggles to maintain consistent food temperature. By utilizing the proposed PCM unit with slightly increased energy during cooking, food can be kept warm for longer periods without the need for continuous external heating.

Keywords: Latent heat, Heat storage, PCM, Paraffin wax and food

1. Introduction

Different products, such as thermos flasks and thermo-steel tiffin boxes, and materials, like phase change materials (PCMs), can effectively store thermal energy [10]. PCMs, readily available at affordable prices, utilize latent heat for thermal storage. When the source temperature increases, the chemical bonds within the PCM weaken, causing a phase change from solid to liquid.

This process, known as an endothermic reaction, absorbs heat from the environment. Once the temperature reaches the paraffin wax's melting point, the wax begins to melt. During this process, the temperature remains constant until the melting is complete. This constant temperature period allows the material to store a large amount of heat with minimal temperature change, resulting in a high storage density.

2. Literature

Tanathep Leungtongkum [1] In refrigeration contexts, the use of phase change materials (PCMs) can lead to improved energy management and enhanced food safety and security. However, the effectiveness of PCMs in different applications often relies on user experience and trial-and-error approaches. This research investigated the use of PCMs in various insulated boxes and refrigerated equipment, including cold storage facilities, domestic refrigerators, display cabinets, and refrigerated trucks. The study analyzed the melting point of the PCM, insulation material, mass of PCM, load, position, external air temperature, and energy consumption. The authors emphasize that important parameters like PCM configurations and wall insulation, previously cited by others, need to be considered together due to their complex interactions and the specific usage conditions. Notably, the relationship between airflow inside the equipment and heat exchange requires further investigation, particularly in relation to the position of the PCMs.

Sakamon Devahastin [2] This study investigated the feasibility of using paraffin wax as a PCM for latent heat storage (LHS) to store excess solar energy and release it during periods of low availability. The research focused on the thermal characteristics of the PCM during charging and discharging cycles. During charging, the effects of various parameters were examined, including inlet air velocities of 1 and 2 m/s and inlet hot air temperatures ranging from 70°C to 90°C. During discharge, the effect of inlet ambient air velocity was considered. Additionally, the study explored how LHS affects the drying kinetics of sweet potatoes and its ability to conserve energy during the drying process. The results showed that a lower inlet ambient air velocity increased the drying rate of sweet potatoes. At an air velocity of 1 m/s, approximately 1920 kJ/min kg-1 of energy was extracted from the LHS, saving 40% of energy. At 2 m/s, 1386 kJ/min kg-1 of energy was extracted, with a 34% energy saving.

Tyagi et al. [13] This review paper examines ways for increasing the thermal conductivity of PCMs in order to improve their heat transfer characteristics. Tyagi et al. investigate several ways, such as the use of nanoparticles, hybrid composites, and PCM encapsulation techniques, to improve heat conductivity while maintaining PCM stability and phase transition features. The review highlights substantial advances in this area and proposes future research objectives to better optimize PCM-based heat storage devices. This comprehensive review provides insights into the recent developments in thermal energy storage using PCMs. The paper discusses various types of PCMs, including organic, inorganic, and eutectic materials, along with their applications in solar energy storage, building heating/cooling systems, and electronics cooling. The author emphasize the importance of PCM selection based on specific application requirements and highlight the need for further research to enhance PCM stability and thermal conductivity.

The examined research papers emphasize the importance of Phase Change Materials (PCMs) in developing heat storage solutions for a variety of applications. While great progress has been achieved in enhancing the thermal characteristics and performance of PCMs, more research is required to address issues such as thermal conductivity enhancement, PCM stability, and cost-efficiency. Continued work in this regard are critical for realizing PCMs' full potential in fulfilling the increasing need for efficient and sustainable heat storage systems.

3. Problem Identification

When people cook food, usually they don't eat it immediately after cooking. But, naturally the cooked food cools down after some time, experts feel that one should avoid reheating food because the chemical change in food due to reheating often leads to food poisoning and food-borne diseases. Also, it increases the energy usage and hence decreasing the efficiency. So, keeping the food hot and fresh for few hours after cooking is extremely important, at least for the food which cannot be made instantly when ever required (such as rice, puddings, etc.). Also reheating of some food such as rice is difficult, it is required to keep it warm after it gets ready.

4. Resources and Materials

A. Catia:

Modelling of the heat storage unit was done using the Catia modelling software. This design was further used as a reference during manufacturing of the product.

B. Cooker:

The most commonly available and used kitchen utensil was selected for the experimental purpose with consideration that user need not to buy a new product for cooking rather the designed product will fit to the already existing product of the kitchen. The cooker of 1.5 litre capacity in which food items such as rice or puddings can be prepared was taken.

C. Heat Storage Unit:

This unit was used for storing the heat during cooking time and utilised it for keeping the food hot after few hours of cooking. It is designed to be coupled to the cooker directly for being compatible for easy application. [19]

D. Phase Change Material (PCM):

Paraffin Wax is the selected Phase change material (PCM) which is filled inside the cavity of the Heat storage Unit [6][7]. Phase change material

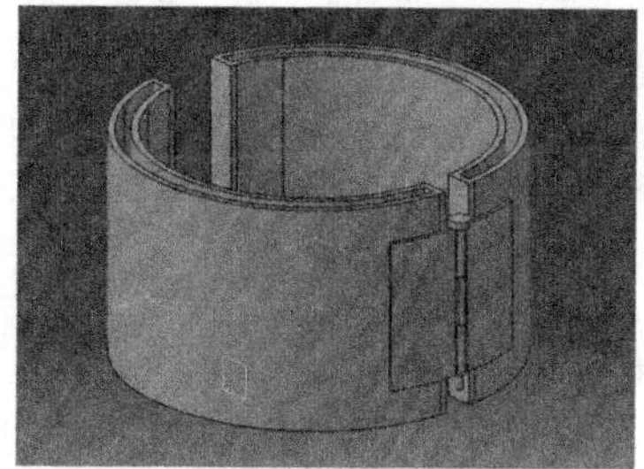

Fig. 12.1 3D catia model of the heat storage unit

Source: Author

Fig. 12.2 Cooker (cooking unit)

Source: Author

(PCM) is a substance which releases/absorbs sufficient amount of thermal energy at the period of phase transition to provide useful heating or cooling. Generally, the phase transition (phase change) takes place between first two fundamental states of matter – solid to liquid or liquid to solid as per the surrounding conditions [20].

4.1 Paraffin Wax as a PCM

There are many classes of phase change materials which are available in the market and used for different purpose [8]. Paraffin Wax is commonly used phase change material.

Reasons behind selecting the Paraffin Wax as PCM for the experiment:

[a] It is chemically stable, although they require closed containers as they undergo the oxidation process when get exposed to oxygen.

[b] They don't show the regular degradation in their thermal properties even after repeated freezing and melting cycles. Paraffin wax have good heat of fusion.

[c] In short, Paraffin waxes are reliable, safe, non irritating in nature, inexpensive and relatively used in a wide range of temperatures.

4.2 Calculations

Dimensions of Cooker

The most commonly available cooker was selected for the experiment, which has following dimensions,

Height, H = 83 mm = 0.083 m

Diameter, D = 140 mm = 0.14 m

Calculation for Heat Transfer through the Cooker

Material of Cooker: Stainless Steel

Overall heat transfer coefficient for S.S. is 20 W/m^{2}K[5]

(Targeted area to be protected or covered using the Heat storage unit i.e., neglecting the area covered by cap of cooker and bottom surface of the cooker only the lateral area of cooker is of interest)

$$\text{Area of Heat Transfer, } A = \pi DH$$

$$= \pi * 0.14 * 0.083$$

$$A = 0.0365 \text{ m}^2$$

[The Pressure cookers were designed at 1.2 bar and the corresponding Temperature for 1.2 bar pressure was of 120°C (from steam table),

Therefore, Temperature at the end of cooking = 120°C

As this study was carried out in the Pune, Maharashtra where the ambient temperature is around 26°C over a larger span of the year,

Therefore, Room or ambient Temperature = 26°C]

$$\text{Heat Transfer, } Q_{Cooker} = UA\,\Delta T$$
$$= 20 * 0.0365 * [120 - 26]$$
$$\therefore \mathbf{Q = 68.62\ KJ}\ \text{(From Cooker)}$$

Calculation for Heat Transfer from the Rice after Cooking was Done

Specific Heat of Rice grains, C_P = 1.22 KJ/ Kg° K [3]

True Density of Rice, ρ = 1412.56 Kg/m^3 [4]

(During the trial with water, 1 litre of water was selected which occupied height of 65mm of the cooker. To maintain the same condition height of 65mm was selected during the trial of rice [16]

i.e., $$V = \pi/4 * 0.14^2 * 0.065 = 0.001\ m^3)$$

Mass of the Rice after cooking,

$$M = \rho * V$$
$$= 1412.56 * 0.001$$
$$M = 1.412\ Kg$$

Heat Transfer through Rice,

$$Q_{Rice} = M\,C_P\,\Delta T$$
$$= 1.412 * 1.22 * [120 - 26]$$
$$\mathbf{Q_{Rice} = 161.92\ KJ}\ \text{(From Rice)}$$

Total Heat Transfer (Q_{Total}),

$$Q_{Total} = Q_{Cooker} + Q_{Rice}$$
$$= 68.62 + 162.92$$
$$\mathbf{Q_{Total} = 231.54\ KJ}$$

Calculation for Required Mass of PCM

Latent Heat of Paraffin wax = 235 KJ/ Kg [6][18]

$$\text{Mass of PCM, } M = \frac{\text{Total Heat Transfer}}{\text{Latent Heat of Paraffin Wax}}$$
$$= \frac{231.54}{235}$$
$$\mathbf{M = 0.985\ Kg}\ \text{(Mass of Paraffin Wax)}$$

Volume for PCM

Density of Paraffin Wax, $\rho = 900\ kg/m^3$ [6]

$$\text{Volume} = \frac{Mass}{Density} = \frac{0.985}{900}$$

$$\text{Volume} = 0.00109\ m^3$$

Dimensions of Heat Storage Unit (Finding Required Outer Diameter)

$$\text{Inner Diameter } (D_i) = 140\ mm = 0.14\ m$$

$$\text{Height } (h) = 83\ mm = 0.083m$$

$$\text{Volume} = \frac{\pi}{4} * (D_0^2 - D_i^2) * h$$

$$0.00109 = \frac{\pi}{4} * (D_0^2 - 0.14^2) * 0.083$$

$$\text{Outer Diameter, } D_o = 0.190\ m$$

D_o = 190 mm (Outer Diameter of the Unit)

So, for putting the required amount of PCM inside the Heat storage unit, on the safer side instead of 190 mm the outer diameter of **200 mm** was selected.

5. Design and Schematic Diagram

The general structure of the Heat Storage system can be divided into three main units: Cooking unit, Heat Storage unit and PCM. These units together formed a working system which was capable of storing huge amount of latent heat.

5.1 The Cooking Unit

The cooker was having dimension of 140 mm diameter, 83 mm height and having 1.5 litre capacity which can be used for making food items such as rice, puddings and other unseasoned.

5.2 The Heat Storage Unit

This unit was the main unit of overall system, basically this unit can be coupled with the cooker and even can be taken out (separated) from the cooking unit as per the requirement. This unit consisted of Paraffin Wax between its cavity. This unit was air tight to not let the PCM went out in the atmosphere after phase change from solid to liquid.

5.3 Phase Change Material (PCM)

At the phase change temperature (PCT), by melting and solidifying process, a PCM can store and releasing larger amount of energy [9]. Heat absorption and release takes place when materials internal structure changes, PCM are referred as latent heat storage (LHS) materials. During the application PCMs act as passive element and henceforth any additional energy source is not required [14][15].

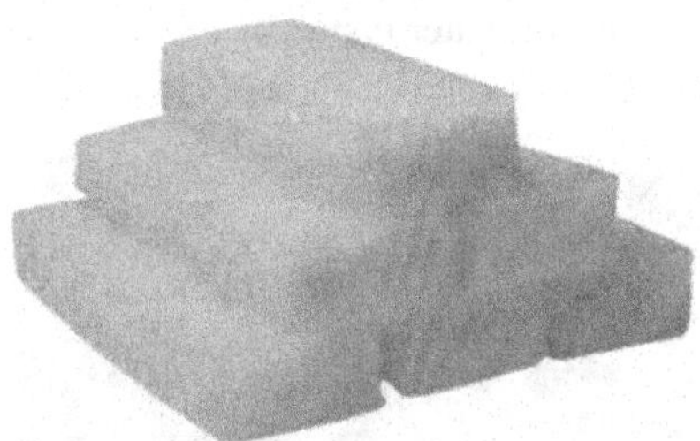

Fig. 12.3 Bricks of paraffin wax [21]

Fig. 12.4 Manufactured heat storage unit

Source: Author

5.4 Paraffin Wax

The properties of paraffin wax [6][7][8]:

Table 12.1 Properties of paraffin wax [6][7][8]

Property	Description
Chemical Formula	C_nH_{2n}
Appearance	White Solid
Nature	Odourless & chemically inert with maximum materials
Melting Temperature	68°C (154°F)
Boiling Point	370°C (698°F)
Density	900 Kg/m^3
Latent Heat	235 KJ/Kg
Specific Heat Capacity	2140 J/Kg-k
Heat of Combustion	42000 J/Kg
Other Properties/ Characteristics	Primary Hydrocarbon in Nature, water repellent, less prone to Chemical reactions, No phase Separation and less corrosive.

6. Experimentation

Experiment No. 1:

1) Without Heat Storage Unit

Experiment was carried out with water inside the cooker and Induction was used for hitting or energy supply purpose (400 W of constant supply was kept during the trial with both the conditions). By taking 1 litre of water inside the cooker the cooking was carried out:

Fig. 12.5 Experiment on cooker without heat storage unit

Source: Author

Fig. 12.6 Experiment after coupling heat storage unit with cooker

Source: Author

Experiment No. 2:

2) With Heat Storage Unit

Experiment was carried out with actual food inside the cooker and same as the case of trial with water, Induction was used for supplying the energy (400 W energy supply was kept during both of trials). After cooking the rice inside the cooker experiments were carried out:

1. Without heat storage unit
2. With heat storage unit

7. Result and Discussion

7.1 Experiment with Water

Table 12.2 Results table of experiment with water

Condition	Time required for 1st Blow	Cooling Down Time
Without Heat Storage Unit	9 min 24 sec	1 hour 42 min
With Heat Storage Unit	10 min 31 sec	2 hour 23 min

Source: Author

1. For the first experiment on cooker for water without heat storage unit at 400W:
   a) The time required till 1st blow cooker valve was around 9 min 24 sec (564 sec)
   b) The time required for it to cool down to 40°C was around 1 hours 42 minutes.
2. For the 2nd experiment after coupling of heat storage unit with cooker for water at 400W:
   a) The time required till 1st blow cooker valve was around 10 min 31 sec (631 sec)
   b) The time required for it to cool down to 40°C was around 2 hours 23 min.

From this result it was observed, by using the heat storage unit the total time required for cooling of the boiled water can be increased around 41 min.

As Wattage for Induction during heating was 400W (same in both cases) it can be concluded that Electricity input was increased by 11.8%, as a result the time it will stay warm is increased by 40.2%.

7.2 Experiment with Rice

Table 12.3 Result table of experiment with cooked rice

Condition	Time required for 1st Blow	Cooling Down Time
Without Heat Storage Unit	14 min 06 sec	33 min
With Heat Storage Unit	15 min 47 sec	61 min

Source: Author

1. For the first experiment on cooker for rice without heat storage unit at 400W:
   a) The time required till 1st blow cooker valve was around 14 min 06 sec (846 sec)
   b) The time required for it to cool down was around 33 minutes.

2. For the 2nd experiment after coupling of heat storage unit with cooker for rice at 400W:
   a) The time required till 1st blow cooker valve was around 15 min 47 sec (947 sec)
   b) The time required for it to cool down was around 61 minutes.

From these results it was observed that, by using the heat storage unit, the total time required for cooling of the cooked rice can be increased around 28 min.

As Wattage for Induction during heating was 400W (same in both cases) it can be concluded that Electricity input was increased by 12%, as a result the time it will stay warm is increased by 84.84%.

8. Conclusion

This paper experimentally demonstrates how the paraffin wax can be used for seeking the aim of heat storage and how it can be utilised it for keeping the food warmer for longer period.

Study with the same input conditions were carried out and the conclusions which can be drawn from these results of study:

1. With little extra time and amount of heat during the cooking time the cooling time for the water can be increased by 41 minutes.
2. 11.8% Electricity input supply is increased but as a result cooling time is increased by 40.2% during the experiment with water.
3. Same for the cooked rice, the cooling time can be increased by 28 minutes and various foods items can be kept fresh for longer time
4. 12% Electricity input supply is increased, as a result cooling time is increased by 84.84% during the experiment with cooked rice. This much effect because the model is designed by considering thermodynamics properties of rice
5. Retaining time is a function of food item to be cooked i.e., as the food to be cooked changes the retaining time will vary. The design is done by considering the rice because rice is the most cooked food item in cooker.
6. Paraffin based phase change materials can be used for the different applications such as pharmaceutical and medical industries, thermal batteries, solar power systems, automobile, transportation, and so on.

Looking into the current market there is no such product available to keep the bulk amount of food warmer for longer time and the proposed technology can be utilized to resolve this issue.

Many catering organizers (in companies and at functions) use the system in which they keep the food vessel in a water and keep supplying the energy to keep that water warm, even though they unable to maintain the food that warmer. By using the proposed technique by supplying little higher energy during cooking, the food can be kept warmer for longer time (need not to supply the energy afterwards as it required in current case mentioned).

9. Future Scope

After carrying out various experiments in future with different PCM materials the most effective PCM can be found for the purpose of retaining the heat of cooked food in kitchen application (considering the economical point of view) With some design changes the more flexible model can be achieved. This unit can be implemented in areas such as catering in big events, in case of food deliveries and in the milk industries [11][17].

Reference

1. Tanathep Leungtongkum, Denis Flick - Insulated box and refrigerated equipment with PCM for food preservation: State of the art, Journal of Food Engineering 317, 110874, 2022
2. Sakamon Devahastin, Saovakhon Pitaksuriyarat - Use of latent heat storage to conserve energy during drying and its effect on drying kinetics of a food product, Applied thermal engineering 26 (14-15), 1705–1713, 2006
3. A Iguaz, MB San Martín, C Arroqui, T Fernández, JI Maté, P Vírseda Thermophysical properties of medium grain rough rice (LIDO cultivar) at medium and low temperatures European Food Research and Technology 217, 224–229, 2003
4. K. R. Bhattacharya, C. M. Sowbhagya, Y. M. Indudhara Swamy Some physical properties of paddy and rice and their interrelations Journal of the Science of Food and Agriculture 23 (2), 171–186, 1972
5. FR Campbell, LR Bourque, R Deshaies, H Sills, MJF Notley In-reactor measurement of fuel-to-sheath heat transfer coefficients between UO2 and stainless steel. Atomic Energy of Canada Ltd., 1977
6. Stella P. Jesumathy, M. Udayakumar & S. Suresh Heat transfer characteristics in latent heat storage system using paraffin wax Journal of mechanical science and technology 26, 959–965, 2012
7. William R Turner, Donald S Brown, Donald V Harrison Properties of paraffin waxes Industrial & Engineering Chemistry 47 (6), 1219–1226, 1955
8. T Kousksou, A Jamil, T El Rhafiki, Y Zeraouli Paraffin wax mixtures as phase change materials Solar Energy Materials and Solar Cells 94 (12), 2158–2165, 2010
9. Pramod B Salunkhe, Prashant S Shembekar A review on effect of phase change material encapsulation on the thermal performance of a system Renewable and sustainable energy reviews 16 (8), 5603–5616, 2012

10. Alure Gowda, NG Kiran Consumer buying behavior towards kitchen Storage Products and Services, A Case Study during Maha savings day at the Big Bazaar, Mandya, Karnataka Asian Journal of Management 10 (2), 135–140, 2019
11. Gregory Atwood Pcm applications and an outlook to the future Phase Change Memory: Device Physics, Reliability and Applications, 313–324, 2018
12. Judith C Gomez High-temperature phase change materials (PCM) candidates for thermal energy storage (TES) applications National Renewable Energy Lab. (NREL), Golden, CO (United States), 2011
13. A. Sharma, V. V. Tyagi, C. R. Chen, D. Buddhi, Review on thermal energy storage with phase change materials and applications, Renewable and Sustainable energy reviews, 13(2) (2009) 318–34.
14. Ind. Eng. Chem. 1929, 21, 11, 1090–1092.
15. J. Sol. Energy Eng. Aug 1984, 106(3): 299–306.
16. Bo He, Viktoria Martin, Fredrik Setterwall Energy 29 (11), 1785–1804, 2004.
17. Patrik Sobolčiak, Haneen Abdelrazeq Applied Thermal Engineering 107, 1313–1323, 2016.
18. MK Rathod, Jyotirmay Banerjee Experimental heat transfer 27 (1), 40–55, 2014.
19. Abhay Dinker, Madhu Agarwal, GD Agarwal Journal of the Energy Institute 90 (1), 1–11, 2017.
20. S Jegadheeswaran, Sanjay D Pohekar Renewable and Sustainable energy reviews 13 (9), 2225–2244, 2009.
21. https://www.etsy.com/hk-en/listing/767326228/paraffin-wax-unscented-pure-diy, dated 14 July 2024

Technologies for Energy, Agriculture, and Healthcare – Shailesh Nikam et al. (eds)
© 2024 Taylor & Francis Group, London, ISBN 978-1-032-98028-7

13

AI-DRIVEN SEIZURE DETECTION USING EEG: A SYNERGISTIC APPROACH WITH DWT, ENTROPY FEATURES AND ANN

Indu Dokare[1]

K. J. Somaiya College of Engineering,
Somaiya Vidyavihar University,
Mumbai, India

VES Institute of Technology,
Mumbai

Sudha Gupta[2]

K. J. Somaiya College of Engineering,
Somaiya Vidyavihar University,
Mumbai, India

Abstract: Integrating Artificial Intelligence (AI) and signal processing techniques has become a promising option for building more efficient and reliable seizure detection systems. Electroencephalography (EEG) is a widely used tool for diagnosing and monitoring neurological conditions, and it is particularly effective in detecting episodes of epileptic seizures. This paper proposes a comprehensive approach to improve the reliability of EEG signals for enhanced seizure detection. In this approach, Discrete wavelet transform (DWT) is used for denoising the signal by thresholding the coefficients. Furthermore, entropy-based features are extracted and fed to the artificial neural network (ANN). The effectiveness of this proposed method is evaluated by experimenting with the CHB-MIT database. Mean square error (MSE) and correlation coefficient have been used to assess the effect of denoising. The accuracy, precision, sensitivity, specificity, F1-score and Area Under the Receiver Operating Characteristic curve (AUC-ROC) score obtained by the classifier for classifying EEG signals into seizure and non-seizure segments are 96.62%, 96.7%, 86.26%, 99.25%, 91.18% and 0.927 respectively.

[1]indu.dokare@somaiya.edu, [2]sudhagupta@somaiya.edu

DOI: 10.1201/9781003596707-13

This work highlights DWT's potential as a useful method for denoising EEG signals for seizure detection along with the discriminative power of entropy-based features. This approach will aid in creating robust reliable systems for the detection of seizures in clinical settings.

Keywords: Epilepsy, Multichannel EEG, DWT, Entropy, ANN

1. Introduction

Seizure detection is a vital aspect of healthcare, particularly in monitoring and treating a neurological disorder like epilepsy (WHO, 2023), (Beghi, 2020). The timely and correct diagnosis of seizure occurrences is critical in improving patient outcomes, guiding treatment methods, and increasing the overall quality of life. The main challenge in epilepsy diagnosis using EEG signals is to detect the significant pattern of seizure activities that are corrupted by the noise components. The source of noise may be due to electrode placement, environmental noise, or artifacts like eyes or muscles. The recurrent occurrence of seizures, due to sudden and abnormal electrical discharges in the brain is a main characteristic of Epilepsy (WHO 2023), (Fisher et al., 2005). There is an enormous amount of variability in seizure patterns among patients (Sanei & Chambers, 2013) and within a patient over time. Due to these seizure pattern variabilities and the interference of different types of noise, it's a great challenge to remove noise without losing the seizure pattern. By offering a frequency-domain view of EEG data, the use of DWT for denoising presents a viable solution to these problems. Numerous studies have reported various techniques for denoising the EEG signal for seizures. However, very few works have reported the use of DWT for denoising EEG signals in seizure detection.

The noise-removing algorithm proposed (Chu et al., 2022) has used an adaptive threshold selection method to change the threshold based on the decomposition layers. The author has experimented with EEG from the Fz channel of Goetz data set 2a and reported the suppression of Gaussian noise. In another approach (Hussain & Qaisar, 2022) adaptive-rate finite impulse response (FIR) filtering for denoising, adaptive-rate DWT-based subbands decomposition, statistical features extraction, mutual information (MI) based feature selection and one-vs-all ensemble classification was implemented for seizure detection. Butterworth Bandpass filter along with the moving average filter was employed for noise removal (Dash et al., 2020). The feature set was obtained using 2D power spectral density, dynamic mode decomposition power, variance and Kartz fractal dimension. Hidden Markov model (HMM) was used as a classifier. The study proposed (Orosco et al., 2016) has employed the Butterworth filter and DWT for decomposition. This work has used the spectral and energy features of each channel as a feature vector. Linear discriminant analysis and neural networks were used for seizure detection. A 48th-order FIR high pass filter was used to remove the low-frequency artifacts and after decomposing the signal using DWT, statistical features of the wavelet coefficients were estimated. For the classification of signals in seizure or non-seizure segments, a support vector machine (SVM) was employed in this implementation (Chen et al., 2017).

The next part of this paper is organized as follows: Section 2 covers the proposed method including DWT-based signal denoising, feature extraction, classifier and performance measures employed. The simulation results with analysis are presented and discussed in section 3. Finally, the conclusions drawn are reported in section 4.

2. Proposed Method

To develop an efficient and accurate seizure detection system, the convergence of AI and signal processing techniques holds the potential to address several challenges faced by conventional methods. This paper proposes a comprehensive approach leveraging the DWT for denoising EEG signals for seizures, in conjunction with entropy-based features and ANN for classifying the EEG signal into seizure and non-seizure segments. The outline of the proposed approach is depicted in Fig. 13.1.

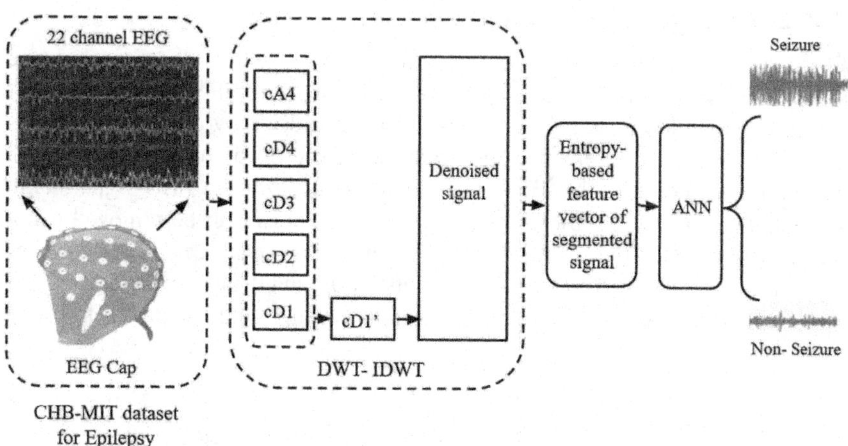

Fig. 13.1 Proposed approach for seizure detection

The main contribution of this work is as follows:

1. DWT is used for denoising the EEG signal which provides a multiresolution analysis of a signal for allowing denoising the signal locally and globally.
2. Entropy-based feature extraction is employed due to its ability to capture irregularity and randomness of seizure patterns in EEG signals.
3. ANN is used to classify the EEG signal into seizure and no-seizure segments.

Considering the seizure pattern variability within a patient over time and even the seizure pattern variability among different patients, this proposed approach provides flexibility to achieve the best performance.

2.1 Dataset Description

One of the widely used multichannel scalp EEG CHB-MIT dataset (Guttag and John, 2010), (Shoeb and Guttag, 2010) available at https://physionet.org is employed in this work for experimentation purposes. This dataset was collected at Children's Hospital Boston (CHB) in association with the Massachusetts Institute of Technology (MIT). The recording was done using the International 10-20 system of EEG electrode placement with a sampling frequency of 256 Hz. The dataset includes recordings of 23 seizure patients, with a text file containing the start and end time of each seizure. The dataset comprises multiple EEG recordings in separate files for each patient. In this research, the experiment is carried out using the dataset of patient 1, who is eleven years old. The dataset for patient 1 contains 7 seizures that occurred at different time intervals spanning 442 seconds in total.

2.2 Denoising using Discrete Wavelet Transform

DWT is a popular tool used for denoising in a wide range of applications because of its ability to represent signals at varied resolutions. The ability of DWT to adjust to signal variability is necessary for managing the diversity of seizure patterns since it allows for the selection of various wavelets, decomposition levels, and thresholds according to the distinctive characteristics of each EEG signal being studied. The wavelet coefficients contaminated by noise can be removed using a thresholding operation (Donoho, 1995). The EEG signal x[n] of length N is decomposed into approximation coefficients $cA_j[n]$ and detail coefficients $cD_j[n]$ at a level j by passing the signal through a high pass g[k] and low pass h[k] filters successively.

$$cA_j[n] = \sum_k h[k] \cdot x[2n-k] \tag{1}$$

$$cD_j[n] = \sum_k g[k] \cdot x[2n-k] \tag{2}$$

The steps for denoising followed in this work are as below:

1. Apply j^{th} level DWT to all 22 channels EEG signal.
2. Estimate noise level by calculating the mean absolute deviation (MAD) (Donoho, 1995) of each detail coefficient and determine a threshold.

$$MAD_j = median\left(\left|cD_j - median\left(cD_j\right)\right|\right) \tag{3}$$

3. Obtain the denoised detail coefficients by applying soft thresholding.
4. Reconstruct the signal by taking inverse discrete wavelet transform (IDWT). Change the threshold factor as per the need.
5. Repeat steps 4, 5, and 6 for different detail coefficients individually or in combination.

2.3 Segmentation and Feature Extraction

The EEG signal from 22 channels is segmented into a fixed segment length of 4 seconds. In this proposed work the sample entropy, permutation entropy (Bandt & Pompe, 2002), and Shannon entropy are used to extract the features of the signal. Entropy measures improve the discriminative power of a classifier in seizure detection tasks. Sample entropy (E_s), permutation entropy (E_p) and Shannon entropy (E_{sh}) are described in equations 4, 5 and 6 respectively.

A template vector of length m, such that $X_m(i) = \{x_i, x_{i+1}, x_{i+2}, \ldots, x_{i+m-1}\}$ and the distance function d[$X_m(i)$, $X_m(j)$] ($i \neq j$) is to be the Chebyshev distance

$$E_s = -\ln\frac{A}{B} \tag{4}$$

A = number of template vector pairs having d [$X_{m+1}(i)$, $X_{m+1}(j)$] $< r$

B = number of template vector pairs having d [$X_m(i)$, $X_m(j)$] $< r$

For each time series, let p be a probability distribution associated with it, where π_i are the frequencies associated with the i possible permutation patterns, therefore i = 1, 2,…, D!, where D is the embedding dimension. The permutation entropy can be determined as

$$E_p = -\sum_{i=1}^{D!} \pi_i \log_2 \pi_i \tag{5}$$

$$E_{sh}(X) = -\sum_{i=1}^{N} p(x_i)\log_2 p(x_i) \tag{6}$$

Where $p(x_i)$ is the probability occurrence of feature values from x_1 to x_N

2.4 Classification

The set of three features is fed to feed forward ANN to classify the signal into seizure and non-seizure segments. The input layer and hidden layer comprise 10 neurons with a ReLu activation function while the output layer contains a neuron with a sigmoid activation function. The model is trained using the Adam optimizer with binary crossentropy loss. The model is validated using 5-fold cross validation method. The effectiveness of the denoising stage for noise removal is assessed by MSE and correlation coefficient. The performance of the classifier is evaluated using accuracy, sensitivity, specificity, precision, F1-score and AUC-ROC score. This will provide the overall system performance for the seizure detection task.

3. Results and Discussion

The proposed method of denoising is experimented on the CHB-MIT database for seizure patients. In this work, FP1-F7, F7-T7, T7-P7, P7-O1, FP1-F3, F3-C3, C3-P3, P3-O1, FP2-F4, F4-C4, C4-P4, P4-O2, FP2-F8, F8-T8, T8-P8, P8-O2, FZ-CZ, CZ-PZ, P7-T7, T7-FT9, FT9-FT10, FT10-T8 these 22 channels of seizure patients are considered for experimentation. The EEG data of patient 1 comprises 7 seizures that occurred at separate times and lasted 7 minutes and 22 seconds in total. The non-seizure signal segment used in this experiment is 28 minutes and 20 seconds. All experiments are performed using Python on 11th Gen Intel(R) Core (TM) i5-11300H with 3.10GHz and 16.0 GB of RAM.

The results are obtained using wavelet Daubechies 4 (db4) decomposed up to level 4 obtaining approximate coefficients cA4 and detail coefficients cD4, cD3, cD2 and cD1. The effect of applying thresholding to each detail coefficient individually and the combined effect of two thresholded detail coefficients for a seizure segment is presented in Fig. 13.2. The correlation coefficient and MSE values obtained show the change in the original and denoised signals as reported in Table 13.1. The high value of the correlation coefficient indicates less amount of change occurred in the denoised signal, while the high value of MSE shows more change occurred in the denoised signal. Table 13.2 shows the performance measures of the classification task. The resulting accuracy, precision, sensitivity, specificity, F1-score and AUC-ROC score range from 96.59% to 96.76%, 96.7% to 98.69%, 84.29 to 86.26, 99.25% to 99.71%, 90.92% to 91.48% and 0.92 to 0.928 for different denoised coefficients, respectively. As demonstrated in Table 13.2, the results achieved by thresholding the coefficients cD1 and cD2 with a threshold factor 2 are superior compared to others.

Though accuracy is a commonly used metric for evaluating classification models, it may not provide a thorough assessment of model performance, especially when

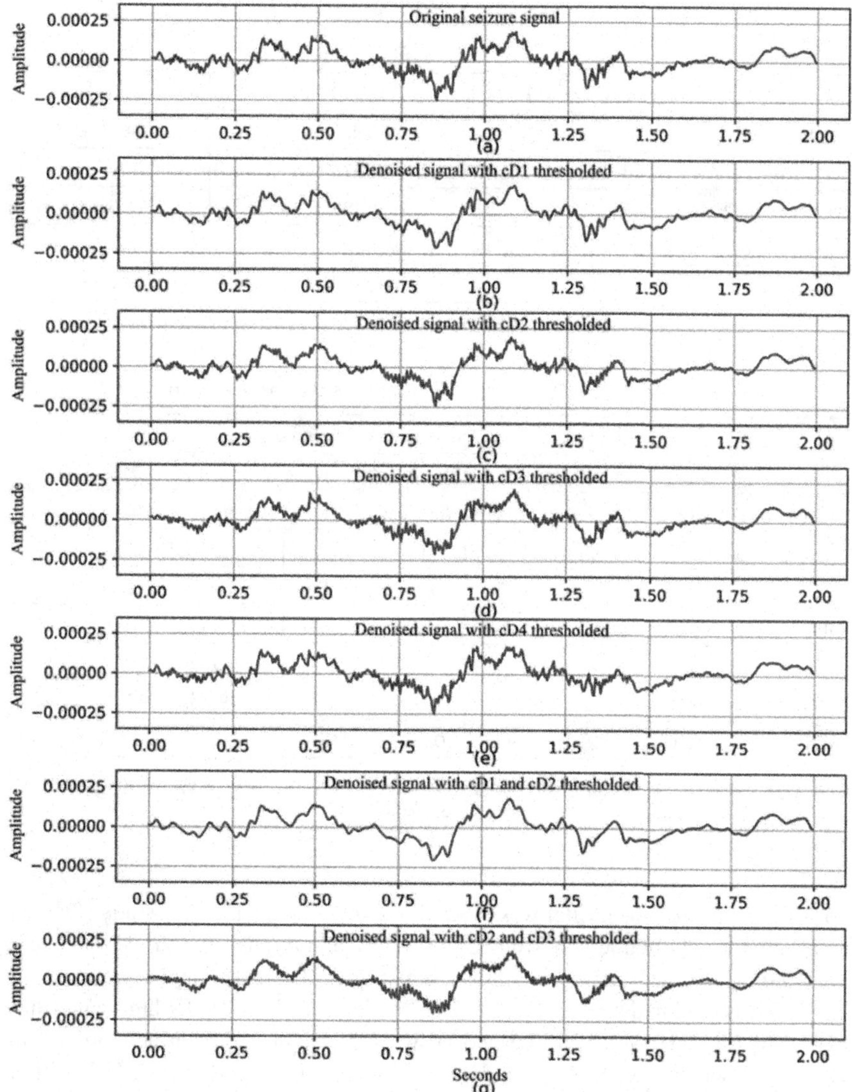

Fig. 13.2 Effect of thresholding the detail coefficients in denoised signal for a seizure segment

the datasets are imbalanced. The work (Hussain & Qaisar, 2022) employed an adaptive-rate FIR filter for denoising and reported an accuracy of 99.38% using

Table 13.1 Performance measures of denoising stage

Reconstructed signal	Correlation coefficient	MSE
cA4, cD4, cD3, cD2, cD1'	0.98654	0.06448
cA4, cD4, cD3, cD2', cD1	0.98198	0.05149
cA4, cD4, cD3', cD2, cD1	0.97735	0.05603
cA4, cD4', cD3, cD2, cD1	0.97885	0.06049
cA4, cD4, cD3, cD2', cD1'	0.96829	0.11597
cA4, cD4, cD3', cD2', cD1	0.95894	0.10755

(cD1', cD2', cD3', and cD4':denoised coefficients of cD1, cD2, cD3, and cD4 respectively)

Table 13.2 Performance measures of classification

	Denoised coefficients							
	cD1'		cD2'		cD3'		cD1' & cD2'	
	T=2	T=4	T=2	T=4	T=2	T=4	T=2	T=4
ACC	96.59	96.62	96.73	96.70	96.67	96.65	96.62	96.76
PRC	98.69	97.14	97.76	97.16	97.00	97.45	96.70	97.92
SEN	84.29	85.83	85.83	86.26	86.26	85.69	86.26	85.83
SPC	99.71	99.36	99.50	99.36	99.32	99.43	99.25	99.54
F1-SCR	90.92	91.14	91.41	91.38	91.31	91.19	91.18	91.48
ROC-AUC	0.920	0.926	0.926	0.928	0.928	0.926	0.927	0.927

(cD1', cD2', cD3', and cD4':denoised coefficients of cD1, cD2, cD3, and cD4 respectively, ACC: Accuracy, PRC: Precision, SEN: Sensitivity, SPC: Specificity,
T: Threshold factor, F1-SCR: F1-Score)

16 features. Butterworth filter was used for the noise removal in the study (Orosco et al., 2016) and obtained an accuracy of 99.9% and a sensitivity of 71.4% using 12 features. An accuracy of 92.30%, sensitivity of 91.71% and specificity of 92.89% was obtained by (Chen et al., 2017) using a 48th-order FIR high pass filter with 9 features. However, this proposed work has achieved accuracy, precision, sensitivity, specificity, F1-score and AUC-ROC score of 96.62%, 96.7%, 86.26%, 99.25%, 91.18% and 0.927 respectively for coefficients cD1 and cD2 thresholded with threshold factor 2. This proposed work employs only 3 features for seizure detection, while the earlier research presented in this paper used 9 to 16 features. Even the use of DWT for denoising allows the selection of various wavelets, decomposition levels, and thresholds according to the distinctive characteristics of EEG signals for diverse seizure patterns.

By leveraging a smaller feature set, the proposed approach achieves a balance between performance accuracy and computational efficiency, making it a viable and a practical solution for seizure detection applications. The results obtained show the effectiveness of this denoising method in the seizure detection task. This proposed method of seizure detection along with DWT for denoising provides flexibility to adjust the noise level. Even the estimated noise can be used to denoise all coefficients. Different wavelet families with different decomposition levels provide another choice for adjusting the threshold. Hence, this proposed method is found to be promising in seizure detection tasks due to inter-patient and intra-patient variability in the seizure patterns. Experimenting with this methodology on various seizure cases across datasets will demonstrate the robustness of this method. More study on creating a comprehensive set of handcrafted features is required that generalize well across diverse datasets.

4. Conclusion

The incorporation of DWT has proven effective in isolating important signal components while minimizing noise, thereby enhancing the quality of EEG signals for subsequent analysis. Entropy-based features provide a powerful tool for quantifying the complexity and irregularity inherent in EEG signals, offering relevant information for discriminating between normal brain activity and seizure occurrences. Moving beyond preprocessing, the incorporation of an ANN for classification has shown promising findings in automating seizure pattern recognition. The ANN, which was trained on the extracted features, has the potential to generalise across varied datasets, making it a powerful tool for real-world seizure detection. The accuracy, precision, sensitivity, specificity, F1-score and AUC-ROC score of 96.62%, 96.70%, 86.26%, 99.25%, 91.18% and 0.927 respectively. This comprehensive technique, which includes denoising, feature extraction, and classification, aids in the construction of precise and reliable seizure detection systems in healthcare. Looking ahead, further development and validation of this integrated method offer great promise for improving the clinical value of EEG-based seizure detection, ultimately benefiting people with Epilepsy and their healthcare providers. The model can be refined by doing additional experiments with various seizure datasets.

References

1. Bandt, C., & Pompe, B. (2002). "Permutation entropy: a natural complexity measure for time series". Physical review letters, 88(17), 174102. https://doi.org/ 10.1103/ PhysRevLett.88.174102
2. Beghi, E. (2020). "The epidemiology of epilepsy". Neuroepidemiology, 54(2), 185–191. DOI: 10.1159/000503831

3. Chen, D., Wan, S., Xiang, J., & Bao, F. S. (2017). "A high-performance seizure detection algorithm based on Discrete Wavelet Transform (DWT) and EEG". *PLOS one*, *12*(3), e0173138. https://doi.org/10.1371/journal.pone.0173138
4. Chu, R., Wang, J., Zhang, Q., & Chen, H. (2022, December). "An adaptive noise removal method for EEG signals". In Journal of Physics: Conference Series (Vol. 2414, No. 1, p. 012007). IOP Publishing. **DOI:** 10.1088/1742-6596/2414/1/012007
5. Dash, D. P., Kolekar, M. H., & Jha, K. (2020). "Multi-channel EEG based automatic epileptic seizure detection using iterative filtering decomposition and Hidden Markov Model". Computers in biology and medicine, 116, 103571. https://doi.org/ 10.1016/j.compbiomed.2019.103571
6. Donoho, D. L., & Johnstone, I. M. (1995). "Adapting to unknown smoothness via wavelet shrinkage". Journal of the American Statistical Association, 1200–1224. DOI: 10.1080/01621459.1995.10476626
7. Fisher, R. S., Boas, W. V. E., Blume, W. (2005). "Epileptic seizures and epilepsy: definitions proposed by the International League Against Epilepsy (ILAE) and the International Bureau for Epilepsy (IBE)". Epilepsia, 46(4), 470–472. https://doi.org/10.1111/j.0013-9580.2005.66104.x
8. Guttag, John. (2010). CHB-MIT Scalp EEG Database (version 1.0.0). PhysioNet. https://doi.org/10.13026/C2K01R.
9. Hussain, S. F., & Qaisar, S. M. (2022). "Epileptic seizure classification using level-crossing EEG sampling and ensemble of sub-problems classifier". Expert Systems with Applications, *191*, 116356. https://doi.org/10.1016/j.eswa.2021.116356
10. Orosco, L., Correa, A. G., Diez, P., & Laciar, E. (2016). "Patient non-specific algorithm for seizures detection in scalp EEG". Computers in biology and medicine, *71*, 128–134. https://doi.org/10.1016/j.compbiomed.2016.02.016
11. Sanei, S., & Chambers, J. A. (2013). *EEG signal processing*. John Wiley & Sons
12. Shoeb, A. H., & Guttag, J. V. (2010). Application of machine learning to epileptic seizure detection. In Proceedings of the 27th international conference on machine learning (ICML-10) pp. 975–982. 2010.
13. World health organization (WHO), February 2023. Epilepsy. https://www.who.int / news-room/fact-sheets/detail/epilepsy

Note: All the figures and tables in this chapter were made by the authors.

Technologies for Energy, Agriculture, and Healthcare – Shailesh Nikam et al. (eds)
© 2024 Taylor & Francis Group, London, ISBN 978-1-032-98028-7

14

Thermomechanical Analysis of Fused Deposition Modeling for ABS Turbomachinery Rotor in Energy Generation

Shrikrishna Pawar*, Dhananjay Dolas

Department of Mechanical Engineering, MGM University, Aurangabad, India

Abstract: Fused deposition modeling (FDM) is a material extrusion-based additive manufacturing (AM) technology used to build various prototypes and functional parts. FDM is a widely used AM process that works with thermoplastic material and is inexpensive relative to other AM technologies. Several research works have experimentally investigated the effect of FDM process parameters on the performance of the parts; however, due to the wider scope of using different materials and design flexibilities, it is imperative to use a numerical simulation approach to speed up the research work in order to reduce the delay and research costs. In this study, the numerical simulation tool Digimat-AM was used to predict the effect of layer thickness and infill density on process-induced defects such as residual stress. A rotor made up of ABS material was used as a specimen for numerical simulation; such rotors are widely used in energy generation sectors. Simulation results were analysed using ANOVA, which indicated that layer thickness had a significant impact on residual stress while the infill density did not influence the output parameter significantly.

Keywords: Additive manufacturing, Energy, Rotor, Turbo machinery, Thermomechanical

*Corresponding author: shrikrishnapawar24@gmail.com

DOI: 10.1201/9781003596707-14

1. Introduction

Additive manufacturing (AM) is a versatile manufacturing method entails the development of complex three-dimensional (3D) objects [1]. There are many applications of AM across different industrial sectors, including power generation, aerospace, and medical [2]. With the advent of material science and relevant technologies, various AM techniques are utilised for manufacturing functioning parts quickly [3]. Due to simplicity, flexibility, cost-effectiveness, and ease of control over the process parameters, fused deposition modeling (FDM) has gained widespread popularity among AM practitioners [4]. In the FDM process, the initial computer-aided design (CAD) model of 3D objects is sliced by slicing software to convert the 3D model details into printing instructions in the form of a tessellation file format (stl) [5]. Viscous polymer extrudate is deposited over the build platform through a small-diameter nozzle. This nozzle moves along a path that is predefined and preprogrammed by motion control codes, referred to as G-codes. One deposited layer of extrudate is followed by the subsequent deposition of the next layers, resulting in the printing of the parts layer-by-layer. [6]

The mechanical behaviour of FDM-built parts has faced challenges due to residual stress. In addition, the FDM process is also affected by the warpage. Another challenge associated with the FDM process is regarding the selection of material for a particular application [7]. Despite the effectiveness of the FDM process, there are many factors that impact the quality and functionality of FDM-built parts. The major control parameters include layer thickness, infill density, infill pattern, shell thickness, raster angle, build orientation, print speed, and temperature, which affect the microstructure and eventually the performance of the parts [8]. However, it is difficult to employ an experimental approach to predict the impact of all process parameters on the quality and performance of FDM-built parts. Also, it is nearly impossible to experimentally investigate the utilisation of materials with varying compositions and blends [9]. In this regard, focus was placed on the utilisation of a simulation approach for investigating the impact of control parameters on local microstructure and then extrapolating the derived result to conclude the macroscopic-level impact [10].

Hussein Azyod et al. [11] investigated the impact of printing parameters on the residual stress of ABS parts manufactured by the FDM process. Three process parameters were selected for developing the numerical solution, which include raster angle, print orientation, and infill pattern. Results indicated that print orientation had a significant impact on the residual stresses generated in the parts. In another study performed by Ans Al Rashid et al. [12], the thermomechanical performance of 3D-printed polyamide-6 composite specimens was simulated.

The influence of different infill densities and infill patterns on warpage, residual stresses, and mechanical response was studied through numerical simulation. These simulation results were compared with experimental results, and the difference between these two results was within an acceptable range of 0.22% to 7.27%. A multiscale (micro and macro) modeling approach was utilised by Sharafi et al. [13] to predict the mechanical behaviour of FDM-manufactured parts. ASTM tensile testing specimens built from two different materials, Acrylonitrile Butadiene Styrene (ABS) and Polyaryletherketone (PAEK), were selected for the simulation study. Digimat software and ANSYS software were used for micro- and macro-level studies, respectively.

In this article, a numerical model is developed using the simulation software Digimat to predict the response of the FDM-build rotor in terms of residual stress. Additionally, the warpage, printing time, and material consumption of each sample are also recorded. Three levels of each process parameter (layer thickness and infill density) are selected for the simulation study. Layer thickness and infill density are significant parameters that can potentially impact complex geometry parts such as rotors. Moreover, these parameters offer the control through adjustment during the manufacturing process, enabling the process optimisation. For focused investigation, only two parameters were selected. The results of the simulation study are statistically analysed using ANOVA to determine the significant parameters and their relevant contribution to impacting the response parameter.

2. Numerical Study for FDM

2.1 Specimens, Material and Design of the Experiment

This study is the primary part of the comprehensive research work planned to experimentally investigate the FDM process for printing the turbo machinery parts. Instead of following an experimental approach for feasibility evaluation, a numerical approach is used due to its benefits, such as time savings and waste minimization. In this study, a specimen rotor, as depicted in Fig. 14.1, was selected a priori in the investigation, as the main focus was not on the design of the parts but on evaluating the feasibility of the FDM process for manufacturing the high-precision parts.

Fig. 14.1 Specimen for numerical simulation

ABS is a thermoplastic material used as a base material in this study owing to its remarkable mechanical properties. The FDM process parameters layer thickness and infill density with three levels were used to simulate the process, as reported in Table 14.1. Simulated experiments are designed with a full factorial approach comprising three levels of layer thickness and infill density. Full factorial approach was selected due to its benefits, such as facilitating the investigation of all potential combinations of the factors and their levels.

Table 14.1 Process parameters and its levels

Parameters with Symbol	Levels		
	1	2	3
Layer thickness LT	0.127	0.254	0.33
Infill density ID (%)	20	60	100

2.2 FDM Process Simulation

Initially, a 3D CAD model of the rotor was converted to STL format and a separate g-code file was generated for each specimen as listed in Table 14.2. A g-code file, a toolpath instruction, was generated using slicer software. Digimat-AM software was used for simulating the FDM process, which involves four major stages: definition, manufacturing, simulation, and results. In the last stage, results are obtained for residual stress, as depicted in Table 14.2. A Rotor specimen as depicted in Fig. 14.1 was examined nine times with varying values of layer thickness and infill density. Resultant residual stress for each experimental run is exhibited in Table 14.2.

Table 14.2 Details of simulation results

Specimen	LT	ID	σ (Mpa)
S1	0.127	20	86.49
S2	0.127	60	87.54
S3	0.127	100	87.54
S4	0.254	20	45.45
S5	0.254	60	46.85
S6	0.254	100	46.85
S7	0.33	20	36.57
S8	0.33	60	36.97
S9	0.33	100	36.97

3. Results and Discussion

Due to rapid heating and cooling during FDM process, geometrical variations occurred that resulted into deflections and ultimately induces the residual stresses between the deposited layers. In this section, results of numerical simulation in terms of maximum residual stresses (σ) for different layer thickness and infill densities are discussed. Details of this simulation results are reported in Table 14.2.

3.1 Residual Stress

Digimat provides stress results in X, Y, and Z directions, as well as for transverse planes. In this study, Von Mises stress values are selected as they represent the overall residual stresses induced in the specimen. ANOVA for maximum residual stress was conducted, and its results are reported in Table 14.3.

Table 14.3 ANOVA for residual stress

Source	DF	SS	MS	P
LT	2	4291.78	2145.89	0.000
ID	2	1.81	0.90	0.026
Error	4	0.34	0.09	
Total	8	4293.93		
Model Summary				
S	R-sq %	R-sq (adj)	R-sq (pred)	
0.292973	99.99	99.98	99.96	

A significant effect of layer thickness is observed on maximum residual stresses. A maximum residual stress of 87.54 MPa was observed for a layer thickness of 0.127 mm. Layer thicknesses of 0.33 mm showed significantly lower maximum residual stress values as compared to other layer thicknesses. A lower maximum residual stress of 36.57 MPa was observed for a layer thickness of 0.33 mm. Higher layer thickness, corresponding to lower residual stress, and lower layer thickness, corresponding to higher residual stress, are due to changes in the number of layers deposited during the manufacturing process. Lower layer thickness requires more layers of extrudate to be deposited for printing the part, whereas higher layer thickness requires fewer layers of extrudate for building the same part. More layers result in an increasing number of cooling and heating times, which ultimately increases the residual stresses induced in the printed parts.

Thinner layer thickness results in higher residual stresses; this can impact the structural integrity and performance of the rotor. Infill density showed relatively lesser impact on the residual stresses; however it is significant parameter which demands further investigation to explore its impact on other mechanical and performance parameters.

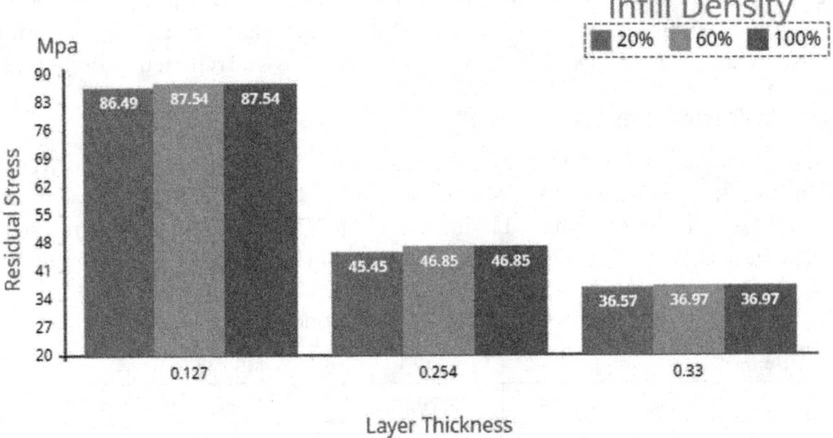

Fig. 14.2 Maximum residual stresses for different process parameters

4. Conclusion

This study presented the numerical modeling of FDM-printed parts using the simulation software Digimat-AM. The effect of layer thickness and infill density on process-induced defects such as residual stress and deflection was investigated. Three levels of each process parameter (layer thickness and infill density) were selected for the simulation study. In addition, printing time, material consumption, and warpage values were recorded. The results of the simulation were statistically analysed using ANOVA. This simulation result exhibited that layer thickness, as compared to infill density, had a significant impact on residual stress and deflection. An increase in the number of layers due to lower layer thickness was the contributing factor for exhibiting higher residual stresses and deflections. The limitation of this study was that it employed only a simulation approach, and physical testing was not covered. The numerical simulation method employed in this study could be an excellent method for performance prediction of FDM parts printed with different materials. The future scope for conducting further research work includes studies regarding the utilisation of different FDM-suitable materials for printing the end-use parts, such as turbo-machinery parts.

References

1. Tofail Syed & Koumoulos, Elias et al., (2017), "Additive Manufacturing: Scientific and Technological Challenges, Market Uptake and Opportunities", Materials Today, Vol. 21, pp 22–37.
2. Ngo Tuan & Kashani Alireza et al., (2018), "Additive Manufacturing (3D Printing): A Review of Materials, Methods, Applications and Challenges", Composites Part B Engineering, Vol. 143, pp 172–196.
3. Committee F42 on Additive Manufacturing Technologies, ASTM International, www.astm.org, accessed on 14 June 2023.
4. Krishnanand, & Taufik Mohammad, (2021), "Fused Filament Fabrication (FFF) Based 3D Printer and Its Design: A Review", In:Deepak, Parhi, D.R.K., Biswal, B.B. (eds) Advanced Manufacturing Systems and Innovative Product Design, Lecture Notes in Mechanical Engineering. Springer, Singapore, pp. 497–505.
5. Guan Yang & Sun Xun et al., (2021), "Development of 3D Printing Entity Slicing Software" China Foundry, Vol.18, pp.587–592.
6. Wang Xin & Jiang M et al., (2016), "3D Printing of Polymer Matrix Composites: A Review and Prospective", Composites Part B: Engineering, Vol. 110, pp. 442–458.
7. C Katia & Cazzato Alberto et al., (2017), "Residual Stress Measurement in Fused Deposition Modeling Parts",Polymer Testing, Vol.58, pp. 249–255
8. Solomon I. & Pandian S et al., (2021), "A Review on the Various Processing Parameters in FDM"' Materials Today: Proceedings, Vol. 37, pp.509–514.
9. Akessa Adugna & Gebisa Aboma et al., (2019), "Numerical Simulation of FDM Manufactured Parts by Adopting Approaches in Composite Material Simulation", IOP Conference Series: Materials Science and Engineering, Vol.700, pp.1–10.
10. Falco Simone & Fogell N. et al., (2022), "Homogenisation of Micromechanical Modelling Results for the Evaluation of Macroscopic Material Properties of Brittle Ceramics", International Journal of Mechanical Sciences, Vol. 220, pp.1–15.
11. Hussein Alzyod, Lajos B, et al., (2023), "Rapid Prediction and Optimization of the Impact of Printing Parameters on the Residual Stress of FDM-ABS Parts Using L27 Orthogonal Array Design and FEA", Materials Today: Proceedings.
12. Al Rashid Ans & Koç M, (2022), "Numerical Simulations on Thermomechanical Performance of 3D Printed Chopped Carbon Fiber-Reinforced Polyamide-6 Composites: Effect of Infill Design", Journal of Applied Polymer Science, Vol.139, pp. 1–13.
13. Sharafi Soodabeh & Santare Michael, et al., (2022), "A Multiscale Modeling Approach of the Fused Filament Fabrication Process to Predict the Mechanical Response of 3D Printed Parts", Additive Manufacturing, Vol. 51, pp. 1–13

Note: All the figures and tables in this chapter were made by the authors.

Technologies for Energy, Agriculture, and Healthcare – Shailesh Nikam et al. (eds)
© 2024 Taylor & Francis Group, London, *ISBN 978-1-032-98028-7*

15

Vibration Analysis of Faults in Rotating Machineries

Shamim Pathan*,
Praseed Kumar, Bipin Mashilkar,
Nilaj Deshmukh, Aqleem Siddiqui
Department of Mechanical Engineering,
Fr. C. Rodrigues Institute of Technology Vashi,
Navi Mumbai, India

Abstract: Vibration in machines can be primarily caused by misalignment and imbalance. These defects may cause rotating systems' components to experience excessive forces, which could cause undesirable vibration. An experimental investigation was conducted to comprehend the alterations in the system's dynamic traits resulting from these faults. An experimental test set up was built to anticipate vibration pattern associated various faults such as shaft misalignment and unbalance. Frequency spectrum is captured using a piezoelectric, a three-axis shear-type accelerometer (Model AC 102-A, Serial Number 66760) paired with the Photon+ (manufactured by Brüel & Kjaer) compact dynamic signal analyzer. The vibration data is acquired with and without faults at various operating frequencies misalignment can be characterized mainly by 2X shaft running speed and unbalance by 1X shaft running speed.

Keywords: Misalignment, Unbalance, Frequency spectrum, Piezo-electric accelerometer, etc.

1. Introduction

When it comes to rotating machinery, shaft misalignment and rotor unbalance are major challenges. M.Xu. et al. developed a theoretical model for understanding

*Corresponding author: shamim.pathan@fcrit.ac.in

DOI: 10.1201/9781003596707-15

the change in behavior of dynamic characteristics of the rotor system, when subjected to faults such as shaft misalignment and unbalance. Misalignment can result in a number of adverse outcomes, such as early bearing failure, greater consumption of energy, excessive seal lubrication leakage, and coupling failure. Harmonic analysis of an Aluminum (Al) shaft rotor bearing system, for predicting the vibration response of a misalignment system was carried out by Hujare et al. Both FEA and experimental results were compared. Parallel misalignment links translational and angular deviations via the stiffness matrix and force vector [AL-Hussain et al.]. Incorrect alignment of shafts using couplings results in significant vibration in rotating machinery. The surge in harmonics due to misalignment can be accurately simulated through Finite Element Method (FEM) analysis [Sekhar, A.S et al.]. The spline joint-flexible coupling-rotor system has found widespread application in the fields of aviation and navigation. Chao Zhang et al. studied the dynamic modeling and analysis of the rotor system with misalignment. Drive shaft misalignment causes spline joint and flexible coupling misalignment in the navigation and aviation industries. As a result, the flexible coupling, rotor, and spline joint system has been widely used. Chao Zhang et al. studied the dynamic modeling and analysis of the rotor system with misalignment. Misalignment between drive shafts can result in misalignment problems within the spline joint and the flexible coupling. Such misalignment exacerbates spline wear and poses risks to the reuse of the spline joint, potentially causing harm [Sekhar, A.S et al.]. Forces and moments will be developed in flexible couplings while accommodating misalignments. F Tuckmantel et al., two methods were compared for modeling the forces and moments produced by a metallic disc coupling when subjected to angular misalignment. The first approach relies on a well-established model that is founded on the linear bending flexure of the disc packs. It assumes that the misalignment effects can be represented by the combined influence of the first four harmonic components. Utilizing the finite element approach, a structural analysis of the connection was performed in the second method. The experimental result was compared with analytical result. Misalignment is one of the reasons for creating additional vibrations when rigid couplings connect shafts [Paolo Pennacchi et al.]. Ideal shaft alignment is very difficult in practical conditions and may present parallel or angular misalignment. If the rotor system is built with fluid – film journal bearing, change in load will alter its dynamic characteristics. A Lees et al. studied the misalignment in rigidly coupled rotors. Harmonics can emerge due to nonlinearities present in fluid film journal bearings or as a result of the kinematic characteristics of couplings. The majority of researchers have endeavored to elucidate the occurrence of harmonic excitation by attributing it to the nonlinearity inherent in either the bearings or the flexible couplings within the system. A Lees et al. introduced a method employing Kalman filtering to identify unbalance within rotor systems. The technique was developed using a mathematical model of the rotor system along with response measurements. M Xu et al. conducted experimentation to verify the theoretical model of rotor imbalance and shaft misalignment. A helical coupling and flexible coupling were used for the experimentation work. According to the experimental data, misalignment and imbalance can be expressed as a function of one or two times the shaft running speed.

2. Design Overview of Experimental Setup

The performance of rotating machinery depends on condition of different components like misalignment, unbalance, looseness etc. An experimental set – up was designed and developed to acquire frequency spectrum by FFT analyzer for diagnosing faults such as misalignment of the shaft and unbalance. Experimental work was performed at different speed so as to study the defect influenced frequencies in rotating system. Figure 15.1 illustrates the experimental arrangement comprising a 0.25 HP DC motor equipped with a variable speed controller, featuring an elongated motor shaft. A bearing is installed on the extended shaft of the motor to capture the vibrations of the driving shaft. A flexible coupling & three identical deep groove ball bearings TR SB 202-10 (one bearing on extended shaft of motor). Three bearing housing were provided, one for supporting the extended part of the motor shaft. On the driven side, the shaft is upheld by two identical ball bearings and spans a length of 350 mm, with a distance of 225 mm between the bearings. The shaft has a diameter of 15.875 mm.

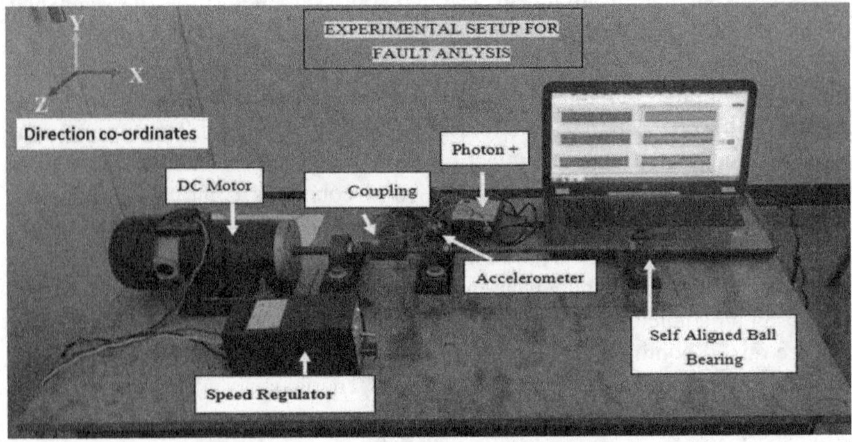

Fig. 15.1 Experimental setup of fault simulator

3. Materials and Methods

The assembly table was supported by four anti-vibration leveling pads, ensuring effective vibration absorption and precise readings. The CAD model of the experimental setup is illustrated in Fig. 15.2, highlighting various components listed below. To create the experimental setup, a steel frame was constructed to support the components. The top of the frame was securely fastened with a plywood sheet, providing a suitable surface for mounting bearings. The motor was supported by a bracket made entirely of mild steel, with an additional bracket attached to the main frame to enhance horizontal alignment. To accommodate bearings of smaller height and achieve proper elevation, the bracket was fabricated at a lower height than the plywood, as depicted in Fig. 15.2. This meticulous design ensures optimal functionality and alignment of the experimental setup.

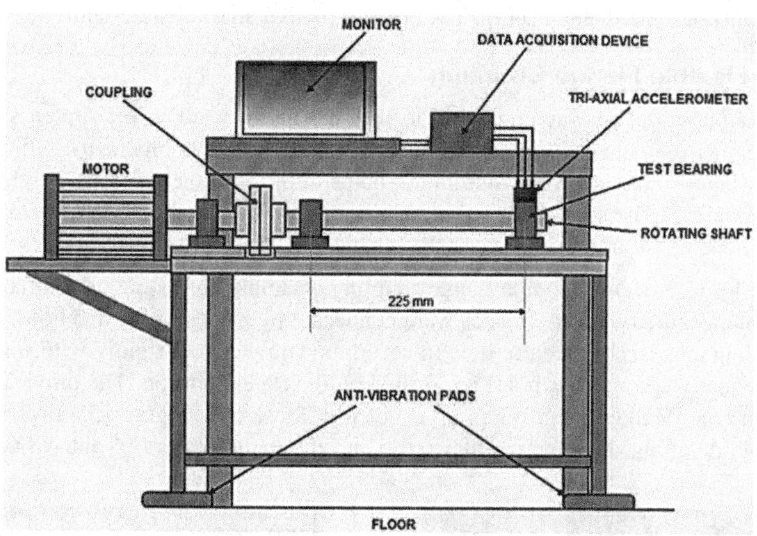

Fig. 15.2 Cad model of the experimental set-up

A square slot was incorporated into the region where the coupling interfaces with the plywood surface, serving to accommodate the coupling's height while maintaining a significantly elevated plywood surface. To provide dedicated spaces, a separate table was allocated for the display device, such as a monitor for a laptop or computer, and the data acquisition device, namely the FFT analyzer. On the top surface of the test bearing, a tri-axial accelerometer was mounted, and data was seamlessly transmitted to the FFT analyzer through connecting wires, as illustrated in Fig. 15.2. Further details and a comprehensive discussion regarding the selection and specifications of various components within the experimental setup are thoroughly covered in subsequent articles.

3.1 Motor

The choice of a DC motor for our current study was primarily motivated by the need for convenient speed variation, a crucial aspect of our investigation. The system's torque requirement, determined by the overall inertia of 5.8×10^{-4} kg/m^2, is estimated at 0.6 N-m. The selected DC motor, with a power rating of 0.25 horsepower, offers a speed range spanning from 250 rpm to 1500 rpm, aligning well with our study's speed control objectives.

3.2 Shaft

A mild shaft of 350 mm length and 15.875 mm diameter was selected to mount all elements of experimental setup. The bearing span of shaft was 225 mm.

3.3 Flexible Flange Coupling

A flexible coupling was employed to link the motor shaft to the driven shaft, serving as the connection between the driver and driven components. Utilizing lightweight materials like aluminum helps minimize the overall weight of the coupling. The selection of the coupling size was based on both the torque requirements for transmission and the dimensions of the driver and driven shafts.

The Fig. 15.3, shows flexible flange coupling assembly consisting of two flanges of same geometry. These flanges were connected by means of nut and bolts. The coupling consists of a centre hole to accommodate the shaft rigidly with flange. The equally spaced four holes are drilled on the flange portion. The dimensions of pin type flexible flange coupling is given in Table 15.1. Figure 15.4 illustrates a table detailing the experimental setup and the incorporation of anti-vibration

Fig. 15.3 Flexible coupling

Fig. 15.4 Table and anti-vibration pads

Table 15.1 Dimensions of flexible coupling

Components	Dimensions (mm)
Shaft diameter	15.875
Hub Dimeter	30
Length of hub	15
Flange diameter	90
Flange thickness	15
Diameter of pin holes on driven shaft	20
Diameter of pin hole on driver shaft = 10 mm	10
Pitch circle diameter	60
Pin diameter	8
Rubber bush inner diameter	10
Rubber bush outer diameter	20

pads. The table was fabricated through a welding process, ensuring a tailored fit for the assembly of the experimental setup, with dimensions carefully determined to match those of the shaft. The table was stabilized by four anti-vibration pads, strategically placed to level the table and prevent misalignment in the vertical direction while also minimizing base vibrations. Alignment was verified using a spirit level.

4. Data Acquisition and Processing

Frequency spectra were obtained for various speeds on the end bearing with the help of a piezoelectric, tri-axial shear-type accelerometer (Type AC 102-A, SI. No. 66760) in conjunction with the ultra-portable dynamic signal analyzer, Photon+ (Brüel & Kjær). This accelerometer records data in three different directions and is placed on top of the bearing block to capture vibration spectra. The gathered data undergoes processing via FFT software, with vibration data subsequently gathered at a computer terminal utilizing the RT Photon+ interface. Measurements were carried out within the frequency domain, and the results are presented as a frequency versus amplitude graph using the RT Photon+ software. The displacement, quantified as vibration amplitude, is measured across 1600 spectral lines within the frequency range of 0-50 Hz. The maximum displacement is confined within a range of 0 mm to 1 mm. Frequency spectra were acquired for 10Hz, 15Hz, and 20 Hz for both healthy and defective bearings on the end bearing. This approach was adopted to acquire vibration spectra specifically for detecting misalignment of shafts.

5. Experimental Investigation of Effect of Parallel Misalignment on Frequency Spectrum

A parallel misalignment of 1mm occurs by horizontally shifting the two bearings on the driven shaft side. Vibration spectra were obtained under both the initially aligned state and the parallel misalignment condition at different frequencies, including 20Hz and 30Hz. The signals were acquired in radial direction, Z-direction (direction coordinates are as shown in Fig. 15.1) for aligned condition and parallel misalignment, since the effect of mentioned fault give prominent vibration in radial directions.

Figure 15.5 illustrates a standard vibration spectrum obtained from the middle bearing, showcasing the aligned shaft position at a frequency of 20 Hz. Conversely, Figure 15.7 depicts the vibration spectrum under parallel misalignment conditions at the same frequency. Similar spectrums were acquired at 30Hz also and are shown in Fig. 15.6 and Fig. 15.8. It can be clearly seen from the vibration spectrum that peak is at 1× for aligned case and for parallel misalignment case the peak is at 1× and 3×. The findings align closely with those documented in the literature by numerous researchers. The vibration spectrum analysis for parallel misalignment reveals the following observations.

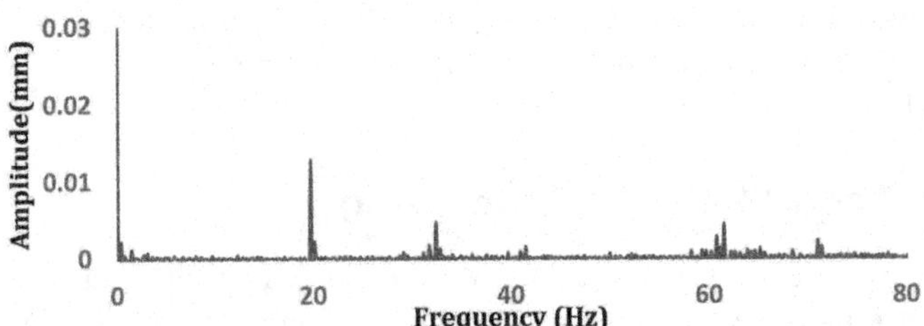

Fig. 15.5 Vibration spectrum for aligned setup on middle bearing at 20Hz

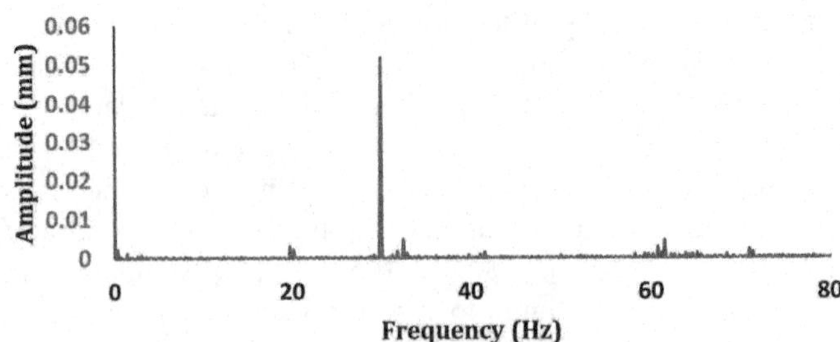

Fig. 15.6 Vibration spectrum for aligned setup on middle bearing at 30Hz

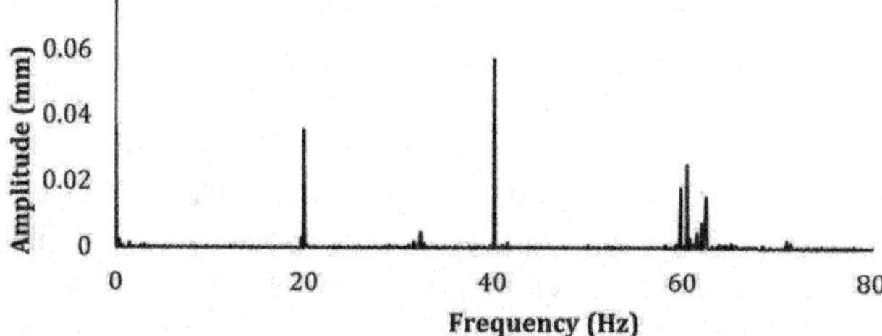

Fig. 15.7 Vibration spectrum with parallel misalignment on motor side bearing at 20Hz

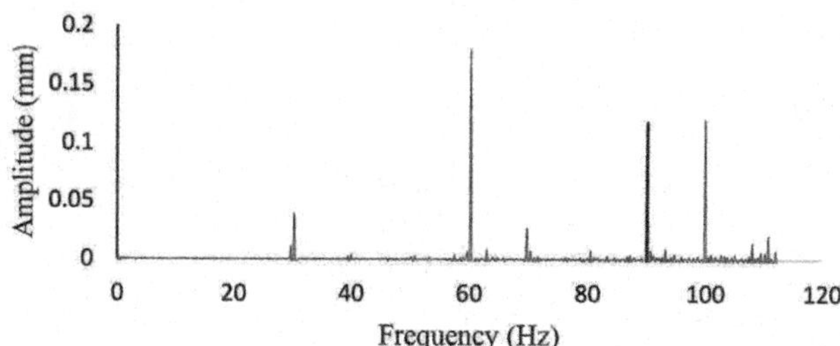

Fig. 15.8 Vibration spectrum with parallel misalignment on motor side bearing at 30Hz

In case of an aligned system vibration spectrum shows a very less vibration in all three directions, it gives small peaks at 1×, 2×, 3× running speed. Parallel misalignment generates a unique vibration spectrum and it depends on type of coupling.

From vibration spectrum it has been observed that dominating frequencies are at 1×, 2× running speed, peak observed at 2× running speed is greater than peak at 1× running speed. Amplitude of machinery vibration is directly proportional to the amount of dynamics forces generated due to misalignment.

6. Experimental Investigation of Effect of Angular Misalignment on Frequency Spectrum

To induce angular misalignment, the end bearing was shifted horizontally by 1 mm within its slot, while the middle bearing remains stationary. Vibration signals were then collected at the middle bearing for both the aligned and angular misaligned configurations using a tri-axial accelerometer. The vibration signals were acquired for angular misalignment in axial direction (x-axis) since angular misalignment shows dominating effect of vibration in axial direction. The frequency spectrum acquired for 20 Hz and 30Hz. Angular misalignment between two shafts linked by a flexible coupling introduces an extra driving force, which has the potential to induce torsional or lateral vibrations. Figure 15.9 and 15.10 shows vibration spectra for aligned and misaligned condition on middle bearing at the frequency of 20Hz. The Fig. 15.10 shows dominating peak at 20Hz, 40Hz, and 60Hz i.e. 1×, 2× and 3× rpm. These findings closely resemble those documented in the literature by several researchers. Vibration severity seen is more in axial direction for angular misalignment.

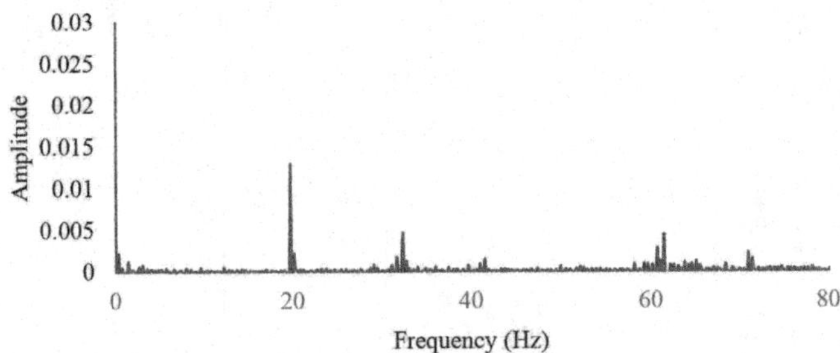

Fig. 15.9 Vibration spectrum for aligned setup on middle bearing at 20Hz

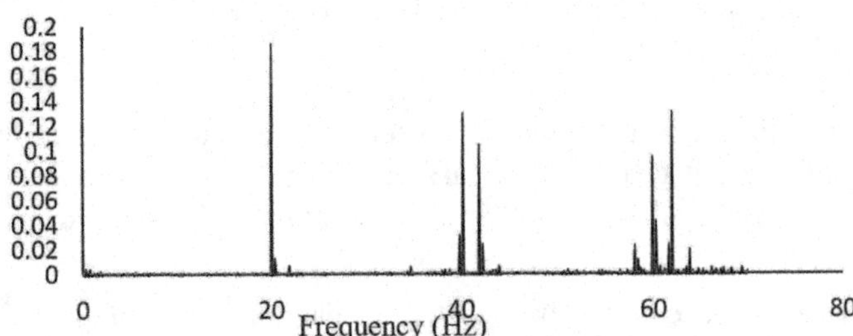

Fig. 15.10 Vibration spectrums with angular misalignment on motor side bearing at 20Hz

Following observation are made from spectrum analysis for angular misalignment:

In scenarios where the system is properly aligned, the vibration spectrum exhibits reduced vibration levels in contrast to a misaligned system. Vibration Spectrum shows that vibration amplitude increases as speed increases. Dominating frequencies reported for angular misalignment are 1×, 2× and 3× running speed and it was acquired in axial direction. Misalignment occurs in a certain direction and as a result, radial forces are not uniform in all direction. Angular misalignment creates bending moments in machine components, which when rotates, exerts a repeating force on machine.

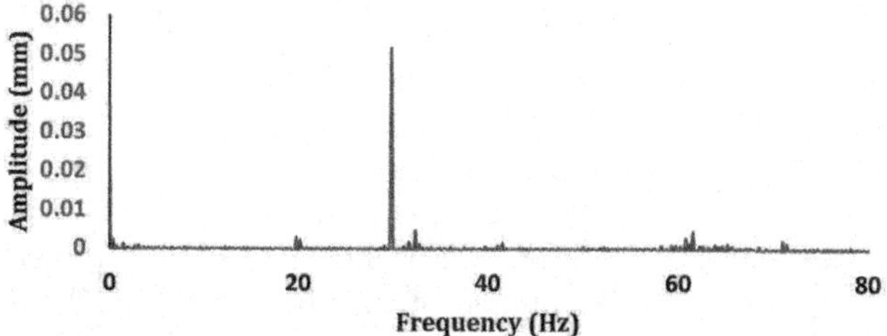

Fig. 15.11 Vibration spectrum with alignment on motor side bearing at 30Hz

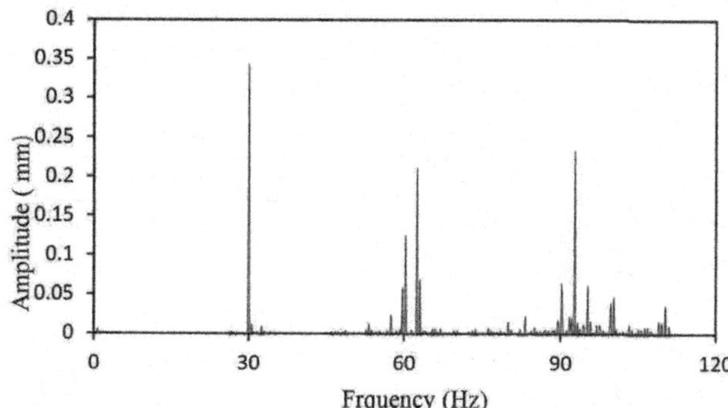

Fig. 15.12 Vibration spectrum with angular misalignment on motor side bearing at 30Hz

7. Experimental Investigation of Effect of Mass Unbalance on Frequency Spectrum

In addition to misalignment, faults stemming from unbalance were investigated using a fault simulator. Rotor unbalance stands out as a primary contributor to vibration in rotating machinery. The experimental configuration comprises a disk affixed at the center of the shaft, with the disk's mass set at 0.4 kg. The disk serves as a platform for attaching additional mass to induce unbalance in the fault simulator. A nut and bolt assembly were utilized as the unbalance mass, allowing for easy removal to balance the system as needed. The unbalance in the system was introduced by attaching mass of 160 grams to disk at a location of 30 mm from the center. Four holes were drilled diametrically opposite on the disk at equidistance 30 mm from centre. The disk was fixed to shaft with grub screw. A Tri-axial transducer and Bruel and Kajer 'Photon+' analyzer has been used for data acquisition. The tri axial sensor was placed on top of bearing blocks to acquire vibration spectrum. Vibration spectra were captured under both balanced and unbalanced conditions at the rotor's operating frequencies of 20Hz and 30Hz, respectively.

A typical vibration spectrum on motor side bearing at a frequency of 20 Hz for balanced and unbalanced case is given in Fig. 15.13 and 15.14 respectively. These figures show peak at 1 × rpm. Vibration spectrums are acquired at 30Hz also for balanced and unbalanced condition and is shown in Fig. 15.15. The vibration spectrum clearly shows that amplitude of vibration increased due to effect of unbalance. The existence of unbalance alters the dynamic characteristics of the system. The frequency response distinctly reveals a rise in amplitude solely at the 1× running speed component.

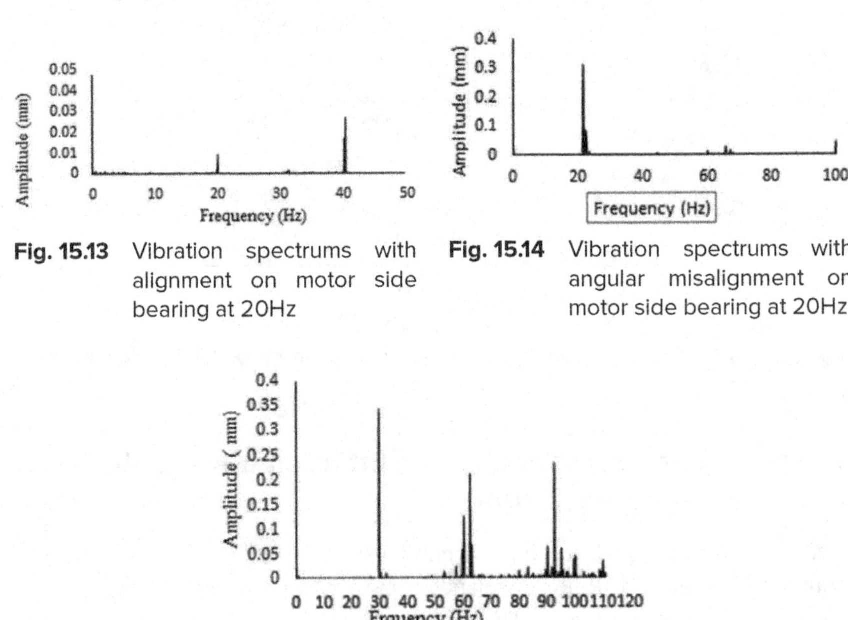

Fig. 15.13 Vibration spectrums with alignment on motor side bearing at 20Hz

Fig. 15.14 Vibration spectrums with angular misalignment on motor side bearing at 20Hz

Fig. 15.15 Vibration spectrums with alignment on motor side bearing at 30Hz

8. Conclusion

Based on the experimental findings obtained through the fault simulator, the following conclusions are derived regarding parallel misalignment, angular misalignment, and shaft unbalance. Parallel misalignment exhibits predominant peaks at 1×, 2×, and 3× RPM, with the peak at 2× surpassing that at 1×. Conversely, shaft unbalance results in a rapid increase in the peak at 1× shaft running speed.

8.1 Conflict of Intersest

The authors declare that they have no known competing financial interests or personal relationships that could have appeared to influence the work reported in this paper.

References

1. M. Xu. and Marangoni, R.D. (1994). Vibration Analysis of Motor-Flexible Coupling-Rotor System Subject to Misalignment and Unbalance Part-I: Theoretical Model and Analysis, J. Sound and Vibr.176(5): 663–679.
2. Hujare, D. P. and Karnik, M. G. (2018). Vibration responses of parallel misalignment in Al shaft rotor bearing system with rigid coupling, Mat. Today: Proceedings, 5(11):23863–23871.
3. AL-Hussain, K.M. and Redmond, I. (2002). Dynamic Response of Two Rotors Connected by Rigid Mechanical Coupling with Parallel Misalignment, J. of Sound and Vibr. 249(3):483–498.
4. Sekhar, A.S. and Prabhu, B.S. (1995). Effects of coupling misalignment on vibrations of rotating machinery, J of Sound and Vibr. 185(4):655–671.
5. Chao, Z., Peng, C., Rupeng., Z, Weifang, C. and Dan, W. (2023). Dynamic modeling and analysis of the spline joint-flexible coupling-rotor system with misalignment, J. of Sound and Vibr.554 (023):117696.
6. Xiaolan, H., Bo Hu, Feitie Zhang, Bing Fu, Hangyang, Li and Yunshan, Zhou. (2018). Influences of spline assembly methods on nonlinear characteristics of spline gear system, Mechanism and Machine Theory, 127: 33–51.
7. Felipe Wenzel da Silva. Tuckmantel. and Katia Lucchesi Cavalca, (2019). Vibration signatures of a rotor-coupling-bearing system under angular misalignment, Mechanism and Machine Theory, 133:559–583.
8. Paolo Pennacchi, Andrea Vanda. and Steven Chatterton, (2012). Nonlinear effects caused by coupling misalignment in rotors equipped with journal bearings, Mechanical Systems and Signal Processing, 30:306–322.
9. Lees, A.W. (2007). Misalignment in rigidly coupled rotors, J. of Sound and Vibr., 305 (1–2):261–271.
10. M. Xu. and Marangoni, R.D. (1994). Vibration Analysis of a Motor-flexible Coupling-Rotor System Subject to Misalignment and Unbalance, Part II: Experimental Validation, J. of Sound and Vibr. 176(5): 1994:681–691.
11. Hanrahan. and Srinivas, PSS., (2009). Vibration analysis of misalignment shaft ball bearing system, Indian j. Sci. and Technology 2(9): 45–50.

Note: All the figures and table in this chapter were made by the authors.

Technologies for Energy, Agriculture, and Healthcare – Shailesh Nikam et al. (eds)
© 2024 Taylor & Francis Group, London, ISBN 978-1-032-98028-7

16

Energy Prediction for Efficient Resource Management in IoT-enabled Data Centres

Sarika Mane[1]

Ph. D. Scholar,
K. J. Somaiya College of Engineering,
Somaiya Vidyavihar University,
Vidyavihar, Mumbai

K. J. Somaiya Institute of Technology,
University of Mumbai

Makarand Kulkarni[2], Sudha Gupta [3]

K. J. Somaiya College of Engineering,
Somaiya Vidyavihar University,
Vidyavihar, Mumbai

Abstract: Internet of Things (IoT) enabled Data Centres (DC), play a vital role in managing and sustaining modern information-driven infrastructure. Accurate energy prediction is the major requirement of the DC for efficient resource management, management of growing internet infrastructure and increasing demand of digital services. The main challenges in the DC are scalability and the cost effectiveness which are dependent on the accurate energy prediction. Hence there is a need of accurate energy prediction. The work proposes energy prediction model with best agreement between predicted and actual values resulting to approximately zero error and robustness for an IoT enabled DC. The feature normalization concept has been used in energy prediction model to enhance the robustness of the different regression models. Proposed work is validated by comparison with the earlier reported work on energy prediction. Robust Linear Regression-Random Sample Consensus (RLR-RANSAC) and Linear Regression (LR) exhibited remarkable performance with RMSE 6.8176×10^{-17} (KWh),

[1]sarika@somaiya.edu, [2]makarandkulkarni@somaiya.edu, [3]sudhagupta@somaiya.edu

DOI: 10.1201/9781003596707-16

7.9565×10^{-17} (KWh) for hourly time span respectively, as compared with earlier reported work. R-squared (R^2) values approaching 1 indicated a near-perfect fit to the data. The proposed approach demonstrated overall performance improvement and can be applicable in IoT enabled DC environment.

Keywords: Data centre, Energy prediction, Energy efficiency, Energy efficient resource management, Internet of things (IoT), Regression models

1. Introduction

Data Centres (DC) are playing the important role to our nation's energy and information infrastructure. Data centres consume 100x powers compared with large commercial building. 40% powers come from energy needed and cooling requirements. As per (Landrum, 2020), worldwide there are 7,500+ datacetres,7.75M servers are installed per year. On an average one data centre uses the equivalent power of 25,000 homes. Such high consumption of the energy demands the proper resource management to manage growing internet user base and increasing demand of digital services DC. For the efficient and sustainable operations of the data centre facilities, energy management and energy prediction are the closely related aspects. Proactive and strategic approach is required for energy prediction. Known future requirements of the energy consumption can help proper energy management. The work by (Ajayi & Heymann, 2021) reported day ahead energy demand prediction using Artificial Neural Network (ANN). (Hsieh et al., 2020) utilized prediction aware virtual machine consolidation strategy for cloud data centres that use less energy. Neural Network (NN) is used for assessment of cooling power performance of the data centre (Shrivastava et al., 2010).The work (Berezovskaya et al., 2020) proposed modelling toolbox to model the soft datacentre using set of building box. This model is capable to examining the performance and energy saving strategies in dynamic mode. Earliest deadline first based energy aware fault tolerant scheduling using AI driven approaches (Marahatta et al., 2021) for cloud DC was designed. Deep Reinforcement Learning (DRL) was used on real data for the cooling optimization and achieved 15% cooling energy savings and 11% cooling cost reduction (Y. Li et al., 2020). (Gao et al., 2020) used deep learning to develop intelligent solutions that enable a cloud DC while effectively handling the unpredictability of renewable energy. Maximized energy efficiency in cloud data centres for virtual machine (VM) consolidation used strong linear regression prediction(L. Li et al., 2019). To understand the DC electricity need, system dynamic model designed for energy prediction from 2016 till 2030. The work by the (Peoples et al., 2011) is focused on optimization of carbon emission associated with the energy consumption. Power management evaluation strategies (Postema & Haverkort, 2018), system wise utilization reduction (Chisca et al., 2015) are observed in the respective mentioned work. Many researchers worked on various techniques, strategies and Artificial Intelligence (AI) algorithm's implementation. In the existing work it is observed that, emphasis on robustness is missing. This motivated to do work for accurate energy prediction along with robustness integrated AI algorithms.

The notable contributions of the proposed work presented in this paper are as follows:

1. Normalization for enhanced accuracy resulting to approximately zero error and robustness has been achieved in this work
2. Energy prediction for IoT-enabled DC using integrated feature normalization in different regression models such as RLR-RANSAC, LR, and RR.
3. A comprehensive enhancement in overall performance has been presented.

The structure of the paper is organized as follows; proposed methodology is presented in section II, succeeded by the results and discussion in section III, and Conclusions in section IV.

2. Proposed Methodology

Implementation of IoT leverages data acquisition in cloud through gateway. By deploying interconnected sensors and the devices data acquisition can be done for the various parameters such as Power (W), Active Energy (KWh), CPU power consumption (%), GPU power consumption (%), etc. IoT facilitates accurate energy predictions through monitoring these parameters. Figure 16.1 adapted from ("The Impact of IoT on the Data Centre Sector," 2017) has depicted the implementation of an IoT scenario in DC for energy prediction. To carry out the energy predictions we used the data server energy consumption dataset (Asanza, 2021) with 184425 entries (0 to 184424) and Data columns (total 16 features) To understand the data in better manner Exploratory Data Analysis (EDA) is carried out. Through heat map correlation analysis is done. Strong correlation is observed between power factor and power (0.76); also power consumption of CPU (0.68), GPU (0.69) and RAM (0.61) are highly correlated with Active Energy (KWh). Active Energy (KWh) refers to the actual energy consumed is a key metric in energy prediction and efficiency in IoT application.

The proposed methodology is as shown in the Fig. 16.2, using different regression models namely A RLR-RANSAC, LR, RR and Lasso regression. Preprocessing was done to remove unnecessary columns. In the data preprocessing averaging and normalization is used. Averaging is used for the aggregation of the data.

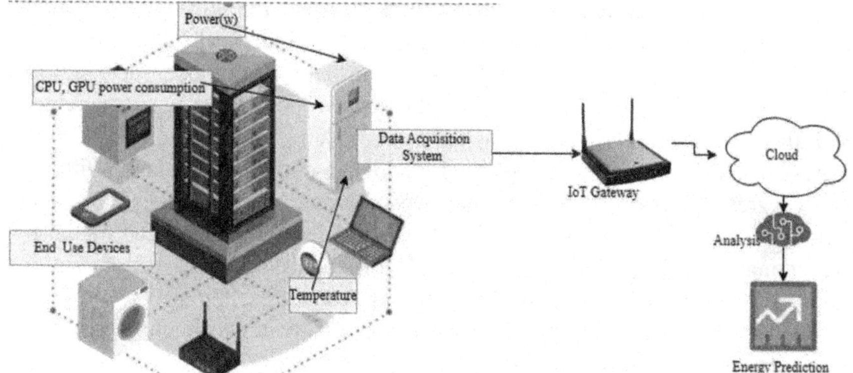

Fig. 16.1 IoT implementation in data centre for energy prediction

Source: Adapted from "The Impact of IoT on the Data Centre Sector." The Irish Advantage, 21 Dec. 2017, www.irishadvantage.com/the-impact-of-iot-on-the-data-centre-sector

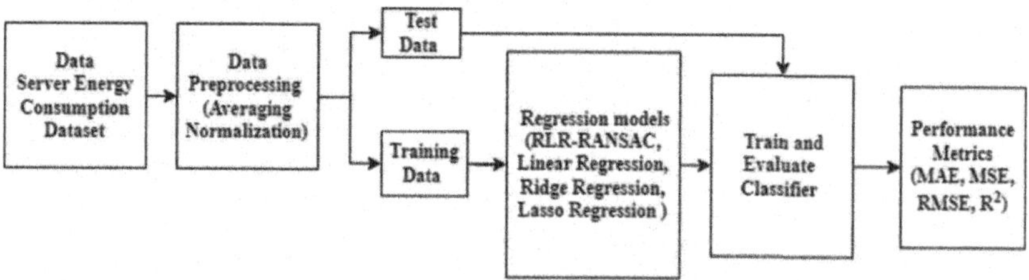

Fig. 16.2 Proposed methodology implementation using various regression models

Source: Author

Normalization reduced biases towards larger-scaled features and promotes numerical stability by ensuring that each feature contributes equally to the training of the regression model. Standard Scalar normalization based on the z score normalization is used. Transformed featuers have mean of 0 and a standard deviation of 1, is given by the equation (1).

$$z = \frac{x - \mu}{\sigma} \tag{1}$$

Where, z is the standardized value, x is the original value of the feature, μ is the mean of the feature and σ is the standard deviation of the feature. Standardization helps to address the challenges related to the scale discrepancies, making the models more accurate, robust, and generalizable. The preprocessed data was then split into training and testing sets and further implementation as shown as in Fig. 16.2.

One of the regression models, RLR-RANSAC is used. RANSAC is a technique used to enhance the robustness of the linear models, specifically in the presense of the outliears. It randomly samples a subset of data, fits a model identifies the inliears based on a predefined threshhold. After repetative process the model with the largest inliers is selected as the final model. Compared to classic linear regression, RLR-RANSAC is less sensitive to data outliers. By robustly fitting a linear regression model, RANSAC improves accuracy in energy prediction. RR's L2 regularization term helps to prevent overfitting and stabilizes the model by preventing excessively large coefficients. Incorporation of the L1 regularization, helps in selecting the subset of the most important features. It lead to a simple and more interpretable model. Thus emphasized essential featuers are important in the enhancement of the accuracy.

For the evaluation of regression models different performance metrics are used such as Mean Absolute Error(MAE), Mean Square Error (MSE), Root Mean Square Error (RMSE) and R-squared (R^2) as per equation (2) and (3) resepectively (Zwingmann, 2022). In reported work RMSE is used for performance measure.

$$RMSE = \sqrt{\frac{\sum_{i=1}^{n}(y_i - \hat{y}_i)^2}{n}} \tag{2}$$

$$R^2 = 1 - \frac{\sum_{i=1}^{n}(y_i - \hat{y}_i)^2}{\sum_{i=1}^{n}(y_i - \bar{y}_i)^2} \tag{3}$$

Where, n is number of data points, y_i is actual (observed) value for ith data point, $\hat{y}_i$ is predicted value for ith data point and $\bar{y}_i$ is the mean of the observed values y_i. Thus feature normalization concept implemented in energy prediction to enhance the robustness of the different regression models. The effectiveness of the implementation is evaluated using comprehensive set of metrics. The following subsection provides the detailed analysis of results.

3. Results and Discussion

Table 16.1, presents different model's comparative analysis over a monthly span and based on various performance metrics. The models are evaluated based on their ability to predict energy consumption in terms of KWh. R^2 represents the proportion of variance between active energy and the time span; the values approaching to 1 represented the better fit.

From Table 16.1, it is demonstrated by both RLR-RANSAC and Linear Regression (LR), with minimal MAE, MSE, and RMSE. Their R-squared values of 1 indicated

Table 16.1 Comparison of different models using various performance metrics

Model	Time Span	MAE (KWh)	MSE $(KWh)^2$	RMSE (KWh)	R^2
RLR-RANSAC	Month	3.7977×10^{-14}	2.4095×10^{-27}	4.9087×10^{-14}	1
LR	Month	4.6653×10^{-14}	3.2818×10^{-27}	5.7287×10^{-14}	1
RR	Month	1.0447	1.8360	1.3550	0.9997
Lasso	Month	0.83263	0.93124	0.9650	0.9998

Source: Author

a perfect fit. It is also observed both could able to identify the underlying patterns in the data. These models are well-known for being straightforward and effective and they make strong arguments for precise monthly energy projections. RR and Lasso Regressions introduced regularization to the models, influencing their performance metrics. Although the error rates of both models remain remarkably low, Lasso performs marginally better than RR in terms of MAE, MSE, and RMSE. These models' regularization methods, whose R-squared values are nearly one, help produce predictions that are stable.

From Table 16.2, RLR-RANSAC algorithm, showed improved forecast accuracy in energy usage. The reported work showed RMSE of 0.0018(KWh) for hourly forecasts and 0.01350(KWh) for daily predictions for LR-Robust linear. Robust Linear -RANSAC performed well, showed RMSE of 6.8176 x10-17 (KWh) for hourly forecasts and 1.6362×10^{-15} (KWh) for daily predictions. Linear Regression also performed well, showed RMSE of 7.9565×10^{-17} (KWh) for hourly forecasts and 1.9095×10^{-15} (KWh) for daily predictions. In comparison with the reported work all error values are less and approximately zero, have resulted enhanced accuracy.

Table 16.2 Comparison of RMSE performance with reported work

Model	Time Span	RMSE (KWh)
LR-Robust linear (Estrada et al., 2022)	Hour	0.0018
	Day	0.01350
RLR-RANSAC (Proposed Approach)	Hour	6.8176×10^{-17}
	Day	1.6362×10^{-15}
	Month	4.9087×10^{-14}
LR (Proposed Approach)	Hour	7.9565×10^{-17}
	Day	1.9095×10^{-15}
	Month	5.7287×10^{-14}

Source: Author

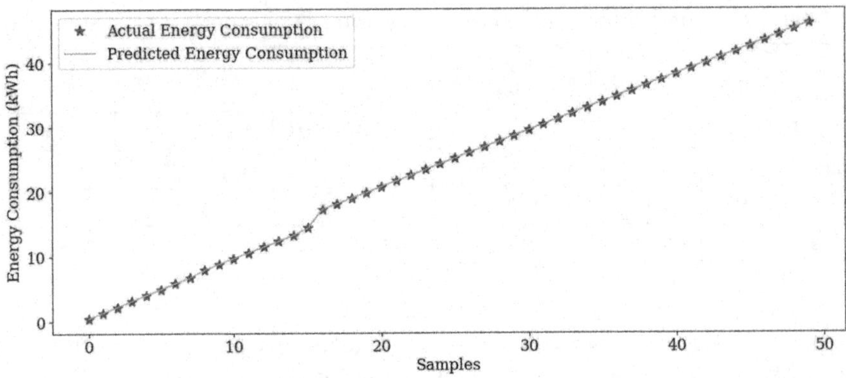

Fig. 16.3 Energy prediction using RLR-RANSAC

Source: Author

Energy Prediction using RLR-RANSAC is as shown in Fig. 16.3. Figure 16.3 showed a nearly perfect fit. Error is approximately 0 so the actual energy consumption is superimposed by predicted energy consumption. But it may also raises concerns over potential over fitting in real time implementation, evaluating the model on new unseen data.

4. Conclusions

The proposed approach with normalization integration implemented with regression models for robustness and accuracy enhancement. The AI plays significant role in addressing energy prediction related aspects. RLR-RANSAC and LR exhibited remarkable performance with RMSE 6.8176×10^{-17} (KWh)· 7.9565×10^{-17} (KWh) for hourly time span respectively, as compared with earlier reported work. In comparison with other models performance RR (1.3550 (KWh)) and Lasso (0.9650 (KWh)) performed well for monthly time span respectively. RMSE values close to the 0 indicated that the models predictions are very accurate and almost indistinguishable from the actual values. RMSE values are consistently extremely small suggested that the models are robust and reliable for predicting energy consumption over the specified time with the errors that are practically negligible. R-squared values approaching 1 indicated a near-perfect fit, able to capture the relations between the variables leading to accurate predictions for all models. Thus integration of normalization with regression models can provide the transformative solution to achieve greater efficiency. It can result in cost effectiveness and scalability along with resource management. In future scope, real time implementation over fitting issue needs to be resolved. Additionally, performance under dynamic, real world condition needed to be resolved using by employing regularization, cross validation, advanced hyperparameter tuning methods.

References

1. Ajayi, O., & Heymann, R. (2021). Data centre day-ahead energy demand prediction and energy dispatch with solar PV integration. *Energy Reports*, *7*, 3760–3774. https://doi.org/10.1016/j.egyr.2021.06.062
2. Asanza, V. (2021). *Data Server Energy Consumption Dataset* [dataset]. IEEE. https://ieee-dataport.org/open-access/data-server-energy-consumption-dataset
3. Berezovskaya, Y., Yang, C.-W., Mousavi, A., Vyatkin, V., & Minde, T. B. (2020). Modular Model of a Data Centre as a Tool for Improving Its Energy Efficiency. *IEEE Access*, *8*, 46559–46573. https://doi.org/10.1109/ACCESS.2020.2978065
4. Chisca, D. S., Castineiras, I., Mehta, D., & O'Sullivan, B. (2015). On Energy- and Cooling-Aware Data Centre Workload Management. *2015 15th IEEE/ACM International Symposium on Cluster, Cloud and Grid Computing*, 1111–1114. https://doi.org/10.1109/CCGrid.2015.141
5. Estrada, R., Asanza, víctor, Torres, D., Bazurto, A., & Valeriano, I. (2022). Learning-based Energy Consumption Prediction. *Procedia Computer Science*, *203*, 272–279. https://doi.org/10.1016/j.procs.2022.07.035
6. Gao, J., Wang, H., & Shen, H. (2020). Smartly Handling Renewable Energy Instability in Supporting A Cloud Datacenter. *2020 IEEE International Parallel and Distributed Processing Symposium (IPDPS)*, 769–778. https://doi.org/10.1109/IPDPS47924.2020.00084
7. Hsieh, S.-Y., Liu, C.-S., Buyya, R., & Zomaya, A. Y. (2020). Utilization-prediction-aware virtual machine consolidation approach for energy-efficient cloud data centers. *Journal of Parallel and Distributed Computing*, *139*, 99–109. https://doi.org/10.1016/j.jpdc.2019.12.014
8. Landrum, J. (2020, July 17). Did You Know These Data Center Facts? *Enabled Energy*. https://enabledenergy.net/insights/did-you-know-these-data-center-facts/
9. Li, L., Dong, J., Zuo, D., & Wu, J. (2019). SLA-Aware and Energy-Efficient VM Consolidation in Cloud Data Centers Using Robust Linear Regression Prediction Model. *IEEE Access*, *7*, 9490–9500. https://doi.org/10.1109/ACCESS.2019.2891567
10. Li, Y., Wen, Y., Tao, D., & Guan, K. (2020). Transforming Cooling Optimization for Green Data Center via Deep Reinforcement Learning. *IEEE Transactions on Cybernetics*, *50*(5), 2002–2013. https://doi.org/10.1109/TCYB.2019.2927410
11. Marahatta, A., Xin, Q., Chi, C., Zhang, F., & Liu, Z. (2021). PEFS: AI-Driven Prediction Based Energy-Aware Fault-Tolerant Scheduling Scheme for Cloud Data Center. *IEEE Transactions on Sustainable Computing*, *6*(4), 655–666. https://doi.org/10.1109/TSUSC.2020.3015559
12. Peoples, C., Parr, G., & McClean, S. (2011). Energy-aware data centre management. *2011 National Conference on Communications (NCC)*, 1–5. https://doi.org/10.1109/NCC.2011.5734700

13. Postema, B. F., & Haverkort, B. R. (2018). Evaluation of Advanced Data Centre Power Management Strategies. *Electronic Notes in Theoretical Computer Science*, *337*, 173–191. https://doi.org/10.1016/j.entcs.2018.03.040
14. Shrivastava, S. K., VanGilder, J. W., & Sammakia, B. G. (2010). *Data Center Cooling Prediction Using Artificial Neural Network*. 765–771. https://doi.org/10.1115/IPACK2007-33432
15. The Impact of IoT on the Data Centre Sector. (2017, December 21). *The Irish Advantage*. https://irishadvantage.com/the-impact-of-iot-on-the-data-centre-sector/
16. Zwingmann, T. (2022). *AI-Powered Business Intelligence*. A book by O'Reilly Media, Inc. ISBN: 9781098111472,45-67

Technologies for Energy, Agriculture, and Healthcare – Shailesh Nikam et al. (eds)
© 2024 Taylor & Francis Group, London, ISBN 978-1-032-98028-7

17

A Simplified Approach for Development of a Syringe Infusion Pump

Abhinav Paniketty[1],
Shruti Jha[2], Sania Ayare[3], Calix Jangul[4]
Student, Department of Biomedical Engineering,
Vidyalankar Institute of Technology Mumbai,
Mumbai University, Mumbai, India
Arunkumar Ram[5], Amol Sakhalkar[6]
Assistant Professor,
Vidyalankar Institute of Technology Mumbai,
Mumbai University, Mumbai, India

Abstract: In patient care, it is important to ensure the efficient delivery of fluids. This ensures the timely delivery of nutrients and medications to critical patients. Syringe infusion pumps have emerged as a promising solution for this purpose. However, the currently available commercial syringe infusion pumps are often prohibitively expensive, rendering them inaccessible to many healthcare facilities, especially for patients in developing nations. This work aims to design and implement a low-cost syringe pump, tailored to the budget constraints of hospitals and clinics in resource-constrained regions. The developed syringe pump utilises a DC motor controlled by a microcontroller board, coupled with a motor driver to enable adjustable drug infusion speed. The mechanism employed to convert rotary motion into linear motion is the lead screw mechanism. The primary objective of this study is to introduce a cost-effective syringe infusion pump, thereby enhancing the overall patient experience and upgrading healthcare infrastructure in tier 3 countries. By accomplishing this objective, we aim to facilitate the efficient and economical delivery of critical medical treatments and

[1]abhinav.paniketty@vit.edu.in, [2]shruti.jha@vit.edu.in, [3]sania.ayare@vit.edu.in, [4]calix.jangul@vit.edu.in, [5]arunkumar.ram@vit.edu.in, [6]amol.sakhalkar@vit.edu.in

DOI: 10.1201/9781003596707-17

fluids to patients, ultimately contributing to improved healthcare outcomes for patients with limited resources.

Keywords: Fluid delivery, Healthcare, Lead screw mechanism, Syringe pump

1. Introduction

The Fig. 17.1 illustrates a standard syringe infusion pump model currently available in the market, highlighting the crucial role of medication and drug delivery in patient care. Precise and reliable administration of medications is essential for ensuring treatment efficacy and patient safety (Islam et al., 2019). Syringe infusion pumps are indispensable tools in medical settings, providing precise and controlled delivery of fluids and medications to patients. They ensure accurate dosing, minimize the risk of medication errors, and enable healthcare professionals to administer treatments with confidence and precision. In our implemented work, we introduce a novel syringe infusion pump designed with a specific goal: to provide reliable and controlled fluid delivery for a wide range of medical treatments, addressing healthcare challenges in developing nations. Our device utilises an Arduino Microcontroller and Stepper Motor for precise fluid delivery within a controlled environment, promising improved patient care and outcomes, reduced costs, and enhanced efficiency.

Fig. 17.1 Syringe infusion pump

Source: Merhi et al., 2019

Through comprehensive experimental testing and evaluation, we assessed the syringe infusion pump's performance, measuring flow rate accuracy, volume delivery and infusion duration. By comparing the advantages of our designed syringe pump against traditional manual drug delivery methods, we emphasise its potential to revolutionise healthcare in resource-limited settings. This innovation not only ensures precise medication delivery but also contributes to bridging the healthcare gap in underserved areas (Ali 2019, 2).

1.1 Research Gap

Our research paper aims to bridge the research gap related to the affordability and accessibility of syringe infusion pumps in healthcare settings, especially in resource-constrained regions. The existing commercial pumps are often prohibitively expensive, limiting their availability. The paper introduces a low-cost syringe pump solution, providing a viable alternative tailored to budget constraints in tier 3 countries. The successful development and validation of this pump address the gap by offering an economical yet reliable option for efficient fluid delivery.

2. Literature Review

The literature surrounding syringe pump innovation and integration underscores the transformative potential of advanced technologies in healthcare delivery. Islam et al., (2019) present a pioneering low-cost smart syringe pump, leveraging wireless and IoT advancements for remote monitoring and control in telemedicine and healthcare applications. Their system architecture enables precise fluid delivery, multiple flow rate control, and wireless operation via a GSM module, promising enhanced healthcare accessibility, particularly in underserved rural areas. Ali (2019) outlines the development of a portable infusion pump designed for administering medication with precision, featuring an alarm system, bubble detection, and adaptability for industrial production. This innovation underscores the critical role of infusion pumps in minimizing medical errors and improving patient care. Markevicius and Navikas (2007) propose the Integrated Syringe Pumps Control System, aiming to automate tasks, reduce errors, and integrate with clinical information systems to enhance medication delivery processes. Highlighting the economic benefits and improved patient care, it offers a promising solution to mitigate human errors in medical settings. Bhavani et al., (2022) present an integration of a semi-automated syringe pump with the Internet of Things (IoT) for real-time monitoring and remote control of fluid flow. Their system, featuring a microcontroller and OLED display, enables remote adjustments based on patient response, reflecting the evolving landscape of syringe pump technology towards greater connectivity and efficiency.

3. Materials and Methods

The Fig. 17.2 provides a comprehensive view of the project's design, framework, and construction. It illustrates the key components involved in the system. The project involves the utilisation of a microcontroller in conjunction with a motor driver to control a DC motor. A long screw shaft is connected to the motor shaft, allowing the rotary motion from the motor to be converted into linear motion. This lead screw is responsible for generating the linear motion necessary for the syringe plunger, enabling the syringe to either expel liquid or draw it into the syringe (ElKheshen et al., 2018). The combination of the microcontroller, motor driver, and lead screw mechanism ensures a reliable and precise fluid delivery system. The Fig. 17.2 serves as a broad overview of the implemented work, giving an idea of how the various components interact and contribute to the successful functioning of the syringe infusion pump.

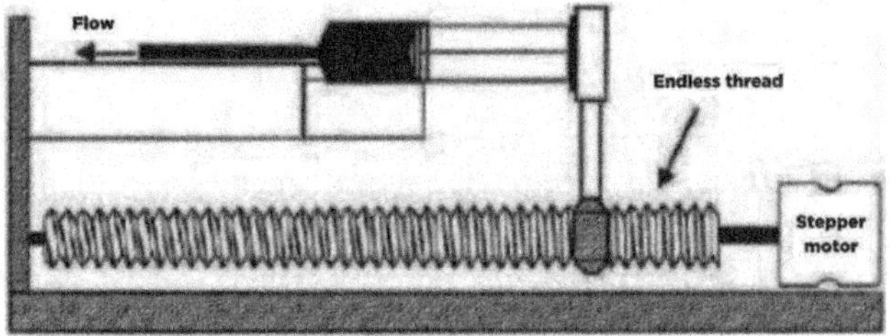

Fig. 17.2 Lead screw mechanism

Source: Assuncao et al. 2014

3.1 Methodology

A syringe infusion pump is used by first loading a syringe containing the medication or fluid to be administered into the pump's syringe holder. The user then programs the pump with the desired infusion rate. Once programmed, the pump delivers the medication or fluid at the specified rate. After the infusion is complete, the syringe is removed, and the pump is reset or prepared for the next infusion as needed.

4. Mechanism Used

The lead screw mechanism plays a crucial role in the syringe infusion pump setup. It acts as the primary means of converting the rotary motion generated by the motor into linear motion required for the syringe plunger's operation (Akash et al., 2016). As shown in Fig. 17.3 the lead screw consists of a screw shaft that is connected to the motor, and its design may include linear shafts supporting it on either side, depending on the project's specific requirements.

The lead screw's characteristics, such as its diameter and pitch, are essential for determining the pump's performance. The pitch, which represents the distance the nut travels along the screw for each full revolution, directly affects the amount of torque required to convert the rotary motion into linear motion (Thanh et al.,

Fig. 17.3 Modelled lead screw mechanism design

Source: Author's compilation

2021). A higher pitch value leads to a lower torque requirement, while a lower pitch value necessitates a higher torque input. In this project, a nut is utilised, which moves axially along the lead screw. This nut is coupled with the syringe plunger or pulls the liquid in and out of the syringe effectively. Overall, the lead screw mechanism forms the backbone of the syringe infusion pump, facilitating precise and controlled fluid delivery through its conversion of rotary motion to linear motion.

5. Components and Modules

As we can see in Fig. 17.4 Arduino Uno is the heart of this setup, which is an open-source microcontroller board based on the ATmega328P chip. Its versatility makes it suitable for users of all levels, from beginners to experts, thanks to its numerous input/output pins and communication interfaces. To link the Arduino Uno to a computer or power source, we use the Uno cable, a USB (Universal Serial Bus) cable that enables data transfer, programming, and power supply.

To control motion, we employ the Nema 17 stepper motor, which is a compact and highly precise motor. It offers accurate positioning and speed control with a step angle of 1.8 degrees per step. We use the A4988 motor driver to control this stepper motor, which supports microstepping for smooth motion and can handle up to 2A of current per coil. To power the setup, we require a 12V adapter. Other mechanical components include the coupling and screw shaft, which allow for the connection of two shafts with some degree of misalignment or axial movement, and the acrylic sheet, a durable and transparent plastic material used for constructing the project's framework.

For input, we utilise a 5x1 keypad that consists of five buttons arranged in a single row, each offering unique functions or values. Finally, to establish electrical connections between components, we use jumper wires, which enable temporary

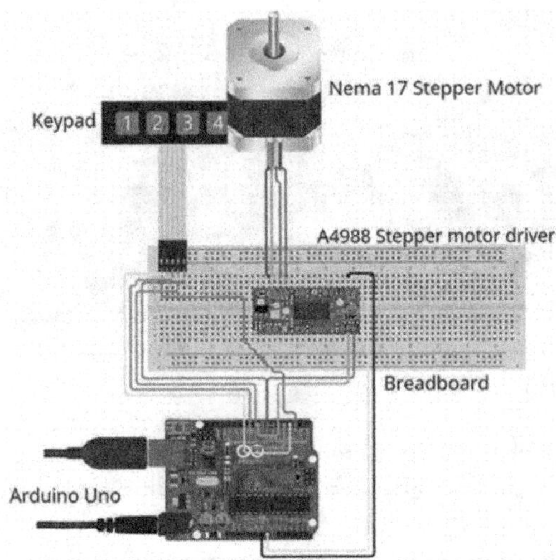

Fig. 17.4 Schematic of syringe pump circuit

Source: Author's compilation

connections on a breadboard or circuit board. Overall, this combination of components provides a comprehensive foundation for building a wide range of interactive and innovative projects, suitable for both beginners and experienced developers alike.

6. Algorithm Overview

The crucial stages in designing any application involve the development of an algorithm. This algorithm provides an overview of the designed system, as it consists of a series of instructions that dictate "what," "when," and "how" to execute a task for a specific input throughout its entire operation (Merhi et al., 2019). The flowchart representing the algorithm is illustrated in Fig. 17.5.

As depicted in the Fig. 17.5, the process of calibrating the code involves several steps aimed at fine-tuning its functionality. The initial step entails the incorporation of three distinct levels of speed within the code, each tailored to suit specific preferences or requirements (Rovere et al., 2016). Once this parameterization is completed, the code is seamlessly integrated into the microcontroller, which serves as the central processing unit for the entire system.

Subsequently, the calibration process is initiated by pressing the designated start button. This action prompts the system to begin the calibration procedure, during which the user is prompted to establish the direction and speed settings in accordance with the intended application. The calibration mechanism serves as a pivotal aspect of ensuring accurate and precise delivery of dosages, making it a crucial step in the overall operation of the device.

Upon commencement of the calibration, the system diligently monitors the dosage administration. If the required dosage is successfully administered within the predefined parameters, the calibration process concludes, and the system transitions to its operational mode. However, should the required dosage not be met within the established parameters, the calibration process recommences. This iterative cycle continues until the desired and accurate amount of dosage is consistently and effectively provided to the patient, ensuring both the safety and efficacy of the medical administration process medical administration process.

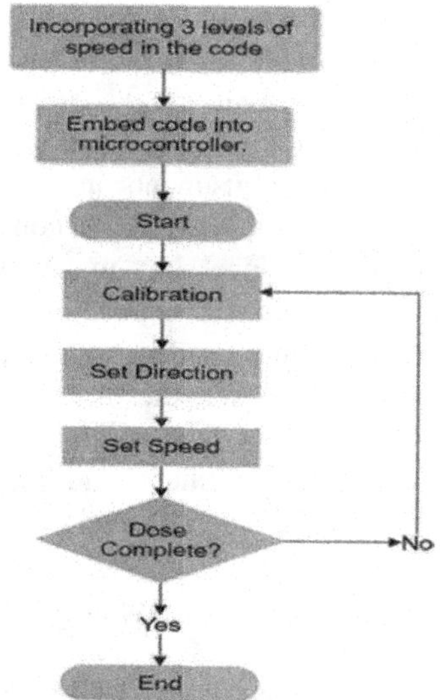

Fig. 17.5 Flowchart of developed system

Source: Author's compilation

7. Results and Discussion

The syringe infusion pump that was developed showcased reliable fluid delivery using a screw lever mechanism and a 12V Stepper motor. The pump was specifically designed to accommodate a 20 ml syringe and its speed was effectively controlled using a motor driver circuit. The inclusion of a keypad input enabled simple manual flow rate adjustments, allowing for precise administration of fluid or medication. The motor driver circuit effectively regulated the motor's speed, ensuring consistent and accurate fluid delivery from the syringe. To maintain the desired flow rates, the pressure applied to the syringe plunger by the motor was carefully calibrated.

The Table 17.1 presents empirical data on Time, Rotation (Speed) and the volume expelled by the syringe pump. This syringe pump has a capacity of pumping 20 ml of fluid at a speed of 1 rotation per second. At this rate, the entire process of pumping 20 ml is completed within a time span of approximately 33.3 minutes. The successful development of the syringe infusion pump brings several advantages to medication administration in clinical settings. The robust screw lever mechanism efficiently translates the Stepper motor's rotary motion into linear motion, ensuring a reliable operation of the syringe plunger. Thanks to the motor driver circuit, precise control over the motor speed is achieved, allowing for accurate adjustments in the flow rate as per the need. The pump's compatibility with 20ml syringe sizes makes it suitable for various clinical applications, aligning with commonly used syringes in medical practice. Moreover, the manual operation flow rate control through the keypad input offers user-friendliness and adaptability, enabling healthcare professionals to easily tailor the medication delivery rate to meet individual patient requirements.

Table 17.1 Observation table

Time (secs)	Speed (rotation/sec)	Volume (ml)
10	10	1.1
10	20	2
10	30	3.1
20	10	2
20	20	4
20	30	6.1
30	10	3.1
30	20	6.1
30	30	9.1

Source: Author's compilation

8. Conclusion and Future Scope

In conclusion, the successful development of the syringe infusion pump showcased its reliability in fluid delivery as depicted in the accompanying Fig. 17.6, making it a promising candidate for clinical settings. The combination of the screw lever mechanism, stepper motor, motor driver circuit, and keypad input sets the stage for future advancements. To enhance safety, it is crucial to incorporate pressure sensors for detecting occlusions and backflow of blood, along with alarms, and to conduct rigorous testing for compliance and reliability. Along with these improvements, the provision of IoT sensors can enable users to control this syringe pump remotely.

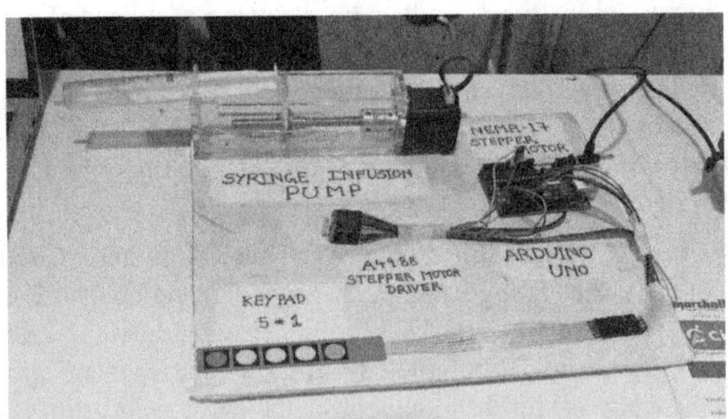

Fig. 17.6 Assembled infusion pump prototype

Source: Author's compilation

References

1. Merhi, N., et al. "An Intelligent Infusion Flow Controlled Syringe Infusion Pump." 2019 31st International Conference on Microelectronics (ICM), Cairo, Egypt, 2019, pp. 48–52, doi: 10.1109/ICM48031.2019.9021516.
2. Islam, M. R., Rusho, R. Z., and Islam, S. M. R. "Design and Implementation of Low-Cost Smart Syringe Pump for Telemedicine and Healthcare." 2019 International Conference on Robotics, Electrical and Signal Processing Techniques (ICREST), Dhaka, Bangladesh, 2019, pp. 440–444, doi: 10.1109/ICREST.2019.8644373.
3. Ali, M. I. "Designing a Low-Cost and Portable Infusion Pump." 2019 4th International Conference on Emerging Trends in Engineering, Sciences and Technology (ICEEST), Karachi, Pakistan, 2019, pp. 1–4, doi: 10.1109/ICEEST48626.2019.8981680.
4. Markevicius, V., and Navikas, D. "Syringe Pumps Integration Influence to Infusion Process Efficiency." 2007 29th International Conference on Information Technology Interfaces, Cavtat, Croatia, 2007, pp. 257–262, doi: 10.1109/ITI.2007.4283780.
5. Bhavani, S., Mishra, S., and Kalpana, R. "Smart Syringe Infusion Pump." International Journal of Innovative Research in Technology, vol. 9, no. 2, pp. 229–234, 2022.
6. Assuncao, R., Barbosa, P., Ruge, R., Guimaraes, Pedro, Alves, Joaquim, Silva, Iroma, and Marques, Maria. "Developing the control system of a syringe infusion pump." 2014, pp. 254–255, doi: 10.1109/REV.2014.6784270.
7. ElKheshen, H., Deni, I., Baalbaky, A., Dib, M., Hamawy, L., and Ali, M. A. "Semi-Automated Self-Monitore – Syringe Infusion Pump." 2018 International Conference on Computer and Applications (ICCA), Beirut, Lebanon, 2018, pp. 331–335, doi: 10.1109/COMAPP.2018.8460462.
8. Akash, K., Sangavi, S., and Venkatesan, M. "Double acting syringe pump using a rack and pinion mechanism — Simulink model." 2016 IEEE International Conference on Computational Intelligence and Computing Research (ICCIC), Chennai, India, 2016, pp. 1–4, doi: 10.1109/ICCIC.2016.7919585.
9. Rovere, S. L., North, M. J., Podestá, G. P., and Bert, F. E. "Practical Points for the Software Development of an Agent-Based Model of a Coupled Human-Natural System." IEEE Access, vol. 4, pp. 4282–4298, 2016, doi: 10.1109/ACCESS.2016.2592418.
10. Thanh, H. T., Do Quang, L., and Ngoc, A. N. "Development of a Low-Delivery-Rate Triple Syringe Infusion Pump for Biomedical Applications." 2021 3rd International Symposium on Material and Electrical Engineering Conference (ISMEE), Bandung, Indonesia, 2021, pp. 277–282, doi: 10.1109/ISMEE54273.2021.9774210.

Technologies for Energy, Agriculture, and Healthcare – Shailesh Nikam et al. (eds)
© 2024 Taylor & Francis Group, London, ISBN 978-1-032-98028-7

18

Blockchain-Based Patient Document Storage and Access

Abhiraj Kale[1], Bhavya Sura[2], Advait Khandare[3], Manav Rupani[4]

Student, Department of Computer Engineering,
KJ Somaiya College of Engineering Mumbai,
Somaiya Vidyavihar University,
Mumbai, India

Deepak Sharma[5]

Vice Principal and Professor,
KJ Somaiya College of Engineering Mumbai,
Somaiya Vidyavihar University,
Mumbai, India

Abstract: In the digital age, a reliable document verification process is crucial, especially for vital records like Electronic Medical Records (EMR) in hospitals. The prevalence of document forgery in the digital landscape challenges the authenticity of such credentials. Blockchain technology offers a decentralized and cryptographically verifiable ledger, serving as a unified platform for secure document storage, sharing, and retrieval. This paper introduces "Doc Chain," a system leveraging blockchain on the Web3.storage platform to simplify EMR verification. By comparing document hashes on the blockchain, it simultaneously verifies patient identity and document authenticity. This innovative approach directly associates documents with individual patients, enhancing security and traceability.

Keywords: Blockchain technology, Ethereum, Web3.storage, Electronic medical records

[1]abhiraj.kale@somaiya.edu, abhirajkale1806@gmail.com; [2]bhavya.sura@somaiya.edu, bhavyasura12@gmail.com; [3]advait.khandare@somaiya.edu, advait.khandare@gmail.com; [4]manav.rupani@somaiya.edu, [5]deepaksharma@somaiya.edu, viceprincipal@engg.somaiya.edu

DOI: 10.1201/9781003596707-18

1. Introduction

The inception of the blockchain concept occurred in 2008 by an individual or group using the pseudonym Satoshi Nakamoto. It served as a foundational element for the Bitcoin cryptocurrency network, effectively addressing the double-spending problem without relying on a trusted central authority or server. This innovation, building upon the earlier work of Stuart Haber, W. Scott Stornetta, and Dave Bayer, has significantly influenced various public cryptocurrencies and has expanded its applications beyond digital currency, acting as a form of payment infrastructure [1].

1.1 Blockchain Architecture

A blockchain comprises a chain of connected blocks, each containing a detailed transaction record akin to a traditional public ledger [1]. Figure 18.1 provides a visual representation of a blockchain, where every block links to its preceding one through a reference termed the previous block hash. Notably, the Ethereum blockchain includes additional hashes of uncle blocks, descendants of a block's ancestors [2]. The initial block in a blockchain is known as the genesis block and lacks a parent block. Below, we explore the internal workings of a blockchain in greater detail.

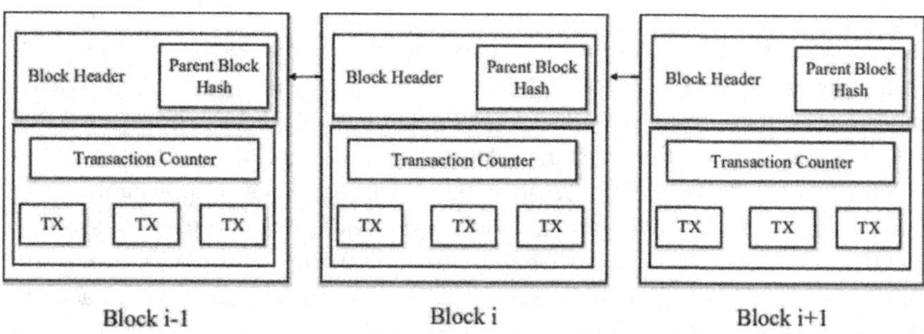

Fig. 18.1 Chaining of blocks

Within the block body, there is a transaction counter and the transactions themselves. The block's transaction capacity depends on various factors, including the block's size and individual transaction sizes. In the realm of blockchain, an asymmetric cryptography system is employed to verify the legitimacy of these transactions [3]. The following section in this paper consists of literature review followed by methodology and approach. After that the paper talks about implementation and the results obtained.

1.2 Problem Definition

Develop an application ensuring the integrity, legitimacy, and authenticity of electronic medical records (EMR). Establish a central repository for accessing these EMRs and implement a distributed, decentralized architecture to enhance data redundancy for improved availability. Create a blockchain for documents, EMRs, and certifications, meeting the distributed and decentralized architecture criteria. Additionally, design a website interface to access records stored on the blockchain.

2. Literature Review

2.1 Blockchain Technology in Certificate Verification: A Comprehensive Review

DocChain underscores the use of IPFS for swift document access and UUID for identity verification [1]. The research explores the crucial security considerations for validating educational certificates on the blockchain [2]. VerifyChain utilizes IPFS, ensuring secure, decentralized file storage with resistance to censorship. The integration of IPFS with blockchain adds an extra layer of security and immutability to the file storage system [3]. The paper offers an overview of recent research in applying blockchain to higher education, presenting experimental outcomes in implementing a system for university transcripts [4]. Blockchain-focused educational initiatives can address contemporary challenges in education, prompting the authors to advocate for a systematic literature review to bridge the gap between these initiatives. The review examines blockchain-based projects, their features in education, and their artistic significance [5]. The incorporation of blockchain in education provides numerous advantages. It establishes a cohesive educational environment, fostering collaboration and resource sharing. Importantly, blockchain guarantees security and intellectual property protection through robust copyright mechanisms [6]. The management of academic diplomas undergoes a transformation with blockchain, shifting the emphasis from institutions to students as document curators. While institutions lose relevance for validation, concerns arise regarding key management and document revocation. Open badges offer a revocation solution, but their broader adoption requires political support and standard enforcement, akin to Blockcerts [7].

3. Methodology and Approach

The chosen methodology utilizes Ethereum as the blockchain platform due to its wide adoption and robust smart contract capabilities.

3.1 Flowchart

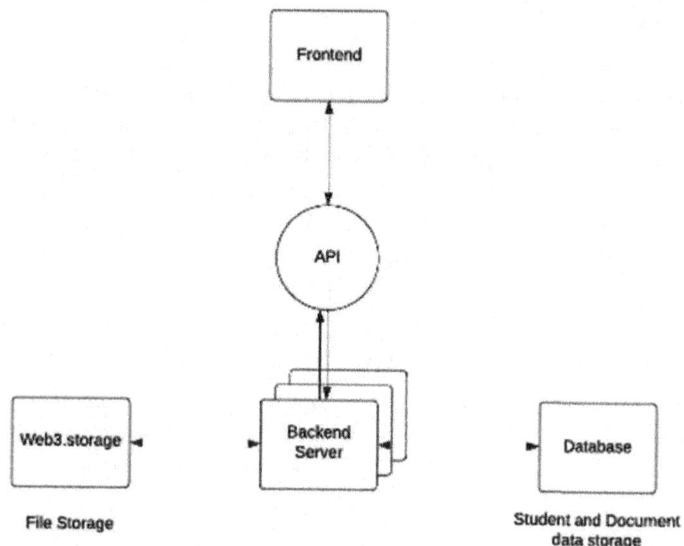

Fig. 18.2 DocChain flowchart

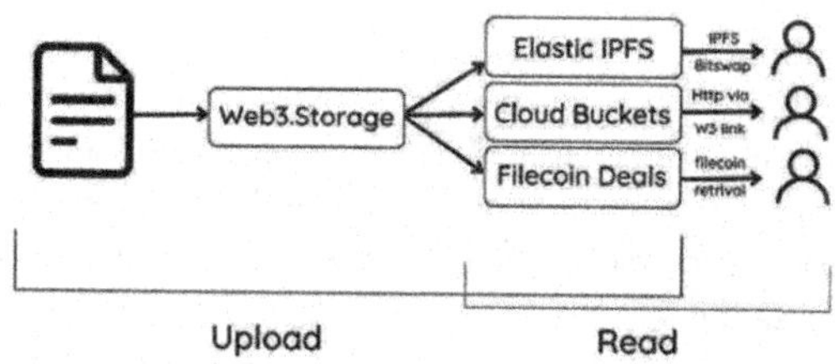

Fig. 18.3 Web3.storage internal structure

3.2 Implementation of Proposed Solution

The implementation plan for the document verification system is designed to be robust and comprehensive.

Smart contracts serve as the cornerstone of the system, facilitating secure storage of hashed document versions and overseeing the verification process. Leveraging Solidity's capacity to facilitate secure transactions on the blockchain ensures the integrity and immutability of the stored data.

During the development and testing phases, the use of Ganache (as shown in Fig. 18.9) as a local Ethereum blockchain simulator provides a controlled environment for testing the system's functionalities. By simulating a decentralized environment with multiple instances for redundancy, developers can thoroughly test the system's resilience and performance before deploying it to the live network.

The implementation plan for the document verification system involves utilizing Web3.storage for storing files securely. Once a file is uploaded, a Content Identifier (CID) is generated, which uniquely identifies the file. This CID will be stored in a MySQL database for easy retrieval and reference (as shown in Fig. 18.5, 18.6 and 18.5.)

The CID serves as a key to access the file stored on Web3.storage through IPFS (InterPlanetary File System). Subsequently, the CID will be inserted into the blockchain, ensuring its immutability and transparency. This insertion into the blockchain adds an additional layer of security and trust to the verification process.

The student will have access to all the documents uploaded by the admin on My Documents Page (Fig. 18.4). Clicking on the link will open the link of the file stored on web3.storage. This link can be shared with whoever the document needs to be shared with.

Users can easily verify the documents by inputting the provided IPFS link into the website (as shown in Fig. 18.8). The website will then compare this CID with the corresponding entry in the MySQL database to authenticate the document's validity.

The website will serve as the primary interface for users to interact with the document verification system. It will be created in Javascript and Bootstrap. Features such as mandatory user authentication and role-based access control will be implemented to safeguard sensitive information and ensure that only authorized individuals can access the system.

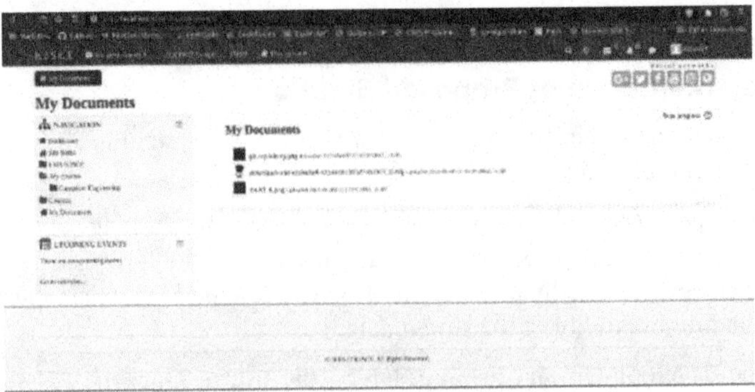

Fig. 18.4 Student's my documents page

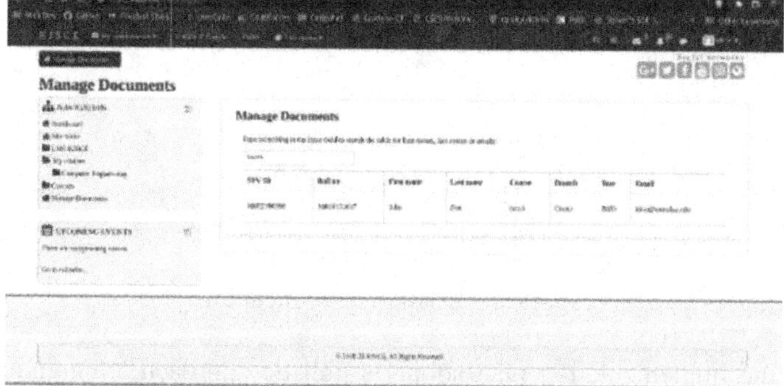

Fig. 18.5 Admin manage documents page

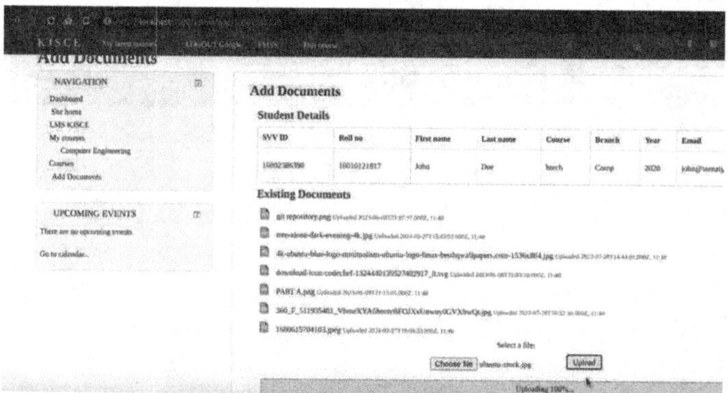

Fig. 18.6 Uploading new document

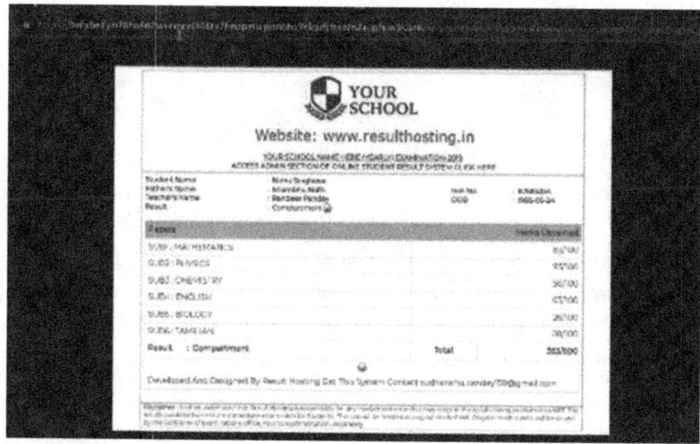

Fig. 18.7 Document uploaded to Web3.storage

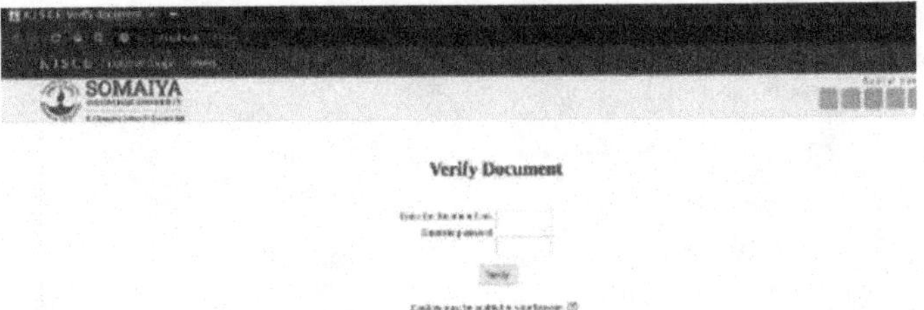

Fig. 18.8 Document verification

Overall, the implementation plan adopts a holistic approach, combining smart contracts, a user-friendly website, Ganache for local testing, and Web3.storage for storing documents.

Fig. 18.9 Document stored on Ganache

4. Results

4.1 Analysis of Doc Chain

A. Data Integrity - Documents on the Ethereum blockchain benefit from its inherent immutability, providing strong resistance to tampering. IPFS maintains data integrity through cryptographic hashes, offering robust protection against tampering. In contrast, centralized databases are vulnerable to insider threats, posing a risk to data integrity. Public cloud storage relies on the security measures of the provider, providing a moderate level of data integrity.

B. Accessibility - web3.storage offers web interface accessibility from anywhere with an internet connection. IPFS may be less user-friendly, requiring knowledge of IPFS tools and nodes. Centralized databases provide user-friendly access but may face downtime or server issues. Public cloud storage is easily accessible but is subject to service availability and potential downtime.

C. Decentralization - web3.storage leverages Ethereum's decentralized network, ensuring redundancy and high availability. IPFS depends on a decentralized network of nodes, guaranteeing data availability. Centralized databases are susceptible to single points of failure due to their centralized nature. Public cloud storage, akin to centralized databases, relies on the infrastructure of the cloud provider.

D. Security - web3.storage utilizes Ethereum's strong security features and smart contract capabilities. IPFS ensures security through cryptographic hashing and peer-to-peer encryption. Centralized databases rely on the database management system and access controls for security. Public cloud storage's security is determined by the cloud provider, often requiring additional encryption and access control configurations.

4.2 Comparative Analysis of Doc Chain with other Solutions

Paper Title \ Parameters	Data Store	Data Fetch	Authenti-cation	Access Control	Record Access	IPFS	Filecoin Storage	Smart Contracts	Access Level	Scalable	Data Validation	Increase Level of Access
Doc chain	☑	☑	☑	☑	☑	☑	☑	☑	☑	☑	☑	☑
Etherdocs [1]	☑	☑	☑	✕	☑	☑	✕	☑	✕	✕	☑	✕
Blockchain Based Framework for Educational Certificates Verification [2]	☑	☑	☑	✕	☑	✕	✕	☑	☑	✕	☑	✕
Verifi-Chain [3]	☑	☑	☑	☑	☑	☑	✕	☑	✕	☑	☑	✕
Blockchain-Based Transcripts for Mobile Higher-Education [4]	☑	☑	☑	✕	✕	✕	✕	☑	✕	✕	☑	☑

Fig. 18.10 Comparative analysis of doc chain with other solutions

5. Conclusion

The proposed solution aims to enhance the security and reliability of university transcripts by developing an innovative blockchain application. This cutting-edge platform will utilize technologies like Ethereum's smart contracts, IPFS, and Solidity programming language to ensure secure document management. Through a distributed and decentralized architecture, the project seeks to minimize data loss risks and ensure continuous access to records. Redundant data storage methods will further bolster system resilience against failures or malicious activities. Security enhancements like role-based access control will provide additional protection. Testing the smart contract systems with Ganache will validate the system's robustness. In essence, this initiative offers a transformative opportunity to modernize how universities manage sensitive student data, providing a more efficient, transparent, and secure approach to maintaining academic records.

References

1. Aniket More, 2023. EtherDocs: A Student Document Management System based on Ethereum Blockchain, Final year project
2. Saleh, Omar & Ghazali, Osman & Rana, Muhammad Ehsan. (2020). Blockchain Based Framework for Educational Certificates Verification. Journal of Critical Reviews. 7. 79–84. 10.31838/jcr.07.03.13.
3. Rahman, Tasfia, Sumaiya Islam Mouno, Arunangshu Mojumder Raatul, Abul Kalam Al Azad, and Nafees Mansoor. 2023. "Verifi-chain: A credentials verifier using blockchain and IPFS." In International Conference on Information, Communication and Computing Technology, pp. 361–371. Singapore: Springer Nature Singapore.
4. Arndt, Timothy & Guercio, Angela. (2020). Blockchain-Based Transcripts for Mobile Higher-Education. International Journal of Information and Education Technology. 10. 84–89. 10.18178/ijiet.2020.10.2.1344.
5. Caldarelli, Giulio, and Joshua Ellul. 2021. "Trusted Academic Transcripts on the Blockchain: A Systematic Literature Review" Applied Sciences 11, no. 4: 1842. https://doi.org/10.3390/app11041842
6. M. Mirzaei, A. Asif and H. Rivaz, Sept. 2020, "Accurate and Precise Time-Delay Estimation for Ultrasound Elastography with Prebeamformed Channel Data," in IEEE Transactions on Ultrasonics, Ferroelectrics, and Frequency Control, vol. 67, no. 9, pp. 1752–1763, doi: 10.1109/TUFFC.2020.2985060.
7. Vimal, S. & Srivatsa, S. (2019). A new cluster P2P file sharing system based on IPFS and blockchain technology. Journal of Ambient Intelligence and Humanized Computing. 10.1007/s12652-019-01453-5.
8. Caldarelli, Giulio, and Joshua Ellul. 2021. "Trusted Academic Transcripts on the Blockchain: A Systematic Literature Review" Applied Sciences 11, no. 4: 1842. https://doi.org/10.3390/app11041842
9. Buterin, V. 2014. "A next-generation smart contract and decentralized application platform." White paper.
10. Satoshi Nakamoto, "Bitcoin: A Peer-to-Peer Electronic Cash System", 2008
11. Patel, Kirtan & Das, Manik. (2019). Transcript Management Using Blockchain Enabled Smart Contracts. 10.1007/978-3-030-36987-3_26.
12. Raaj Anand Mishra, Anshuman Kalla, An Braeken, Madhusanka Liyanage,
13. Privacy Protected Blockchain Based Architecture and Implementation for Sharing of Students' Credentials, Information Processing & Management, Volume 58, Issue 3, 2021, 102512, ISSN 0306–4573, https://doi.org/10.1016/j.ipm.2021.102512.

Note: All the figures in this chapter were made by the authors.

Technologies for Energy, Agriculture, and Healthcare – Shailesh Nikam et al. (eds)
© 2024 Taylor & Francis Group, London, *ISBN 978-1-032-98028-7*

19

Rotating Drum *Cashew* Nut Roaster in Commercial Domestic Oven-Toaster Griller

Amit A. Deogirikar[1]

Assistant Professor,
Department of Agricultural Engineering,
College of Agriculture, Dr BSKKV, Dapoli, District Ratnagiri,
Maharashtra, India

Anagha Amit Deogirikar[2]

Manorath Cashew Delicacy,
Gimhavane, Dapoli, District Ratnagiri,
Maharashtra, India

Abstract: Flavoured cashews (*Anacardium Occidentale*) are preferred for consumption as snacks. In Konkan region, many types of flavoured cashews are available which are locally manufactured. But roasted cashews are not manufactured in this region as commercially available cashew roasters are of high costs and capacities are not suitable for this region. Considering this, a small capacity cashew roasting unit was developed using existing grilling features of the commercially available domestic Prestige Make Oven Toaster Griller (OTG) (210 V AC, 50Hz 1380 W Model POTG 19PCR). A perforated drum of 750 g cashews capacity was developed to mount on grilling rod of oven. Keeping temperature of 210 °C as constant and changing time of roasting (30, 45 and 55 min), three different samples of salted roasted cashews (T_1, T_2 and T_3) were subjected to sensory evaluation. Cashew nuts roasted for 45 min showed maximum sensory score 9.0. Based on this score; 45 min roasting time at 210 °C temperature for 750 g of cashew showed most acceptable salted roasted cashew nuts. This device has potential to solve the need of local cashew nut processing industry to produce roasted cashew nuts.

Keywords: Commercial oven toaster griller, Cashew nuts, Perforated drum

[1]amitdeogirikar@gmail.com, [2]anaghaadeogirikar@gmail.com

DOI: 10.1201/9781003596707-19

1. Introduction

The cashew has a very good medicinal value (Bes-Rastrollo et al., 2009; Jamshidi et al., 2021; Rathore, 2021). It is a good source of calories (553 calories per 100 g), soluble dietary fiber, vitamins, minerals, health-promoting phyto-chemicals, monounsaturated-fatty acids like oleic, and palmitoleic acids. It is favoured as snacks (Blazdell, 2000). Apart from its all medicinal values and advantages, its taste is favoured by the consumers as raw, chocolate or flavour coated, spice coated, salty or salted roasted. The roasted cashew has better medicinal value than the raw one (Griffin & Dean, 2017; Olatidoye, 2021). The moderate consumption of roasted cashew can help to reduce cholesterol, weight, regulate blood pressure, can be an alternative source of protein to fatty meat (Kim et al., 2017). Roasted cashew nut kernels contained more varieties and higher concentrations of aroma volatile compounds than raw kernels (Liao et al., 2019).

Cashew has been an important crop in Konkan region of Maharashtra state (Manerikar et al., 2022). Even though Konkan is producer of cashew nuts, roasted cashew are not produced here. The cashew kernel roasters are available on commercial scale. These roasting machines are designed to roast cashew nuts at a high temperature in salt and give them a perfect golden brown color.

A batch type roasting machine is an automated machine that has capacity to produce the best cashew Kernels for table top snack. Capacity 50 kg / 70 kg / 100 kg per batch. Heat can be provided via Gas or Electric Heaters. The cost of such roasting machines ranges from USD 3500 to 5000 for 200 to 300 kg/h capacity which is not affordable to cashew processors in this region. The production capacity of this machine is also beyond the demand of roasted cashew in local market. The demand of roasted cashew nuts is less in this region and it changes according to tourist season.

To have a solution of small capacity salted cashew roaster, a salted cashew nut roaster unit for roasting 1 kg per hour (6 – 8 kg per day) was developed for small entrepreneurs and Agro tourism centres using commercially available domestic oven toaster griller (OTG). There is wide range of temperature (150 to 200 °C) and time (7 to 20 min) for roasting cashew nuts to golden brown colour. The roasting time and temperature required for preparing salted roasted cashew nuts with developed unit was fixed through filler trials. It was around 35 to 55 min at 210 °C for 750 g cashew nuts in developed unit. The roasting time was to standardized to get acceptable roasted cashew nuts. This was done with experimentation of roasting cashew nuts at 210 °C for 35, 45 and 55 min. Sensory evaluation of these samples was done to standardize roasting time.

2. Methodology and Model Specifications

Oven Toaster Griller: The available commercial Prestige Make 19 L Oven-Toaster-Griller (OTG) (210 V AC, 50Hz 1380 W Model: POTG 19PCR) was used to develop small capacity salted cashew nut roaster (Fig. 19.1 and 19.2). The mechanism of rotating (in vertical plane) square rod present in the OTG for grilling is used to mount a perforated SS drum. The rotating speed of motor was kept unaltered i.e. 2.5 rpm.

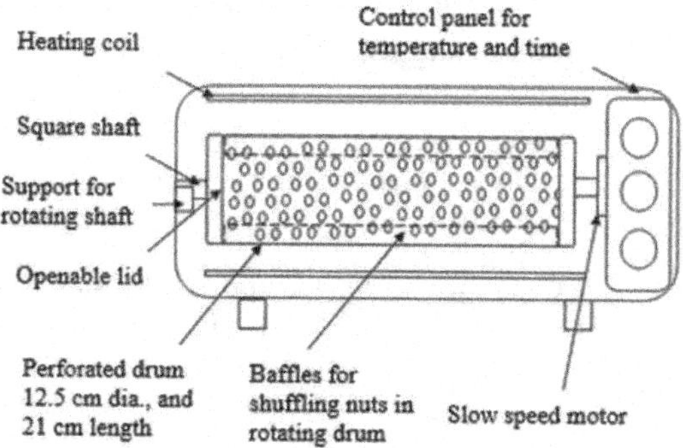

Fig. 19.1 Front view of oven-toaster-griller with mounted drum

Perforated SS drum: The perforated drum was made up of food grade stainless steel sheet of 16 gauge to mount in OTG with the help of available square rod provided for grilling purpose (Fig. 19.1 and Fig. 19.2) (diameter 12.5 cm, length 21 cm and weight 215 g). It (Fig. 19.3) is a removable attachment so as to enhance utility of OTG. One end of the drum is fixed with a stainless steel lid while other

Fig. 19.2 Perforated drum mounted in commercial Oven-Toaster-Griller

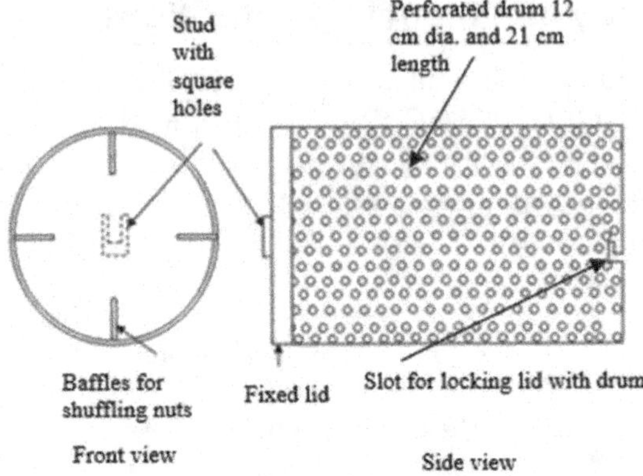

Fig. 19.3 Views of the open drum

end is having 'twist and open/lock' type lid. The lids are having square holes to accommodate a shaft of OTG (Fig. 19.4) where usually food stuff is hanged for roasting purpose while it rotates about horizontal axis. After locking open-able lid of drum, rod can be comfortably inserted centrally through drum. Four equally distributed baffles are provided along its circumference parallel to its length for shuffling nuts filled in it (Fig. 19.5). The drum can easily be placed in or removed through OTG by a handle provided with OTG. Perforated drum mounted on square rod and mounting arrangement made in its lid is shown in Fig. 19.6.

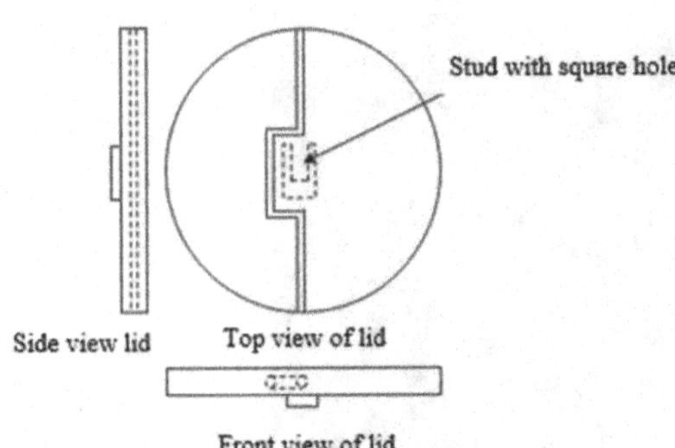

Fig. 19.4 Open-able drum lid and drum

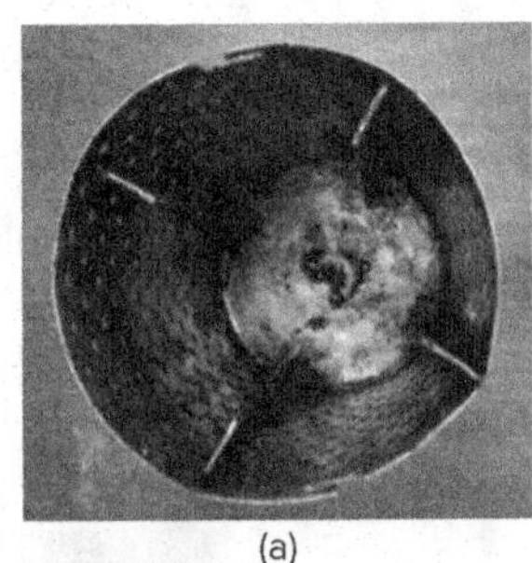
(a)

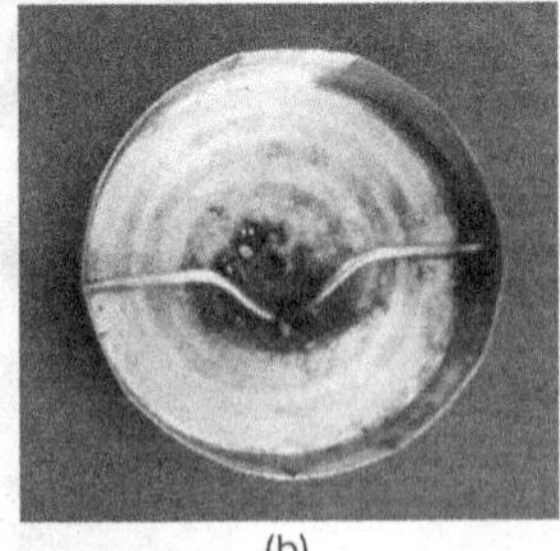
(b)

Fig. 19.5 Internal structure of a. drum and b. lid

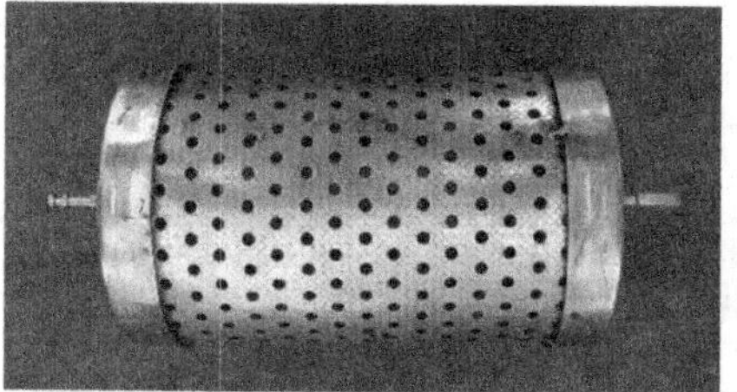
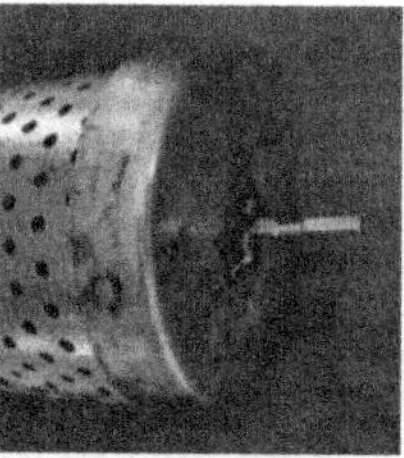

Fig. 19.6 Perforated drum mounted on square rod and mounting arrangement made in its lid

3. Experimentation for Roasting Cashew Nut with Developed Drum for OTG

A batch of 750 g of cashew nuts was soaked in one litre of saturated salt solution for 5 min. These soaked and drained cashew nuts were filled in perforated drum and locked its lid. Mounted the drum with rod for rotating in preheated oven which was set at 210 °C temperature. The temperature 210 °C was standardized with filler trials. Operated OTG for 35 (T1), 45 (T2) and 55 (T3) min at 210 °C keeping rotary motor on. After the set time, removed roasted cashew nuts from drum. The roasted cashew nuts were allowed to cool without using fan and sealed it in air tight plastic bags to maintain its crispiness or crunchiness (Fig. 19.7).

Fig. 19.7 Roasted cashews prepared with developed roaster

3.1 Sensory Evaluation

The sensory evaluation of roasted cashew nuts prepared at 210 °C and roasted for 35, 45 and 55 min was carried out to standardize roasting time. The sensory attributes were evaluated by a panel of 25 untrained judges using 9 points hedonic scale. Hence, three samples of cashew nuts roasted for 35, 45 and 55 min were evaluated by a panel of 25 members in two different sessions. Accordingly, total number of readings were 3 samples × 2 sessions × 25 panelists = 150. The points and parameters of scale were: 1) Dislike extremely, 2) Dislike very much, 3)

Dislike moderately, 4) Dislike, 5) Neither like nor dislike, 6) Like slightly, 7) Like moderately, 8) Like very much, and 9) Like extremely. The sensory attributes or descriptors evaluated were flavor, aroma, color and overall acceptability of roasted cashews.

4. Results and Discussion

Perforated drum for Oven-Toaster-Griller (OTG) for cashew nut roasting: The perforated drum as shown in Fig. 19.1 to 19.4 was fabricated to suite in OTG of specification given above. This small unit could meet the need of roasted cashew nut preparation at low processing capacity and investment price. Oven-Toaster-Griller can be used for other intended purpose too. The total cost of the unit is under USD 120 (Rs 10000). The attachment of stainless steel perforated drum can give consumers option to prepare roasted cashews at home.

4.1 Sensory Evaluation

The sensory evaluation was done using XLSTAT 2023.3.0.1415 with Product characterization (Vidal et al., 2020). The descriptors i.e. flavour, aroma, color and acceptability indicated that T2 i.e. cashew nuts roasted at 210 °C temperature for 45 min showed the most favourable results (Fig. 19.8). The summary statistics is given in Table 19.1. Economics of developed unit kin given in Table 19.2.

Table 19.1 Summary statistics

Variable	Observations	Min.	Max.	Mean	Std. deviation
Flavour	150	3.0	9.0	5.340	2.133
Aroma	150	4.0	9.0	6.520	1.812
Color	150	3.0	9.0	6.427	2.277
Acceptability	150	3.0	9.0	6.273	2.013

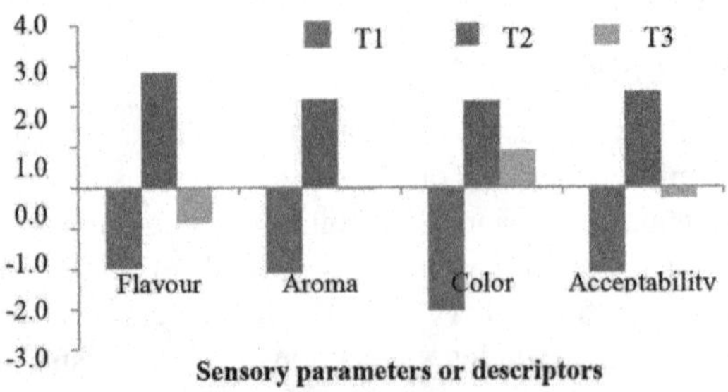

Fig. 19.8 Sensory evaluation of roasted cashews

Table 19.2 Economics of developed unit

S. N.	Parameters	Cost
1	Commercial OTG = USD 4.8 (Rs 4000) and Developed d rum = USD 18 (Rs 1500)	USD 66 (Rs 5500)
2	Cost of repair and maintenance of OTG @ 5 % per annum (20 days per month and 8 h per day = 1920 h per year)	5% of USD 4.8 = USD 0.24 (5% of Rs 4000 = Rs 200) Approx. 0.24/1920 = USD 0.000125/h or USD 0.001/day (Rs 0.1/h or Rs 0.8 /day)
3	Electricity charges	USD 0.12/kg (Rs 10/kg)
4	Assumed electricity charges for roasting 10 kg per day	USD 1.2/day (Rs 100/day)
5	Cost of 10 kg raw cashew	USD 84.5 (Rs 7000)
6	Rate of Indian labour	USD 6/day (Rs 500/day)
7	OTG depreciates in 5 years	USD 4.8/5/20 = 0.05/day (Rs 4000/5/20 i.e. Rs 40/day)
8	Total cost of producing 10 kg roasted cashew nuts	1.2+6+0.05+0.001 = USD 7.251 (100+500+40+0.8 = Rs 640.8)
9	Production cost of 10 kg roasted cashew nuts including raw nus	USD 84.5 + 7.251 = 91.751 (Rs 7000 + 640.8 = 7640.8)
10	Estimated market value of roasted cashew nuts @ USD 17/kg (Rs 1400/kg)	USD 170 (Rs 14000)
11	Net income per day	USD 170 – 91.751 = 78.251 (Rs 14000 – 7640.8 = 6359.2)
12	Considering roaster operated for 20 days a month, net returns in next 5 years	20 × 12 × 5 × 78.251 = USD 93901.2 (20 × 12 × 5 × 6359.2 = Rs 7631040)

The projected income (USD) for five years as per percentage of profit set is shown in Fig. 19.9.

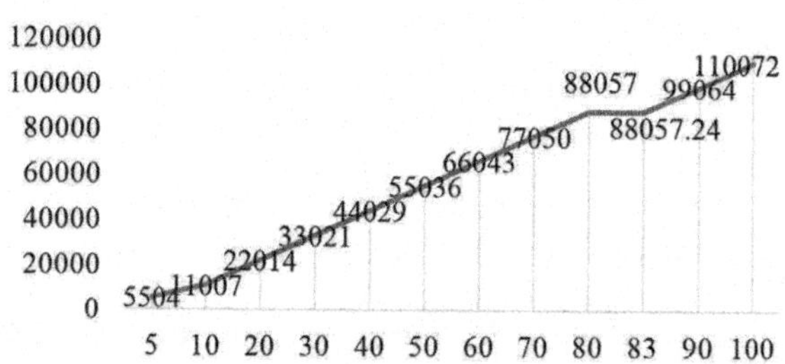

Fig. 19.9 Projected income (USD) in five years as per percentatge of profit set

5. Conclusion

The cashew nuts roasted at 210 °C temperature for 45 min using developed unit for the domestic Oven-Toaster-Griller showed the highest acceptability in respect to flavour, color and aroma.

Patent Granted: No. 453703 dated 11/10/2019

References

1. Bes-Rastrollo, M., Wedick, N. M., Martinez-Gonzalez, M. A., Li, T. Y., Sampson, L., & Hu, F. B. (2009). Prospective study of nut consumption, long-term weight change, and obesity risk in women. *The American Journal of Clinical Nutrition*, *89*(6), 1913–1919. https://doi.org/10.3945/ajcn.2008.27276
2. Blazdell, P. (2000). The mighty cashew. *Interdisciplinary Science Reviews*, *25*(3), 220–226. https://doi.org/10.1179/030801800679251
3. Griffin, L. E., & Dean, L. L. (2017). Nutrient Composition of Raw, Dry-Roasted, and Skin-On Cashew Nuts. *Journal of Food Research*, *6*(6), 13. https://doi.org/10.5539/jfr.v6n6p13
4. Jamshidi, S., Moradi, Y., Nameni, G., Mohsenpour, M. A., & Vafa, M. (2021). Effects of cashew nut consumption on body composition and glycemic indices: A meta-analysis and systematic review of randomized controlled trials. *Diabetes & Metabolic Syndrome: Clinical Research & Reviews*, *15*(2), 605–613. https://doi.org/10.1016/j.dsx.2021.02.038
5. Kim, Y., Keogh, J., & Clifton, P. (2017). Benefits of Nut Consumption on Insulin Resistance and Cardiovascular Risk Factors: Multiple Potential Mechanisms of Actions. *Nutrients*, *9*(11), 1271. https://doi.org/10.3390/nu9111271
6. Liao, M., Zhao, Y., Xu, Y., Gong, C., & Jiao, S. (2019). Effects of hot air-assisted radio frequency roasting on nutritional quality and aroma composition of cashew nut kernels. *LWT*, *116*, 108551. https://doi.org/10.1016/j.lwt.2019.108551
7. Manerikar, S., Kshirsagar, P., Thorat, V., Torane, S., Sawant, P., & Dhunde, A. (2022). *Trends in area, production and productivity of cashew in Konkan region of Maharashtra state: An economic analysis*. The Pharma Innovation Journal 2022.
8. Olatidoye, O. P. (2021). *Effect of temperature and time combinations on colour characteristics, mineral and vitamin content raw and roasted cashew kernel. 554.*
9. Rathore, G. (2021). An analysis of health benefits of cashew nuts. *Asian Journal of Multidimensional Research*, *10*(11), 709–715. https://doi.org/10.5958/2278-4853.2021.01110.1
10. Vidal, N. P., Manful, C. F., Pham, T. H., Stewart, P., Keough, D., & Thomas, RaymondH. (2020). The use of XLSTAT in conducting principal component analysis (PCA) when evaluating the relationships between sensory and quality attributes in grilled foods. *MethodsX*, *7*, 100835. https://doi.org/10.1016/j.mex.2020.100835

Note: All the figures and tables in this chapter were made by the authors.

Technologies for Energy, Agriculture, and Healthcare – Shailesh Nikam et al. (eds)
© 2024 Taylor & Francis Group, London, ISBN 978-1-032-98028-7

20

Sugarcane Crop Acreage Estimates for Bagalkot and Belagavi Districts of Karnataka, India

Sunil Kumar Jha
University of New England, Armidale, NSW 2351, Australia

Vrushali Deshmukh*
K.J. Somaiya Institute of Applied Agricultural Research, Sameerwadi, Bagalkot district, India

Anupriya Jha
Tata consultancy services, Bengaluru, India

Priyanka Kumari
NIIT, Ranchi, India

Abstract: Sugarcane acreage estimation is important for farmers and the sugar industry's decision-makers for scheduling harvest as well as for agricultural field management. Mapping the sugarcane crop using remote sensing is a challenging task due to the diverse cropping patterns and complex phenology of sugarcane crops planted in three different seasons in a year. The newly launched Sentinel-2 satellite by the European Space Agency and free availability of the image's high temporal and spatial resolution on cloud platforms like Google Earth engine has opened up a renewed opportunity to carry out crop acreage estimates with improved accuracies on a real-time basis. The present study was carried out with the objective of sugarcane area estimates for the Bagalkot and Belagavi districts of Karnataka. The Sentinel 2 images between March to May 2020 were used for the study. Maximum Likelihood Classifier (MLC), a supervised classification algorithm was used for sugarcane classification. Results showed overall accuracies of 93.58% and 93.20% for sugarcane crop classification for the Bagalkot and

*Corresponding author: deshmukh.vrushali@somaiya.edu

DOI: 10.1201/9781003596707-20

Belagavi districts, respectively. These predicted values were validated with the Govt of Karnataka data.

Keywords: Crop acreage, Maximum likelihood classification, Remote sensing, Sugarcane

1. Introduction

India's prominence in sugarcane farming is reflected in its global ranking, securing the second position worldwide in both production volume and farming area, following Brazil (DES, 2017). Following Uttar Pradesh and Maharashtra, Karnataka holds the third spot as the largest sugarcane producer in India (Sharma et al., 2019).

In the past, traditional field surveys were the primary method for sugarcane area estimation; however, Virnodkar et al. (2022) and Panigrahy et al. (2009) advocate for efficient and more cost-effective remote sensing technology. For example, Jha et al. (2022) demonstrated that sugarcane yield estimation can occur at various scales, from local levels like sugar mills to regional ones such as districts. With Google Earth Engine (GEE), users gain a cloud-based platform that facilitates easy access to and processing of vast datasets. This includes free satellite imagery from sources like the Sentinel-2 remote sensing satellite, alongside advanced pixel-based classification tools designed for effective crop mapping (Shelestov et al., 2017). Therefore, this paper seeks to investigate the effectiveness of employing the GEE platform in classifying multi-temporal satellite imagery for sugarcane crop mapping, addressing the felt need to estimate sugarcane crop acreage at finer resolutions. This finer-resolution data will significantly aid the sugar industry in better planning of inventories and harvest schedules to attain improved sugar recoveries, ultimately benefiting both the farmers and the sugar industry. The study's main objectives encompass (1) identifying sugarcane crops from other crops or natural vegetation in the Bagalkot and Belagavi districts of Karnataka using Sentinel-2 imagery, (2) developing a sugarcane yield prediction model, and (3) estimating sugarcane crop acreage and production at the cadastral level.

2. Methodology

2.1 Study Area

Acreage assessments were performed at the Bagalkot and Belagavi districts of Karnataka state. The geographical limits of the study area are depicted in

Fig. 20.1. Bagalkot experiences an average annual rainfall of 562 mm, while Belagavi receives around 1200 mm annually.

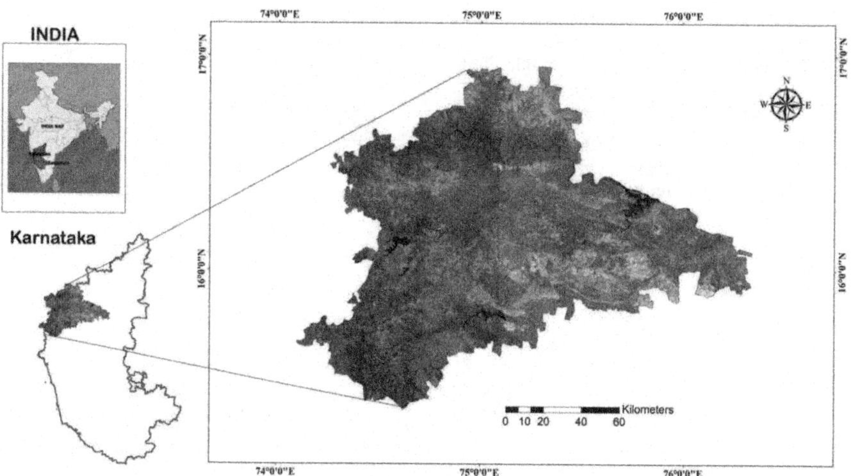

Fig. 20.1 False colour composite image of bagalkot and belagavi districts

Source: Author

2.2 Data used for Area and Production Estimation

The field data collected from various locations within the two districts include information such as crop phenology, cultivar, date of planting, season, coordinates of the field (latitude and longitude), and yield data. In Karnataka, the sugarcane growing seasons consist of adsali, pre-seasonal, and seasonal, with the harvesting period extending from October to April. For sugarcane acreage estimation, 700 samples were used for classification n. The reference dataset was established through field surveys, recording ground truth values with a Global Positioning System (GPS) instrument. This field-sampled dataset was then split into a 7:3 ratio for training and testing purposes.

2.3 Satellite Image Processing

Composite Images generated from Sentinel-2 (S2) imagery within the Google Earth Engine (GEE) were examined in this study. Specifically, cloud-free S2 reflectance imageries obtained from GEE over the Bagalkot and Belagavi districts (refer to Fig. 20.1). For the temporal composition, 5-day composites spanning from March to May 2020 were chosen to enhance temporal resolution. These S2 images are readily available at GEE, which is derived from Level L1C orthorectified scenes, utilizing computed Top of Atmosphere (TOA) reflectance. Each composite comprises all scenes within 5 days, with the most recent pixel positioned at the top. Only four spectral bands (Near Infrared, Red, Green, and Blue) out of 13 spectral bands of Sentinel-2 images were used for sugarcane acreage estimation.

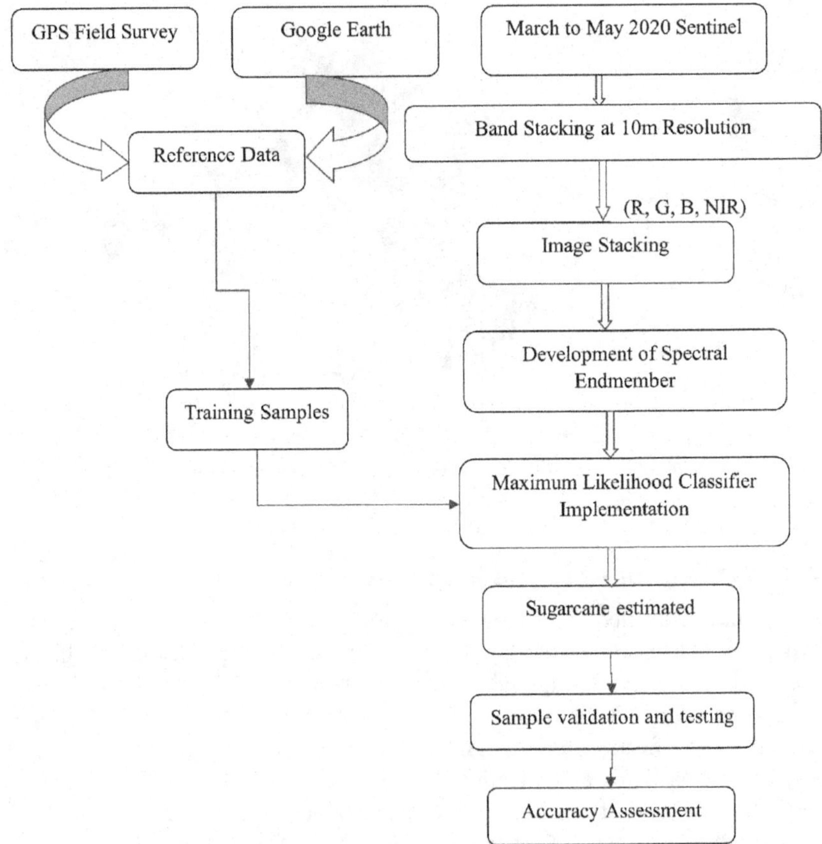

Fig. 20.2 Sugarcane crop classification and production estimation

Source: Author

There are numerous supervised classification methods available, including Maximum Likelihood Classification (MLC), parallelepiped, Mahalanobis distance, Neural Network, Support Vector Machine, and Spectral Angle Mapper. In this study, Maximum Likelihood Classification is employed, which operates under the assumption that the statistics for each class in each band follow a normal distribution and computes the probability of a given pixel belonging to a particular class. Subsequently, each pixel is assigned to the class with the highest probability, known as the maximum likelihood.

2.4 Sugarcane Area Estimation

Sugarcane crop classification includes a process comprising of stacking sentinel-2 cloud-free imagery and training it with sugarcane and other grown land use land covers. A supervised classification approach was utilized for sugarcane across the study area collected by field staff of our institute. Signatures of other land cover features such as built-ups, forests, barren lands, waterbodies, and crop classes were identified using crop calendar, visual interpretation, tone, texture, pattern, and association from the satellite image. The MLC method evaluates the likelihood of a pixel's membership in a specific class (Rao, 2002). This approach assumes the data from training sets (signatures) to follow a normal distribution, enabling the calculation of mean vector and covariance matrix for the spectral cluster of each brightness value category (Lillesand and Kiefer, 2000). Leveraging these principles, the statistical probability of a pixel's digital number (DN) value regarding each land use/cover category was computed. Subsequently, each pixel was allocated to the class demonstrating the highest probability value.

3. Results and Discussion

3.1 Acreage Estimation

The supervised classification on the multi-spectral dataset successfully distinguished between vegetation and non-vegetation classes. Utilizing ground truth sites collected by sugar mill officials, we adopted a supervised classification technique specifically for vegetation classes. The collected pixel profiles for sugarcane and other land covers facilitated accurate classification. The classification results for sugarcane crops in the Belagavi and Bagalkot regions using the MLC are shown in Figs. 20.3 and **20.4**, respectively. The MLC method demonstrated high accuracy in identifying sugarcane fields.

The results of the sugarcane crop classification using MLC classifier were compared with the Government of Karnataka survey data which is shown in Fig. 20.5 and 20.6.

The overall accuracy of sugarcane crop classification in the Belagavi district reached an impressive 93.58%, as depicted in Table 20.2. This accuracy was calculated by Congalton's method (1991), dividing the total number of accurately classified areas by the total number of areas utilized in the classification. For Belagavi, the user's accuracy for correctly identifying sugarcane fields out of all actual sugarcane fields was 95.57%, while the producer's accuracy for accurately classifying sugarcane fields out of all areas classified as sugarcane was 90.07%.

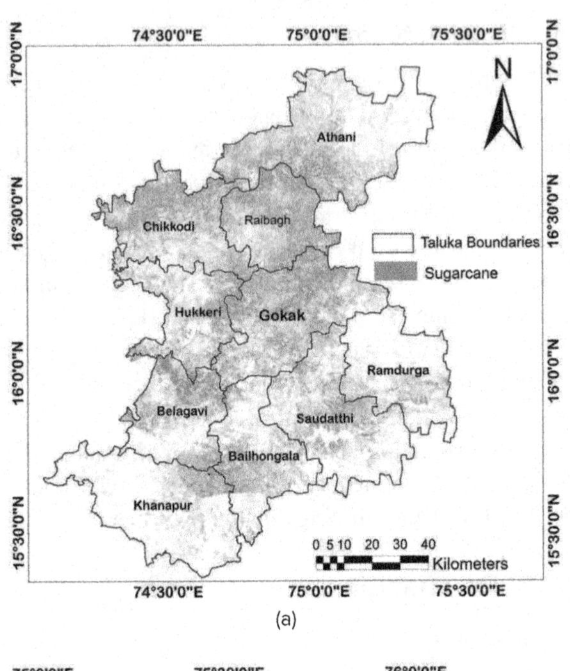

(a)

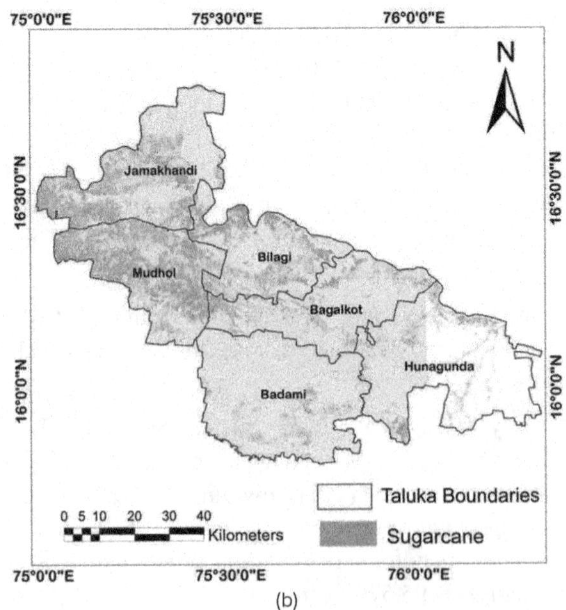

(b)

Fig. 20.3 (a) Sugarcane acreage map for Belagavi and (b) Bagalkot district

Source: Author

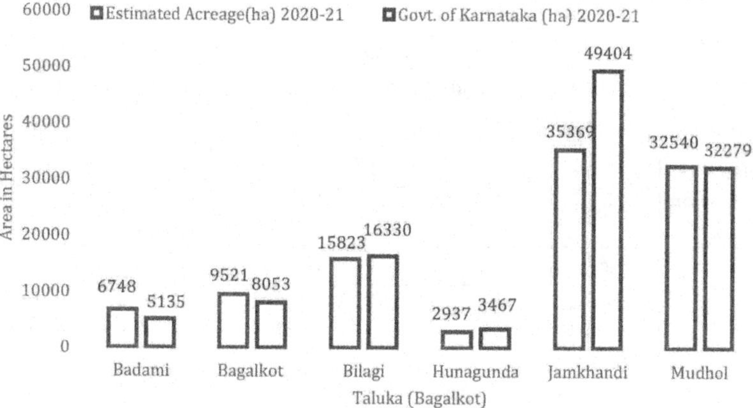

Fig. 20.4 Comparison of sugarcane acreage by taluka in Bagalkot district (KIAAR and Government of Karnataka, 2020-21)

Source: Adapted from DAFW

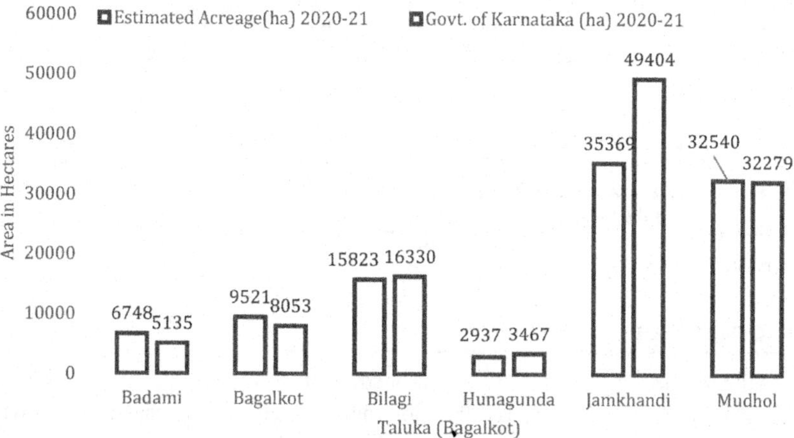

Fig. 20.5 Comparison of sugarcane acreage by taluka in Belagavi district (KIAAR and Government of Karnataka, 2020-21)

Source: Adapted from DAFW

Conversely, in Bagalkot, the user's accuracy for sugarcane fields was slightly higher at 93.81%, with a producer's accuracy of 92.92%. Additionally, Bagalkot exhibited a user's accuracy of 92.57% for correctly identifying non-sugarcane areas and a producer's accuracy of 93.50% for accurately classifying non-sugarcane areas. Overall, Bagalkot achieved an accuracy of 93.20% for all classified areas as shown in Table 20.3.

Table 20.1 Accuracy assessment for belagavi district

		Sample Data			
	Class	**Sugarcane**	**Other**	**Row Total**	**User's Accuracy (Error of Inclusion)**
Remote Sensing Classification	Sugarcane	344	38	382	344/382*100 = 90.05%
	Other	5	283	288	283/288*100 = 98.26%
	Column Total	349	321	670	
	Producer's Accuracy (Error of Exclusion)	344/349*10 = 98.57%	283/321*100 = 88.16%		**Overall Accuracy:** (344+283)/670*100 = 93.58%

Source: Author

Table 20.2 Accuracy assessment for Bagalkot district

		Sample Data			
	Class	**Sugarcane**	**Other**	**Row Total**	**User's Accuracy (Error of Inclusion)**
Remote Sensing Classification	Sugarcane	197	13	210	197/210*100 = 93.81%
	Other	15	187	202	187/202*100 = 92.57%
	Column Total	212	200	412	
	Producer's Accuracy (Error of Exclusion)	197/212*100 = 92.92%	187/200*100 = 93.50%		**Overall Accuracy:** (197+187)/412*100 = 93.20%

Source: Author

4. Conclusion

In conclusion, the study effectively demonstrates the potential of utilizing remote sensing technology, particularly the Sentinel-2 satellite imagery coupled with the Google Earth Engine platform, for accurate and efficient sugarcane acreage estimation. The study, focusing on Bagalkot and Belagavi districts of Karnataka, utilized Sentinel-2 imagery from March to May 2020 and applied MLC for sugarcane classification. The obtained results revealed impressive overall accuracies of 93.58% and 93.20% for sugarcane crop classification in the respective districts. These accuracies were validated against the Government of Karnataka's survey data. It's important to note that while Maximum Likelihood Classification (MLC) yielded superior accuracy in this study, there are several other methods available for classifying sugarcane. Further exploration of techniques such as Support Vector Machines (SVM) or Random Forests could be beneficial, especially with a larger number of training points for sugarcane and other land use/land cover (LULC) classes. This could potentially enhance the accuracy and reliability of sugarcane acreage estimation in diverse agricultural fields. The classified sugarcane map generated from this study holds practical utility beyond mere acreage estimation. For instance, in the event of natural calamities such as floods or droughts, this map can serve as a valuable tool for disaster response and management, aiding in assessing the extent of damage to sugarcane crops and guiding relief efforts. Additionally, for efficient resource management, including water and fertilizer allocation, and for facilitating insurance assessments, this detailed information on sugarcane distribution proves invaluable to farmers, cane suppliers, and mill managers, enabling them to monitor and manage crops effectively.

Acknowledgments

The authors acknowledge support from Shri Samir Somaiya (Chairman and MD, Somaiya group), sugarcane farmers of the region, and Godavari Bio-refineries Ltd.

References

1. Congalton, R. G. 1991. "A review of assessing the accuracy of classifications of remotely sensed data." Remote Sensing of Environment 37 (1): 35–46.
2. Jha, Sunil Kumar, Virupakshagouda C. Patil, B. U. Rekha, Shyamal S. Virnodkar, Sergey A. Bartalev, Dmitry Plotnikov, Evgeniya Elkina, and Nilanchal Patel. "Sugarcane Yield Prediction Using Vegetation Indices in Northern Karnataka, India." University Journal of Agricultural Research 10, no. 6 (2022): 699–721.
3. Lillesand, T. M., and R. W. Kiefer. 2000. Remote Sensing and Image Interpretation. New York: John Wiley & Sons.
4. Panigrahy, R. K., S. S. Ray, and S. Panigrahy. 2009. "Study on the utility of IRS-P6 AWIFS SWIR band for crop discrimination and classification." Journal of the Indian Society of Remote Sensing 37: 325–333.
5. Rao, P. V. K., V. V. Rao, and L. Venkatraman. 2002. "Remote Sensing: A Technology for Assessment of Sugarcane Crop Acreage and Yield." Sugar Tech 4 (3&4): 97–101.
6. Sharma, N., Saxena, S., Dubey, S., Choudhary, K., Sehgal, S., & Ray, S. S. "Analysis of Sugarcane Acreage and Yield Estimates Derived from Remote Sensing Data and Other Hybrid Approaches under FASAL Project." ISPRS - International Archives of the Photogrammetry, Remote Sensing and Spatial Information Sciences, 2019
7. Shelestov, A., M. Lavreniuk, N. Kussul, A. Novikov, and S. Skakun. 2017. "Exploring Google Earth Engine platform for big data processing: Classification of multi-temporal satellite imagery for crop mapping." Frontiers in Earth Science 17.
8. Virnodkar, S., V. P. Jha, and S. Patil. 2022. "Multi-Sensor Approach for Sugarcane Classification Using Deep Convolutional Neural Network." The Interciencia 47: 53–81.
9. Xie, H., Y. Q. Tian, J. A. Granillo, and G. R. Keller. 2007. "Suitable remote sensing method and data for mapping and measuring active crop fields." International Journal of Remote Sensing 28 (2): 395–411. https://doi.org/10.1080/01431160600702673.

Technologies for Energy, Agriculture, and Healthcare – Shailesh Nikam et al. (eds)
© 2024 Taylor & Francis Group, London, ISBN 978-1-032-98028-7

21

Design and Development of Wilkinson Power Divider for Wifi Application

Prathamesh Bagal[1], Aditya Mishra[2], Makarand Kulkarni[3], Nilesh Lakade[4]

K.J Somaiya College of Engineering, Vidyavihar, Mumbai

Abstract: This paper presents the design and simulation of a 3-port Equal-Split Wilkinson Power Divider specifically for Wi-Fi applications operating at 2.4 GHz. This component is critical for ensuring uniform signal distribution across multiple devices within a network. Its compact design and precise power division make it ideal for various applications, particularly in the agriculture, healthcare, and energy management sectors. In agriculture, it supports IoT devices in precision farming, enabling effective data collection and analysis for improved crop management. In healthcare, it ensures reliable communication and data transfer for medical devices and telemedicine services within hospital networks. In Energy management, it aids smart grid technologies by providing consistent signal distribution to monitoring and control devices, thus enhancing energy efficiency and grid reliability. The Wilkinson power divider distinguishes itself from other dividers by offering superior isolation and minimal insertion and reflection losses. This paper leverages PathWave-ADS software to design and simulate an equal-split 3-port Wilkinson power divider tailored to Wi-Fi frequencies for developing a power divider that meets the stringent demands of modern agricultural, healthcare, and energy systems, ensuring robust and reliable signal distribution.

Keywords: Wilkinson power divider, IoT, Equal-Split, PathWave-ADS, Signal distribution, Wifi application

[1]prathamesh.bagal@Somaiya.edu, [2]aditya.mishra@somaiya.edu, [3]Makarandkulkarni@Somaiya.edu, [4]nilesh.lakade@Somaiya.edu

DOI: 10.1201/9781003596707-21

1. Introduction

The Wilkinson power divider is known for its compact design and vast area of application, Among the other types of power dividers available, the Wilkinson power divider offers better isolation alongside the insertion and reflection loss. In this paper, we have designed and simulated an equal-split 3-port Wilkinson power divider tailored for wifi frequency using Pathwave ADS software. The objective is to design a 3-port equal-split Wilkinson power divider tailored for wifi frequency. Wilkinson power dividers have versatile applications across agriculture, healthcare, and energy sectors. In agriculture, Wilkinson Power Dividers facilitate equal distribution of sensor signals for precision farming, enabling efficient data collection and analysis to optimize crop management practices. In healthcare, these devices support biomedical imaging systems by splitting and directing signals for diagnostic procedures, enhancing patient care and medical research. Additionally, in the energy sector, Wilkinson Power Dividers are integral components of smart grid systems, enabling balanced power distribution and efficient energy management for sustainable infrastructure development. This adaptable nature of Wilkinson Power Dividers makes them indispensable tools across diverse fields

2. Design and Simulation

For the design of the desired Wilkinson power divider, a Quarter wave transformer line ($\lambda/4$) topology has been employed, with a characteristic impedance of 1.414Z0. Additionally, an isolation resistor of 2Zo (100Ω) is included. The Quarter wave line introduces a 90-degree phase shift, ensuring equal amplitude and phase splitting of the input signal into the two output ports, while also providing impedance matching across the input and output ports (refer to Fig. 21.1 and Fig. 21.2). The goal of the design is to achieve all S-parameters, including the reflection and isolation parameters, to be below -15 dB, and the transmission parameters within the range of -2.8 to -3.2 dB, ideally -3 dB.

The substrate design is composed of FR4 (flame-resistant) Material with the height of the substrate being 1.6 mm and the thickness of the conductor, which is copper being 0.035 mm.

3. Simulation Results and Discussion

On Simulating the design at wifi frequency (2.4 GHz), we get the following results:

Reflection loss: S_{11} = -48.389 dB; $S_{22} = S_{33}$ = -15.433 dB;

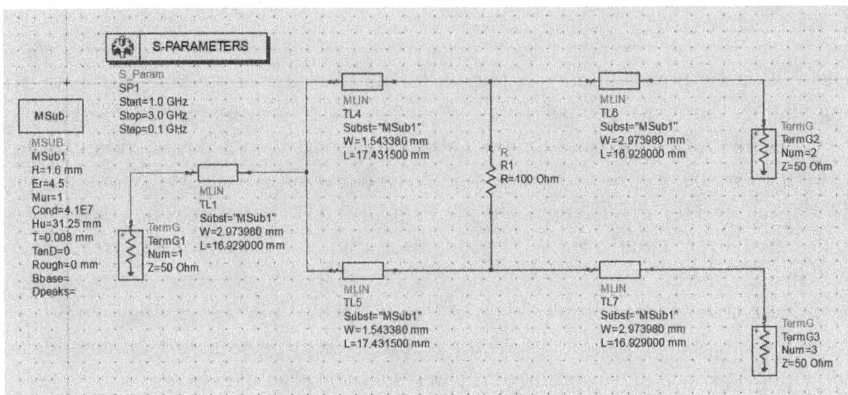

Fig. 21.1 Schematic design of equal power split wilkinson power divider

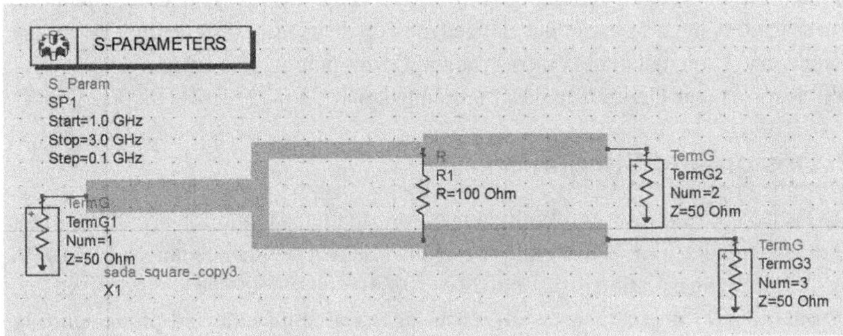

Fig. 21.2 Layout design of equal power split wilkinson power divider

Insertion loss: $S_{12} = S_{21} = S_{13} = S_{31} = -3.031$ dB;

Isolation loss: $S_{32} = S_{23} = -15.484$ dB.

In Fig. 21.3, it is evident that the values are matched and with decent isolation between the two output ports, facilitated by the presence of the isolated SMD resistor. The reflection coefficients S_{11}, $S_{22,}$ and S_{33}, as well as the isolation coefficients S_{23} and S_{32}, are all below -15 dB. The designed power divider operates in the narrowband at 2.4 GHz, with a power division ratio of 1:2. The transmission coefficients, S_{21}, S_{12}, S_{31}, and S_{13}, demonstrate almost perfect power division at -3.031 dB, falling within the accepted range of -2.8 to -3.2 dB.

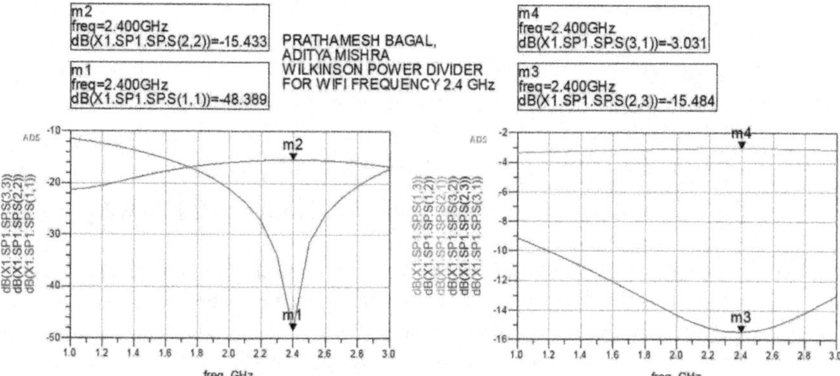

Fig. 21.3 Simulation results on ADS (Figure A. Reflection coefficients (S_{11}, S_{22}, S_{33}) and Figure B. Isolation Coefficients (S_{23} , S_{32}) and Insertion loss(S_{21}, S_{12}, S_{31}, S_{13}))

4. Conclusions

The design and simulation of the equal-split Wilkinson power divider shows outstanding performance across critical parameters. The reflection parameters S_{11}, S_{22}, and S_{33} are recorded at -48.389 dB and -15.433 dB respectively, demonstrating effective impedance matching and minimal signal reflection at the input and output ports. Additionally, the isolation parameters S_{23} and S_{32} are at -15.484 dB, well below -10 dB, indicating minimal signal leakage between ports, which is essential for maintaining signal integrity and reducing interference. Moreover, the insertion parameters S_{21}, S_{12}, S_{31}, and S_{13} are measured at -3.031 dB, falling within the acceptable range of -2.8 to -3.2 dB. This ensures balanced signal splitting across all ports. The consistent performance across these parameters confirms that the design objectives of the 3-port equal-split Wilkinson Power Divider for Wi-Fi frequencies at 2.4 GHz have been successfully met. Future advancements could involve incorporating a Discrete Ground Structure (DGS) into the ground plane of the Wilkinson power divider PCB. This addition would function as an LC filter, to prevent the unwanted harmonics from appearing. By implementing a Discrete Ground Structure, the power divider's performance could be significantly enhanced, making it more suitable for applications requiring high-frequency stability. This improvement would not only refine signal integrity but also expand the device's applicability across various fields.

Acknowledgment

Research reported in this presented work was supported by K.J. Somaiya College of Engineering, A constituent College of Somaiya Vidyavihar University.

References

1. D. Pozar, Microwave Engineering, 3rd ed. Hoboken, New Jersey: John Wiley & Sons Inc. pp. 308–361, 2005.
2. M. C. Farah and F. Salah-Belkhodja, "Design of Wilkinson Power Divider for Mobile and WLAN Applications," in Proceedings of the International Conference for Pioneering and Innovative Studies, June 5–7, 2023.
3. Prof. G. Kalpanadevi and S. J. Keerthanaa, "Design of Wilkinson Power Divider for Mobile and WLAN Applications," IOSR Journal of Electronics and Communication Engineering (IOSR-JECE), pp. 57–60, e-ISSN: 2278–2834, p-ISSN: 2278–8735.
4. Hazeri Ali Reza, "An ultra wideband Wilkinson power divider", International Journal of Electronics, vol. 99.4, pp. 575–584, 2012.
5. M. A. Maktoomi and M. S. Hashmi, "A Performance Enhanced Port Extended Dual-Band Wilkinson Power Divider", IEEE Access, vol. 5, pp. 11832–11840, 2017
6. Elftouh, Hanae & Bakkali, Moustapha & Touhami, Naima & Zakriti, Alia & Mchbal, Aicha & Dkiouak, Aziz. (2020). The Unequal Wilkinson Power Divider 2:1 for WLAN Application. Procedia Manufacturing. 46.777–781. 10.1016/j.promfg.2020.04.004
7. Faraz Ahmed Shaikh, Sheroz Khan, Ahm Zahirul Alam, Mohamed Hadi Habaebi, "Design and Analysis of 1-4 Wilkinson Power Divider for antenna array feeding network", 2018 IEEE International Conference on Ubiquitous Wireless Broadband (ICUWB), pp. 1–4, 2018.
8. Renu Kumrai, Vishal Acharya "Design and Simulation of Two way Power Divider in S Band", 2017 11th European Conference on Antennas and Propagation (EUCAP), pp. 3079–3081, 2017.

Note: All the figures in this chapter were made by the authors.

Technologies for Energy, Agriculture, and Healthcare – Shailesh Nikam et al. (eds)
© 2024 Taylor & Francis Group, London, ISBN 978-1-032-98028-7

22

CFD Analysis of Thin Wall Pump Impeller for Sustainable Manufacturing and Energy Management

Rajesh Mandale[1], Sangita Bansode[2], Ramola Sinha[3]
Research Scholar KJSCE SVU,
Associate Professor KJSCE SVU

Abstract: Complex parts of nearly any alloy can be made easily by sand casting which uses expendable sand mould. The casting defects leads to the wastage of energy which is required to melt the metal and remelt the defective part for further use. However, for effective energy utilization we can predict the most probable region for failure due to hot spots and hot tears. The non-uniform heat transfer rate and improper directional solidification is the main reason for these defects. A thin-walled pump impeller structure made up of aluminium alloy (LM6) is taken for study as its casting is difficult due to thinner wall thickness. This work opens the new avenue for the prediction of hot spots and hot tears using commercially available software ANSYS FLUENT (R2 2023). For the simulation of solidification, the solution of the transient non-linear heat transfer problem is considered. It considers the effect of wide range of pouring temperature on heat transfer rate. A case study is illustrated to explain the approach.

Keywords: Aluminium alloy (LM6), Hot spot, Hot tear, Simulation, Optimization, Ansys fluent

[1]rajesh.mandale@somaiya.edu, [2]sangeetabansode@somaiya.edu, [3]ramolasinha@somaiya.edu

DOI: 10.1201/9781003596707-22

1. Introduction

A wide range of complex structures can be manufactured by sand casting and semi open type pump impeller is one among them. Due to three directional solidification of casting all the regions do not solidify at the same rate. The cooling of casting is affected by geometry and the mould design. The guidelines and design principles are developed over many years pertaining to casting primarily through practical experience. But new analytical methods, process modelling, and computer aided design and manufacturing techniques are coming into wider use and improving productivity and quality of casting. This technique finally results in significant cost reduction and energy saving due to reduction in high rejection rate or failure during operation. Hot tears are formed in the casting due to reduced strength of metal at high temperature and differential cooling rates. Hot spot is the last solidifying point in the casting which is the most probable location for shrinkage related defects. The effect of heat transfer and fluid flow dominates each other in case of casting of thin wall structures. Hence CFD simulation of thin wall impeller is carried out under the light of process parameters such as geometrical attributes, velocity and temperature. In order to reduce the high rejection rates and low yield to avoid the wastage of energy to melt the metal and remelt the defective part for further use. The analysis of casting defects like hot spot and hot tears in a thin wall pump impeller structure is carried out in this work. The opacity of mould is the major obstacle in the study of solidification of casting. The progress of metal flow and solidification can be effectively studied by fluent simulation instead of carrying out costly conventional trial and error methods. The gating and feeding system design can be iteratively improved by simulation for giving high yield.

2. Literature Review

Simulation techniques help in visualising the mould filling and solidification of casting. It is a powerful tool to predict defects and improve the casting design thereby increasing productivity. But as far as realistic considerations are concerned experimental routes have an edge over conventional method for design and development of mould and to arrive at optimum processes parameters. It can be tracked back from the work of B. Ravi (2013) that simulation of complete casting process is a convenient way of mould design and analyse the effect of different process parameters. Nusselt number (Nu) can be used to predict the nature of flow; if it exceeds 100 then the flow of molten metal will be turbulent. Study conducted by Samson Dare et al. (2023) revealed that rapid sand casting (RSC) is an emerging field in which mould making is replaced by printing mould package as per design. Experimental validation done by Samir Chakrawarti et al.

(2023) revealed the effect of pouring velocity on air porosity. Porosity was found minimum when pouring velocity is around 5000 mm/sec and almost remained constant irrespective of pouring temperature. B. Ravi ((2008) mentioned that in most of the casting the path of molten metal in entire mould and gating system is turbulent. The major purpose of the gating system is to reduce the turbulence though it can't be eliminated. The defects in casting may be because of solid inclusion, gases entrapment and incomplete filling. Jin-Wu Ka et al. (2021) used skeletal sand mould, made by 3D printing, which are 54% lighter than conventional sand mould. Experimental case study conducted by Muhammad Aslam Khan et al. (2022) made use of lean implementation for performance improvement of casting process. The focus of study was on the use of lean framework in sand casting process using software SOLID CAST for identifying probable defects. Mark Jolly et al. (2022) made a study on modelling of defects in aluminium cast product by using commercially available software MAGAMSOFT. The different probable defects and the related reasons is mentioned in the work. Yiwei Dong et al. (2020) targeted prediction of shrinkage in hollow thin-walled casting by back propagation (BP) neural network correlating cooling rate with structural parameters. The coupling mechanism between complex structure was used to develop BP neural network using software PROCAST. The results were compared with actual casting which had shown same trends as that of simulation results. Mayur Sutaria et al. (2012) made the use of Level Set method (LSM) for capturing moving boundaries. LSM is an Eulerian computational technique. The study revealed feed path provide better visualization of feeding mechanism which plays major role in quality and yield optimization in industries.

Findings of literature review are summarised as; Numerical simulation of casting is useful in improving quality and minimizing product cost and scrap. The following rules may promote directional solidification and reduces shrinkage stresses and distortion. Casting walls shall not have any abrupt changes but be connected by smooth transitions. Transition between casting walls of different thickness must be smooth. Avoid local metal accumulation and massive elements. All these points will help in producing casting right at first time thereby enhancing the productivity and avoid undue expenditure of energy in producing defective castings. However above-mentioned studies do not shed any light on avoiding hot spots and hot tears in casting of thin wall impeller structures. As observed from literature survey, very less work has been done in this area. Therefore, first simulation work is carried out to observe heat flux, wall shear stress, Nusselt number, heat transfer coefficient. These heat transfer parameters are related to hot spots and hot tears therefore the simulation results will be validated experimentally in near future.

3. Research Methodology

Simulation experiments are conducted to study solidification of thin wall impeller structure of a semi open type pump impeller used in medium capacity hydraulic pumps. An impeller is a driven rotor used to increase the pressure and flow of a fluid. The selected impeller consists of six thin wall blades with a base plate. Foundry visits were arranged at pump impeller manufacturers at Anil foundry works Dombivali (E) and Kiran Founders, Ghatkopar(W), for insight and to get the input to carry out simulation experiments. Discussion was useful to understand the most probable locations of hotspots and hot tears usually encountered in the casting of thin wall structures. It was observed in both the foundries that the most probable location of these defects is in the thin wall blade section. These finding are in perfect agreement with the findings of literature review as mentioned by B. Ravi (2013). However, the scope of this work is limited to simulation of casting of thin wall impeller in ANSYS FLUENT (R2 2023) to study the directional solidification and predict the most probable locations of hot spots and hot tears. Simulation results reduce number of trial casting and helps in focusing scope of experimental work. Validation of simulation results experimentally is essential and is the future scope of this work. In this research work, a CAD model of the impeller is prepared as shown in Fig. 22.1. As the geometry and loading are symmetric about the same plane thus in axisymmetric component like impeller, simulation can be carried out by considering a portion of the actual structure. Half geometry of model is considered for simulation as it reduces memory requirement and a simulation run time. In symmetrical model finer mesh can be used resulting more accurate solution. The mesh model of the geometry is as shown in Fig. 22.2.

Fig. 22.1 CAD model of impeller using solid works (Tip diameter 297 mm, hub diameter of 45 mm, blade thickness 15 mm)

The impeller used in medium capacity pump has backward curved vanes with exit blade angle 22^0. It is capable of generating a head of 40 m. The material aluminium alloy (LM6), used for impeller exhibits exclusive corrosion resistance, and super castability property. Aluminium LM6 has mainly 10-13% Si and 0.6 % Fe, 0.5% Mn, 0.2 % Tn with balance of aluminium. In this simulation work, properties of material LM 6 has been used as input parameter as shown in Table 22.1.

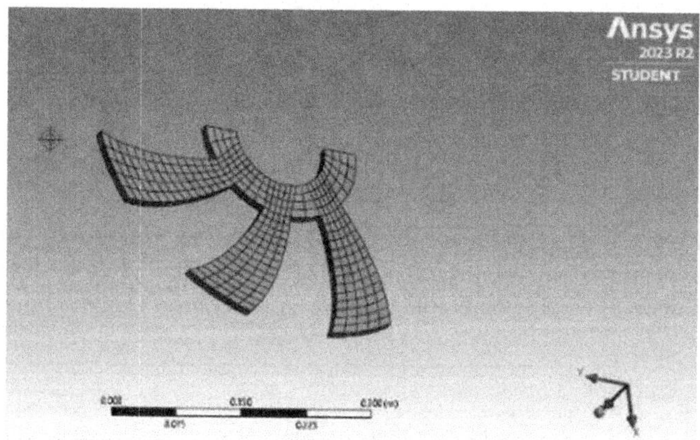

Fig. 22.2 Half geometry used for Mesh Model

Table 22.1 Thermo-physical properties of aluminium alloy LM6

Sr. No	Property	Symbol	Value
1	Specific heat	Cp	896 J/Kg. K
2	Thermal conductivity	K	204.2 W/m. K
3 4	Viscosity Density	μ ρ	0.00553 Kg/m-s 2707 Kg/m^3

Different combinations of pouring temperature are taken by keeping pouring velocity, and the blade thickness constant as given in Table 22.2. Pouring temperature varies between1000-1200K, thickness is 0.015 m and velocity is 0.002 m/s. The melting point of aluminium alloy (LM6) varies between 900K to 953K depending on its composition. Temperature 1000K is slightly above melting point of LM6 but below its fusing temperature1250K, hence higher temperature is kept 1200K for simulation.

Table 22.2 Design of simulation experiments

Simulation Experiment No.	Pouring Temperature (K)	Blade Thickness. (m)	Inlet Velocity (m/s)
1	1000	0.015	0.002
2	1100	0.015	0.002
3	1200	0.015	0.002

Continuity equation, momentum equation and energy equation are the three pillars of CFD simulation.

Conservation of mass (continuity equation)

$$\frac{\partial}{\partial t}(\alpha_q \rho_q + \dot{\nabla} \alpha_q \rho_q \overrightarrow{v_q}) = \sum_{p=1}^{n} (\dot{m}_{pq} - \dot{m}_{qp}) + S_q \tag{1}$$

If continuity equation is satisfied, then flow is possible. Continuity equation in CFD plays an important role to ensure the stability, accuracy, physical validity of simulation by enforcing principle of mass conservation trough computational domain.

Momentum equation

$$\sum_{p=1}^{n} \overrightarrow{R_{pq}} = \sum_{p=1}^{n} K_{pq} (\overrightarrow{v_p} - \overrightarrow{v_q}) \tag{2}$$

Momentum equation can give the idea regarding evolution of velocity flow field under the action of various forces

Conservation of energy equation

$$\frac{\partial}{\partial t}\left(\alpha_q \rho_q h_q + \dot{\nabla} \alpha_q \rho_q \overrightarrow{u_q} h_q \right) = \alpha_q \frac{\partial p_q}{\partial t} + \overline{\overline{T}}_q : \nabla \overrightarrow{u_q} - \nabla \overrightarrow{q_q} + S_q +$$

$$\sum_{p=1}^{n} (Q_{pq} \dot{m}_{pq} h_{pq} - \dot{m}_{qp} h_{qp}) \tag{3}$$

Energy equation is capable of simulating the heat transfer mechanism as well as the temperature distribution in the domain. It is subjected to Dirichlet boundary conditions with minimum cooling temperature 300 K.

4. Results and Discussion

CFD analysis of selected impeller blade has been carried out by varying pouring temperature during casting by keeping thickness of the blade and inlet velocity constant. Simulation results obtained are shown in Table 22.3.

Table 22.3 Simulation results showing impact of change in temperature

Simulation Experiment No.	Heat Transfer Coefficient (W/m^{2}K)	Specific Enthalpy (J/Kg)	Heat Flux (W/m^2)	Nusselt Number Nu	Wall Shear Stress (Pascal)
1	0.05303	102504.9	570.3463	2.194	763.334
2	4.342	459026.3	3729.62	4.512	744.07
3	4.70042	813785.8	3992.535	12.44	744.25

The high value of shear stress may reduce the strength of thin blade wall and leads to formation of hot spots and hot tears. The blade may fail the quality test, or it may fail during operation. Increased stresses may cause casting to wrap and develops crack in the casting. In spite of this, uniformity of temperature distribution is noticed in counters of temperature in all experiments. The values of Nu for experiments are below 100 indicating absence of turbulence in molten metal resulting in comparatively low shear stress. Figure 22.3 depicts related sample contours of wall shear stress for given condition in Experiment 1, this indicates even distribution of shear stress. Temperature contours shown in Fig. 22.4 indicates even distribution of temperature. Contours of scaled residuals for experiment 1 indicated no fluctuations in velocity and they are vanished which is the reason for low momentum fluctuation and reduced shear stress. For this type, temperature distribution is uniform due to uniform cooling of casting. Hence

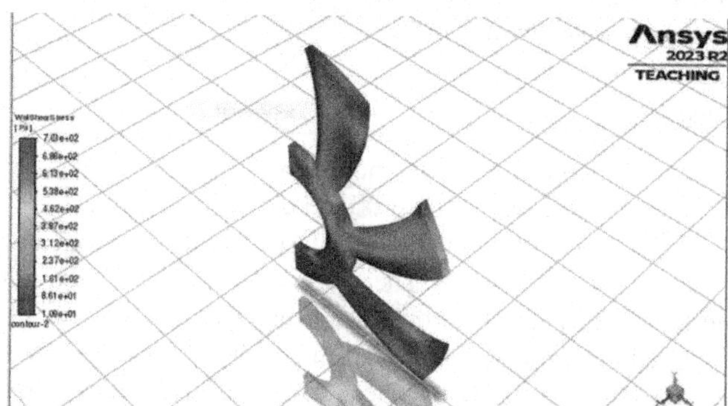

Fig. 22.3 Wall shear stress contours for Experiment 1

Fig. 22.4 Temperature contours for Experiment 1

experiment 1, with reduced shear stress is selected as proper combination to be recommended for production of casting avoiding hot spots and hot tears. Formation of hot spots is mainly due to faster cooling of casting area than surrounding area. They can be avoided by proper cooling of casting. But these regions of higher stress are most susceptible to formation of hot spots hence such regions shall fall in the pouring basin or in risers which will be chipped of in the end.

Another simulation approach using casting solidification software is carried out. In this work probable location of hot spots has been obtained by using Autocast software from E foundry IIT Mumbai portal as shown in Fig. 22.5. The hot spots are indicated by yellow color located at hub as well as blades. However, the hub portion at the center will be removed by machining.

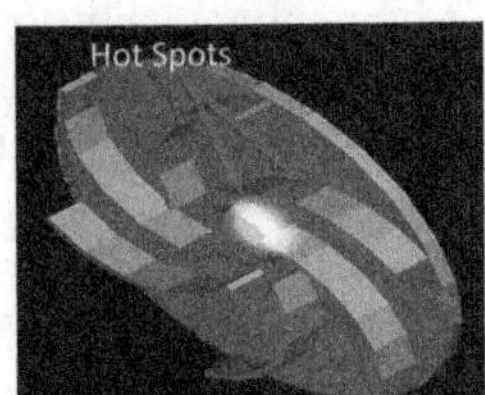

Fig. 22.5 Probable location of hot spots defect using Auto cast software

5. Conclusions

It can be concluded from results as shown in Table 22.3 that when pouring temperature of metal increases from 1000-1200K for given blade thickness and assumed feed velocity, heat transfer coefficient, specific heat, heat flux and Nusselt number increases. However, the wall shear stress decreases slightly with increase in temperature. The correct selection of pouring temperature, for given metal pouring velocity and thickness of thin wall impeller structure plays an important role in producing defect free casting. These parameters, if selected properly, will play an important role in distribution of shear stress and temperature in the thin blade and predict the possible region for failure due to hot spot and hot tears. Finally, it results in minimizing the wastage of energy in producing a defect free casting. These results can be used as an input for design of feed path, placement of runner and risers.

Acknowledgement

We sincerely acknowledgeAnil foundry works, Dombivali (E) and Kiran Foundars, Ghatkopar (W) for providing access to their foundry infrastructure and required necessary details for conducting research work.

References

1. B. Ravi, Casting Design and Simulation Software: 25-Year Perspective and Future", Transactions of 61st Indian Foundary Congress 2013.

2. Samson Dare, Oguntuyi, Kasongo Nyembwe. Challenges and recent progress on the application of rapid sand casting for part production: a review" The International Journal of Advanced Manufacturing Technology(2023) 126:891–906. https://doi.org/10.1007/s00170-023-11049-1
3. Samir Chakravarti, Swarnendu Sen. An investigation on the solidification and porosity prediction in aluminium casting process" Journal of Engineering and Applied Science (2023) 70:21. https://doi.org/10.1186/s44147-023-00190-z
4. Jin-wu Ka , Hao-long Shangguan.Cooling. Control for Castings by Adopting Skeletal Sand MouldDesign.China Foundary Research and Development. Vol.18 No.1 (January 2021). https://doi.org/10.1007/s41230-021-0150-7
5. Muhhamad Aslam Khan, Muhhamad Khurram Ali (2022) Muhhamad Sajid. Lean Implementation Framework: A Case of Performance Improvement of Casting Process. IEEE Access.Digital Object Identifier 10.1109/ACCESS.2022.3194064
6. Mark Jolly, Laurens Katgerman .Modelling of defects in aluminium cast products. Journal of Progress in Materials Science. 123 (2022) 100824
7. Yiwei Dong,Weiguo Yan, Zongpu Wu Modelling of shrinkage characteristics during investment casting for typical structures of hollow turbine blades.(2020). The International Journal of Advanced Manufacturing Technology.110:1249–126. https://doi.org/10.1007/s00170-020-05861-2
8. Mayur Sutaria, Vinesh H. Gadaa, Atul Sharma, Ravi B., "Computation of feed paths for casting solidification using level-set-method" Journal of Materials Processing Technology 212 (2012) 1236–1249.

Note: All the figures and tables in this chapter were made by the authors.

Technologies for Energy, Agriculture, and Healthcare – Shailesh Nikam et al. (eds)
© 2024 Taylor & Francis Group, London, ISBN 978-1-032-98028-7

23

Comparative Analysis and Ensemble-Driven Improvement in Depression Detection on Reddit Posts

Aditya Pai[1],
Prathamesh Powar[2], Rohan Sharma[3]
Students, KJSCE IT
Snigdha Bangal[4], Leena Sahu[5]
Assistant Professors, KJSCE IT

Abstract: Depression is a prevalent mental health issue that impacts a vast number of people worldwide. Early detection and intervention are essential for effective therapy, and social media data has become an important resource for identifying early signs of depression. In this work, we assess multiple machine learning techniques and introduce an improved method for Reddit anonymous text post sadness detection. We evaluate the performance of each method using a carefully selected dataset based on relevant research. We find that, with an accuracy of 96.57%, the stacked ensemble technique outperforms individual algorithms. Moreover, we demonstrate the effectiveness of merging several algorithms, making a major contribution to the field of early depression identification.

Keywords: Depression detection, Ensemble learning, Reddit posts, Linguistic patterns, Sentiment analysis

[1]pai.a@somaiya.edu, [2]prathamesh.powar@somaiya.edu, [3]rohan20@somaiya.edu, [4]snigdha.b@somaiya.edu, [5]l.sahu@somaiya.edu

DOI: 10.1201/9781003596707-23

1. Introduction

Depressive illnesses have a major effect on the world at large, affecting not just emotional stability but also cognitive performance and general well-being. The accuracy limitations of traditional diagnostic tools highlight how crucial it is to recognize problems early and take appropriate action. But in the middle of all of these difficulties, a glimmer of optimism appears from the enormous social media pool.

Overall wellbeing, cognitive performance, and emotional resilience are all significantly impacted by depressive illnesses. But conventional diagnostic techniques suffer from accuracy issues, which emphasizes the importance of prompt detection and action. Social media data has been a potentially useful resource for investigating algorithmic identification of depression indications in recent times. In instance, Reddit provides an enormous amount of textual data that has been anonymized and might be utilized to identify sadness.

In this work, we investigate the effectiveness of a stacked ensemble learning method for identifying depressive-symptomatic comments on Reddit articles. In order to interpret the complex linguistic patterns linked to depressive expressions, our work makes use of the advantages of eight different base learners: RF, AdaBoost, Decision Tree, KNN, MLP, SVM, and Logistic Regression.

Classifiers are:

Naive Bayes: A classifier, based on the probabilistic assumption of independence of features that is effectively used when working with text classification tasks.

Logistic Regression: Binary classification model using statistical method chosen due to its interpretability and applicability in distinguishing depressed and non-depressed people.

ADA Boost: Ensemble learning method wherein performance of the model is iterated and fine-tuned over misclassified data which is known for its robustness and resistance to overfitting.

Decision Tree: A hierarchical model that reflects intricate connections in data and gives insight into decision making processes.

K Nearest Neighbors: A flexible nonparametric similarity measure between individuals that adapt to different kinds of data distributions.

Multi-Layer Perceptron (MLP): An artificial neural network well suited to learning complex patterns, a powerful tool capable of recognizing subtle and non-linear relations.

Support Vector Machine (SVM): A general model utilizing hyperplanes for sorting data whether either linear or non-linear relationships are present.

Random Forest: An ensemble approach constructing multiple decision trees, known for enhanced accuracy and robustness in handling noisy data.

Our solution is designed with teens and young adults in mind, taking into account their prevailing social media usage habits. In addition, the COVID-19 epidemic, which resulted in quarantine restrictions for individuals, emphasizes the value of social media as a platform for expressing annoyance and obtaining assistance.

2. Literature Survey

Tao et al. , introduced an innovative ensemble binary classifier using Quality of Life (QoL) scales to improve machine learning's performance in identifying depressed cases, achieving a high F1 score (0.976) and low misclassification rate (4%). This ensemble approach leverages diverse data processing and modeling techniques to effectively capture depression risk factors.

Safa et al. demonstrate the convergence of mental health analysis, social media, and machine learning, exploring predictive approaches for mental disorders from online content with a focus on auto-detection frameworks, highlighting challenges in integrating social media data with surveys.

Avinash et al. focus on optimizing feature selection methods for sentiment analysis, comparing their efficacy against machine learning classifiers and ensemble methods. It investigates characteristic word extraction, word co-occurrence analysis, and employs classifiers like SVM and Naive Bayes for sentiment classification in product reviews. Ensemble techniques, including Voting and Bagging, are introduced to enhance feature extraction, aiming to improve accuracy and reliability in sentiment analysis classifiers on benchmark datasets.

Currently, Murarka et al. have increasingly been utilizing social media data for mental health exploration, initially focusing on Twitter and later shifting attention to Reddit. Various techniques, including classical NLP methods, regression models, neural MTL models, and CNNs, have been employed in prior studies to analyze language patterns and sentiments.

Jingfang et al. propose a hybrid machine learning approach to detect depression in individuals through social media content. By combining wrapper, filter, and embedded methods, along with base learners like naive Bayes and support vector machines,it achieves a 90.27% accuracy in identifying users with depression. These results suggest the hybrid method's potential to enhance the accuracy of depression detection based on social media activities.

M. M. Tadesse et al. explore the relationship between mental health and language expression, identifying patterns in individuals with depression, including increased use of negative words, first-person pronouns, and simpler sentence structures.

Utilizing social media platforms like Facebook, Reddit, and Twitter, the study highlights the CLEF eRisk competition's focus on early detection of depression or anorexia through Reddit data, employing various machine learning techniques and feature engineering for improved accuracy.

R. Qayyum et al. present a novel framework for diagnosing mental health issues through social media text analysis, employing hierarchical classification with RoBERTa and comparing it with baseline deep learning approaches.

Hong-Han Shuai and colleagues present a two-phase framework, Social Network Mental Disorder Detection, employing binary SVMs for identifying Cyber-Relationship Addiction, Net Compulsion, and Information Overload. The approach attains a commendable 90% accuracy by tackling feature extraction challenges and integrating multi-source data from diverse online social networks

3. Dataset and Methodology

In conducting this study, a carefully curated Reddit dataset [9] was employed, with a specific emphasis on research related to depression detection. This dataset comprises anonymized user comments, each labeled as either indicative of depression or not. The different features included in this dataset are Reddit posts, LIWC features and number of negative words. The labeling process is informed by well-defined criteria, ensuring a robust and accurate representation of the mental health context within the dataset. The dataset enables the evaluation of these models in capturing the linguistic expressions associated with depression, thereby enhancing the study's accuracy in understanding mental health cues within social media discourse.

The proposed approach as seen in Fig. 23.1 employs a piled ensemble literacy frame, which uses the collaborative method of multiple base learners to achieve superior overall delicacy. The study incorporates 8 different machine learning algorithms as base learners.

This above diagram shown in Fig. 23.1 outlines the process of creating a stacking classifier for classification. The process begins with data loading and preprocessing, followed by data splitting for training and testing.Although we initially engaged in feature selection for the dataset, the absence of noticeable improvements led us to discard this approach. Consequently, we proceeded with the stacking classifier to enhance our model's performance. Text vectorization is then performed before the data is classified using various models such as Naive Bayes, Logistic Regression, ADA Boost, Decision Tree, KNN (k=3), Multi-Layer Perceptron, SVM, and Random Forest. The results are then displayed and evaluated using the stacking classifier.

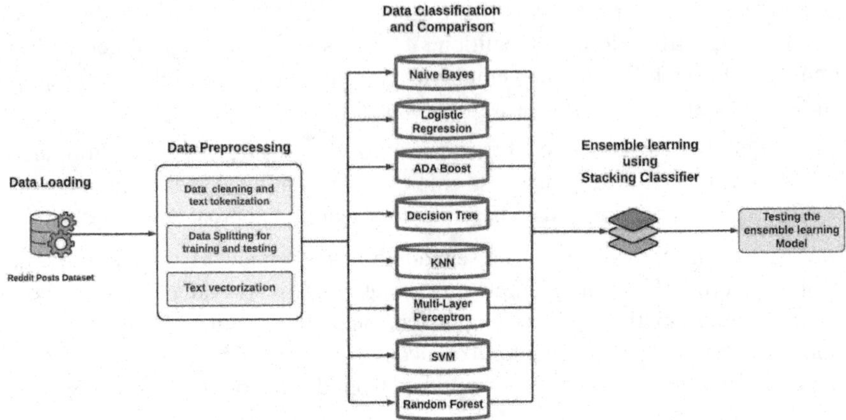

Fig. 23.1 Process flow diagram

Stacking is an **ensemble learning technique** that combines diverse base models with a meta-model to optimize the combination of their predictions. Experimenting with various base model combinations and meta-models enhances performance. It involves three key components:

Base Models: Diverse individual models, such as decision trees, support vector machines, or neural networks, making predictions on input data.

Meta-Model: Also known as a combiner or blender, it takes predictions from base models as input features and produces the final prediction using methods like logistic regression or linear regression.

Training Process: It comprises a two-stage process. First, base models are trained on the initial dataset, and then their predictions serve as input for the meta-model, which is trained on the same dataset.

In this, each base learner specializes in different aspects of the underpinning data patterns, contributing to a further comprehensive understanding of the complex connections within the dataset. The addition of different algorithms, like decision trees, SVM, enhances the model's capability to generalize well on different datasets. Through the strategic combination of these varied models, the piled ensemble leverages their individual strengths hence forming a high-performance system.

4. Performance Parameters

Accuracy: Accuracy measures the correct predictions out of total predictions performed

$$\text{Accuracy} = \frac{\text{Correct Predictions}}{\text{Total Predictions}} \tag{1}$$

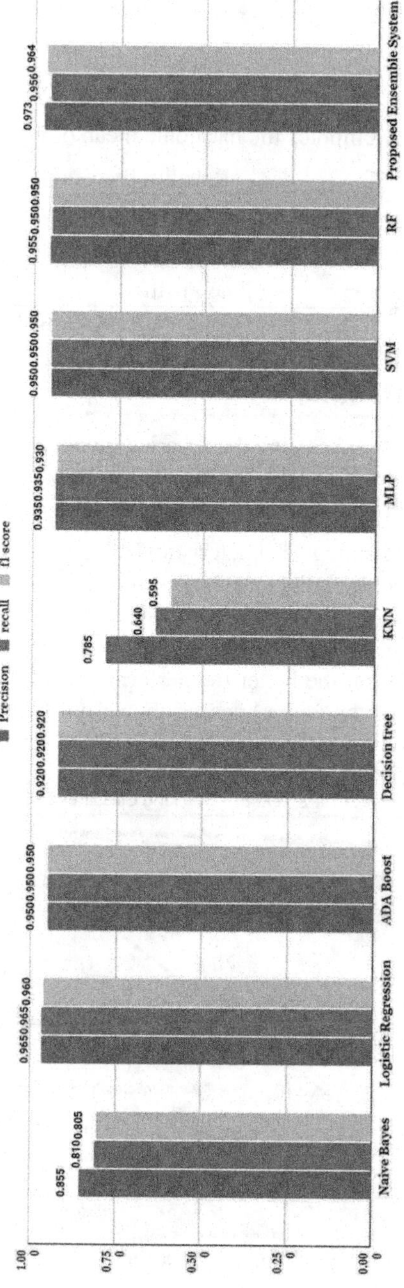

Fig. 23.2 Comparison of the performance metrics of all the algorithms

Precision: Precision measures the capability of a system to produce only relevant results.

$$\text{Precision} = \frac{\text{True Positives}}{\text{True Positives} + \text{False Positives}} \tag{2}$$

F-Measure: F-Measure computes the harmonic mean of precision and recall.

$$\text{F} - \text{Measure} = 2 * \frac{\text{Recall} * \text{Precision}}{\text{Recall} + \text{Precision}} \tag{3}$$

Recall: Recall is the ability of a system to make correct positive predictions.

$$\text{TPR} = \frac{\text{True Positives}}{\text{True Positives} + \text{False Negatives}} \tag{4}$$

5. Experimental Results

The Stacking Classifier, synthesizing these varied models, significantly enhanced overall predictive capabilities. This collective strength signifies a substantial advancement in the field of depression detection through social media analysis.

The ROC (Receiver Operating Characteristics) Plot focuses on the classifier's ability to differentiate between the classes. A model's performance is determined by the trade-off between an FPR (False Positive Rate) on the X-axis and a TPR (True Positive Rate) on the Y-axis. The closer the curve of a model is to the upper left corner of the ROC plot, the better the performance of the model. According to the AUC (Area Under the Curve) scores shown on the plot in Fig. 23.3, the Stacking Classifier performs the best, followed by Logistic Regression and SVM.

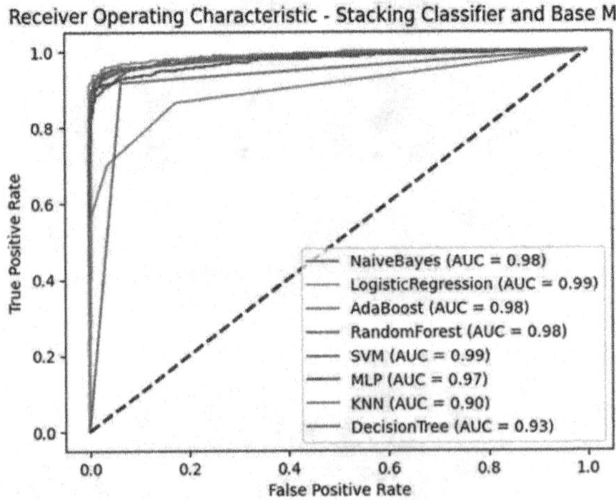

Fig. 23.3 Algorithm performance ranking: AUC scores on ROC plot

The Kappa statistic is a strong indicator of inter-rater agreement and thus provides an excellent testimony to the reliability of our stacked ensemble method. As seen in the Fig. 23.4, 0.93 Score indicating near-perfect alignment between what the model predicts and expert annotations shows that it can make reliable, reliant outcomes.

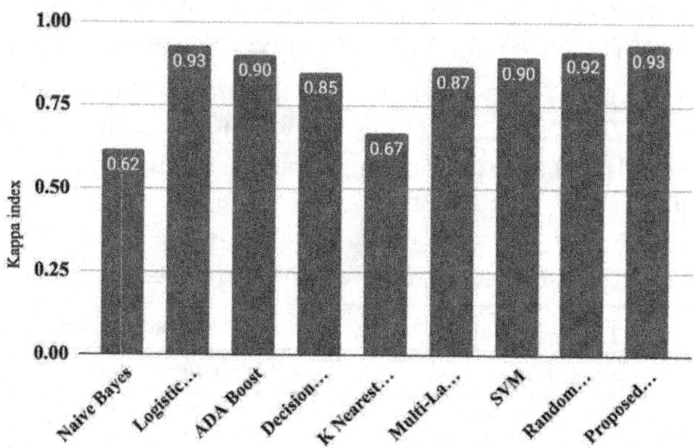

Fig. 23.4 Kappa statistic comparison for All algorithms

Table 23.1 Algorithm performance spectrum across F1-score, precision, and recall

Classifier Performance Comparison			
	Precision	**Recall**	**F1 score**
Naive Bayes	0.855	0.810	0.805
Logistic Regression	0.965	0.965	0.960
ADA Boost	0.950	0.950	0.950
Decision tree	0.920	0.920	0.920
KNN	0.785	0.640	0.595
MLP	0.935	0.935	0.930
SVM	0.950	0.950	0.950
RF	0.955	0.950	0.950
Proposed Ensemble System	**0.973**	**0.956**	**0.964**

The evaluation of our depression detection models yielded insightful metrics, detailed in Table 23.1, focusing on precision, recall, and F1 score for each classifier. The Stacking Classifier emerged as the top performer, boasting a precision of 97.3%, a recall of 95.6%, and an impressive overall F1 score of 96.4%. This ensemble method, integrating eight diverse classifiers, demonstrated collective strength in effectively identifying depression indicators within Reddit post comments.

Our ensemble technique effectively harnessed the unique strengths of individual base learners, including Logistic Regression, ADA Boost, SVM, and Random Forest. This resulted in exceptional F1 scores consistently surpassing 93%. Notably, Logistic Regression stood out with a precision and recall of 96.5%, highlighting its effectiveness in discerning depressed content.

With the help of Table 23.2 we can see that the proposed ensemble system has better accuracy of 96.57% than the present systems as it offers a comparison of the different models and their performance from the existing systems as presented in those research papers.

Table 23.2 Comparison of different models based on its accuracy

Study	Dataset Used	Methods Used	Accuracy (%)
Tao et al. [1]	NHANES dataset	Ensemble classifier	95.4
Murarka, Ankit et al. [4]	Reddit Posts and title dataset form 13 subreddits	BERT and ROBERTA classifiers	85
Jingfang Liu et al. [5]	Textual Dataset with depressed and non-depressed users	Naive Bayes with SVM	90.27
R. Qayyum et al. [7]	Reddit Posts dataset	BERT and ROBERTA transformer models	84
Proposed Ensemble Approach	Reddit Posts dataset	Stacking Classifier	96.57

These findings underscore the potential of ensemble learning and the strategic combination of diverse algorithms to elevate the accuracy and reliability of depression detection models. The robust individual performances of base learners, combined with the collective strength of the ensemble, provide a promising foundation for future research in mental health screening through social media analysis.

6. Conclusion and Future Scope

Our novel stacked ensemble learning approach surpasses previous benchmarks, achieving a remarkable 97% accuracy in classifying Reddit post comments indicative of depression. This success lies in the synergy of diverse learners. Each base learner, from Naive Bayes to Random Forest, excels at capturing distinct linguistic patterns of depression expression. The stacking architecture then mitigates the limitations of individual models. While a single learner might struggle with outliers or complex data interactions, the ensemble leverages collective wisdom to overcome these challenges. The final model learns from the base learners' mistakes, enhancing its generalizability and ability to adapt to subtle textual features.

This multi-layered architecture fosters robustness. Prediction errors from one learner don't propagate through the system, preventing cascading failures. This inherent redundancy is crucial for the sensitive nature of mental health assessment.

These promising results unlock exciting avenues for utilizing social media data in mental health understanding:

Enriched Data Fusion: Integrating user profiles, posting behavior, and social interactions alongside Reddit comments could further enhance accuracy and completeness.

Interpretable Models: While the stacked ensemble achieves high accuracy, delving into its internal workings is crucial. Employing interpretability techniques can reveal the linguistic markers and patterns associated with depression expression in social media data.

By pursuing these research directions, we can advance the field of mental health screening through social media, paving the way for effective tools for early depression detection and intervention, ultimately benefiting both individuals and society as a whole.

References

1. Tao, X., Chi, O., Delaney, P.J. et al. Detecting depression using an ensemble classifier based on Quality of Life scales. Brain Inf. 8, 2 (2021).
2. Safa, Ramin & Edalatpanah, S A. & Sorourkhah, Ali. (2023). Predicting mental health using social media: A roadmap for future development.
3. Madasu, Avinash & Elango, Sivasankar. (2020). Efficient feature selection techniques for sentiment analysis. Multimedia Tools and Applications. 79. 1–23. 10.1007/s11042-019-08409-z.
4. Murarka, Ankit & Radhakrishnan, Balaji & Ravichandran, Sushma. (2020). Detection and Classification of mental illnesses on social media using RoBERTa.
5. Liu, Jingfang & Shi, Mengshi. (2022). A Hybrid Feature Selection and Ensemble Approach to Identify Depressed Users in Online Social Media. Frontiers in Psychology. 12. 10.3389/fpsyg.2021.802821
6. M. M. Tadesse, H. Lin, B. Xu and L. Yang, "Detection of Depression-Related Posts in Reddit Social Media Forum," in IEEE Access, vol. 7, pp. 44883–44893, 2019, doi: 10.1109/ACCESS.2019.2909180.
7. R. Qayyum, H. Afzal, K. Mahmood and N. Iltaf, "Detection and Analysis of Mental Health Illness using Social Media," 2023 International Conference on

Communication Technologies (ComTech), Rawalpindi, Pakistan, 2023, pp. 34–41, doi: 10.1109/ComTech57708.2023.10165143.

8. H. -H. Shuai et al., "A Comprehensive Study on Social Network Mental Disorders Detection via Online Social Media Mining," in IEEE Transactions on Knowledge and Data Engineering, vol. 30, no. 7, pp. 1212–1225, 1 July 2018, doi: 10.1109/TKDE.2017.2786695.
9. Low, D. M., Rumker, L., Torous, J., Cecchi, G., Ghosh, S. S., & Talkar, T. (2020). Natural Language Processing Reveals Vulnerable Mental Health Support Groups and Heightened Health Anxiety on Reddit During COVID-19: Observational Study. Journal of medical Internet research, 22(10), e22635.
10. Babu, N.V., Kanaga, E.G.M. Sentiment Analysis in Social Media Data for Depression Detection Using Artificial Intelligence: A Review. SN COMPUT. SCI. 3, 74 (2022).
11. Sumathi, Ms & B., Dr. (2016). Prediction of Mental Health Problems Among Children Using Machine Learning Techniques. International Journal of Advanced Computer Science and Applications. 7. 10.14569/IJACSA.2016.070176.
12. Adrian Benton, Margaret Mitchell, and Dirk Hovy. 2017. Multi-task learning for mental health using social media text.
13. J. Wolohan, M. Hiraga, A. Mukherjee, Z. A. Sayyed, and M. Millard, "Detecting linguistic traces of depression in topic-restricted text: Attending to self-stigmatized depression with nlp," in Proc. 1st Int. Workshop Lang. Cognition Comput. Models, 2018, pp. 11–21.
14. Y. Tyshchenko, "Depression and anxiety detection from blog posts data," Nature Precis. Sci., Inst. Comput. Sci., Univ. Tartu, Tartu, Estonia, 2018.
15. S. C. Guntuku, D. B. Yaden, M. L. Kern, L. H. Ungar, and J. C. Eichstaedt, "Detecting depression and mental illness on social media: An integrative review," Current Opinion Behav. Sci., vol. 18, pp. 43–49, Dec. 2017.
16. R. A. Calvo, D. N. Milne, M. S. Hussain, and H. Christensen, "Natural language processing in mental health applications using non-clinical texts," Natural Language Eng., vol. 23, no. 5, pp. 649–685, 2017
17. M. Cao and Z. Wan, "Psychological Counseling and Character Analysis Algorithm Based on Image Emotion," in IEEE Access, doi: 10.1109/ACCESS.2020.3020236.
18. A. Benton, M. Mitchell, and D. Hovy. (2017). "Multi-task learning for mental health using social media text." [Online]. Available: https://arxiv.org/abs/1712.03538
19. Fidel Cacheda et al. "Analysis and Experiments on Early Detection of Depression." In: CLEF (Working Notes) 2125 (2018).
20. Sijia Wen. "Detecting Depression from Tweets with Neural Language Processing". In: Journal of Physics: Conference Series. Vol. 1792. 1. IOP Publishing. 2021, p. 012058.
21. Anu Priya, Shruti Garg, and Neha Prerna Tigga. "Predicting anxiety, depression and stress in modern life using machine learning algorithms". In: Procedia Computer Science 167 (2020), pp. 1258–126

Note: All the figures and tables in this chapter were made by the authors.

Technologies for Energy, Agriculture, and Healthcare – Shailesh Nikam et al. (eds)
© 2024 Taylor & Francis Group, London, ISBN 978-1-032-98028-7

24

Numerical Investigation on the Effect of Inclined Delta Flow Obstructions on the Thermo-hydraulic Performance of an Air Heater

Bronin Cyriac[1]
Research Scholar,
Department of Mechanical Engineering,
K. J. Somaiya College of Engineering,
Mumbai

Akash Bidwaik[2]
Research Scholar,
Department of Mechanical Engineering,
K. J. Somaiya College of Engineering,
Somaiya Vidyavihar University,
Mumbai

Siddappa S. Bhusnoor[3]
Professor,
Department of Mechanical Engineering,
K. J. Somaiya College of Engineering,
Somaiya Vidyavihar University,
Mumbai

Abstract: The efficiency of the air heater (AH) is often low because of the poor heat transfer properties of air. This paper numerically investigates the impact of inclined delta flow obstructions (IDFO) on the thermal and hydraulic performance of AH using Computational Fluid Dynamics (CFD). The CFD model used for the study is validated by comparing the CFD results for AH without obstructions with

[1]bronin.c@somaiya.edu, [2]akash.bidwaik@somaiya.edu, [3]siddappabhusnoor@somaiya.edu

DOI: 10.1201/9781003596707-24

the results obtained from the standard empirical correlations. The results show that the IDFO significantly improves the performance of air heaters compared to non-inclined delta obstructions (NIDFO). It was observed from the results that the use of IDFO results in a Nu enhancement of 2.42 - 3.33 with a corresponding f augmentation of 4.65-6.72 compared to an AH without obstruction. This results in a improved Thermal Enhancement Factor (TEF) ranging from 1.35 to 1.57, signifying a significant improvement in AH efficiency.

Keywords: Air heater, Thermo-hydraulic performance, Convective heat transfer, Friction factor, Thermal enhancement factor (TEF)

1. Introduction

The search for novel and effective methods to conserve energy has escalated recently due to the rising demand for power and the depletion of conventional energy sources (Saxena, Varun, and El-Sebaii 2015). AH is an essential component in solar energy collection systems. AH is widely used in numerous applications like drying agricultural products, space heating, seasoning wood, and curing industrial products (Yadav and Thapak 2016). The major drawback of traditional AH is its poor heat transfer performance, mainly due to the low heat transfer coefficient of air (Cyriac and Bhusnoor 2023a). Researchers have suggested many solutions to improve the efficiency of AH systems, such as using flow obstructions or obstacles, jet impingement, surface modifications, fluid additives, and extended surfaces (Acharya and Pise 2017; Yu et al. 2011; Cyriac and Bhusnoor 2023b; Cyriac Bronin and Bhusnoor 2023).

Various studies have examined how flow obstructions affect the thermal and hydraulic characteristics of an AH. Kumar and Layek 2022 examined the efficacy of winglet-shaped ribs in improving the performance of AH. Baissi et al. 2019 investigated the use of longitudinally-curved delta obstruction to improve the performance of AH. Their study shows that the curvature configuration improved the thermal and hydraulic characteristics of the AH. Jain et al. 2019 proposed fractured inclined ribs for the performance improvement of AH. Their research demonstrated that the configuration of slanted ribs reduces frictional losses, reducing the overall energy consumption in the AH. Promvonge et al. 2011 examined the simultaneous use of rib and delta winglet geometries and the impact of these arrangements on Nu, f, and the TEF. Ebrahim Momin, Saini, and Solanki 2002 used ribs (V type) to improve the efficiency of AH.

The influence of obstruction inclination of a delta flow obstruction on AH performance is rarely studied. This paper numerically investigates the effect of IDFO on AH performance. The introduction of obstructions in the air duct aims to induce transverse and longitudinal vortices, enhancing heat transfer by promoting better mixing and turbulence in the air. The ensuing disruption of the stagnant boundary layer and increased heat transfer coefficient contribute to more efficient heat exchange, addressing the limitations associated with conventional air heaters. By enhancing heat transfer, IDFOs have the potential to significantly reduce energy consumption in AH systems, contributing to advancements in energy conservation for agricultural applications. This can lead to increased efficiency and potentially higher crop yields by enabling more precise control over drying processes and improved space heating in greenhouses and storage facilities.

2. Materials and Methods

A rectangular channel of 1200 mm × 300 mm × 50 mm, as shown in Fig. 24.1, was considered for performance assessment. A computational domain featuring a single row of obstructions in the longitudinal (P_l) of 100 mm was considered for the flow modelling of AH with obstructions. The relative obstruction longitudinal pitch (height of obstruction to the longitudinal pitch of obstruction), $P_l/e = 4$, and relative obstruction transverse pitch (transverse pitch of obstruction to the base width of obstruction), $P_t/b = 2$, relative obstruction height (obstruction height to the height of AH), $e/H = 0.5$ were kept constants for all the geometries. IDFO considered for this study is inclined 45^0 to the inlet, as shown in Fig. 24.2.

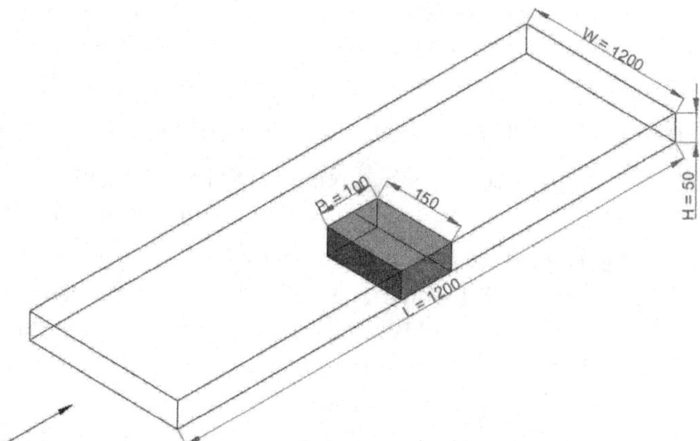

Fig. 24.1 Rectangular channel along with the computational domain used for the numerical study

The ICEM CFD 2019 R3 was used to discretise the computational domain using an unstructured tetrahedral mesh that was improved with a prism mesh at the boundary. A constant heat flux (800 W/m^2) was considered at the base of the computational domain. One side of the domain was subjected to a symmetry boundary condition, while the opposite and top sides were modelled as walls. The thermophysical characteristics of air were assumed to remain constant, aligned with the inlet air temperature of 300 K. ANSYS Fluent 2019 R3 was used to solve conservation and turbulence models in the computational domain. The flow was simulated as turbulent and incompressible, using the k-ε turbulence model, COUPLED algorithm and QUICK scheme to model AH. The airflow through the AH was assumed to be steady, turbulent, incompressible and fully developed. Also, heat losses from the system were neglected for the analysis. The flow was considered converged when the residuals of continuity, velocity, and turbulence dissipation terms reached 10^{-3}. For the energy equation, convergence was acknowledged upon reducing the residual to 10^{-6}.

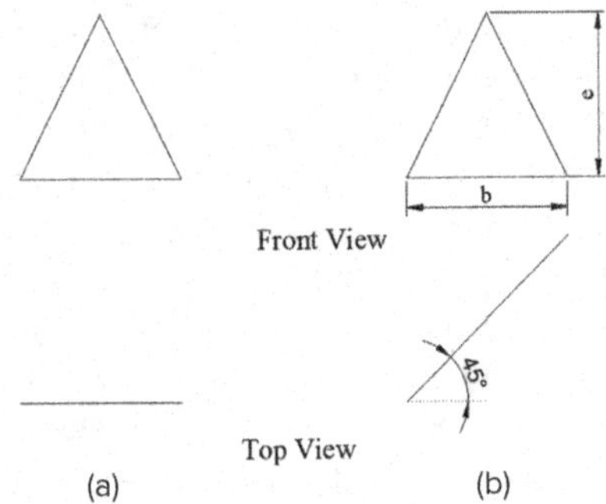

Fig. 24.2 Front and top view of a (a) non-inclined and (b) inclined delta flow obstruction

The performance parameters obtained for the AH without obstructions at various Re were compared with the corresponding values derived from standard empirical correlations (Equations 1 and 2) for a smooth duct to validate the CFD model used for the numerical study.

$Nu = 0.023\ Re^{0.8}\ Pr^{0.4}$ Dittus–Boelter correlation (Rohsenow et al. 1998) (1)

$f = 0.085\ Re^{-0.25}$ Modified Blasius correlation (Kakaç, Ramesh K. Shah, and Win Aung 1987) (2)

The thermo-hydraulic performance parameter, TEF, provides an overall evaluation of the AH, considering both thermal and hydraulic characteristics. The TEF is defined as (Webb, Eckert, and Goldstein 1971; Han, Park, and Lei 1985),

$$TEF = \left(\frac{Nu}{Nu_o}\right)\left(\frac{f}{f_o}\right)^{\frac{-1}{3}} \tag{2}$$

Where Nu and Nu_o are the Nu of the roughened and plane surface, respectively, and f and f_o are the f of the roughened and smooth surface, respectively.

3. Results and Discussions

The thermo-hydraulic performance analysis of a smooth AH without flow obstructions, with non-inclined and with inclined delta flow obstruction for various *Re* (5000, 10000, 15000 and 20000) are discussed in the following subsections. The obtained numerical results from the Computational Fluid Dynamics (CFD) of the AH without obstruction were validated using standard correlations.

3.1 Validation of the CFD Model

Figures 24.3 and 24.4 illustrate the impact of Reynolds number (*Re*) on the performance parameters - Nusselt number (*Nu*) and friction factor (*f*) of the smooth air heater (AH), for Re in the the range of 5000 to 20000. As Reynolds

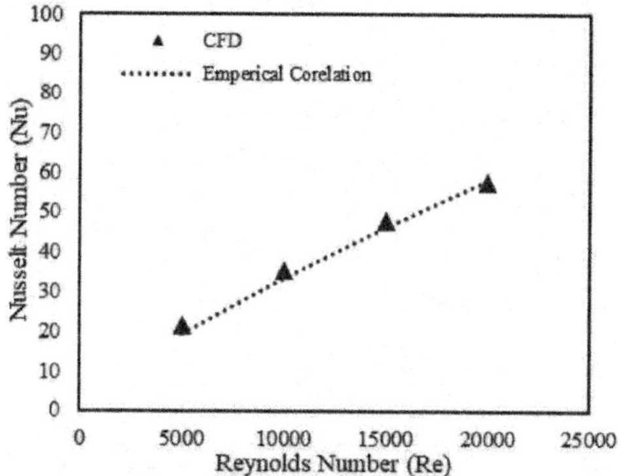

Fig. 24.3 Effect of flow rate (Re) on thermal performance parameter (Nu) of air heater without obstructions

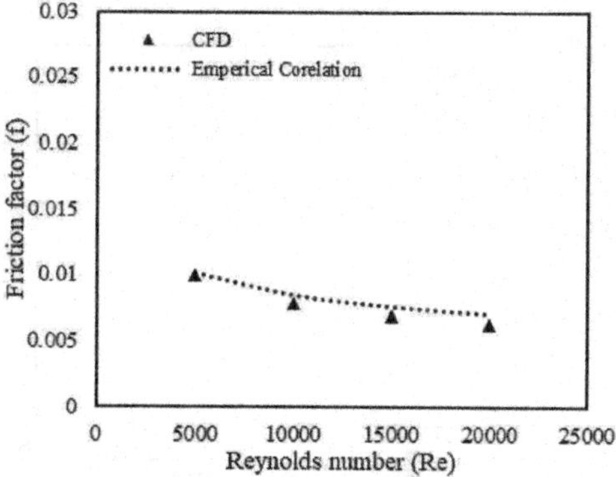

Fig. 24.4 Effect of flow rate (Re) on hydraulic performance parameter (f) of air heater without obstructions

number increases, *Nu* exhibits a rising trend, indicating enhanced convective heat transfer due to increased turbulence at higher flow rates. The '*f*' exhibits a decreasing trend with increasing *Re*, indicating a reduction in fluid flow resistance against the surface. This behaviour is consistent with expectations, as higher flow rates lead to a thinner boundary layer, resulting in lower drag forces and decreased flow resistance. There is substantial agreement on this with empirical correlation results for the smooth duct, which ensures the accuracy of the CFD model in predicting the thermal and hydraulic performance parameters (*Nu* and *f*).

3.2 Performance Analysis of Air Heater with Delta Flow Obstructions

The *Nu* and *f* of the AH without obstructions (smooth), with NIDFO and IDFO at various *Re*, are shown in Fig. 24.5 and Fig. 24.6, respectively.

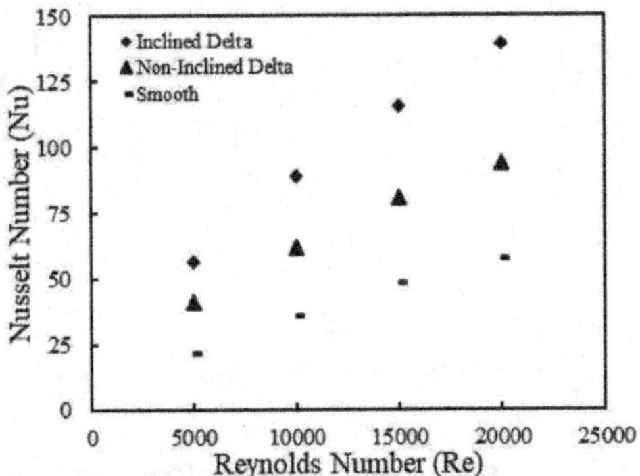

Fig. 24.5 Effect of inclined and non inclinded flow obstructions on Nu of the AH at various *Re*

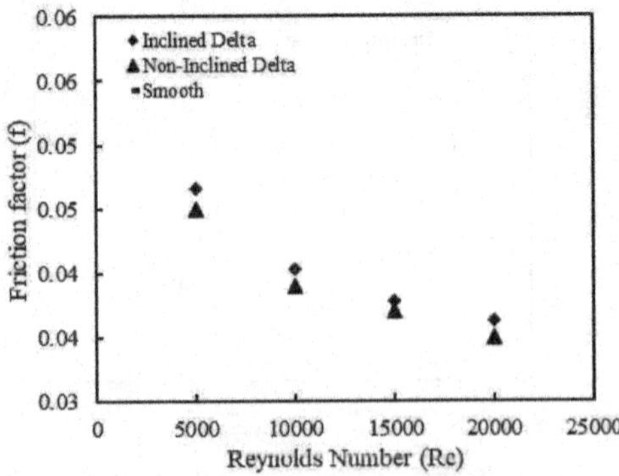

Fig. 24.6 Effect of inclined and non inclinded flow obstructions on *f* of the AH at various *Re*

As Reynolds number (*Re*) increases, the air heater's Nusselt number (*Nu*) demonstrates an increasing trend for all the arrangem,ents of air heater with and without flow obstructions. This behaviour can be attributed to the increased turbulence intensity at higher flow rates. IDFO produces better *Nu* enhancement than NIDFO in the entire range of *Re*. The f values for all obstructions decrease with the increase in Re, as illustrated in Fig. 24.6. With the increase in Reynolds number (*Re*), the friction factor (*f*) consistently decreases for all the arrangem,ents of air heater with and without flow obstructions. This is due to the reduced residence time of air on the heated surface at higher flow rates. IDFO produces higher *f* enhancement than NIDFO in the entire investigated range of *Re*. The use of IDFO results in a *Nu* enhancement ($Nu_{\text{with obstruction}}$ / $Nu_{\text{without obstruction}}$) of 2.42-3.33 with a corresponding *f* augmentation ($f_{\text{with obstruction}}/f_{\text{without obstruction}}$) of 4.65-6.72 compared to an AH without obstruction.

Figure 24.7 shows the effect of *Re* on the TEF of the AH without obstructions (smooth), AH with NIDFO and IDFO.

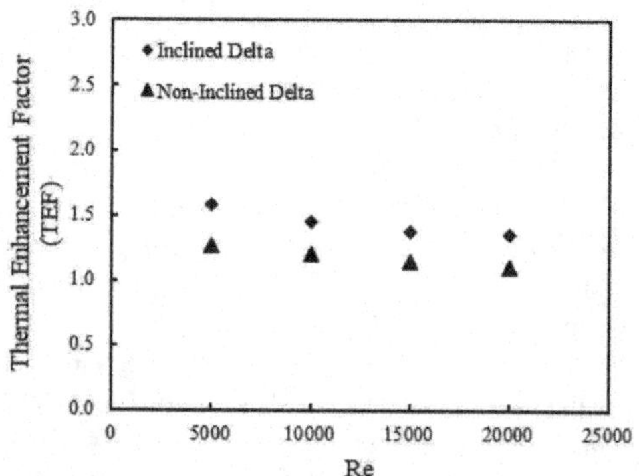

Fig. 24.7 Effect of inclined and non inclinded flow obstructions on TEF of the AH at various *Re*

The inclined delta obstructions produce a TEF varying from 1.35 to 1.57, compared to 1.1 to 1.27 for non-inclined delta flow obstructions. It signifies an improved performance of the AH with IDFO.

4. Conclusions

The numerical investigation reveals that introducing inclined delta flow obstructions in the AH substantially enhances thermo-hydraulic performance. The inclined configuration (IDFO) promotes better mixing and turbulence, increasing convective heat transfer and reducing friction. Comparative analysis with non-inclined delta obstructions (NIDFO) highlights the superior performance of inclined configurations across the range of Reynolds numbers studied. The Thermal Enhancement Factor (TEF) demonstrates a significant improvement, reaching a maximum of 1.57 at Reynolds number 5000. These findings underscore the potential of IDFO as an effective strategy for improving the effectiveness of AH, contributing to energy conservation and sustainable thermal systems.

References

1. Acharya, Anil, and Ashok Pise. 2017. "A Review on Augmentation of Heat Transfer in Boiling Using Surfactants/Additives." *Heat and Mass Transfer/Waerme- Und Stoffuebertragung* 53 (4): 1457–77.
2. Baissi, M. T., A. Brima, K. Aoues, R. Khanniche, and N. Moummi. 2019. "Thermal Behavior in a Solar Air Heater Channel Roughened with Delta-Shaped Vortex Generators." *Applied Thermal Engineering*, no. August 2018: 113563.
3. Cyriac Bronin, and Bhusnoor Siddappa S. 2023a. "Numerical Investigation of Heat Transfer Performance of an Air Heater with Delta Flow Obstruction." *Materials Today: Proceedings* 72: 1246–52.
4. Cyriac Bronin, and Bhusnoor Siddappa S. 2023b. "Thermal and Hydraulic Characteristics of an Air Heater with Modified Delta Flow Obstructions." *E-Prime - Advances in Electrical Engineering, Electronics and Energy*, March, 100147.
5. Cyriac Bronin, and Bhusnoor Siddappa S. 2023. "Performance Optimisation of an Air Heater with Delta Flow Obstructions: A Taguchi Approach." In *Proceedings of International Conference on Intelligent Manufacturing and Automation*, edited by Vijaya Kumar N. and Raina Amool A Vasudevan Hari and Kottur, 621–30. Singapore: Springer Nature Singapore.
6. Ebrahim Momin, Abdul Malik, J. S. Saini, and S. C. Solanki. 2002. "Heat Transfer and Friction in Solar Air Heater Duct with V-Shaped Rib Roughness on Absorber Plate." *International Journal of Heat and Mass Transfer* 45 (16): 3383–96.
7. Han, J. C., J. S. Park, and C. K. Lei. 1985. "Heat Transfer Enhancement in Channels with Turbulence Promoters." *Journal of Engineering for Gas Turbines and Power* 107 (3): 628–35.
8. Jain, Sheetal Kumar, Ghanshyam Das Agrawal, Rohit Misra, Prateek Verma, Sanjay Rathore, and Doraj Kamal Jamuwa. 2019. "Performance Investigation of a Triangular Solar Air Heater Duct Having Broken Inclined Roughness Using Computational Fluid Dynamics." *Journal of Solar Energy Engineering, Transactions of the ASME* 141 (6): 1–11.
9. Kumar, Amit, and Apurba Layek. 2022. "Evaluation of the Performance Analysis of an Improved Solar Air Heater with Winglet Shaped Ribs." *Experimental Heat Transfer* 35 (3): 239–57.
10. Promvonge, P., C. Khanoknaiyakarn, S. Kwankaomeng, and C. Thianpong. 2011. "Thermal Behavior in Solar Air Heater Channel Fitted with Combined Rib and Delta-Winglet." *International Communications in Heat and Mass Transfer* 38 (6): 749–56.

11. Saxena, Abhishek, Varun, and A. A. El-Sebaii. 2015. "A Thermodynamic Review of Solar Air Heaters." *Renewable and Sustainable Energy Reviews* 43: 863–90.
12. Webb, R. L., E. R.G. Eckert, and R. J. Goldstein. 1971. "Heat Transfer and Friction in Tubes with Repeated-Rib Roughness." *International Journal of Heat and Mass Transfer* 14 (4): 601–17.
13. Yadav, Anil Singh, and Manish Kumar Thapak. 2016. "Artificially Roughened Solar Air Heater: A Comparative Study." *International Journal of Green Energy* 13 (2): 143–72.
14. Yu, Wenhua, David M. France, Jules L. Routbort, and Stephen U.S. Choi. 2011. "Review and Comparison of Nanofluid Thermal Conductivity and Heat Transfer Enhancements." 432–60.

Note: All the figures in this chapter were made by the authors.

Technologies for Energy, Agriculture, and Healthcare – Shailesh Nikam et al. (eds)
© 2024 Taylor & Francis Group, London, ISBN 978-1-032-98028-7

25

Modelling and Experimental Studies on Performance Evaluation of Energy Efficient Minichannel Shell and Tube Heat Exchanger

Akash S. Bidwaik[1],
Iqbal Muzawar[2], Bronin Cyriac[3]
Research Scholar,
Department of Mechanical Engineering,
K. J. Somaiya College of Engineering,
Somaiya Vidyavihar University,
Mumbai, India

Shailesh R. Nikam[4]
Associate Professor,
Department of Mechanical Engineering,
K. J. Somaiya College of Engineering,
Somaiya Vidyavihar University,
Mumbai, India

Siddappa S. Bhusnoor[5]
Professor, Department of Mechanical Engineering,
K. J. Somaiya College of Engineering,
Somaiya Vidyavihar University,
Mumbai, India

Abstract: Due to the rapid increase in industrialization and depletion of fossil fuels, there is a need to save energy. Heat exchangers play a crucial role in efficient energy transmission in the industry. The significant approach for increasing energy efficiency is making them compact. Minichannels are crucial for improving shell and tube heat exchangers (STHXs) performance. This study explored the

[1]akash.bidwaik@somaiya.edu, [2]iqbal.m@somaiya.edu, [3]bronin.c@somaiya.edu, [4]shailesh.n@somaiya.edu, [5]siddappabhusnoor@somaiya.edu

DOI: 10.1201/9781003596707-25

impact of minichannels on the performance metrics of a STHX under various test conditions (cold water inlet temperature = 26.4°C, hot water inlet temperature ranging from 35°C to 55°C, flow rates for tube side: 4 to 14 lpm and the shell side: 2 to 7 lpm) using both experimental and modelling techniques. From the observation it is revealed that the predicted performance metrics obtained through MATLAB® closely aligned with the experimental results, showing a negligible error of approximately 4%. Furthermore, it was also observed that the results of the current experiment show an increase of about 2.7% when compared to the results published in the literature using conventional STHXs.

Keywords: Minichannel, Heat exchanger, Energy conservation

1. Introduction

It is necessary to give priority to make efficient energy utilization in industrial operations (Holik et al., 2021). To reduce energy expenses and consumption, researchers are devoting more time to studying methods for improving heat transfer(Klemeš et al., 2020). Reducing heat exchanger size passively and lowering pumping power for flow maintenance are two important tactics (Bhattacharyya et al., 2022). These methods lower energy, initial investment, operational costs, and waste while improving thermal and hydrodynamic performance and conserving materials and space. More than 40% of thermal systems employ STHXs, which have several benefits including cost effective, low installation cost, efficient heat transmission, and robustness against high pressure (Bell, 2004). We may improve compactness by using passive methods to improve STHX performance and comply with energy and space-saving objectives. This is accomplished by altering geometric parameters, such as hydraulic diameter and flow channel length. As a result of lower CO_2 emissions, we may utilize less harmful fluids, save money, and have less environmental effects (Del Col et al., 2010).

Although small flow channels enhance heat transmission, they also increase pressure drop, which is an undesirable hydrodynamic impact. In heat exchangers using circular tubes, numerous interesting phenomena occur when the flow channels are smaller. Like the square of the diameter, the flow rate increases as lower the tube diameter (D). The surface area used for transferring heat, on the other hand, increases only in a linear manner with diameter. It may be observed that the ratio of the tube's internal volume to its heat transfer surface area varies inversely with its diameter. In other words, the ratio represents the compactness of the heat exchanger (Kandlikar, 2007).

The objective of this study is to introduce minichannels in STHX for the performance improvement. This study included the theoretical (using MATLAB® Simulation) and experimental study under various temperature ranges and flow rates and comparing the results with the conventional STHXs to determine and evaluate the performance of a minichannel STHX,

2. Methodology

The methodology employed to evaluate the performance of minichannel STHX under varying flow conditions is studied theoretically (MATLAB®) and experimentally is discussed in the following sub-sections.

2.1 Theoretical Studies

A theoretical investigation has been undertaken, utilizing water as the working fluid for both the tube and shell sides. The analytical study involved varying the hot water temperature (from 35°C to 55°C), cold water temperature (26.4°C) and flow rates on the tube side (4 to 14 lpm) and shell side (2 to 7 lpm). The detailed analysis methods for estimating the performance metrics of the minichannel STHX are discussed in the following sub-sections.

Calculation of Tube Side Heat Transfer Coefficient of Minichannel Shell and Tube Heat Exchanger

The tube-side heat transfer coefficient of a mini-channel STHX is calculated using the following empirical correlations.

$$\text{Nusselt's Number, } Nu_t = \frac{(f/2)Re_b Pr_b}{1.07 + 12.7\left(\frac{f}{2}\right)^{\frac{1}{2}}(Pr_b^{\ 2/3} - 1)} \quad \text{(Petukhov 1970)} \qquad (1)$$

$$\text{Friction factor, } f(turbulent) = (1.58\, lnRe - 3.28)^{-2} \qquad (2)$$

$$\text{Tube side Heat Transfer Coefficient, } h_i = \frac{Nu_t \times k_h}{d_i} \qquad (3)$$

Calculation of Shell Side Heat Transfer Coefficient of Minichannel Shell and Tube Heat Exchanger

The following empirical correlation is used to determine the shell-side heat transfer coefficient of a mini-channel STHX.

$$\text{Shell Side Heat Transfer Coefficient, } h_o = h_{id}\, J_c\, J_l\, J_b\, J_s\, J_r \qquad (4)$$

Calculation of Performance Characteristics of Minichannel Shell and Tube Heat Exchanger

The performance metrics of minichannel STHX such as overall heat transfer coefficient, effectiveness and heat recovery can be calculated by using the following process

Overall Heat Transfer Coefficient for clean surface,

$$U_c = \frac{1}{\frac{d_o}{d_i \times h_i} + \frac{d_o ln(d_o / d_i)}{2k} + \frac{1}{h_o}} \tag{5}$$

Effectiveness, $$\varepsilon = \frac{2}{1 + c^* + \left[(1 + c^{*2})^{0.5} \frac{1 + exp[-NTU(1 + c^{*2})^{0.5}]}{1 - exp[-NTU(1 + c^{*2})^{0.5}]} \right]} \tag{6}$$

Heat Recovery, $Q = C_h \times (T_{h_i} - T_{h_o})$ (7)

2.2 Mathematical Studies using MATLAB

The virtual thermal-fluids system presented in this paper can be described as a set of steady-state energy balance equations that define the heat transfer process occurring in a STHX. The governing equations are configured to accept user-defined inputs such as fluid inlet temperature, fluid flow rates, length of heat exchanger, outer & inner diameter of tube as well as number of tubes in heat exchanger and determine the output of the system such as Convective heat transfer coefficient inside and outside, overall heat transfer coefficient, heat transfer rate, exit temperature for the two fluids which are involved in the process without entering any thermo-physical property of fluid (i.e. water) Moreover, the response of the system is analysed in real-time with the use of MATLAB® software

2.3 Experimental Studies

The experimental setup consists of two water tanks with a capacity of 252 litres each and a minichannel STHX with 30 tubes inside the shell as shown in Fig. 25.1. Each tube has3 mm outer diameter, 2 mm inner diameter and flow length of 364 mm and the shell's inner diameter is set at 80 mm, with baffles featuring a 25% cut. Hot water from a tank enters the shell, flowing through the tubes, while cold water from another tank circulates around the shell. The end housings have fittings for temperature sensors, and there are four "K" type thermocouples to measure inlet and outlet fluids temperatures. The hot water is heated in a tank with an electric heater, controlled by a thermostat up to 65°C. A gear pump circulates the hot water through the exchanger, and flow valves in the service unit adjust the flow of both hot and cold fluids.

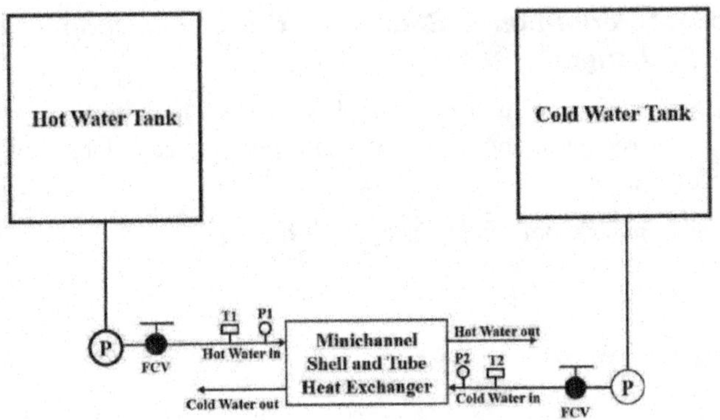

Fig. 25.1 Schematic representation of the experimental setup

3. Results and Discussion

The performance parameters (overall heat transfer coefficient, heat recovery and effectiveness) of minichannel STHX were evaluated experimentally and theoretically (using MATLAB® simulation). The inclusion of minichannels in STHXs notably enhances the performance parameters as discussed below.

Figure 25.2 illustrates the dynamic relationship between the overall heat transfer coefficient and Reynolds number (Re), showcasing their interplay. It has been

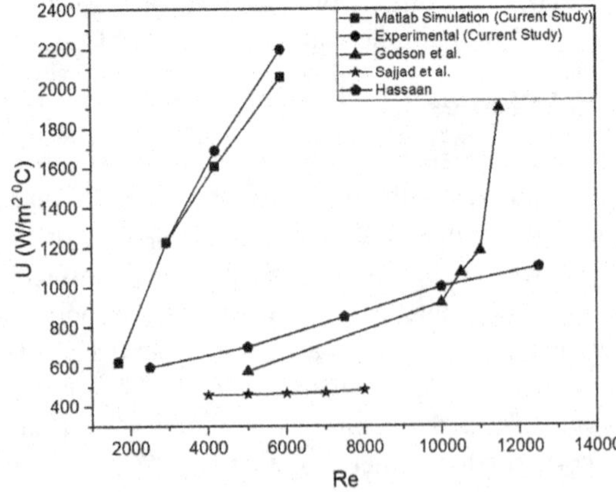

Fig. 25.2 Comparative analysis between overall heat transfer coefficient (U) and Reynolds number (Re)

noted that a rise in Reynolds number leads to an enhancement in the overall heat transfer coefficient and the highest overall heat transfer coefficient reported by Godson et al. was ~1990 W/m^2 °C at tube side Re = 11,500, it was obtained in the current study at Re ≅ 4900. Similarly, the overall heat transfer coefficient reported by Hassaan was ~ 1100 W/m^2 °C at Re = 12,500, it was obtained in the current study at Re ≅ 2400.

Figures 25.3 and 25.4 show the variations in rate of heat transfer (Q) and effectiveness (%) respectively with respect to Re. The experimental investigation

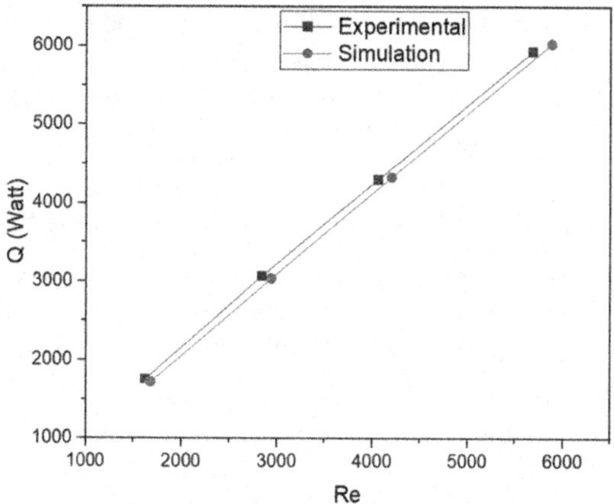

Fig. 25.3 Variation of heat transfer rate (Q) with tube side Reynolds No. (Re)

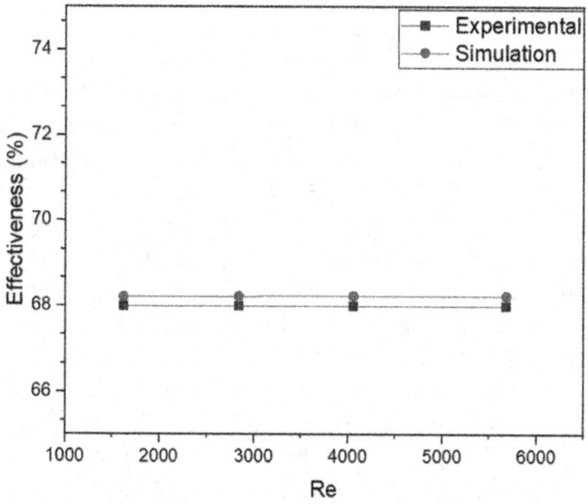

Fig. 25.4 Variation of effectiveness (%) with tube side Reynolds No. (Re)

reveals that the maximum recorded heat transfer rate reaches 6030 W at a Reynolds number of 5876, and the associated effectiveness is 68%. Due to the smaller hydraulic diameter, there is an increase in the performance parameters. The predicted performance metrics obtained through MATLAB® closely aligned with the experimental results, showing a negligible error of approximately 4%

4. Conclusions

The use of minichannels in STHXs is one of the best options to enhance the performance metrics of a heat exchanger. Decreasing the hydraulic diameter of tubes enhances the rate of heat transfer, effectiveness, and overall convective heat transfer coefficients of the heat exchanger. Therefore minichannels in STHXs present a viable approach to effective energy and material utilisation in many industrial applications, highlighting the need to take thermal parameters into account.

The conclusions drawn from the present study are listed below,

- The mini-channel's experimental overall heat transfer coefficient showed dependency on the Reynolds number. Observations indicate that within the range of Reynolds numbers from 1600 to 5900, there is a discernible rise in the overall heat transfer coefficient.
- Observational findings revealed that the maximum experimental overall heat transfer (2199 $W/m^2\,°C$), heat transfer rate (6030 W) and the corresponding effectiveness (68%) obtained at Re = 5876 because of an increase in surface area due to small hydraulic diameter of the tubes.

References

1. Holik, M., Živić, M., Virag, Z., Barac, A., Vujanović, M., & Avsec, J. (2021). Thermo-economic optimization of a Rankine cycle used for waste-heat recovery in biogas cogeneration plants. Energy Conversion and Management, 232(February). https://doi.org/10.1016/j.enconman.2021.113897.
2. Klemeš, J. J., Wang, Q. W., Varbanov, P. S., Zeng, M., Chin, H. H., Lal, N. S., Li, N. Q., Wang, B., Wang, X. C., & Walmsley, T. G. (2020). Heat transfer enhancement, intensification and optimisation in heat exchanger network retrofit and operation. Renewable and Sustainable Energy Reviews, 120(October 2019). https://doi.org/10.1016/j.rser.2019.109644.
3. Bhattacharyya, S., Vishwakarma, D. K., Srinivasan, A., Soni, M. K., Goel, V., Sharifpur, M., Ahmadi, M. H., Issakhov, A., & Meyer, J. (2022). Thermal performance enhancement in heat exchangers using active and passive techniques: a detailed review. In Journal of Thermal Analysis and Calorimetry (Vol. 147, Issue 17). Springer International Publishing. https://doi.org/10.1007/s10973-021-11168-5.

4. Bell, K. J. (2004). Heat exchanger design for the process industries. Journal of Heat Transfer, 126(6), 877–885. https://doi.org/10.1115/1.1833366
5. Del Col, D., Cavallini, A., Da Riva, E., Mancin, S., & Censi, G. (2010). Shell-and-tube minichannel condenser for low refrigerant charge. Heat Transfer Engineering, 31(6), 509–517. https://doi.org/10.1080/01457630903409738
6. Kandlikar, S. G. (2007). A roadmap for implementing minichannels in refrigeration and air-conditioning systems - Current status and future directions. Heat Transfer Engineering, 28(12), 973–985. https://doi.org/10.1080/01457630701483497.
7. Godson, L., Deepak, K., Enoch, C., Jefferson Raja, B. R., & Raja, B. (2014). Heat transfer characteristics of silver/water nanofluids in a shell and tube heat exchanger. Archives of Civil and Mechanical Engineering, 14(3), 489–496. https://doi.org/10.1016/j.acme.2013.08.002.
8. Sajjad M., Ali H., Kamran M. (2020). Thermal-hydraulic analysis of water based ZrO2 nanofluids in segmental baffled shell and tube heat exchangers, Therm. Sci. 24 1195–1205. https://doi.org/10.2298/TSCI180615291S
9. Petukhov, B. S. (1970). Heat Transfer and Friction in Turbulent Pipe Flow with Variable Physical Properties. Advances in Heat Transfer 6 (C): 503–64. https://doi.org/10.1016/S0065-2717(08)70153-9

Note: All the figures in this chapter were made by the authors.

Technologies for Energy, Agriculture, and Healthcare – Shailesh Nikam et al. (eds)
© 2024 Taylor & Francis Group, London, ISBN 978-1-032-98028-7

26

VISION-BASED AUTONOMOUS PATH FOLLOWER DRONE

Anoushka Bhat[1]
Student,
K. J. Somaiya College of Engineering,
Somaiya Vidyavihar University,
Mumbai

Kartik Patel[2]
Assistant Professor,
K. J. Somaiya College of Engineering,
Somaiya Vidyavihar University

Abstract: This research aims to develop a visual tracking system tailored for Unmanned Aerial Vehicle (UAV) with a specific focus on applications in precision agriculture. The primary objective is to orchestrate precise path tracking, enabling the UAV to navigate and adhere to predefined trajectories with a high accuracy within agricultural landscapes. The proposed methodology employs an innovative vision-based line tracking technique, using a simple camera sensor for path tracking through image processing. After image acquisition, recorded images are wirelessly transmitted to a Ground Control Station (GCS) where the image is processed in real-time. The GCS calculates the present yaw measurement and distance between the agricultural path and quadrotor, which is crucial for precise agricultural navigation. This calculated information is then relayed to the onboard controller of the quadrotor. The outcome of this study from the visual tracking system suited for UAVs holds significant implications for precision agriculture. Within agricultural settings, the high accuracy of 5 pixels and precision demonstrated in contour formation play a crucial role in crop monitoring. The system's response time for precise navigation directions emerges as a key achievement, enabling the drone to adapt its course swiftly within milliseconds of detecting changes in the track's position. Under diverse lighting conditions, the system showcased

[1]anoushka.b@somaiya.edu, anoushka0826@gmail.com; [2]kartik@somaiya.edu

DOI: 10.1201/9781003596707-26

consistent performance, maintaining contour accuracy with negligible variations in brightness, including scenarios with up to 80% illumination changes. These tangible results not only underline the immediate applicability of the visual tracking system in aerial surveillance, search and rescue missions, but also illuminate its potential to revolutionize agricultural practices. The commitment to continual refinement in design and algorithmic efficacy, evident in these results, establishes this technology as a cornerstone in the evolving landscape of precision agriculture and UAV applications. This innovative framework combines advanced image processing with UAV technology, specifically designed for precision agriculture. It shows potential in enhancing path tracking for precise agricultural practices, highlighting the transformative impact of advanced technology in modern farming.

Keywords: Autonomous drone, Line following algorithm, Image processing, OpenCV (Open computer vision), Agriculture technologies

1. Introduction

The surge in drone usage is attributed to cost-effectiveness, lightweight design, and enhanced safety features. Drones leverage autonomous flight capabilities through the integration of advanced sensing technologies, including Lidar, Global Positioning System (GPS), or optical flow sensors, ensuring optimal stability and precision during operation such as in (Granillo & Beltrán, 2018; Shao et al., 2019). These UAVs serve diverse purposes, such as surveillance, reconnaissance, etc. Despite the widespread use of GPS, its limitations in indoor environments and payload restrictions for heavier sensors have prompted exploration into alternative solutions such as vision-based image processing using cameras for precise path tracking (Liu et al., 2020; Ceppi, 2020), aligning with Agriculture Technologies.

The main objective of the paper is to contribute to the intersection of innovative technology and agriculture by developing an autonomous quadcopter drone using computer vision to follow a predefined coloured line on the ground. This aligns with the principles of precision agriculture, which aims to optimize crop yields, reduce input usage, and enhance efficiency through accurate and targeted interventions. The autonomous drone system comprises of two key components: an image processing unit driven by a Raspberry Pi and a drone propulsion system featuring a Pixhawk flight controller board. The image processing unit, powered by a Raspberry Pi, utilizes Python and the OpenCV library, interprets visuals from a USB camera, identifying a predefined colour track, and sending real-time serial signals for drone movement. The propulsion system includes advanced components like the Pixhawk 2.4.8 flight Controller, GPS module, telemetry, Electronic Speed Controller (ESC), motors, and a transmitter-receiver system for control. Using image processing and real-time tracking indicates a reliance on data analysis techniques, such as segmentation of the captured images, showcasing an application of technology for agricultural monitoring and analysis. The autonomous capabilities of the drone, especially its ability to follow a predefined path, can be associated with the broader theme of automation in agriculture. Integrating hardware components, including Pixhawk flight controller, cameras, and sensors, showcases the use of advanced tools and techniques for smart agriculture.

Beyond agriculture, the project extends its applications to environmental monitoring, traffic management, and defense, offering benefits such as ecosystem mapping, wildlife protection, real-time traffic updates, and military surveillance. This alignment with the broader vision of Agriculture Technologies underscores the project's significance in modernizing agricultural practices and contributing to smart automation.

Rest of the paper is structured as follows. Section-2 reviews the extant literature. Section-3 describes the system Design and Methodology. Section-4 discusses the Implementation. Section-5 summarises the result.

2. Literature Review

Zou et al. (2022) developed a GPS-independent visual tracking system for UAVs, specifically for tracking and landing on a stationary mobile robot. Using Tello EDU equipped with Light Emitting Diode (LED) of various colours, the study efficiently separated areas based on distinct hues and luminosities in images. The RGB (Red, Green, Blue) to HSV (Hue, Saturation, Value) colour space conversion facilitated precise colour representation, achieving autonomous tracking and landing capabilities using optical flow sensors. Choi et al. (2011) introduced a navigation strategy for UAVs to track and land on a moving target using an on-board vision sensor. Determining the target's position and speed through image processing, the drone established its location in the inertial frame. A dual-velocity guidance system, based on geolocation information and UAV status, facilitated target approach, and monitoring without relying on GPS signals. The use of optical flow positioning in indoor environments and a guiding law for integrating target and drone velocities demonstrated adaptability in speed computation during simulations.

Granillo et al. (2018) proposed an economical alternative to GPS or Inertial Navigation System (INS) for drone trajectory using the Lucas-Kanade (LK) optical flow algorithm. This method efficiently provided real-time drone speed, displacement direction, and distance from a target. The study emphasized challenges in trajectory control with velocity data and suggested improvements in image conditioning, such as using the optical flow algorithm for monitoring building trajectories and natural light in indoor environments.

In their work, Shao et al. (2019) utilizes the Parallel Tracking and Mapping (PTAM) method for target tracking in quadrotor UAVs, addressing resource limitations in traditional visual-guided systems. Their novel method, Simultaneous Localization and Mapping (SLAM) algorithm employs a monocular vision sensor, focusing on geometric constraint equations of moving target feature points to enhance the robustness and accuracy of depth estimation algorithms. This approach integrates dynamic target tracking, monocular SLAM placement, and target feature detection for robust UAV visual positioning and tracking in small quadrotors. Liu et al. (2020) developed a method enabling a drone to accurately monitor and land on a mobile. They utilized a Proportional-Integral-Derivative (PID) controller adjusted using the Ziegler-Nichol technique, fine-tuned with experimental data. The mobile robot's embedded system (Up Board), running Linux Ubuntu and using OpenCV to process images, can send the Tello EDU drone a control command (SDK 2.0) via Wi-Fi using User Datagram Protocol (UDP) protocol.

Ceppi (2020) employed vision-based control and image processing to create a line-tracking algorithm for a small-scale drone. Real-time tests revealed altitude sensitivity. The Model-in-the-Loop phase used, involved evaluating the developed model in a simulated environment before implementing it in real-time hardware testing. The control system, built using MATLAB and Simulink, incorporated computer vision, image processing, and aerospace functionalities. The line-tracking algorithm involved image processing of the downward-facing camera's output, followed by user-defined noise reduction functions. Custom and pre-existing functions from the computer vision toolbox were used to eliminate noise from the binary image, including path detection and line angle analysis.

Garcia-Olvera et al. (2020) transformed a Parrot Bebop 2 quadcopter into a line follower UAV using Wi-Fi communication and the Robot Operating System (ROS). Their adaptive system efficiently positions the camera, processes images rapidly, and integrates information for UAV-based target tracking and identification system. Utilizing neural networks, moving background processing, and consensus-based target tracking algorithms, the system is tailored for embedded implementation. Geographic information from a Geographic Information System (GIS) enhanced the context and provided environment details and geolocation information, showcasing the effectiveness of the proposed target tracking and recognition framework.

3. Methodology

Key hardware integrations played a pivotal role in crafting a robust autonomous drone. The avionic system encompassed crucial components such as the Flight Controller, ESC, frame, propellers, battery, power distribution board, receiver, GPS, and Telemetry System. The Image Processing System was equipped with an onboard USB camera and a Raspberry Pi. The Pixhawk Flight Controller renowned for its open-source nature, modularity, and support for various communication protocols, served as the central hub. This detailed table provides specific specifications for each hardware component used in the drone's development.

Table 26.1 Hardware specifications

Component	Specification and Description
Flight Controller	Custom-developed initially with STM32 microcontroller and MPU6050 Inertial Measurement Unit (IMU); Upgraded to Pixhawk 2.4.8 Flight Controller
Electric Motors	A2212 Brushless DC (BLDC) motor with a rating of 1000kv having 11100 RPM (Revolutions per minute) at full throttle
Propellers	Two-blade 8*4.5 inches propellers
Frame	All components of the avionic system are installed on this.
Power Distribution Boards (PDB)	To distribute the power from the battery to all the components, including the motors and the flight controller
ESC	Controls and regulates the speed of an electric motor
Power Distribution Board	Efficient distribution of power to components
Battery	Reliable battery for powering the drone, 2200mAh 3 cell orange battery.
Receiver	A FlySky FS-iA6B RF 2.4GHz 6 Channel output with iBus port receiver Used for communication with the Pixhawk Flight Controller.
GPS Module	Provides precise location awareness
Telemetry System	FlySky FS-T6 6-Channel Transmitter-Receiver for seamless real-time data transmission
Onboard USB Camera	Real-time image processing for vision-based navigation
Raspberry Pi	Enables real-time image processing and integration for image interpretation

Source: Author's compilation

This comprehensive drone system strategically utilized Pixhawk's high-precision IMU sensors for precise attitude stabilization crucial for line tracking. The

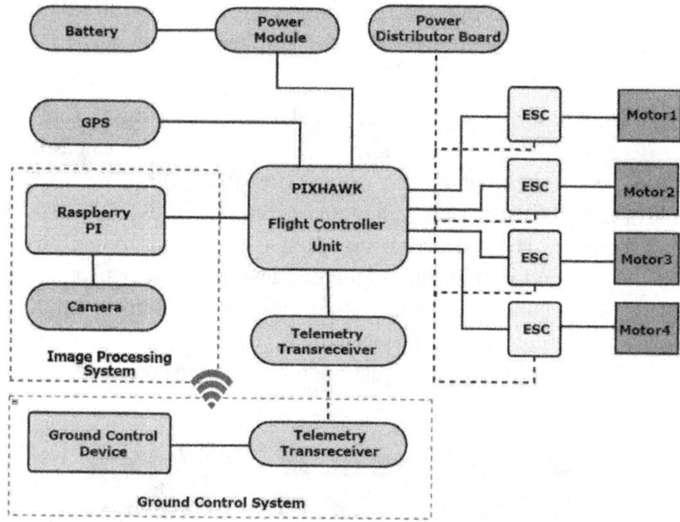

Fig. 26.1 Block diagram of proposed system with Pixhawk

Source: Author's compilation

Raspberry Pi enabled real-time image processing, interpreting predefined coloured tracks. The telemetry system ensured continuous data transmission, allowing for remote monitoring and control. This system exemplified the synergy between Pixhawk's versatility and carefully selected hardware components, showcasing the feasibility of autonomous flight and navigation based on computer vision and sensor inputs.

3.1 Logic of the Algorithm

The image processing algorithm employed in real-time using Python and OpenCV for identifying a predefined coloured track follows a sequential series of steps. Computer vision algorithms, part of the image processing software, enable real-time analysis, encompassing tasks like object detection, tracking, and mapping. Initially, the camera captures each frame using the VideoCapture() function, for enhanced processing accuracy and to ensure a standardized video feed, the frame is resized to 500 × 375 pixels with a 4:3 aspect ratio. Subsequently, the frame undergoes a conversion from the BGR colour space to the HSV colour space, facilitating better isolation of the specified colour. A red colour filter is applied using specific HSV range to create a binary mask. Through masking technique, all colours in the frame are suppressed except for the desired colour, effectively highlighting the track. The masking process is crucial for accurate identification of the track, particularly in dynamic environments. The algorithm then identifies contours within the isolated, red-coloured frame, and the largest contour is computed, and it's bounding rectangle is computed.

The centroid of this identified contour is calculated, serving as a pivotal point for further decision-making. The fundamental approach involves partitioning the screen into seven vertical segments, designating the central one as the pivotal segment. It is imperative that the calculated centroid consistently resides within this central segment. The horizontal direction is then determined based on the position of the centroid within this segmented layout. The corresponding signals generated for drone control are transmitted through serial communication. This iterative process repeats for each subsequent frame captured by the camera, ensuring continuous and adaptive adjustments to the drone's movements based on the perceived track information.

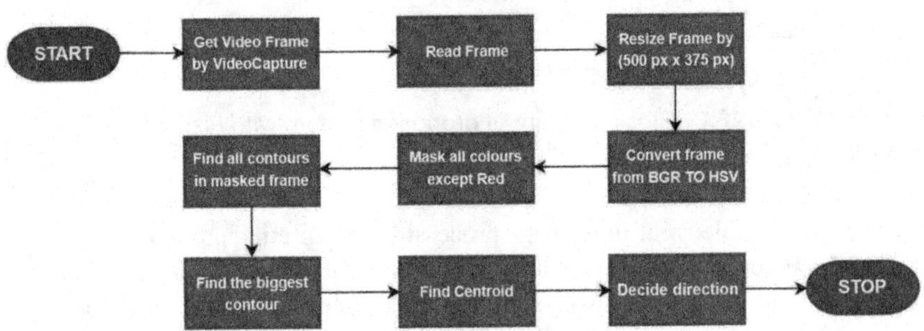

Fig. 26.2 Block diagram of image processing

Source: Author's compilation

The following aspects were taken into consideration while designing the image processing system:

Parameter Tuning: The colour threshold parameters were fine-tuned in the HSV space from 0 to 255 to optimize the algorithm's sensitivity to the red-coloured track. Adjustments were made iteratively, considering lighting conditions. Specifically, the hue, saturation, and value thresholds were calibrated to achieve reliable colour isolation, enhancing the accuracy of the track detection algorithm.

Frame Rate and Real-time Processing: The algorithm operates at a frame rate of 30 frames per second (FPS), enabling real-time analysis of each captured frame. To achieve this, we optimized the algorithm for efficiency, leveraging parallel processing capabilities of the Raspberry Pi. This ensures the drone can make prompt adjustments to its trajectory based on the latest track information.

Integration with Drone Control: The centroid's position, is translated into control signals for the drone's flight controller. Serial communication, using the Universal Asynchronous Receiver Transmitter (UART) protocol, facilitates seamless integration, ensuring quick and precise adjustments in the drone's flight path.

Computational Resources: The algorithm was implemented on Raspberry Pi with a quad-core ARM Cortex-M4F processor. Computational efficiency was a key consideration, and optimizations, such as using hardware-accelerated libraries for image processing tasks, were implemented. This allowed the algorithm to meet real-time processing requirements while ensuring efficient utilization of the available computing resources.

4. Implementation

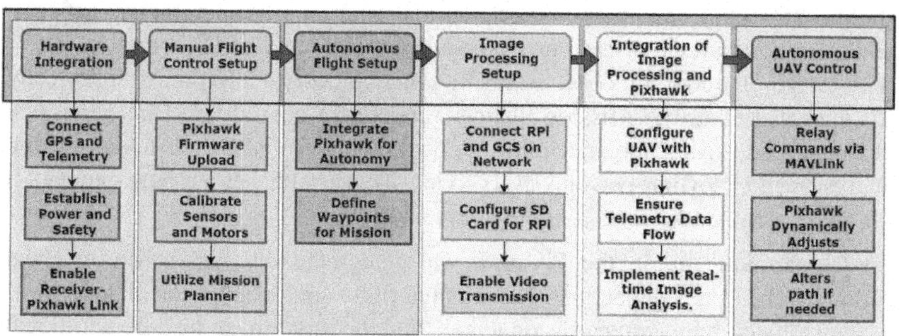

Fig. 26.3 Workflow of implementation

Source: Author's compilation

4.1 Configuration of Pixhawk Flight Controller and other Components

The integration of the components in the autonomous line follower drone involved meticulous connections to the Pixhawk Flight Controller, showcasing a comprehensive avionic system. The GPS module uses Inter-Integrated Circuit (I2C) connection, utilizing Serial Data (SDA) and Serial Clock (SCL) pins for data transfer, ensuring accurate location awareness with a 5V power supply. The Telemetry System interconnects with the Pixhawk, via pins for power, communication, and grounding, enabling real-time data transmission and remote monitoring.

The Pixhawk Flight Controller is linked to power components and safety features. The Power Module Connections include pins for voltage and current, ensuring a stable +5V power supply. Safety measures involve a +3.3V powered safety switch interfacing with LED Safety and Safety pins. The receiver communication employs Pulse Position Modulation (PPM) input pins on the Pixhawk board, while the Pixhawk establishes connections with a Raspberry Pi, facilitating power, data transmission, and grounding. This interconnected hardware setup forms the foundation for both manual and autonomous flight control.

Fig. 26.4 Assembled drone

Source: Author's compilation

The Pixhawk Flight Controller underwent firmware upload and configuration using the Mission Planner from ArduPilot for manual flight control phase. Calibration processes including axis, accelerometer, compass, and radio calibration, ensured precise control. Motor and ESC calibration, along with transmitter configuration, results in a meticulously configured system ready for manual flight. The Mission Planner serves as a central interface, reflecting changes made using the transmitter and allowing for real-time telemetry log analysis and FPV (first-person view) driving. This establishes a robust foundation for subsequent autonomous flight implementation.

For autonomous flight, the Pixhawk was integrated into the hardware setup following the same connections as in manual flight control. Mission planning was executed by defining waypoints on a map, and the mission, along with additional commands, was uploaded to the Pixhawk. The entire process, from hardware setup to autonomous flight, was systematically conducted, ensuring the Pixhawk flight controller's successful integration into the autonomous line follower drone.

4.2 Implementation of Image Processing

The Raspberry Pi and the GCS were connected to the same Network or Local Area Network (LAN) for communication. The Raspberry Pi's SD card is configured with the same network used by the GCS. A Virtual network computing (VNC) facilitated video transmission viewer, requiring the IP address of the Raspberry Pi. This initiates the connection, revealing the remote desktop on the GCS.

4.3 Integration of Image Processing and Pixhawk

This entailed configuring the UAV with Pixhawk, connecting it to an onboard camera, and transmitting video output to a computer for image processing. Pixhawk continuously communicates telemetry data to the GCS, providing essential information for precise analysis such as GPS location, altitude, and orientation. Results from image processing are sent back to Pixhawk, empowering UAVs with autonomous capabilities.

The finely tuned image processing algorithm, executed on Raspberry Pi, enables the drone to recognize and respond to predefined visual cues. Commands are relayed through the Micro Air Vehicle Communication Protocol (MAVLink) communication protocol in ArduPilot. Pixhawk dynamically adjusts flight path based on processed data, establishing seamless integration between image processing and UAV control. Upon detecting an obstacle, the Pixhawk can alter its path by transmitting new commands to the control system. This autonomy extends to instructing motor and control surfaces for real-time adjustments in trajectory. This enhances UAV autonomy and safety, particularly in challenging terrains or obstructed environments.

5. Results

Designed for precision agriculture, the image processing system provides real-time guidance to the drone for optimal field monitoring and management. As the camera scans the predefined track in an agricultural field, the image processing algorithm forms contours around the identified region indicated in green in the below figure, and the centroid of the largest contour is determined indicated in yellow in the below figure. The result of the image processing system guides the drone to position itself over the track by giving precise directions. If the track or for example the crop is positioned to the right of the centroid or central region, a "turn right" message is generated, indicating that the drone needs to adjust its

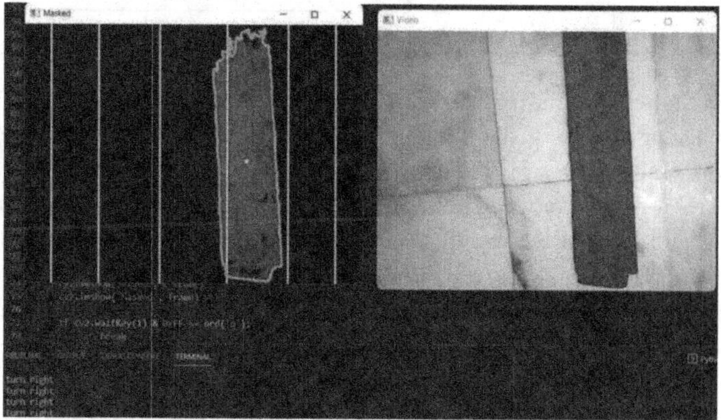

Fig. 26.5 Image processing system Output1

Source: Author's compilation

direction to the right to align the track within the central region. When the centroid falls within the central region, the output shown is "keep going," indicating that the drone is successfully following the track and it should continue the same path.

Conversely, if the track is located to the left of the central region, a "turn left" message is displayed. This signals that the drone should turn towards the left to bring the track back to the central region. These messages serve as real-time guidance for the drone's navigation based on the position of the track relative to the central region.

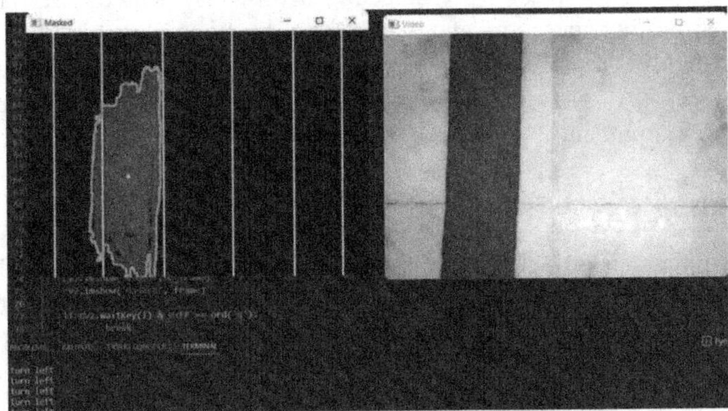

Fig. 26.6 Image processing system Output2

Source: Author's compilation

Table 26.2 Performance metric table

IoU	Centroid Accuracy	Spatial Accuracy	GPS Accuracy
85%	5px	2 cm	1 m

The metric table summarizes the key performance metrics used to evaluate the accuracy and precision of the image processing algorithm. Here the Intersection over Union (IoU), measures the overlap between the detected bounding box i.e., predicted region and annotated bounding box i.e., ground truth region. By dividing the intersection area of the two bounding boxes by their union area, it determines the accuracy of object localization. Higher IoU value indicates a greater overlap and, as a result, greater object detection accuracy. Centroid accuracy metric assesses the accuracy of detected objects' centroid localization relative to their ground truth positions. A smaller deviation in centroid position indicates greater accuracy in object localization, which is critical for precise crop identification and tracking in agricultural fields. Spatial accuracy measures the precision of detected bounding boxes in relation to the actual boundaries of the objects being detected. This metric quantifies how closely the detected regions match the ground truth regions, providing information about the algorithm's ability to accurately delineate crop and other agricultural features. The GPS accuracy metric measures the precision of geographic positioning for detected objects. It assesses the ability of GPS coordinates to accurately represent the spatial location of crops within agricultural fields. This metric is especially important in applications involving large-scale crop monitoring and management, where precise geospatial data is required for decision-making. Overall, the system demonstrates the following features:

Accuracy and Precision in crop monitoring: The contour formation exhibited a high degree of accuracy, successfully outlining the predefined track. This accuracy is crucial for precise monitoring of crop health and identification of potential issues.

Response Time for precise Navigation: The system showcased rapid response times, with the drone adjusting its course within milliseconds of detecting changes in the track's position. This quick response contributes to the system's effectiveness in real-time navigation for precision agriculture tasks.

Impact of Environmental Factors: The system exhibited robust performance under varying lighting conditions; however, challenges arose when the track surface texture changed significantly. Future enhancements may focus on improving the system's adaptability to diverse track surfaces or agricultural terrains.

Dynamic Track Following: The system demonstrated exceptional adaptability to dynamic changes in the track's position. As the track moved within the camera's field of view, the drone consistently adjusted its trajectory, ensuring continuous and accurate track following.

The system's outcomes not only demonstrate immediate applicability in realms such as aerial surveillance and search and rescue missions but also underscore the commitment to continual refinement in design and algorithmic efficacy within the dynamic landscape of this technological domain.

6. Conclusion

This research endeavour has culminated in the successful development of an autonomous path follower drone, integrating state-of-the-art components, sophisticated algorithms, and meticulous design considerations. The collaboration of the Pixhawk flight controller with a carefully curated set of components has resulted in a drone boasting advanced flight control capabilities, exhibiting stability, precision, and adaptability within the agriculture landscape. Similarly, the amalgamation of Pixhawk's open-source, modular design with computer vision algorithms for real-time image analysis has proven to be instrumental in achieving enhanced autonomy and safety for the drone. The seamless communication between Pixhawk and the image processing software contributes to a streamlined and efficient decision-making process, further solidifying the system's practicality. The modular nature of Pixhawk not only ensures adaptability within the agricultural sector but also allows for potential expansions and upgrades, paving the way for incorporating cutting-edge technologies or refining existing functionalities. This adaptability positions the integrated system as a foundation for continued research and development in the rapidly evolving field of autonomous UAVs.

Despite the integration of the advanced components and sophisticated algorithms, the Pixhawk-based drone system has certain limitations. One notable limitation is the reliance on environmental conditions and sensor accuracy. Factors such as adverse weather conditions, varying terrain, and inconsistent lighting can affect the drone's performance and reliability. Additionally, while the use of additional sensors like sonar or Light detection and ranging (LiDAR) can aid in obstacle detection and avoidance, there may still be scenarios where unexpected obstacles are encountered, leading to potential collisions. Moreover, the integration of advanced algorithms for autonomous navigation and object tracking, while promising, may introduce complexities in terms of computational resources and real-time processing requirements, potentially limiting the drone's responsiveness in dynamic environments. Addressing these limitations will be crucial for further advancements in ensuring the reliability and adaptability of autonomous UAV systems in agricultural settings.

In future, the precision and stability of the drone's line-following capabilities can be enhanced by integrating object tracking algorithm such as Mean-Shift or Kalman filter. For instance, the drone's speed and direction can be altered in real-time using feedback control or deep learning algorithms. Adding obstacle recognition and avoidance skills could boost the drone's adaptability. Leveraging deep learning algorithms can enable the drone to learn and adapt to various track configurations. Neural networks can be trained on diverse datasets to recognize patterns and optimize the drone's response to different track conditions. Furthermore, specialized payloads can be added onto the drone to extend its utility. For example, attaching a water-spraying mechanism can enable the drone to perform tasks such as agricultural irrigation or firefighting in remote areas. This enhances the drone's versatility, allowing it to undertake a broader range of tasks beyond line following. Looking forward, the potential applications of this autonomous path follower drone are vast, ranging from precision farming operations to crop monitoring. The fusion of cutting-edge technology and meticulous engineering showcased in this project serves as a testament to the possibilities within the realm of autonomous aerial systems. In essence, this research endeavour not only contributes to the current discourse on autonomous drones but also underscores the significance of interdisciplinary collaboration, persistent refinement, and a forward-looking approach in the ever-evolving landscape of unmanned aerial vehicle technology.

References

1. Zou, J.-T.; Dai, X.-Y. The Development of a Visual Tracking System for a Drone to Follow an Omnidirectional Mobile Robot. Drones 2022, 6: 113. https://doi.org/10.3390/drones6050113
2. Choi, J.H.; Lee, W.-S.; Bang, H. Helicopter Guidance for Vision-based Tracking and Landing on a Moving Ground Target. In Proceedings of the 11th International Conference on Control, Automation and Systems, Gyeonggi-do, Korea, 26–29 October 2011: 867–872.
3. Granillo, O.D.M.; Beltrán, Z.Z. Real-Time Drone (UAV) Trajectory Generation and Tracking by Optical Flow. In Proceedings of the 2018 International Conference on Mechatronics, Electronics and Automotive Engineering, Cuernavaca, Mexico, 26–29 November 2018: 38–43.
4. Shao, Y.; Tang, X.; Chu, H.; Mei, Y.; Chang, Z.; Zhang, X. Research on Target Tracking System of Quadrotor UAV Based on Monocular Vision. In Proceedings of the 2019 Chinese Automation Congress, Hangzhou, China, 22–24 November 2019: 4772–4775
5. Liu, R.; Yi, J.; Zhang, Y.; Zhou, B.; Zheng, W.; Wu, H.; Cao, S.; Mu, J. Vision- guided autonomous landing of multirotor UAV on fixed landing marker. In Proceedings of the 2020 IEEE International Conference on Artificial Intelligence and Computer Applications, Dalian, China, 27–29 June 2020: 455–458
6. Ceppi, Paolo (2020): Model-based Design of a Line-tracking Algorithm for a Low-cost Mini Drone through Vision-based Control. University of Illinois at Chicago. Thesis. https://doi.org/10.25417/uic.14134460.v1
7. García-Olvera, David & Hernández-Godínez, Armando & Nicolás-Trinidad, Benjamín & Cuvas-Castillo, Carlos. (2020). Line follower with a quadcopter. Pädi Boletín Científico de Ciencias Básicas e Ingenierías del ICBI. 7. 30–36. 10.29057/icbi.v7i14.4727.
8. Wang, Shuaijun, Fan Jiang, Bin Zhang, Rui Ma and Qi Hao. "Development of UAV-Based Target Tracking and Recognition Systems." IEEE Transactions on Intelligent Transportation Systems 21 (2020): 3409–3422.

Technologies for Energy, Agriculture, and Healthcare – Shailesh Nikam et al. (eds)
© 2024 Taylor & Francis Group, London, ISBN 978-1-032-98028-7

27

KISAN SEVAK: A COMPREHENSIVE MOBILE-BASED DIAGNOSIS SYSTEM FOR SUSTAINABLE AGRICULTURE

Chirag Sharma,
Anirudha Ta, Tejas Pundlik,
Uditi Sinha, Sarfaraz Shaikh, Atharva Tambe
Student, K.J. Somaiya College of Engineering,
Somaiya Vidyavihar University
Swati Mali[1], Archana Gupta[2]
Assistant Professor,
K.J. Somaiya College of Engineering,
Somaiya Vidyavihar University

Abstract: Plant ailments commonly stem from pests, insects, and pathogens, potentially causing a considerable reduction in productivity on a large scale unless effectively controlled in a timely manner. As a result, a significant portion of harvest is lost due to crop diseases, primarily because of delayed and informal treatment. This paper aims to tackle this issue by developing an integrated system for farmers and experts. The solution is designed to employ a machine learning model for timely and precise diagnosis with treatment recommendations for crop diseases. It ensures a proactive approach to crop health management. The machine learning model is backed up by experts. Additionally, it provides agriculture-relevant news information, like weather, based on the farmer's location. The proposed solution achieves better results in terms of accuracy, efficiency of diagnosis and accessibility.

Keywords: Crop disease detection, Custom ResNet50, Mobile and web based solution, ResUNet, Sustainable agriculture

Corresponding author: [1]swatimali@somaiya.edu, [2]archana.gupta@somaiya.edu

DOI: 10.1201/9781003596707-27

1. Introduction

A significant portion of crop harvest is lost to diseases such as leaf rust, blue-green mold, transit rot, anthracnose, stem end rot and other such diseases. Lack of timely expert guidance heightens the issue, leading to substantial overall harvest losses. According to Roberts (2006), on average, 12% of maize, barley, and soya bean crops are affected, while 24% of potatoes and groundnuts are lost to diseases. According to Oerke (2006), wheat and cotton face 50-80% infection rates, resulting in annual economic losses of around 40 billion dollars.

To tackle this problem, we propose a unified solution supported with a mobile application and a website. The mobile application allows farmers to submit the diseased crop image to the model, which is backed by crop health experts. Visualization tools, such as choropleth maps, provide insights, and the platform includes weather predictions based on the farmer's location. This comprehensive solution aims to address the lack of available expertise and improve the detection and treatment of crop diseases, reducing annual losses.

The paper is structured to ensure a coherent presentation. It begins with an introduction that provides an initial overview, followed by a section that delves into a literature survey, offering insights into existing research. The methodology of the system is then detailed. Results are discussed in another section, followed by future scope in the next section and the final section concludes the paper.

2. Literature Survey

Extensive research has been carried out to thoroughly explore diverse domains related to crops and their disease Some of the relevant initiatives are listed here in Table 27.1.

Table 27.1 Comparison of different papers [3][4][5]

Features	Wang et al., 2023	Anwarul et al., 2023	Sharma et al., 2020
Model Used	Ultra Lightweight Efficient Net	Convolutional Neural Network (CNN)	Convolutional Neural Network (CNN)
Dataset Used	Plant Village	Plant Village	Plant Village
Number of classes covered	38	15	–
Results	Accuracy: 98%	Accuracy: 94%	Accuracy: 90%

Hota and Verma (2022) advocate the use of the Interpretive Structuring Model to promote smart agriculture in India. They emphasize the crucial role of digital technology integration for sustainability and farmer well-being, despite existing challenges in digital agriculture adoption.

The work done by Abbasi et al. (2022) propose a smart agriculture system for Agriculture 4.0 to address growing agri-food demands. Their review of crop farming trends explores digital technologies, particularly in open-air farms, with applications mostly in the prototypical phase. The study identifies hurdles to digitization in agriculture, offering insights into the current status and future prospects.

In their study, Wang et al., (2023) introduce a highly efficient network tailored for identifying plant disease and pest infections from images, particularly suitable for scenarios with constrained computational resources. Their method emphasizes achieving precise classification of diseases and pests while ensuring minimal model complexity.

Sharma et al., (2020) used data augmentation techniques with a convolution based neural network. The study used a plant village dataset with 38 classes achieving a commendable accuracy of 90.32%.

Given that farming actively employs Machine Learning techniques, Anwarul et al., (2023) introduced a web-based application designed to streamline the process of uploading images for identification purposes. Their model, utilizing a Convolutional Neural Network architecture comprising seven convolutional layers, was trained on the Plant Village dataset, which encompasses various plants such as potatoes, bell peppers, and tomatoes.

Despite available insights, there is a noticeable absence of a solution that incorporates both disease prediction and an expert system. The proposed project seeks to integrate these two components, along with agricultural news, weather forecasting, and a discussion platform for farmers to engage in crop-related topics.

3. Methodology

This paper adopts an approach wherein farmers capture and upload images of diseased crops to the machine learning model from the mobile application, as shown in Fig. 27.1. The model promptly provides a suspected diagnosis and prescription to the farmer. In cases of low accuracy in the suspected diagnosis, the crop image, along with the farmer's description, is forwarded to an expert through the dedicated website for expert consultations. The expert's diagnosis is then communicated back to the farmer. The model is further trained using the uploaded image and the expert diagnosis received for the crop. The ResUNet model is used for the segmentation of the diseased part of the crop, while the ResNet50 model is used for the prediction of the disease..

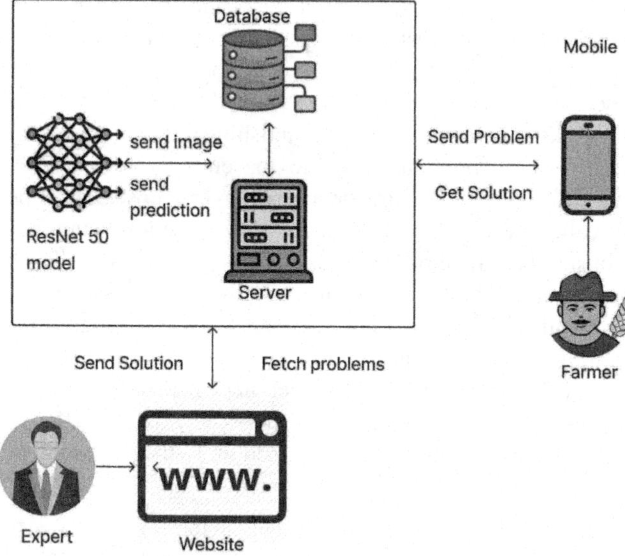

Fig. 27.1 Overview of proposed solution

Source: Author

4. Dataset

The dataset utilized for training the machine learning model is published at Kaggle by Ali and belongs to Mohanty (2016).The images in the dataset are utilized for both testing and training in disease classification. Additionally, the dataset includes masks used for training and testing segmented leaf images. It comprises 54,303 RGB images depicting healthy and diseased crops, along with masks designed for the segmentation of diseased crop leaves. The dataset encompasses a total of 38 classes.

For each class, the dataset exhibits a wide variety of diseased leaves, and the provided masks are precise, making them ideal for image segmentation. The image dataset has been divided, allocating 80% for training purposes and 20% for testing. To enhance the model's generalization capacity and to fight class imbalance problems, image augmentation techniques such as rotation, flipping, and shearing have been applied to the images. The images in the dataset are of resolution 250 by 250.

5. Machine Learning Model

The prediction of diseases in the model occurs in two stages. First the image is passed to a ResUNet model for segmenting the diseased leaves from the rest of the background in the image. After which the segmented image is passed on to a custom ResNet50 model. ResUNet consists of residual blocks arranged in two layers. These two layers are joined together with the help of bridges. The residual blocks have convolutions of stride 2 that reduce the dimensions of the image by half. The skip links help in alleviating the vanishing gradient problem by allowing the gradients from the previous block to the current block.Thus the residual blocks help in capturing fine image details. In layer 2 upscaling is performed of the reduced feature vectors produced in encoding. The upscaling is done to help the blocks construct the final binary mask of the segmented leaf. The bridge takes feature vectors from layer 1 and adds them to the layer 2 feature vector. This is done as the upscaling of the feature vector results in loss of details thus adding the gradients from the corresponding residual block helps in capturing finer details. The final output is a binary mask of the segmented leaf. Bitwise and operation between the mask and the original image helps in extracting the diseased leaf from the background. Now the image is passed on to the ResNet50 model where the image classification is done.

The machine learning model used is the ResNet 50 model (Fig. 27.2) that contains skip links. Each block of the model consists of convolution layers with batch normalization followed by ReLU and Max pooling layer. The convolutional layer helps in extracting features from the image and the batch normalization layer helps in normalizing the input to prevent the issue due to too much internal and external covariate shift. The max pooling layer helps in picking up important features from the feature list and also pick up location invariant features.

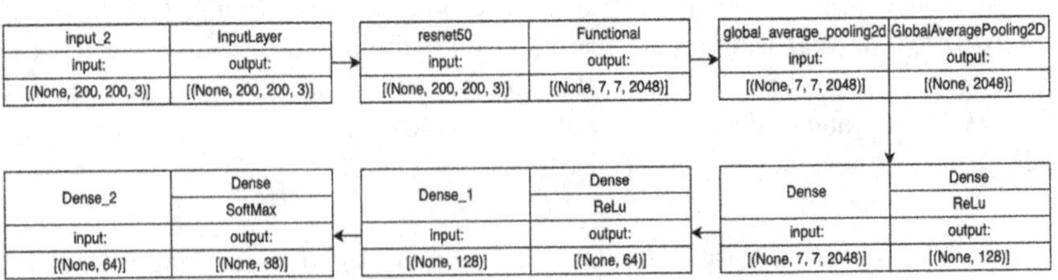

Fig. 27.2 Layers of custom ResNet50

Source: Author

The top layer of the model consists of a global max pooling layer followed by two fully connected dense layers. Both the layers have ReLU as the activation function since it helps in non linearity and is easy to compute. The output layer consists of 38 neurons representing the classes in the dataset.

The input layer takes images in the dimensions (200, 200, 3). Which is then connected to a preprocessing layer after which we have a layer that adds bias to the input image. Finally, the image goes to the ResNet 50 model which gives the predicted output class of the image.

6. Farmer Mobile Application

The Home Section allows farmers to submit photos of suspected plant diseases. After processing through the ResNet50 Model, the system provides a prompt reply to the farmer.

The Community Section encourages farmer interaction through posts with likes and dislikes, powered by WebSocket for real-time updates. In replies, farmers access the history of crop issues with expert or model-generated suggestions. The Settings Section allows farmers to modify personal details and enable location-based push notifications using the Publisher/Subscriber model. There is a Map Section that showcases nearby cold storage facilities and trending diseases, while the News Section keeps farmers updated on agricultural news and real-time weather conditions with a 7-day forecast in India.

The expert selection model retrieves five specialists based on crop category, sending crop images to a panel consisting of three newcomers and two seasoned professionals. Expert ratings are influenced by farmer upvotes, encouraging continuous improvement and knowledge sharing.

7. Expert Website

The platform enables experts to address farmers' crop disease submissions, where experts can categorize farmers' crop disease submissions based on their status—whether they are completed, pending, or encompassing all submitted problems. Experts provide their diagnosis and prescription which is stored in the database with the total expert responses. On reaching 5 responses, the most frequent diagnosis is chosen.

The database incorporates a scheduled task to monitor the duration since a problem was submitted. If the elapsed time exceeds 5 days, the issue is reassigned to a larger pool of experts. This process repeats once. Similarly, it applies if there is a tie among expert diagnoses or if there is no clear winner. If a problem remains unresolved for 10 days, it is opened to all experts for voting and prescribing. After a 3-day voting period, the prescription with the highest number of upvotes becomes the designated solution for the farmer. Expert selection is tailored to those with expertise in the specific crop category.

Maps provide information on proximate retailers for farmers by taking farmers location and querying it in the database to get a list of retailers. The locations of the retailers are marked on map along with relevant information regarding the retailer

8. Results

The customized ResNet50 model demonstrates outstanding accuracy in both its training set, consisting of 43,442 images, and its testing set, comprising 10,860 images. Figure 27.3 showcases an impressive accuracy rate of 95.24%, coupled with a corresponding loss of 0.511, as illustrated in Fig. 27.4. It is crucial to emphasize that the images within the plant village dataset are captured under ideal conditions, and it is important to acknowledge that accuracy may diminish in real-world scenarios. In order to evaluate overfitting, the model undergoes exposure to grayscale images as well as images with varying aspect ratios and scales. The accuracy exhibits a decrease to 85%, indicating the presence of some overfitting, although the decline is not significant.

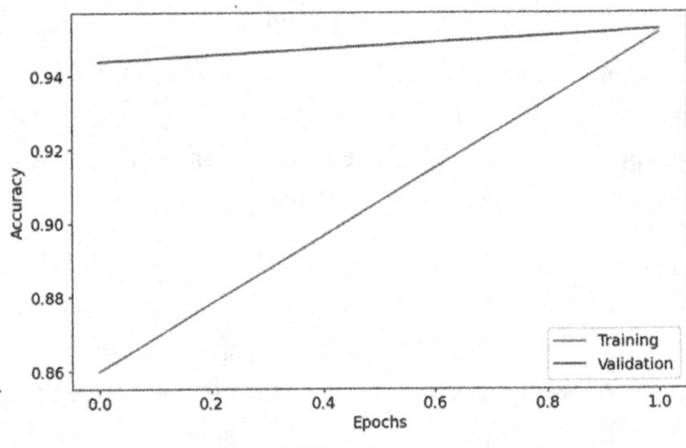

Fig. 27.3 Accuracy

Source: Author

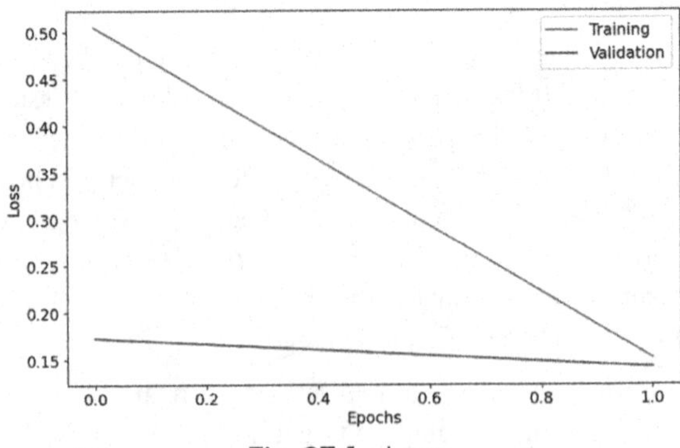

Fig. 27.4 Loss

Source: Author

Table 27.2 Results comparison [3][4][5]

Features	Wang et al., 2023	Anwarul et al., 2023	Sharma et al., 2020	Proposed Model
Accuracy	98%	94%	92%	95.4%
Speciality	Uses Spatial pyramid pooling with Residual Depthwise convolution	Uses 7 layers of Convolution neural network with 4 pooling layers & 2 fully connected layer	Uses Convolution, max pooling and dense layers	Uses max pooling and fully connected layers on top of ResNet50
Mobile, Web app integration	No	No	No	Yes

This study proposes a model with 95% accuracy and provides both prediction and expert diagnosis, as outlined in Table 27.2. While the model's accuracy is slightly lower compared to Wang et al. (2023), it compensates with faster training due to reduced complexity. The model also has a quicker response time, making it well-suited for integration with mobile applications.

9. Conclusion

In conclusion, the potential enhancements include implementing real-time chat for direct farmer-expert communication. The application could also benefit from multi-lingual support to reach a wider global audience. Incentivizing experts to participate and soliciting regular feedback from farmers will further enhance the app. Insights/trend analysis on diseases based on location, seasons, crops are some of the more advanced features that can be added to the application for farmers.

The presented innovative system has a commendable accuracy of 95.24%, along with quick response time from the model, empowering farmers through intelligent crop health monitoring and diagnosis. This integrated system merges machine learning, expert consultations, and a community-driven knowledge-sharing platform. This approach enhances the accessibility of crucial information and also provides farmers with the tools they need to navigate the dynamic landscape of agriculture effectively.

References

1. Oerke, E-C. "Crop losses to pests." *The Journal of Agricultural Science* 144, no. 1 (2006): 31–43. https://doi.org/10.1017/S0021859605005708.

2. Roberts, Michael James. *The value of plant disease early-warning systems: a case study of USDA's soybean rust coordinated framework.* No. 18. USDA Economic Research Service, 2006.
3. Wang, Beibei, Chenxiao Zhang, Yanyan Li, Chunxia Cao, Daye Huang, and Yan Gong. "An ultra-lightweight efficient network for image-based plant disease and pest infection detection." *Precision Agriculture* (2023): 1–26. https://doi.org/10.1007/s11119-023-10020-0.
4. Anwarul, Shahina, Manya Mohan, and Radhika Agarwal. "An Unprecedented Approach for Deep Learning Assisted Web Application to Diagnose Plant Disease." *Procedia Computer Science* 218 (2023): 1444–1453. https://doi.org/10.1016/j.procs.2023.01.123.
5. Sharma, Ritesh, Sujay Das, Mahendra Kumar Gourisaria, Siddharth Swarup Rautaray, and Manjusha Pandey. "A model for prediction of paddy crop disease using CNN." In *Progress in Computing, Analytics and Networking: Proceedings of ICCAN 2019*, pp. 533–543. Singapore: Springer Singapore, 2020. https://doi.org/10.1007/978-981-15-2414-1_54.
6. Abbasi, Rabiya, Pablo Martinez, and Rafiq Ahmad. "The digitization of agricultural industry–a systematic literature review on agriculture 4.0." *Smart Agricultural Technology* 2 (2022): 100042. https://doi.org/10.1016/j.atech.2022.100042.
7. Hota, Jyotiranjan, and Virendra Kumar Verma. "Challenges to Adoption of Digital Agriculture in India." In *2022 International Conference on Maintenance and Intelligent Asset Management (ICMIAM)*, pp. 1–6. IEEE, 2022. https://doi.org/10.1109/ICMIAM56779.2022.10147002.
8. Ali, Abdallah. 2019. PlantVillage Dataset https://www.kaggle.com/datasets/abdallahalidev/plantvillage-dataset/ (accessed December 15, 2023).
9. Mohanty, SP. 2016. PlantVillage-Dataset. https://github.com/spMohanty/PlantVillage-Dataset (accessed December 15, 2023).

Technologies for Energy, Agriculture, and Healthcare – Shailesh Nikam et al. (eds)
© 2024 Taylor & Francis Group, London, ISBN 978-1-032-98028-7

28

A Trident Shaped EBG Integrated Monopole Antenna for Wearable Applications

Monika Budania[1],
Bharati Singh[2], Vandana Satam[3]

Department of
Electronics and Telecommunication Engineering,
K J Somaiya College of Engineering,
Somaiya Vidyavihar University,
Vidyanagar, Mumbai

Abstract: A compact trident shaped EBG integrated monopole antenna is proposed for 2.45 GHz ISM band. The overall size of the proposed EBG integrated antenna is $50 \times 30 \times 4$ mm^3. The simulated impedance bandwidth of 50 MHz (2.43-2.48 GHz) is achieved on tissue layer model. The realized gain value for the EBG integrated antenna is 0.134 dBi and the free space efficiency is above 60%. The antenna is kept at a minimum distance of 4 mm away from the body as compared to literature where this distance is greater than 5 mm. The simulated SAR values are 0.114 W/Kg over 1 g tissue and 0.218 W/Kg over 10 g tissue for 0.1 W input power and it follows the FCC limit making it suitable for wearable applications.

Keywords: Compact, EBG, SAR, Tissue model, Wearable antenna

1. Introduction

With the rapid growth in the wearable technology and evolution of 6G networks the utilization of consumer wearable devices has increased. In near future one can

[1]monika.budania@somaiya.edu, [2]bhartisingh@somaiya.edu, [3]vandanam@somaiya.edu

DOI: 10.1201/9781003596707-28

expect to carry number of devices and sensors that constantly communicate with each other for different applications. To realize this vision a wearable antenna plays a pivotal role. The antennas used for on body communication facilitates bio-telemetry to transmit the biological information collected from sensors while the in- body antennas can be used for other medical applications like endoscopy, cancer treatment etc. [1]. The wearable antennas must possess specific characteristics like lightweight, cost- effective, low-maintenance, easy to integrate without the need for complex installations, addressing antenna performance during movements etc. Further, the electromagnetic radiation from the wearable antenna requires a specific absorption rate (SAR) level that should comply the safety limits as set by the international bodies. According to the European standard, the SAR value should be below 2 W/kg averaged over 10 g tissue and under 1.6 W/Kg over 1 g tissue [2].

In recent years, many efforts have been taken to overcome the challenges in wearable antenna design. A triangular slotted CPW fed monopole antenna backed by an 4 × 4 artificial magnetic conductor (AMC) array is presented in [3]. The proposed antenna achieves low SAR values but the separation distance between human body model and the antenna is very large (15 mm). The authors in [4] proposed a compact dual- band wearable antenna on RO3003 substrate with size of 41 × 44 mm^2. The SAR value at 2.4 GHz frequency is 0.955/0.571 W/kg for 1 g/10 g of human tissue and 0.478 /0.127 W/kg for 5.8 GHz respectively. The gain is narrow and the distance between antenna and body for SAR evaluation is not mentioned.

In [5] a low-profile EBG integrated wearable textile antenna for Ultra-Wideband (UWB) healthcare applications is presented. The proposed antenna demonstrates robust performance for bending conditions but at larger (10 mm) gap positioning between the antenna and body; also, the radiation pattern is deteriorated at higher frequency. The authors in [6] proposed a dual-band CPW fed T-structured antenna at 2.4 GHz/5.2 GHz for on-body applications. With EBG, the SAR values are reduced but the gain obtained on tissue layer model is negative. A broadband high-gain antenna based on a non-uniform metasurface (MS) is proposed in [7]. The measured average gain is 7.23 dBi. There is discrepancy in resonating frequency due to fabrication error. It has been observed that the performance of the antenna is greatly improved with the help of metamaterial structures but the overall profile of the antenna is increased due to large size of metasurfaces. In this work an effort has been made to overcome the above shortcoming and a compact EBG integrated wearable antenna design is proposed. Section-2 presents the design and analysis of the proposed antenna using CST studio software, Section-3 describes the antenna performance over EBG in free space, Section-4 presents the

EBG integrated antenna performance on tissue layer model and SAR evaluation, Section-5 concludes the proposed work.

2. Proposed Antenna Design

Initially a printed microstrip monopole antenna is designed (see Fig. 28.1) with a partial ground plane using equations (eq.1-eq.6). The proposed antenna is designed on FR-4 substrate having a thickness of 1.6 mm. The value of dielectric constant is 4.3. The size of the antenna is 42.45 × 33.96 mm^2. The antenna resonates with a wide bandwidth of 97 MHz (2.81-3.78 GHz) and the gain is 2.1 dBi. The radiation pattern is omnidirectional. The free space efficiency is 80%. It is found that the SAR values are too high (15.7 W/Kg over 10 g tissue).

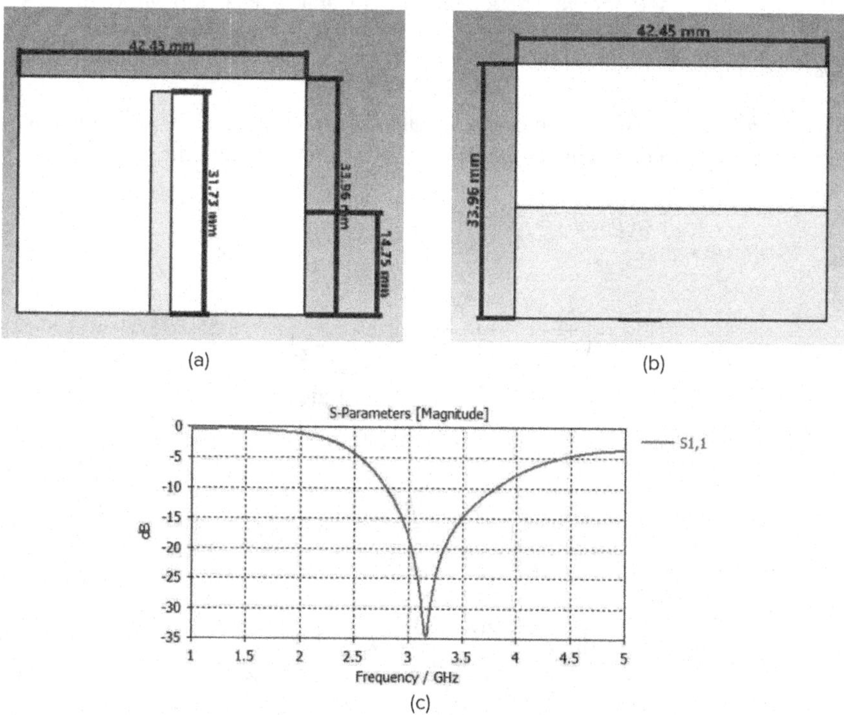

Fig. 28.1 Microstrip monopole antenna (a) Top (b) Bottom view (c) S_{11} vs Frequency plot

Source: Author

Hence, the design is further modified to improve the performance. In the next step a trident shaped monopole antenna is designed (see Fig. 28.2). The overall size of this design is compact 40 × 20 mm^2 and the impedance bandwidth obtained is 160

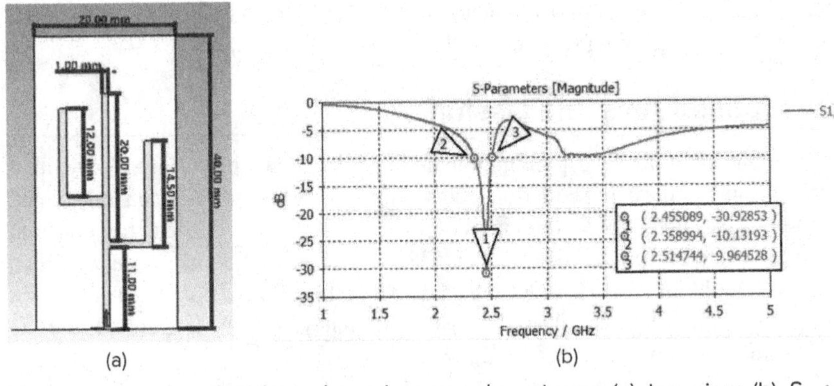

(a) (b)

Fig. 28.2 Proposed trident shaped monopole antenna (a) top view (b) S_{11} vs Frequency Plot

Source: Author

MHz (2.35 to 2.51 GHz) with centre frequency of 2.45 GHz. The two side stubs are added for impedance matching as well as for miniaturization.

$$L(patch) = \frac{\lambda}{4} = \frac{c}{4f\sqrt{\varepsilon_{eff}}} \tag{1}$$

$$\varepsilon_{eff} = \left(\frac{\varepsilon_r + 1}{2}\right) + \left(\frac{\varepsilon_r - 1}{2}\right)\left(1 + \frac{12h}{w}\right)^{-0.5} \tag{2}$$

W (substrate) = 2.5 L, L (substrate) = 2 L (3)

L (partial ground) = L (feedline) (4)

$$\text{W}(\text{feedline}) = \frac{7.48h}{e^{\left(z_0 \frac{\sqrt{\varepsilon_r + 1.41}}{87}\right)}} - 1.25 * t \tag{5}$$

$$L(\text{feedline}) = \frac{\phi(\pi/180)}{k_0 \times \sqrt{\varepsilon_{eff}}}, \quad k_0 = 2\pi \text{f/c} \tag{6}$$

{ε_{eff} = Effective dielectric constant, f = Resonance frequency, C = free space velocity, Z_0 = characteristic impedance, K_0 = electrical length in degree}

After the parametric study for the width of the patch it is noted that with decrease in the width the impedance matching improves and the bandwidth reduced. There is no change in the frequency as well as on impedance matching with the change in the armlength of the stub. Changing the position of the stub effects impedance matching due to change in current density at that point. The SAR values obtained when simulated on tissue layer model is 30.2 W/Kg over 1 g tissue and 12.3 W/Kg over 10 g tissue for 0.1 W input power. To reduce SAR values and further improve the gain of the proposed antenna EBG surface is designed.

3. Free Space Performance of the Antenna Over EBG

A rectangular ring EBG unit cell is proposed of size 2.2×2.8 mm^2. The 0° phase reflection is obtained at 2.45 GHz and the frequency range is 1.2 GHz – 3.7 GHz. The permittivity and permeability values are extracted from S parameter and it is found that both the values are negative. Thus, the proposed EBG design falls under the category of double negative left-handed metamaterial. The proposed antenna is placed over 18×5 array of EBG with separation gap of 4 mm in order to provide proper isolation and avoid short circuits and to eliminate mismatches. In simulation software air layer is inserted with dielectric of 1 to represent the

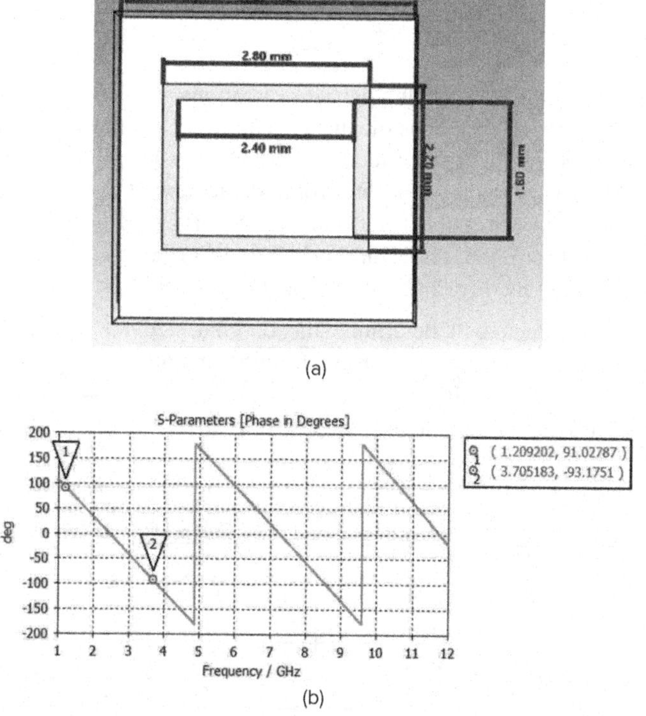

Fig. 28.3 EBG unit cell (a) Rectangular ring EBG unit cell (b) Reflection-phase graph

Source: Author

gap. It is observed that the EBG integrated antenna have narrow bandwidth as compared to antenna alone but the proposed EBG integrated antenna shows robust performance against frequency detuning (see Fig. 28.4).

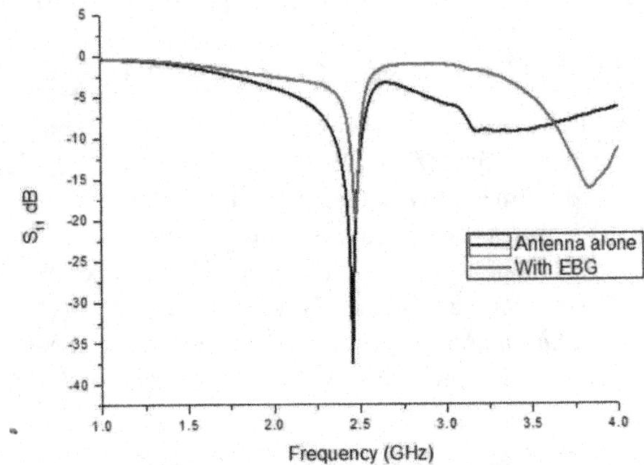

Fig. 28.4 S_{11} vs frequency plot

Source: Author

The EBG surface helps in reducing the back radiations and thus lowering the SAR values. It is also observed that as the distance between antenna and EBG Surface increases the back lobe further reduces also the resonating frequency shifts. Ideally this distance should be $\lambda/4$ but from practical point of view keeping such large separation is not feasible as it will create difficulties in incorporating the antenna onto body surface. The radiation pattern of the antenna with and without EBG is as shown in the Fig. 28.5. In H plane the pattern is circular.

A study was conducted to determine the optimal size for an EBG array by evaluating various array sizes and determining maximum value of gain and efficiency. Upon examination of the results (Table 28.1), it was found that the 18 × 5 EBG array exhibited the highest gain (4.3 dBi) and 90% total efficiency.

Table 28.1 Effect of EBG array sizes on antenna performance

EBG array size	Frequency (GHz)	Max Gain (dBi)	Total Efficiency (%)
14 x 1	2.45	-0.34	35
14 x 3	2.45	0.41	38
14 x 5	2.45	0.44	40
5 x 4	2.45	0.40	38
18 x 1	2.45	2.9	80
18 x 5	2.45	4.3	90

Source: Author

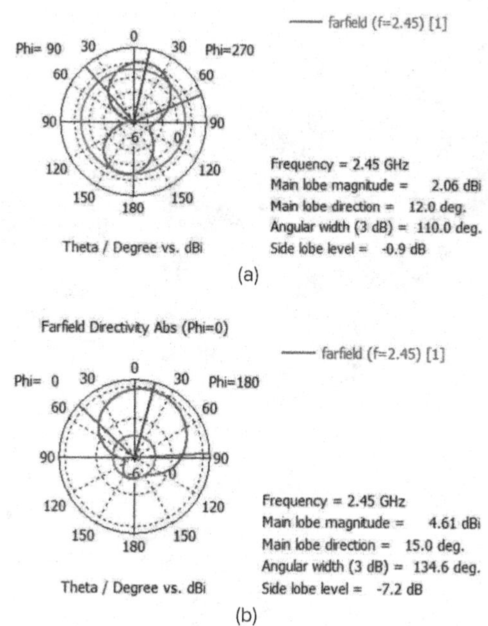

Fig. 28.5 Radiation pattern in E-Plane (a) without EBG (b) with EBG

Source: Author

4. Performance of the EBG Integrated Antenna on Tissue Layer Model

The performance of the proposed EBG integrated antenna is determined on the human body tissue layer model as shown in Fig. 28.6 (a). The four layers of tissue model consists of a 2 mm-thick skin layer, a 4 mm fat layer, a 8 mm muscle layer, and a 13 mm bone layer. The overall size of tissue layer model is 150×150 mm^2. The material properties of the tissues (refer Table 28.2) are determined from the

Table 28.2 Material properties of tissue layer model

Tissue Layer	Epsilon	Electric Cond. (s/m)	Density Rho (Kg/m^3)	Thermal Cond. (W/K/m)	Heat capacity (kJ/K/kg)	Thickness (mm)
Skin	42.92	1.561	1100	0.3	3.5	2
Fat	5.285	0.102	900	0.201	2.5	4
Muscle	52.791	1.705	1080	0.5	3.5	8
Bone	18.49	0.82	1850	0.41	1.3	13

Source: CST studio human tissue library

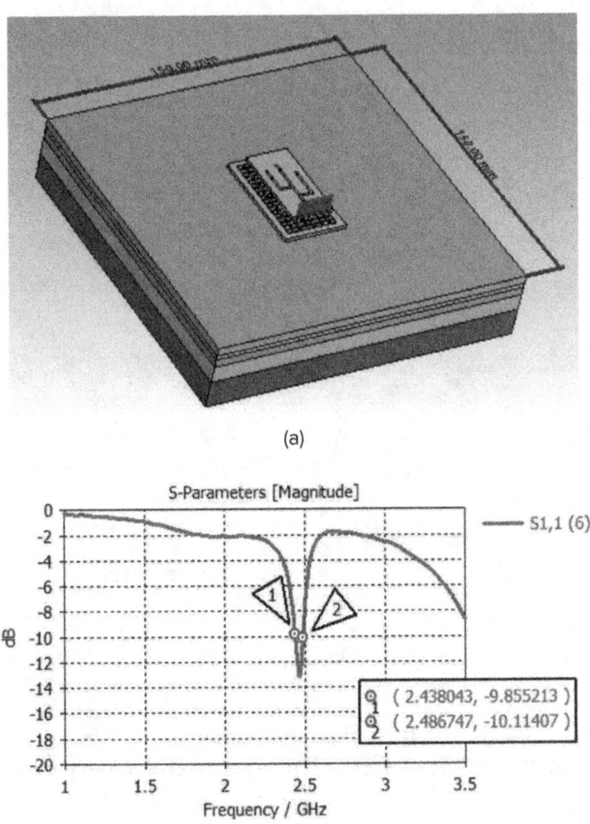

Fig. 28.6 (a) EBG integrated antenna on tissue layer model (b) S_{11} vs frequency plot

Source: Author

CST Studio human tissue library. A 4 mm air gap is set between the antenna and the tissue model to represent any worn clothes. The frequency range covered by EBG integrated antenna is 2.43-2.48 GHz. The impedance matching is affected due to highly dielectric tissue layer.

The maximum SAR value achieved from numerical simulations is 0.218 W/Kg over 10 g tissue and 0.114 over 1 g tissue for a gap of 4 mm between antenna and skin layer as compared to larger 5-15 mm gaps usually considered in literature. The SAR evaluations is based on the IEEE C95.3 standards and the power input given is 0.1W.

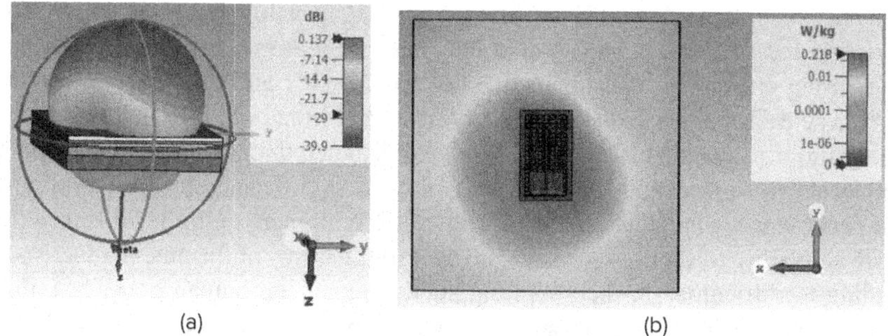

(a) (b)

Fig. 28.7 (a) 3-D Farfield realized gain (b) SAR Evaluation over 10 g tissue

Source: Author

Table 28.3 Comparison between presented work and literature work on EBG integrated wearable antennas in free space

Ref.	Freq. (GHz)	Gain (dBi)	Efficiency (%)	SAR (W/Kg) 10 g tissue	Size (mm^2)	Gap b/w antenna & tissue model (mm)
[08]	2.45/3.65	4.25/7.35	97/88	0.65	55.7 × 52.2	5
[03]	3.5/5.8	9.07/7.66	-0.78/-0.155 (dB)	0.02/0.08	86 × 86	3
[09]	2.45	7.07	> 75	0.244 (over 1 g)	68 × 38	-
[10]	2.45	7.3	80	0.09	35 × 35	5
This work	2.45	4.3	90	0.218	50 × 30	4

Source: Author

5. Conclusion

The design of a wearable antenna is challenging due to many constraints and effects of the human body. This work presents a compact EBG integrated trident shaped monopole antenna for wearable applications at ISM 2.4 GHz. The antenna geometry consists of a simple printed monopole antenna with partial ground plane. The two stub structures are used for impedance matching and miniaturization. The size of the proposed antenna alone is much compact (40 × 20 mm^2). The gain achieved is 2.06 dBi and bandwidth obtained is 160 MHz. The partial ground plane structure in the antenna results in back radiations leading to higher SAR values which is undesirable for wearable applications. To enhance the performance and reduce SAR values an EBG surface is being used. The proposed EBG size is compact as compared to literatures. The proposed EBG integrated antenna integrated shows robust performance against frequency detuning. The performance of the antenna is also enhanced in terms of gain and efficiency after using EBG. The gain in increased upto 4.3 dBi and efficiency is around 90%. The SAR values achieved are very low (0.114/0.218 W/Kg) and within the safety limits as per standards (1.6/2 W/Kg). It is noted that the integrated distance between EBG plane and main antenna plays a significant role in determining the performance characteristics of the antenna. With increase in this distance the SAR values are further reduced but the overall thickness of the antenna increases leading to difficulties in incorporating such antennas on human body. In future work the prototype will be fabricated and the verification of results will be carried out.

References

1. Karthikeyan, S., Gopal, Y.V., Kumar, V.G.N. and Ravi, T., 2019, October. Design and analysis of wearable antenna for wireless body area network. In IOP Conference Series: Materials Science and Engineering (Vol. 590, No. 1, p. 012022). IOP Publishing.
2. Ali, U., Ullah, S., Kamal, B., Matekovits, L. and Altaf, A., 2023. Design, analysis and applications of wearable antennas: A review. IEEE Access.
3. El Atrash, M., Abdalla, M.A. and Elhennawy, H.M., 2019. A wearable dual-band low profile high gain low SAR antenna AMC-backed for WBAN applications. IEEE Transactions on Antennas and Propagation, 67(10), pp.6378–6388.
4. Musa, U., Shah, S.M., Majid, H.A., Mahadi, I.A., Rahim, M.K.A., Yahya, M.S. and Abidin, Z.Z., 2023. Design and analysis of a compact dual-band wearable antenna for WBAN applications. IEEE Access, 11, pp.30996–31009.
5. Sam, P.J.C., Surendar, U., Ekpe, U.M., Saravanan, M. and Satheesh Kumar, P., 2022. A low-profile compact EBG integrated circular monopole antenna for wearable medical application. In Smart Antennas: Latest Trends in Design and Application (pp. 301–314). Cham: Springer International Publishing.
6. Dam, T.H., Le, M.T., Nguyen, Q.C. and Nguyen, T.T., 2023. Dual-Band Metamaterial-Based EBG Antenna for Wearable Wireless Devices. International Journal of RF and Microwave Computer-Aided Engineering, 2023.
7. Gao, G., Meng, H., Geng, W., Zhang, B., Dou, Z. and Hu, B., 2021. Design of a wide bandwidth and high gain wearable antenna based on nonuniform metasurface. Microwave and Optical Technology Letters, 63(10), pp.2606–2613.
8. Zhang, K., Vandenbosch, G.A. and Yan, S., 2020. A novel design approach for compact wearable antennas based on metasurfaces. IEEE Transactions on biomedical circuits and systems, 14(4), pp.918–927.
9. Abbasi, M.A.B., Nikolaou, S.S., Antoniades, M.A., Stevanović, M.N. and Vryonides, P., 2016. Compact EBG-backed planar monopole for BAN wearable applications. IEEE Transactions on Antennas and Propagation, 65(2), pp.453–463.
10. Kiani, S., Rezaei, P. and Fakhr, M., 2021. A CPW-fed wearable antenna at ISM band for biomedical and WBAN applications. Wireless Networks, 27, pp.735–745.

Technologies for Energy, Agriculture, and Healthcare – Shailesh Nikam et al. (eds)
© 2024 Taylor & Francis Group, London, ISBN 978-1-032-98028-7

29

Check Valve Simulation through Resource-Efficient FEA-CFD Coupled Analysis

Keshav Anand Kabra*,
Somrick Das Biswas
Student, Dept. of Mechanical Engineering,
K J Somaiya College of Engineering,
Vidyavihar, Mumbai, India

Siddappa S. Bhusnoor
Professor, Dept. of Mechanical Engineering,
K J Somaiya College of Engineering,
Vidyavihar, Mumbai, India

Abstract: This research presents a cost-effective method for FEA-CFD coupled analyses, tailored for low-cost hydraulic valve design. The approach departs from traditional computational interfaces, integrating manual calculations, static FEA, and finely tuned static CFD simulations to streamline the process for enhanced resource efficiency. Despite sacrificing some fidelity, the strong correlation between spring displacement and fluid pressure in the FEA module, serving as inputs for the CFD module, is established. The results demonstrate significant alignment between theoretical, simulation, and real-world values, confirming the accuracy of the method. This methodology addresses specific challenges in hydraulic valve simulation, contributing to efficient techniques for essential components in diverse industrial contexts within machinery and controls.

Keywords: Computational optimization, FEA-CFD analysis, Valve design, Machinery and controls

*Corresponding author: keshav.kabra@somaiya.edu

DOI: 10.1201/9781003596707-29

1. Introduction

Flow control valves play a vital role in hydraulic systems, efficiently managing fluid dynamics in machinery and controls. This research focuses on a nuanced exploration of these valves, utilizing sophisticated Finite Element Analysis (FEA) and Computational Fluid Dynamics (CFD) coupled analysis for simulation. The core objective is to unravel the technical intricacies of flow control valves and optimize their behaviour for enhanced performance using less computational techniques.

Within the expansive realm of hydraulic systems, flow control valves emerge as mechanical maestros, directing fluid flow within circuits. Beyond simple facilitation or restriction of flow, they possess the capability to shut down a line, redirect pressurized fluid, or finely regulate flow rates in specified regions. The fluid medium, typically oil, navigates through these systems, with the spool acting as the linchpin in controlling its movement—a focal point in this investigation.

The research delves beyond traditional valve functionalities, aiming to unravel complexities through simulation and analysis. Operated manually or automatically, these valves respond to various triggers, including physical, mechanical, pneumatic, hydraulic, or electrical stimuli [2]. The study transcends conventional valve operations, shedding light on nuanced dynamics that can be simulated and optimized through advanced numerical methods like FEA and CFD.

Among the array of valves, the check valve takes centre stage. Known by various names—non-return valve, reflux valve, retention valve, foot valve, or one-way valve—it embodies the concept of allowing fluid flow in a single direction. Understanding check valve behaviour becomes paramount in comprehending the broader dynamics of flow control valves, and this research scrutinizes their role in detail. Check valves, characterized by two openings in the body for fluid entry and exit, operate autonomously. Unlike counterparts, they lack external control mechanisms, devoid of valve handles or stems [4]. Despite apparent simplicity and small size, check valves play a significant role in numerous applications, even in common household items. This paper aims to unravel the intricacies of check valves within the broader canvas of flow control valve simulations.

The research extends to the simulation aspect, emphasizing resource-efficient FEA-CFD coupled analysis. By amalgamating these analytical approaches, the study provides a comprehensive understanding of fluid dynamics and structural behaviour in flow control valves. The goal is to contribute to advancements in hydraulic system design and optimization, elevating efficiency and reliability.

2. Literature Review

The foundation of our research is established upon a thorough examination of pertinent research papers that have significantly contributed to the comprehension and enhancement of valve behaviour. This literature survey encapsulates key insights from a diverse array of studies, each addressing critical aspects of structural and flow analysis in different types of valves. Egure-Hidalgo and Aburto Barrera's [1] seminal work focuses on the Structural Analysis of Ball Valves, employing the Finite Element Method (FEM). Their insights align seamlessly with our research goal of optimizing valve behaviour through FEM.

Saravanan and Mohanasundara Raju's [2] research delves into the Structural Analysis of Non-Return Control Valves, emphasizing the optimization of critical components. This focus on component optimization is pertinent to our objective of efficiency enhancement. Their findings offer valuable considerations for achieving optimal performance in valves. Filo, Lisowski, and Rajda [3] explored the Design and Flow Analysis of an Adjustable Check Valve using Computational Fluid Dynamics (CFD) methods. Emphasizing the improvement of flow characteristics and reduction of pressure losses, their work aligns with a parallel pursuit in our study. Understanding their methodologies contributes to our efforts to enhance flow characteristics in valve design. Skousen's work [4] provides foundational knowledge on check valves, offering insights into valve selection criteria and common problems. This fundamental understanding aids our research by providing context for selecting appropriate valve types based on specific application requirements. Skousen's insights contribute to a robust foundation for our valve design considerations. Wu et. al [5] present an alternative approach by utilizing CFD simulation for a pressure control valve. This aligns with our numerical methods-based strategy. Their work offers valuable perspectives on the application of CFD in analysing and optimizing pressure control valves, complementing our FEA-CFD coupled analysis approach. Davis and Stewart's research [6] validates CFD tools for predicting primary control valve performance, a key aspect in our emphasis on FEA-CFD coupled analysis. Boqvist's investigation [7] into swing check valves using CFD provides insights into dynamic characteristics and fluid-structure interactions. This enriches our understanding of valve dynamics, contributing to the overall analysis of complex fluid systems.

3. Methodology and Model Specifications

Traditional FEA-CFD coupled analysis involves an interface between FEA and CFD modules, exchanging data after each time step of the CFD run, leading to a high computational load. In their approach, they mitigate this computational burden by substituting the time-intensive computational interface with a discretized method.

The methodology begins by sacrificing a degree of fidelity to enhance computational efficiency. The process initiates in the FEA module, leveraging a combination of manual calculations and static FEA to establish a relationship between spring displacement or compression and fluid pressure. This relationship is crucial for understanding the behaviour of the valve, as the spring compression exhibits a linear correlation with the applied pressure. By varying parameters such as spring material properties, wire diameter, stiffness constants, and length, they construct a comprehensive framework for the analysis of multiple valves.

The step lengths chosen in Fig. 29.1 represent their engineering judgment as to what positions of the ball and spring assembly are most important for analysis. Design of Experiment (DOE) could be used as a feasible tool to balance computational load with desired fidelity.

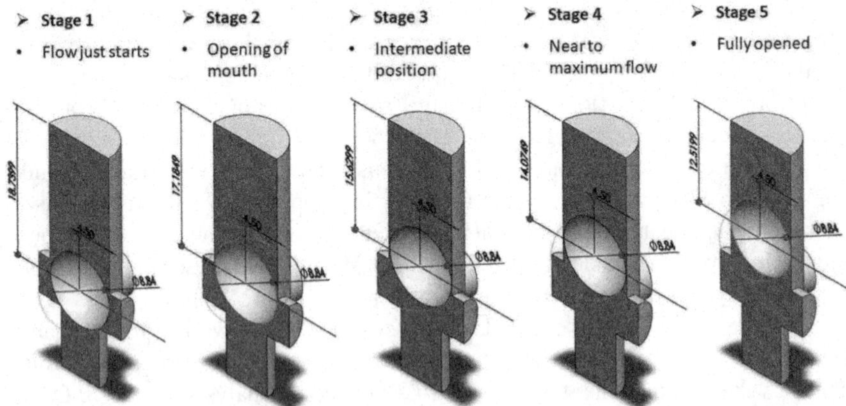

Fig. 29.1 Different stage formation of fluid region

Subsequently, they transition to CFD simulations, where specific valve opening areas are selected to calculate key parameters such as flow rate as a function of fluid pressure, minimum cracking pressure, and from Fig. 29.2 spring compression as a function of fluid flow rate. The use of static CFD simulations proves computationally efficient, especially given the compact dimensions of the valve (less than 30 mm in length and approximately 25 mm in diameter). These small sizes enable the utilization of fine tetrahedral meshes without imposing significant computational demands. Figure 29.3 depicts the velocity vectors of fluid in different stages of opening of valve.

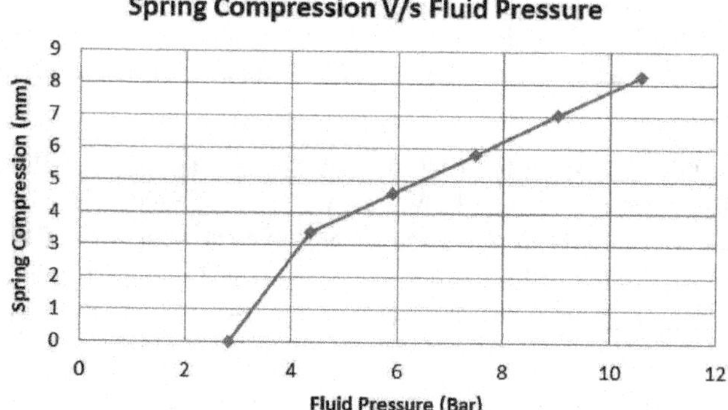

Fig. 29.2 Spring compression V/s Fluid pressure

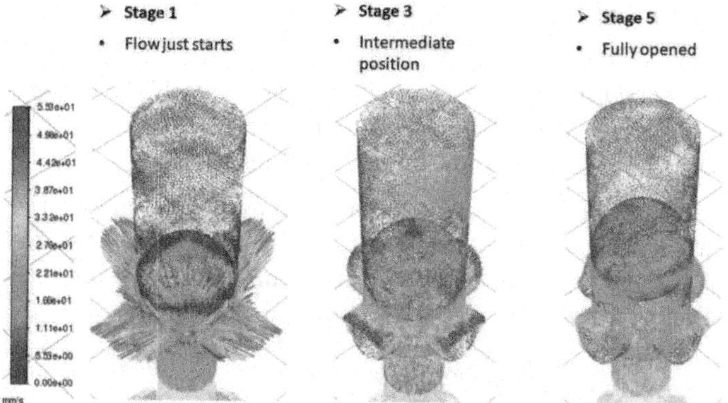

Fig. 29.3 Velocity vectors of fluid in different stages

The discretized approach involves multiple static CFD simulations, strategically chosen to cover a range of operating conditions. Post-convergence on these runs, they interpolate between computational data points to obtain a comprehensive understanding of the valve's behaviour. Notably, their analysis reveals that this method exhibits remarkable similarity with data available in published literature, affirming its efficacy in capturing the intricacies of complex systems.

4. Results

This innovative approach provides a cost-effective computational process for simulating complex systems. By strategically combining manual calculations with FEA and static CFD simulations, they streamline the simulation process, making it accessible for designing and analysing valves with limited computational resources. The results obtained through this methodology demonstrate its potential as a reliable and economical tool for simulating and optimizing complex engineering systems.

The application of the resource-economical methodology in simulating FEA-CFD coupled analyses for the design of low-cost valves has yielded compelling results. The results are presented through a scatter plot that showcases three distinct lines representing Theoretical, Simulation, and Real-World values, the latter sourced from existing literature. Notably, the alignment and proximity of these lines affirm the efficacy and accuracy of the proposed methodology.

The Theoretical line represents the anticipated values derived from the physics-based calculations between spring displacement and fluid pressure. These values serve as a benchmark, providing a theoretical foundation for the expected behavior of the valve under various conditions.

The Simulation line depicts the results obtained through the discretized approach, involving a combination of manual calculations, static FEA, and static CFD simulations. The strategic selection of operating conditions and fine tetrahedral meshes has contributed to the precision of the simulated values.

The experimental line consists of values adopted from literature [8]. These serve as the real metric for their comparisons. As Fig. 29.4 shows, their results can capture the real-world dynamics at a fraction of the computational load. The simulated values are within ~10% of the real-world results from literature.

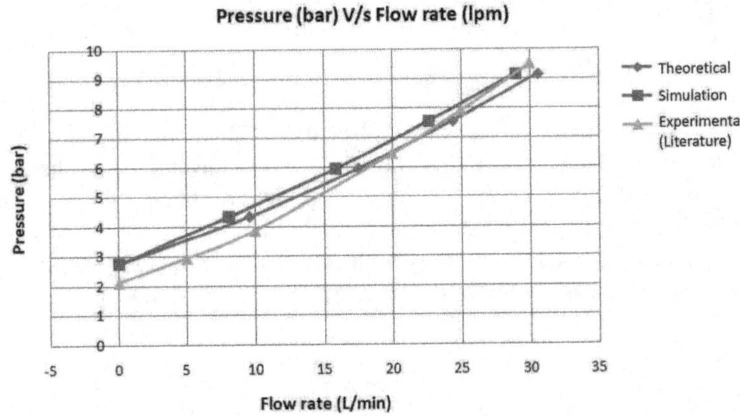

Fig. 29.4 Pressure v/s flow rate plot of valve

5. Conclusion

This research unveils a cost-effective approach for simulating FEA-CFD coupled analyses in the design of low-cost valves. The method, replacing the traditional computational interface with a manually discretized approach, demonstrates exceptional accuracy and efficiency. The alignment between Theoretical, Simulation, and Real-World values attests to the precision in capturing valve intricacies. By integrating manual calculations with FEA and static CFD simulations, the approach streamlines the simulation process, enabling the design and analysis of valves with limited computational resources. The robustness of the methodology underscores its potential significance in the domain of machinery and controls. This research opens a promising avenue for future studies in the realm of efficient mechanical design within this crucial domain.

References

1. Egure-Hidalgo, M., Aburto-Barrera, J. M., Torres-San Miguel, C. R., Martinez-Reyes, J., & Romero-Ángeles, B. (2019, June 27). Structural Analysis by Finite Element Method in Ball Valves to Improve Their Mechanical Properties. Advanced Structured Materials, 175–185. https://doi.org/10.1007/978-3-030-20801-1_13
2. K. G. Saravanan, & N. Mohanasundara Raju. (2015, April 22). Structural Analysis of Non-Return Control Valve using Finite Element Analysis. International Journal of Engineering Research And, V4(04). https://doi.org/10.17577/ijertv4is040889
3. Filo, G., Lisowski, E., & Rajda, J. (2021, April 16). Design and Flow Analysis of an Adjustable Check Valve by Means of CFD Method. Energies, 14(8), 2237. https://doi.org/10.3390/en14082237
4. Skousen, P. L. (1998, January 1). Valve Handbook. McGraw-Hill Professional Publishing.
5. Wu, D., Li, S., & Wu, P. (2015, September). CFD simulation of flow-pressure characteristics of a pressure control valve for automotive fuel supply system. Energy Conversion and Management, 101, 658–665. https://doi.org/10.1016/j.enconman.2015.06.025
6. Davis, J. A., & Stewart, M. (2002, August 19). Predicting Globe Control Valve Performance—Part I: CFD Modeling. Journal of Fluids Engineering, 124(3), 772–777. https://doi.org/10.1115/1.1490108
7. Swing check valve. (2004, August). World Pumps, 2004(455), 8. https://doi.org/10.1016/s0262-1762(04)00285-8
8. CVH0.S08 - VIS HYDRAULICS SRL - PDF Catalogs Technical Documentation Brochure. https://pdf.directindustry.com/pdf/vis-hydraulics-srl/cvh0s08/232900-998377.html

Note: All the figures in this chapter were made by the authors.

Technologies for Energy, Agriculture, and Healthcare – Shailesh Nikam et al. (eds)
© 2024 Taylor & Francis Group, London, ISBN 978-1-032-98028-7

30

A Smart Healthcare Companion with Tesseract OCR and KNN Integration

Nishant Kathpalia[1]
Department of Computer Engineering,
SIES GST, Nerul, India

Gabriel Nixon Raj[2]
Department of Computer Engineering,
SIES GST, Nerul, India

Madhavan Venkatesh[3]
Department of Computer Engineering,
SIES GST, Nerul, India

Abstract: Conversational Artificial Intelligence (AI) systems have emerged as transformative tools for enhancing user interactions across various industries. Our aim is the development and implementation of a conversational AI in the Medical Sector. Our objective is to streamline user interactions by leveraging AI-driven conversations that intelligently identify and present highly relevant information, reducing time consumption and eliminating irrelevant results. In the medical field, our conversational AI engages users in natural language conversations, leveraging NoSQL databases and cloud hosting for data storage. By asking targeted questions and using machine learning models, it efficiently provides users with highly relevant medical information. This streamlined, tech-driven approach reduces search efforts, making healthcare information more accessible and user-friendly. Our goal is to transform healthcare delivery by offering timely, precise, and individualized responses to medical queries.

Keywords: Adverse drug reaction (ADR), Computer vision, Conversational AI, Embedding, Indexing, Vector searching

[1]nishanttkathpalia@gmail.com, [2]gabriel.nixonraj@gmail.com, [3]maddyparker2002@gmail.com

DOI: 10.1201/9781003596707-30

1. Introduction

In the wake of unparalleled demographic shifts, India stands at the center of a population explosion, challenging its healthcare infrastructure and necessitating innovative approaches to address the evolving needs of its citizens. Since 1950, India's population has burgeoned by over 1 billion individuals, reaching an estimated 1.4 billion people surpassing the combined populations of Europe and rivaling that of the entire Americas. Remarkably, this demographic surge continues to propel India's population upward, diverging from the demographic trends observed in other populous nations like China. The challenge is not only in the sheer numbers but also in the escalating population density, a metric that underscores the demand for healthcare resources. In 2024, India's population density has risen to 438.58 people per square kilometer, marking a 0.92% increase from the previous year. [1] This trend follows a consistent upward trajectory, with each passing year witnessing incremental growth, reflecting the urgency for scalable healthcare solutions. Compounding the complexity, India has not conducted a comprehensive census since 2011, making real-time population data elusive. This information gap poses a unique challenge to healthcare initiatives, requiring innovative and adaptive approaches to cater to the health needs of a dynamic and ever-expanding population. In response to this imperative, beyond simply addressing medical queries, this conversational medical AI seeks to bridge the knowledge gap, providing personalized health insights and fostering proactive health management among India's diverse and expansive population. This paper aims to leverage data-driven insights to empower individuals, enhance health literacy, and contribute to the overall well-being of a nation undergoing transformative demographic shifts.

1.1 Rise in the Prevalence of Diseases in India

In recent years, India has witnessed a noticeable surge in the number and diversity of diseases, posing a multifaceted challenge to the nation's healthcare landscape. Factors such as rapid urbanization, changing lifestyles, environmental degradation, and increased global connectivity have contributed to the escalation of both communicable and non-communicable diseases. [2] Infectious diseases, including respiratory infections and vector-borne illnesses, persist as significant public health concerns, necessitating vigilant preventive measures and responsive healthcare systems. Simultaneously, the prevalence of non-communicable diseases, such as cardiovascular disorders, diabetes, and cancer, has seen a concerning uptick, often linked to lifestyle choices and an aging population. This complex interplay of various health challenges underscores the need for adaptive healthcare strategies, innovative medical technologies, and a robust public health infrastructure to effectively address the evolving disease landscape in India.

1.2 Preventable Diseases due to Inadequate Planning

The global landscape of health is marked by a multitude of diseases, some of which, with more prudent planning, could have been averted or mitigated. Effective disease prevention often hinges on the intersection of informed public health policies, community education, and strategic infrastructure development. The ability to prevent diseases goes beyond the realm of medical treatments; it involves addressing root causes, understanding socio-economic factors, and implementing measures that foster healthier living conditions. This broad perspective highlights the importance of comprehensive planning to proactively identify and tackle potential health challenges before they escalate.

1.3 The Importance of Boosting Medical Literacy

Enhancing people's awareness about the medicines they use is of utmost importance. With diverse healthcare practices and an abundance of medications available, it becomes crucial for individuals to have a deeper understanding of their prescribed drugs. Medications are central to managing health conditions, but the lack of awareness about their proper usage, potential side effects, and interactions is a common challenge. By empowering individuals with comprehensive knowledge about their medicines, we not only promote informed decision-making but also encourage active participation in personal healthcare.

2. Statistics

Boston University's School of Public health surveyed the misuse/overuse of antibiotics in different states in India conducted by Shaffi Fazaludeen Koya.

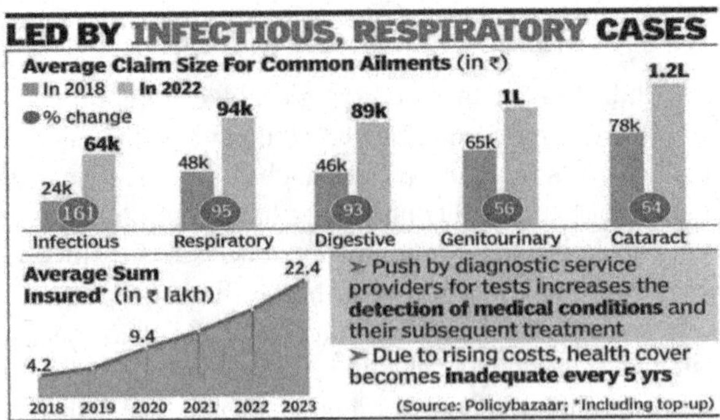

Fig. 30.1 Rise in medical conditions

Source: Times of India (2023)

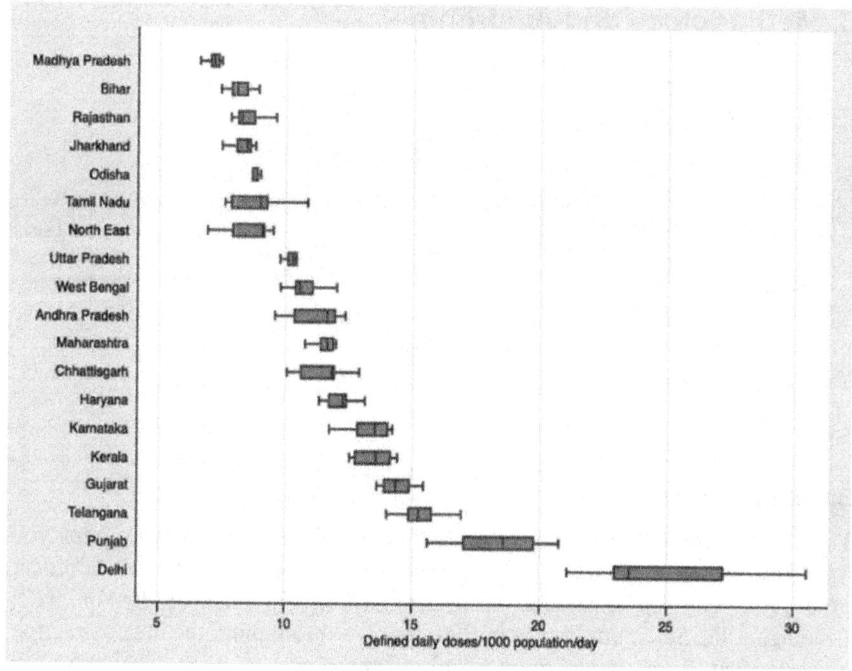

Fig. 30.2 Antibiotic misuse in india

Source: Boston University (SPH,2022)

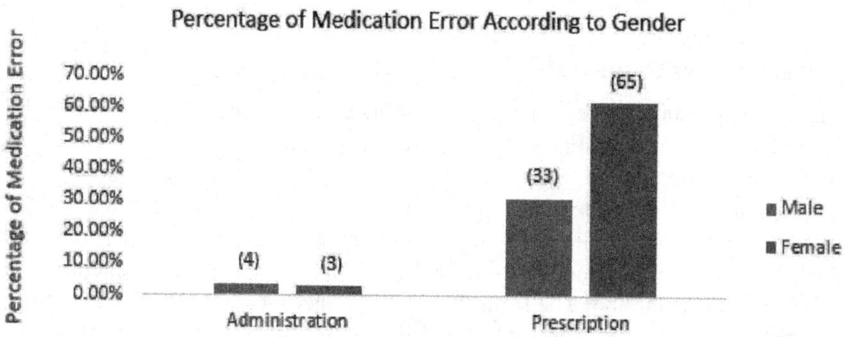

Fig. 30.3 Percentage of medication error

Source: Indian Journal of Pharmacy and Pharmacology

3. Methodology and Algorithm

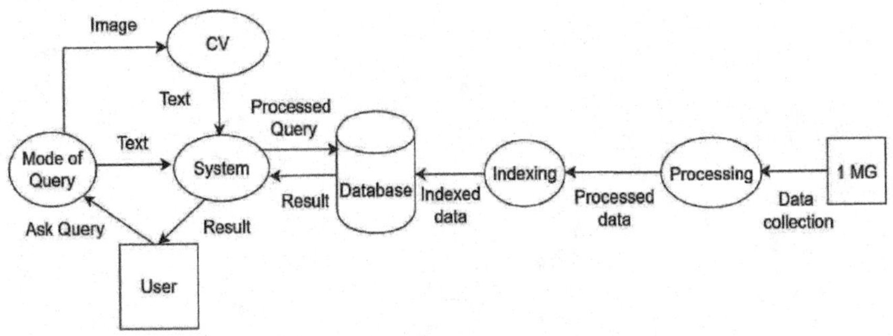

Fig. 30.4 Flowchart of working model

Source: Author

3.1 Data Collection

This paper presents a comprehensive elucidation of the methodologies employed for data acquisition, annotation, and preprocessing essential for the development of a proficient conversational AI system within the medical domain. [12] Leveraging the Selenium library in Python for web scraping, the data extraction is performed from 1mg.com, a platform managed by Tata. The extracted information encompasses various facets including the generic name of medicines, drug classification, marketer details, therapeutic uses, advantages, potential side effects, usage instructions, mechanism of action, and information on adverse drug interactions.

3.2 Data Preprocessing

The data acquisition process involves gathering and structuring information in the form of key-value pairs. Each piece of drug-related information is encapsulated as a NoSQL document. Subsequently, this data is transformed into natural language text. [11] The textual representation is then converted into a vector, a crucial step for vector similarity search operations. For this purpose, the 'E5-small-V2' transformer model from Hugging Face is employed, generating a vector of dimension 384. The resulting vector is incorporated into the document with the designated key "embeddings." Finally, the entire document is stored in the MongoDB Atlas database.

3.3 Natural Language Processing Models

1. **T5 model:** Google's Text-To-Text Transfer Transformer (T5) model is a groundbreaking language model developed by Google Research,

distinguished by its unique text-to-text framework. Unlike conventional models designed for specific natural language processing (NLP) tasks, T5 approaches various tasks as text generation problems. The framework involves both input and output treated as text, allowing for remarkable adaptability across a spectrum of NLP tasks. T5 undergoes a two-step process—pretraining on an extensive and diverse corpus to grasp language structure and semantics, followed by fine-tuning on specific tasks using task-specific data. The model's adaptability stems from its encoder-decoder architecture, which is based on transformers. [10] T5's encoder processes input text, and the decoder generates the corresponding output text, enabling the model to capture intricate relationships within the data. Tokenization is employed to represent input and output as sequences of tokens, each corresponding to a word or subword in a predefined vocabulary. T5 has demonstrated exceptional performance in various tasks such as translation, summarization, and question answering, showcasing its versatility and effectiveness in the realm of text generation.

2. **E5-small-V2:** The 'E5-small-V2' model is a transformer-based neural network architecture developed by Hugging Face, specifically designed for generating vector embeddings from textual data. This model plays a crucial role in converting textual information into a numerical representation suitable for tasks like vector similarity searching. With a focus on efficiency, the 'E5-small-V2' model utilizes a transformer architecture, a type of neural network known for its success in natural language processing tasks. In the context of vector embedding, the 'E5-small-V2' model takes textual input, such as drug-related information in the medical domain and transforms it into a numerical vector with a dimensionality of 384.[3] This transformation captures the semantic nuances and contextual information present in the text, facilitating meaningful comparisons and similarity measurements. The resulting vector, often referred to as an embedding, is then seamlessly integrated into the data structure as a key-value pair. In this case, the key is labeled as "embeddings," providing a clear reference to the numerical representation generated by the model.

3.4 Indexing

Indexing in MongoDB Atlas Vector Search is a critical aspect of optimizing performance for similarity searches on vector data. In this context, vector search typically involves searching for vectors that are similar to a given query vector, which is prevalent in applications like similarity matching for embeddings generated by language models or other numerical representations. To enable efficient vector searches, MongoDB Atlas employs index structures that facilitate the rapid retrieval of similar vectors. The indexing process involves creating a specialized index on the vector field within the MongoDB collection. This index allows MongoDB to organize and structure the vector data in a way that accelerates search operations based on similarity metrics.

3.5 Querying

The process involves the transformation of input natural language queries into vector embeddings through the utilization of the 'E5-small-V2' model. MongoDB Atlas then employs the pre-built vector index to quickly identify and retrieve vectors in the collection that are most similar to the provided query vector. [9] Vector searching is done in Atlas through the application of KNN (K-Nearest Neighbors) algorithm. KNN uses similarity functions like cosine or dot product to find k vectors closest to query vector. The similarity metric used in the search depends on the chosen indexing method and is typically based on distance calculations in the vector space. This approach utilizes vector embeddings to represent the underlying semantic information, facilitating a more sophisticated and contextually meaningful search in the database.

3.6 Computer Vision

In our application, we employ computer vision methodologies, specifically leveraging the Tesseract OCR engine, to extract textual information from images, particularly those containing details related to medicines. Users input images that may include labels, instructions, or other text relevant to the medical domain. Before text extraction, we apply preprocessing techniques to optimize the image for recognition. [4] Tesseract OCR is then integrated into our workflow, processing the preprocessed image to convert visual text into machine-readable text. Postprocessing steps refine and validate the results, ensuring accuracy and reliability. The extracted text then follows the same querying process as textual input such as vector embedding, searching and retrieving relevant results. Beyond proficiency in English or the ability to read medical information, individuals can effortlessly access accurate details about their medications. Tesseract OCR's language-agnostic capabilities ensure that users, irrespective of linguistic backgrounds, can benefit from our system's functionalities. Moreover, the system's ability to provide comprehensive medication details proves especially advantageous for the older demographic. Given the prevalence of memory lapses and forgetfulness about dosage instructions among the elderly, the convenience of capturing an image of the medicine box and promptly accessing all relevant information emerges as invaluable. The integration of computer vision, notably Tesseract OCR, thus becomes a key solution that addresses the nuanced challenges posed by linguistic diversity and memory issues in the dynamic and diverse population of India.

4. Softwae

4.1 Django and ReactJS

In crafting our innovative system, we seamlessly integrated the capabilities of ReactJS and Django to develop a robust and user-friendly website. React, a JavaScript library for building interactive user interfaces, played a pivotal role in the frontend development. [5] Its component-based architecture allowed us to create a dynamic and responsive user interface, enhancing the overall user experience. Leveraging ReactJS virtual DOM, we achieved efficient updates and rendering, ensuring smooth interactions for users engaging with our website.

In the backend, Django, a high-level Python web framework, provided a solid foundation for server-side development. Django's model-view-controller (MVC) architecture streamlined the development process, enabling us to organize our code systematically. The seamless communication between React on the frontend and Django on the backend facilitated a cohesive and integrated web solution.

4.2 MongoDB Atlas

In handling the extensive dataset acquired through web scraping from the 1mg site, we strategically employed MongoDB Atlas as our NoSQL database to store and manage diverse medical information.[6] MongoDB Atlas is a multi-cloud database service by the same people that built MongoDB. Atlas simplifies deploying and managing your databases while offering the versatility you need to build resilient and performant global applications on the cloud providers of your choice MongoDB's document-oriented structure proved instrumental in accommodating the flexible and evolving nature of the scraped data, allowing us to store information in a JSON-like format. By utilizing MongoDB collections and documents, each representing a distinct medicine or related details, we achieved efficient data organization and retrieval. The schema-less design of MongoDB facilitated seamless integration of the scraped data, offering adaptability to varying attributes and structures. With support for nested arrays and subdocuments, MongoDB provided a comprehensive and structured representation of the information.

4.3 Tesseract

Tesseract, an open-source optical character recognition (OCR) engine, plays a pivotal role in our system's ability to extract meaningful information from images of medicine boxes. [7] Leveraging machine learning and neural network techniques, Tesseract is trained to recognize and convert text embedded in images into machine-readable text. The process begins with the input image such medicine labels, prescription, etc. containing textual. Tesseract then employs a combination of image processing, feature extraction, and pattern recognition algorithms to identify and interpret characters within the image. The trained model discerns text patterns and intelligently converts them into a digital format. It retrieves important information such as medicine name, dosage, manufacturer, etc. and converts it into machine readable text. Once the text is extracted, it serves as a key for querying our NoSQL database, facilitating the retrieval of relevant data about the medicine.

4.4 AWS

For the seamless deployment of our program, we turned to Amazon Web Services (AWS) to leverage its cloud infrastructure and services. AWS provided us with a scalable and reliable environment to host our application, ensuring high availability and performance. [8] We utilized AWS Elastic Beanstalk for the deployment of our web application, taking advantage of its managed services to handle infrastructure provisioning, auto-scaling, and load balancing.

5. Result

We present a medical chatbot that combines text and image input to provide comprehensive medical information. The chatbot seamlessly integrates both text and image input modalities, enhancing user experience and accommodating diverse user preferences. Leveraging vector search techniques, our chatbot excels in retrieving precise and relevant information from a vast database. For text input, users can articulate queries naturally, and our chatbot, powered by a sophisticated E5-small-V2 model, converts these queries into vector embeddings, enabling efficient semantic searching. The chatbot extends its utility to image input, employing computer vision techniques and Tesseract OCR to extract relevant text from medicine-related images. The integration of vector search not only ensures the accuracy of information retrieval but also enriches user interaction by capturing semantic nuances. Our results showcase the chatbot's effectiveness in delivering accurate and contextually relevant medical information, bridging the gap between

Table 30.1 Performance measures of tesseract OCR in comparison with other models

Performance Meaasure	Kraken	OpenCV	Tesseract OCR
Word Accuracy	85%	80%	86%
Character Accuracy	90%	75%	94%
Line Accuracy	80%	75%	85%
Sentence Accuracy	82%	78%	88%

Source: Author

textual and visual inputs in a seamless and user-friendly manner. This innovation signifies a significant step forward in the realm of medical conversational AI, demonstrating the potential to revolutionize information access and dissemination within the healthcare domain.

Table 30.2 Embedding model in comparison with other models

Rank	Model	Embedding Dimensions	Sequence Length	Average (56 datasets)	Classification Average (12 datsets)	Clustering Average (11 Datasets)	Ranking average (4 datasets)
1	E5-large-v2	1024	512	62.25	75.24	44.49	86.83
2	Instructor-xl	768	512	61.79	73.12	44.74	86.62
3	Instructor-large	768	512	61.59	73.86	45.29	85.89
4	E5-base-v2	768	512	61.5	73.84	43.8	85.73
5	E5-large	1024	512	61.42	73.14	43.33	85.94
5	Text-embedding-ada-002	1536	8191	68.99	70.93	45.9	84.89
6	E5-base	768	512	60.44	72.63	42.11	84.09
7	E5-small-v2	384	512	59.93	72.94	39.92	84.67

Source: Author

6. Conclusion

Many individuals in India may not have adequate knowledge about medications, their uses, and potential side effects. Accessing reliable and timely medical information can be challenging for individuals. Patients may have questions about symptoms, medications, or treatment options. Lack of awareness and health education can contribute to preventable health issues. Deaths in India, due to adverse drug reactions (ADRs) are estimated to be around 400,000 annually in 720,000 ADRs. Chatbots can educate users about the precautions, side effects and other information of generic medicines, helping to dispel misconceptions. By seamlessly integrating both text and image inputs, chatbots can cater to a diverse range of user interactions, offering a holistic solution for accessing medicine-related information. With the increasing adoption of technology and mobile devices, medical chatbots can offer remote healthcare support. This is especially beneficial for individuals who may not have easy access to healthcare facilities or during situations like the COVID-19 pandemic. In essence, our medical chatbot, with its integrated vector search functionality, stands as a testament to the potential of combining natural language processing and computer vision for enhanced user interaction.

References

1. Hwang, T. H., Lee, J., Hyun, S. M., and Lee, K. 2020. "Implementation of Interactive Healthcare Advisor Model Using Chatbot and Visualization." In 2020 International Conference on Information and Communication Technology Convergence (ICTC), 452–455. IEEE. October.
2. Anas, D., Suhas, D., Manthan, T., and Harsh, M. 2019. "AI Based Healthcare Chatbot System Using Natural Language Processing." Doctoral dissertation, Thesis, St. John College of Engineering and Management, Palghar.
3. Srivastava, P., and Singh, N. 2020. "Automatized Medical Chatbot (Medibot)." In 2020 International Conference on Power Electronics & IoT Applications in Renewable Energy and Its Control (PARC), 351–354. IEEE. February.
4. Wang, H., Zhang, Q., and Yuan, J. 2017. "Semantically Enhanced Medical Information Retrieval System: A Tensor Factorization Based Approach." IEEE Access 5: 7584–7593.
5. Safi, Z., Abd-Alrazaq, A., Khalifa, M., and Househ, M. 2020. "Technical Aspects of Developing Chatbots for Medical Applications: Scoping Review." Journal of Medical Internet Research 22 (12): e19127.
6. Dharwadkar, R., and Deshpande, N. A. 2018. "A Medical Chatbot." International Journal of Computer Trends and Technology (IJCTT) 60 (1): 41–45.
7. Kazi, H., Chowdhry, B. S., and Memon, Z. 2012. "MedChatBot: An UMLS Based Chatbot for Medical Students."
8. Nagarhalli, T. P., Vaze, V., and Rana, N. K. 2020. "A Review of Current Trends in the Development of Chatbot Systems." In 2020 6th International Conference on Advanced Computing and Communication Systems (ICACCS), 706–710. IEEE. March.
9. Okonkwo, C. W., and Ade-Ibijola, A. 2021. "Chatbots Applications in Education: A Systematic Review." Computers and Education: Artificial Intelligence 2: 100033.
10. Luo, B., Lau, R. Y., Li, C., and Si, Y. W. 2022. "A Critical Review of State-of-the-Art Chatbot Designs and Applications." Wiley Interdisciplinary Reviews: Data Mining and Knowledge Discovery 12 (1): e1434.
11. Alam, L., and Mueller, S. 2021. "Examining the Effect of Explanation on Satisfaction and Trust in AI Diagnostic Systems." BMC Medical Informatics and Decision Making 21 (1): 178.
12. Hsu, I. C., and Yu, J. D. 2022. "A Medical Chatbot Using Machine Learning and Natural Language Understanding." Multimedia Tools and Applications 81 (17): 23777–23799.

Technologies for Energy, Agriculture, and Healthcare – Shailesh Nikam et al. (eds)
© 2024 Taylor & Francis Group, London, ISBN 978-1-032-98028-7

31

Analysis of the CNN Models Performance to Detect Handwriting Difficulties

Nisha Ameya Vanjari[1], Prasanna Shete[2]

K J Somaiya Institute of Technology,
K J Somaiya College of Engineering,
Somaiya University

Abstract: Handwriting difficulties may be the start of specific learning disability like dysgraphia. Dysgraphia is a disability that impairs a person's ability to communicate symbols and words in writing. It has a detrimental effect on students' academic performance and general well-being. Early intervention for those in need can be facilitated by expanding the availability of dysgraphia testing to a wider audience through the use of automated processes. Raising awareness of the issue of dysgraphia and its impact on society is another goal of this paper. In order to detect handwriting impaired by dysgraphia, we used a deep learning approach in this research. We assembled a dataset of handwritten to accomplish this goal. To determine whether handwriting is impacted by dysgraphia, we have used a deep learning algorithm like VGG16, ResNet50 and CNN1, and we get accuracies 72%, 62% and 84% respectively from pre-trained models. All are based on feed forward approach.

Keywords: Convolution neural network, Hyper parameter tunning, Handwriting

1. Introduction

The neurodevelopmental diseases known as specific learning disorders (SLD), also referred to as learning disabilities, are typified by enduring challenges with

[1]nvanjari@somaiya.edu, [2]prasannashete@somaiya.edu

DOI: 10.1201/9781003596707-31

one of the three core skills—reading, writing, and math. That being said, SLD is incurable. SLD is listed as one of the disabilities in India under the Rights of Persons with Disability Act of 2016, yet diagnosing and screening for SLD are still difficult tasks. For the assessment, a range of instruments are used, each having pros and cons. "NEVADA Today Report" says it will be helpful to detect the symptoms before a pupil enters third or fourth grade. Numerous academics are utilizing various fields such as image processing, artificial intelligence, and deep learning to study handwriting analysis.

The signs can be predicted with the help of several deep learning approaches. Creating automated and trained models with a significant reduction in human contact is the main goal of deep learning. Deep learning systems can be made more accurate by using training, improving algorithms, and organizing layers. By adding more data points and utilizing data clusters, these models can become more precise. For a very long period, artificial neural networks were the most widely used type of computer model. They have strong learning algorithms that allow them to swiftly and efficiently integrate new data into their systems. Crucially, this network is a "feed-forward" neural network since it only transmits data from its neurons to its deeper layers.

Rest of the paper is structured as follows. Section-2 literature review, Section-3 methods and material, Section-4 model used, Section-5 Results/Findings. Section-6 summarisation of the paper.

2. Literature Review

Iza Sazanita et al. (2009) used handwriting pattern analysis and artificial neural networks (ANNs) to identify dysgraphia symptoms. OCR and ROI, or bounding box segmentation, are employed. Less dataset issues were brought up by the author, which reduced the accuracy of their model (which has a 50%–70% range). Neo, Chin Chea et al. (2012) concentrated on handwriting challenges and experimented with different stroke configurations, line alignments, and curves. How to identify handedness is demonstrated by A. Pentel (2017) using variations in typing. The data collecting context is included in the author setup method. The questionnaire was created to examine the connections between handedness and personality traits rather than keystroke dynamics. The characteristics required to determine emotions, age, and gender are retrieved by the author. S. Rosenblum et al. (2017) developed a tool in this work to assess the data and produce details for additional investigation. The author previously worked on a study in which third-grade dysgraphic students would be evaluated and automatically detected. To gather data for this study, they used writing pads. An effective method for identifying handwritten alphabets from scanned image and created datasets of handwritten data, such as MNIST, is provided by U. Munir et al. (2019). Character extraction methods were employed in this procedure. This experiment yielded an accuracy of 98.9%. Z. Dankovičová et al. (2019) concentrate on attribute extraction identification and processing, as well as dysgraphia, a handwriting issue. Adaptive boosting, SVM, and RF were utilized as machine learning approaches. The success rate of the RF technique is 67.1%. According to M. H. Alkawaz et al. (2020), it consists of interfaces for handwritten data that are converted into binary values in pixels and shown online in matrix form. The recognition accuracy was assessed by contrasting it with an environment consisting of 1155 and 35 pixels. The range of 81.48% to 85.40% is straight stroke alphabet identity detection accuracy. The identification test results for curves, strokes, and characters were 92.08%. According to a study by Al-Mahmud et al. (2021), the author employed CNN for classification and recognised the pattern with 98.94% accuracy on the MNIST dataset, which contains typical handwritten data. S. Surana et.al (2022) The author concentrated on machine learning algorithms that may be used to identify the digits and extract handwritten characters from photographs. The author of this study prepared the image cluster for detection using image segmentation, and then OCR was applied. Model hyperparameter adjustment is used in an attempt to reach the objective. As per Agarwal et.al and Vanjari et.al(2023), many CNN models, such as basic CNN, VGG16 and Resnet50, have been the subject of our research. Interestingly, this network forwards information from its neurons to its deeper layers, making it a "feed-forward" neural network.

According to the paper's survey, we came to several important conclusions, including the following: CNN will work on alphabet recognition; dataset issues and model accuracy are the problems. Machine learning algorithms such as RF and SVM could be used for classification but need to polish it to improve the accuracy. Some survey done on the basis of questionnaire or feedbacks and some are done using alphabet stroke analysis. We can also use deep learning models and using these considerations, we tested CNN models on available dataset to categorize the image into certain handwriting difficulty classifications. Mainly we are working on improvement of models in terms of hyper parameter tunning.

3. Methods and Material

The following procedures must be taken into account for CNN-based handwriting analysis: picture acquisition to classification. In this instance, the pre-processed dataset that we are referring to was gathered from two sources: the NIST Special Database and the Seberang Jaya Primary School dataset, which is publicly accessible on Kaggle. Following flow diagram will give more glance about the proposed system.

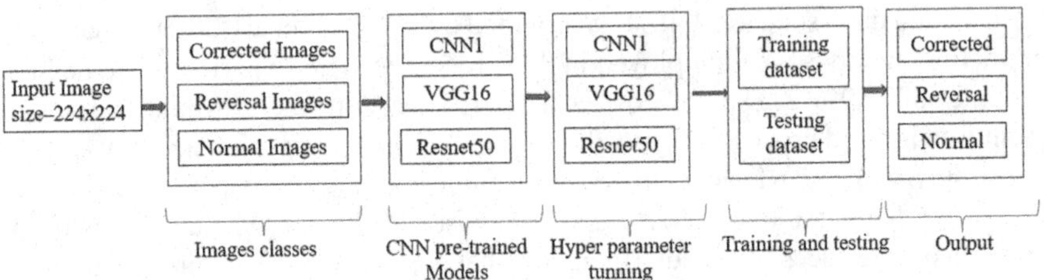

Fig. 31.1 Flow diagram of pre-trained model and hyper parameter tunned model training and testing

Source: Author

As the three main classes in the dataset are reverse, normal, and corrected. In this case, reverse is the mirror image of the alphabets. If the child exhibits dysgraphia symptoms, the model should select the reverse class, which will indicate the symptoms. We did comparison between 2 strategies like pre-train model accuracy and hyper-parameter tunned model accuracy.

4. Model Used

We are using 3 basic model like CNN1, VGG16 and Resnet50 for experimentation.

1. **CNN1 -** CNN architecture contains 1 convolution layer, 1 pooling layer, 2 dense and 1 dropout layer. We can do the architectural changes if required. Total 51075 trainable parameter used for experimentation.
2. **VGG16 -** VGG16 is a 16-layer deep neural network and an extensive network with a total of 138 million parameters and even by today's standards, it's enormous.

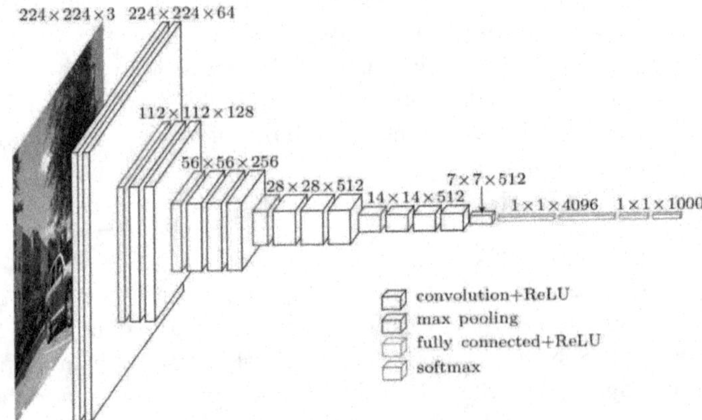

Fig. 31.2 VGG16 architecture

Source: https://medium.com/nerd-for-tech/vgg-16-easiest-explanation-12453b599526

A VGG network consists of small convolution filters and has 3 fully connected layers and 13 convolutional layers.

3. **Resnet50** - In the 2015, He Kaiming, Zhang Xiangyu, Ren Shaoqing, and Sun Jian introduced ResNet, an abbreviation for Residual Network, a particular kind of CNN.

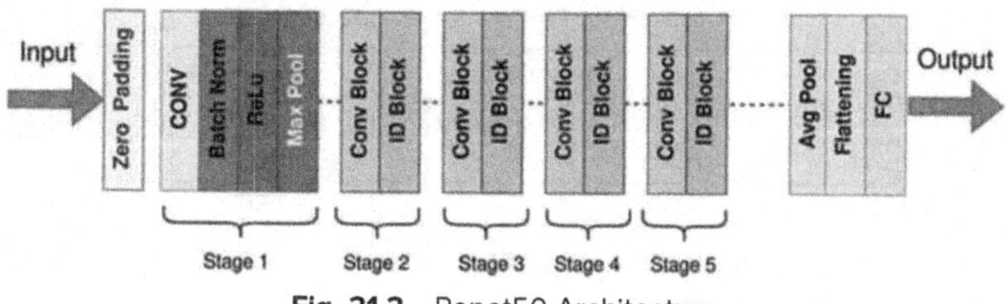

Fig. 31.3 Renet50 Architecture

Source: https://medium.com/@arashserej/resnet-50-83b3ff33be7d

48 convolutional layers, 1 MaxPool layer, and 1 average pool layer make up the 50-layer convolutional neural network ResNet-50. Remaining blocks are stacked to create networks in residual neural networks, a subset of ANN.

5. Results

Following table will give more clarity on VGG16, Resnet50 and CNN1 model analysis. VGG16 consisting of convolutional layers, max-pooling layers, and fully linked layers but ResNet50 includes the concept of residual learning. It contains residual blocks, allowing the network to learn residual functions. This allows for more effective training of deeper networks. The convolution layer of VGG16 is essentially composed of 3x3 convolutional filters with stride 1 and the same padding. The architecture is distinguished by its simplicity, which use simple filters repeatedly to learn hierarchical properties. ResNet50 also uses 3x3 convolutional filters, but it includes residual connections that skip one or more layers. This helps to mitigate the vanishing gradient problem and makes it easier to train very deep networks. In the pooling layer, VGG16 uses max pooling to reduce the spatial dimensions of the input feature maps, whereas softmax is used for classification. In ResNet50, global average pooling is used before the fully linked layers, reducing the spatial dimensions to 1x1 prior to final classification. Rectified Linear Unit (ReLU) activation functions are employed throughout the structures to add nonlinearity.

While these architectures have certain similarities, their key differences are their depth, the usage of residual connections in ResNet50, and the amount of parameters, which can affect their performance on various tasks. Above Table 31.1, shows the result with default setting like LR=0.0001 and split is 80:20 and we have tried 10 epochs for evaluation. Now, we tried to improve the VGG16 and Resnet50 model for better performance and for that reason we have targeted some parameter like splitup and dropout ratio.

Table 31.1 Performance metrics

Sr. No	Model	Pre-trained Model performance metrics			
		Precision	Recall	Accuracy	F1-score
1	VGG16	70.44	71.13	72.50	0.72
2	ResNet50	62.95	60.11	62.50	0.62
3	CNN1	84.61	83.07	84.38	0.84

Source: Author

Table 31.2 Accuracy and loss analysis

Sr No.	Model	Epochs	Splitup – 80:20				Add global pooling			
			Train Acc	Train Loss	Val Acc	Val Loss	Train Acc	Train Loss	Val Acc	Val Loss
1	VGG16	1	0.68	0.62	0.69	0.61	0.69	0.60	0.69	0.60
5		5	0.67	0.60	0.68	0.61	0.60	0.60	0.68	0.61
10		10	0.63	0.56	0.67	0.58	0.64	0.52	0.67	0.54
1	Resnet50	1	0.51	1.07	0.63	0.89	0.52	1.03	0.62	0.89
5		5	0.65	0.79	0.70	0.73	0.65	0.80	0.71	0.74
10		10	0.68	0.75	0.73	0.67	0.69	0.75	0.74	0.68
1	CNN1	1	0.84	0.37	0.93	0.18	0.49	0.88	0.53	0.88
5		5	0.93	0.16	0.96	0.11	0.50	0.84	0.62	0.80
10		10	0.95	0.11	0.97	0.08	0.53	0.80	0.62	0.77
Sr No.	Model	Epochs	Splitup – 70:30				Add global pooling			
			Train Acc	Train Loss	Val Acc	Val Loss	Train Acc	Train Loss	Val Acc	Val Loss
1	VGG16	1	0.69	0.60	0.67	0.55	0.69	0.60	0.69	0.60
5		5	0.65	0.60	0.66	0.55	0.60	0.60	0.68	0.61
10		10	0.64	0.51	0.66	0.53	0.64	0.52	0.67	0.54
1	Resnet50	1	0.51	1.07	0.63	0.89	0.51	1.03	0.62	0.89
5		5	0.65	0.79	0.70	0.73	0.65	0.80	0.71	0.74
10		10	0.68	0.75	0.73	0.67	0.68	0.75	0.74	0.68
1	CNN1	1	0.85	0.33	0.93	0.18	0.45	0.99	0.52	0.98
5		5	0.93	0.16	0.96	0.11	0.57	0.93	0.60	0.90
10		10	0.95	0.11	0.97	0.09	0.58	0.90	0.61	0.87

Source: Author

But as per observation, it does not show tremendous change inVGG16 and Resnet50 model performance and also for CNN1, these parameter changes shows more loss and less accuracy. For CNN1, Max pooling may be more efficient than global pooling in the area of handwriting categorization, where the spatial arrangement of strokes and characters is significant. Max pooling aids in the preservation of local details and the spatial relationships between features, both of which are essential for effectively identifying handwriting patterns and differentiating between characters.

For VGG16 and Resnet50, a critical step in maximizing a learning model's performance is hyperparameter tuning. A hyperparameter's capacity to regulate a particular facet of the training process can have a big impact on how well the model generalizes to new, untested data like learning rate where the number of steps in the optimization process depend on it. Another parameter we have consider is Dropout. Dropout is a regularization strategy that, by randomly removing a portion of neurons during training, helps avoid overfitting. The model can be made more capable of generalizing to new data by properly adjusting the dropout ratio. Underfitting can result from a dropout ratio that is too high, while overfitting cannot be sufficiently reduced by a dropout ratio that is too low. Dataset Split is also impacting on performance of model. The model's performance evaluation is influenced by how your dataset is divided into training, validation, and test sets. Selecting an appropriate split guarantees that the model is trained on a sizable and varied dataset, validated on a different set to fine-tune hyperparameters, and tested on a second set to evaluate generalization.

6. Conclusion

We have experimented deep learning algorithms to identify written work hampered by dysgraphia or not. To achieve this, we put together a handwritten dataset. Using deep learning algorithms such as VGG16 and ResNet50, we are able to assess whether dysgraphia affects handwriting with accuracy rates of 72% and 70%, respectively but it does not show degraded accuracy after hyper parameter tunning. So, to maximize model performance and guarantee appropriate generalization to new data, hyperparameter adjustment is crucial. The particulars of the dataset and the model's complexity determine the proper settings for the hyperparameters. As "one size does not fit for all", just like CNN1 need different strategy to improve the performance of model. It may be improved by doing architectural changes in model as important features should get attended and we can use improved CNN model instead of deeper CNN models.

References

1. Agarwal, B., Jain, S., Beladiya, K., Gupta, Y., Yadav, A. S., & Ahuja, N. J. (2023). Early and Automated Diagnosis of Dysgraphia Using Machine Learning Approach. SN Computer Science, 4(5). https://doi.org/10.1007/s42979-023-01884-0
2. Vanjari, N., & Shete, P. (2023). CNN-based cognitive impairment prediction using handwriting recognition and analysis. In 2023 International Conference on Innovation and Intelligence for Informatics, Computing, and Technologies (3ICT) (pp. 408–412). Sakheer, Bahrain.https://doi.org/10.1109/3ICT60104.2023.10391633
3. Vanjari, N., Patil, P., & Sharma, S. (2019). Interactive Web Based Design for Learning Disabled Children. In 2019 IEEE 5th International Conference for Convergence in Technology (I2CT) IEEE. doi:10.1109/i2ct45611.2019.9033620
4. Al-Mahmud, A., Tanvin, & Rahman, S. (2021). Handwritten English character and digit recognition. In 2021 International Conference on Electronics, Communications and Information Technology (ICECIT) (pp. 1–4). IEEE. https://doi.org/10.1109/ICECIT54077.2021.9641160
5. Munir, U., & Öztürk, M. (2019). Automatic character extraction from handwritten scanned documents to build large scale database. In 2019 Scientific Meeting on Electrical-Electronics & Biomedical Engineering and Computer Science (EBBT) (pp. 1–4). IEEE. https://doi.org/10.1109/EBBT.2019.8741984
6. Surana, S., Pathak, K., Gagnani, M., Shrivastava, V., T. R, M., & Madhuri, S. (2022). Text extraction and detection from images using machine learning techniques: A research review. In 2022 International Conference on Electronics and Renewable Systems (ICEARS) (pp. 1201–1207). IEEE. https://doi.org/10.1109/ICEARS53579.2022.9752274
7. Isa, I. S., Rahimi, W. N. S., Ramlan, S. A., & Sulaiman, S. N. (2019). Automated detection of dyslexia symptom based on handwriting image for primary school children. Procedia Computer Science, 163, 440–449. https://doi.org/10.1016/j.procs.2019.12.127
8. Neo, C. C., Su, E. L. M., Khalid, P. I., & Yeong, C. F. (2012). Method to determine handwriting stroke types and directions for early detection of handwriting difficulty. Procedia Engineering, 41, 1824–1829. https://doi.org/10.1016/j.proeng.2012.08.110
9. Alkawaz, M. H., Seong, C. C., & Razalli, H. (2020). Handwriting Detection and Recognition Improvements Based on Hidden Markov Model and Deep Learning. In 2020 16th IEEE International Colloquium on Signal Processing and Its Applications (CSPA) (pp. 106–110). doi:10.1109/CSPA48992.2020.9068682.
10. Dankovičová, Z., Hurtuk, J., & Feciľak, P. (2019). Evaluation of Digitalized Handwriting for Dysgraphia Detection Using Random Forest Classification Method. In 2019 IEEE 17th International Symposium on Intelligent Systems and Informatics (SISY) (pp. 149–154). doi:10.1109/SISY47553.2019.9111567.

Technologies for Energy, Agriculture, and Healthcare – Shailesh Nikam et al. (eds)
© 2024 Taylor & Francis Group, London, ISBN 978-1-032-98028-7

32

Innovations in Crop Health Monitoring: A Deep Dive into Disease Detection

Nilkamal More*
Head of Department,
Department of Information Technology,
K.J. Somaiya College of Engineering,
Somaiya Vidyavihar University

Prathik Chadaga,
Parthsarthi Singh, Faiz Abbas Syed,
Suraj Mohan, Gautam Bulusu
Student,
Department of Information Technology,
K.J. Somaiya College of Engineering,
Somaiya Vidyavihar University

Abstract: The agricultural sector, particularly in countries like India, grapples with challenges such as low productivity, unpredictable weather, and crop diseases, impacting farmers' profitability and food security. This research focuses on employing image processing for timely disease detection. Various technologies, including IoT sensors, AI, data analytics, and cloud computing, have been integrated into smart farm architecture to enhance farming practices. The system's image processing, utilizing deep learning, aids in the early detection of diseases, crucial for crop protection. This disease detection system significantly contributes to precision agriculture, addressing key challenges faced by farmers. This work details the system's design, development, and presents results from a pilot study conducted with farmers in India.

Keywords: Deep learning, Agriculture

*Corresponding author: neelkamalsurve@somaiya.edu

DOI: 10.1201/9781003596707-32

1. Introduction

Indian farmers confront a multitude of challenges, including harsh weather changes, soil erosion, fluctuating crop yields, and the looming threat of plant diseases. Plant disease diagnosis is complex, often challenging even for experts due to the diversity of cultivated plants. An automated system for disease detection could aid agronomists and farmers, especially in regions lacking agricultural infrastructure. This study seeks to create and assess Convolutional Neural Network (CNN) structures to automate the detection of plant diseases, utilizing crop leaf images obtained from both controlled laboratory settings and real-world field conditions. Subsequent sections detail our motivation, dataset, algorithm, results, and future research directions.

2. Motivation

1. **Agricultural Productivity and Sustainability:** Farmers strive for optimal productivity and resource sustainability. The work provides intelligent recommendations for crop selection, management practices, and resource utilization, fostering sustainable farming and increased productivity[7].
2. **Crop Disease Management:** The system utilizes advanced image processing and machine learning for timely detection and classification of crop diseases, enabling prompt actions to minimize losses and promote healthier crops[9].
3. **Limited Access to Information:** The work addresses the information gap by offering a user-friendly mobile application, providing easily

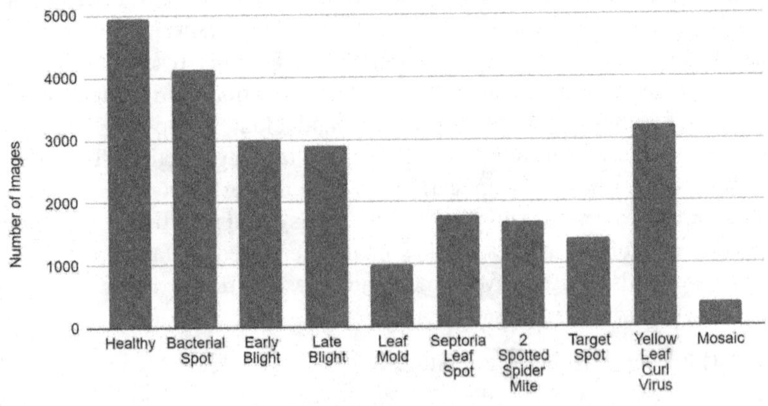

Fig. 32.1 Disease classification

accessible insights for informed decision-making, irrespective of location or technological literacy.

4. **Empowering Smallholder Farmers:** The system's user-friendly interface ensures accessibility for farmers of varying technological proficiencies, helping reduce the digital divide in the agricultural sector.

3. Literature Survey

Deep learning techniques play a pivotal role in enhancing crop disease detection methods, contributing to increased crop quality and farmers' profitability, thereby boosting the overall economy. The literature extensively covers this topic [10,11,12,13,14,15]. [10] introduced a predictive model based on CNN for the classification and image processing of paddy plants.In the domain of disease detection in rice fields, [11] utilized a CNN. Typically, researchers opt for convolutional neural networks with four to six layers to classify various plant species. [12] employed a CNN, coupled with transfer learning techniques, to categorize, identify, and segment various plant diseases. Additionally, [13] develops a web-based application for crop disease detection, aiding farmers in making informed decisions, and [14] evaluates machine learning and deep learning models for predicting fungal illnesses on crops, highlighting SVM, decision trees, and Naïve Bayes as top performers. Numerous techniques have been proposed for accurately identifying and classifying plant infections. Some utilize traditional image processing methods that require manual feature extraction and segmentation, known as handcrafted approaches [15].

4. Dataset

The dataset employed in this plant disease prediction research is comprehensive, centering around three major crops: bell peppers, potatoes, and tomatoes. Within the bell pepper category, the dataset includes 997 images portraying bacterial spots and 1478 images of healthy bell peppers. Regarding potatoes, the dataset comprises 1000 images each of early-blight and late-blight, along with 952 images of healthy potatoes. For tomatoes, the dataset is comprehensive and includes various diseases such as bacterial spot, septoria leaf spot, early-blight, late-blight, leaf mold, two-spotted spider mite, target spot, yellow leaf curl virus, mosaic, and 1519 images of healthy tomatoes. Adapted from the Plant Village Dataset, this subset strategically allocates data into training (70%), testing (20%), and validation (10%) directories, forming a foundational resource for developing and evaluating robust plant disease classification model.

The presented flowchart illustrates the systematic procedure employed in our work for implementing plant disease detection. Initially, we sourced a relevant dataset from Kaggle and meticulously assigned class labels to each image. Following this, we organised the images and subjected them to preprocessing tailored to the specific diseases under consideration. Augmentation techniques were applied to enhance the dataset. Following preprocessing, the dataset was divided into separate sets, comprising training, testing, and validation subsets. This partitioning is crucial for both training the model and evaluating its performance.

To construct an effective plant disease detection model, we introduced additional layers atop pre-existing transfer learning frameworks, thus satisfying our performance criteria.

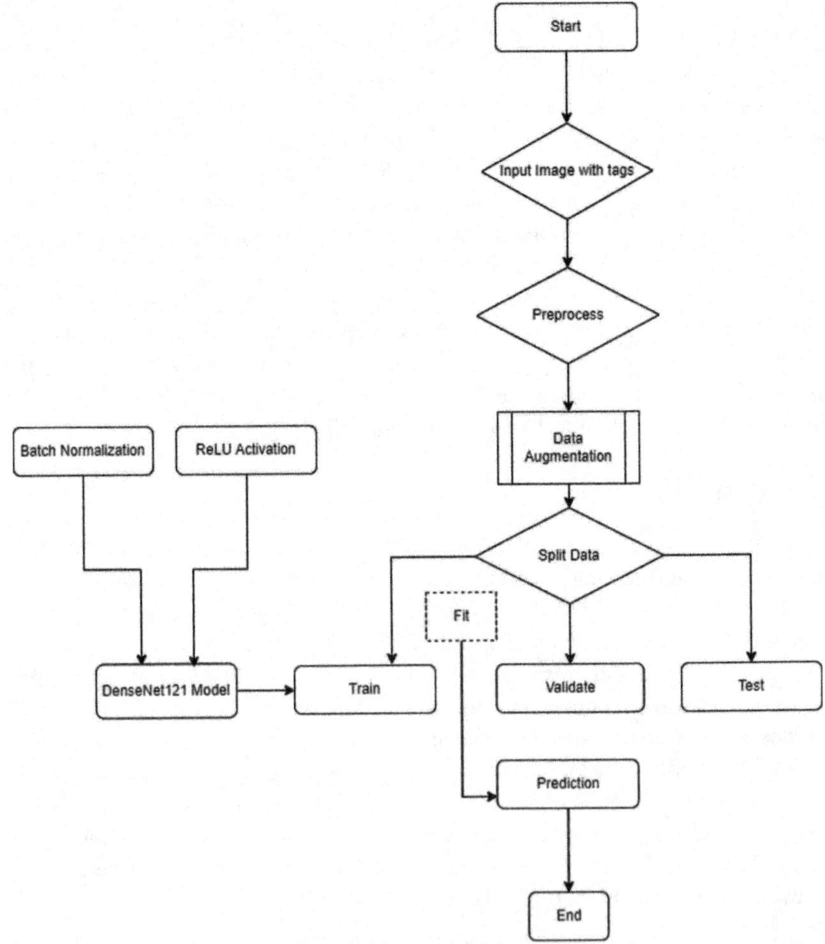

Fig. 32.2 Overview of the proposed system

5. Algorithm

The algorithms used in the study are :

Convolutional Neural Networks (CNNs): Using the DenseNet121 architecture, a Convolutional Neural Network (CNN), our study efficiently detected plant diseases. This CNN leverages convolutional layers to automatically learn hierarchical features, showcasing its prowess in image classification tasks.

DenseNet121 Architecture: DenseNet, short for Densely Connected Convolutional Networks, introduces dense connectivity, where each layer directly receives input from all preceding layers. This design enhances feature reuse, alleviates the vanishing gradient issue, and optimizes parameter utilization.

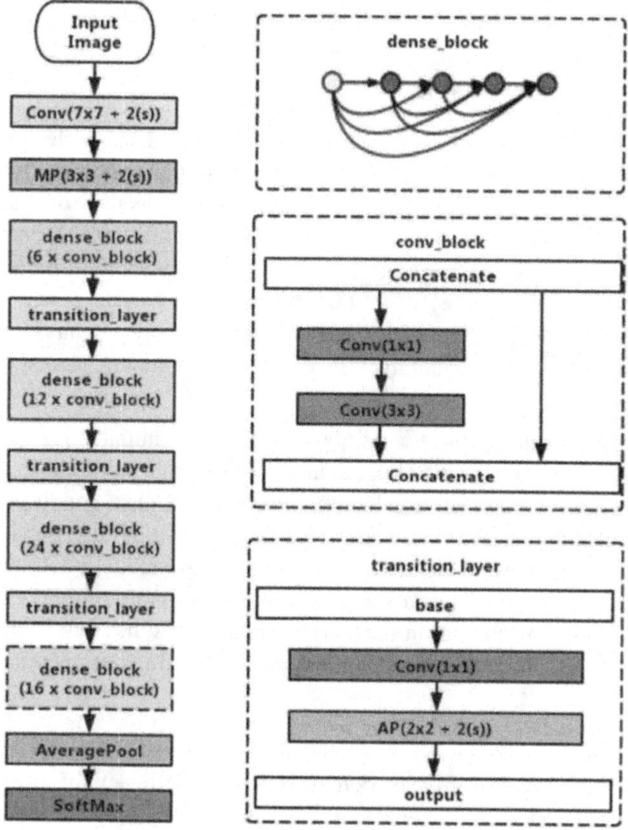

Fig. 32.3 Architecture of DenseNet 121

DenseNet121, with its 121 layers, has shown outstanding performance across different image recognition tasks.

Data Augmentation: To overcome the limited labelled datasets for plant diseases, we applied data augmentation techniques. These involve random transformations like rotation, flipping, and scaling on training images. This strategy diversifies the dataset, enhancing the model's resilience and generalisation capacity.

Mathematical Representation

For each dense block i:

For each layer j within the block

$$Xij = H(X(i-1)j \oplus X(i-1)(j-1) \oplus \ldots \oplus X(i-1)1 \oplus X(i-2))$$

Where , H(·) is the composite function representing bottleneck layer operations (BN, ReLU, Conv(1x1), BN, ReLU, Conv(3x3)), ⊕ is the concatenation operation.

6. Result and Discussion

In our detailed exploration and evaluation of crop disease detection using the DenseNet121 model on the PlantVillage dataset, we observed significant achievements in performance metrics, which are reflective of the model's robustness and effectiveness. Over the span of 15 epochs, the model demonstrated a remarkable capability in learning and classifying images with high precision, achieving an overall accuracy of 98.60% and a low loss rate of 0.0454. These metrics indicate the model's adeptness at correctly identifying various plant diseases from the dataset and its efficiency in minimizing errors during the learning process.

The impressive accuracy rate of 98.60% exceeds the initially reported overall accuracy of 96.35% on the PlantVillage dataset, highlighting the model's exceptional capability to differentiate between healthy and diseased plant conditions across a wide range of crops and disease types. The low loss value of 0.0454, observed during the training phase, indicates the model's effective reduction of the disparity between predicted outcomes and actual labels. The adoption of the DenseNet121 architecture was instrumental in these accomplishments, notably in tackling the vanishing gradient problem, a prevalent obstacle in training deep neural networks.

7. Conclusion

In conclusion, this research has demonstrated the efficacy of deep learning techniques, in the accurate detection of diseases in key agricultural crops like bell peppers, potatoes, and tomatoes. Through analyzing a comprehensive dataset of disease-affected crop images, our model has excelled in identifying and classifying plant health issues with remarkable proficiency. The implications of these findings are profound, suggesting a significant stride towards addressing one of the most pressing challenges in agriculture: ensuring crop health and productivity. By achieving high accuracy in disease detection, our model not only presents a viable solution to enhance crop management but also paves the way for leveraging advanced technologies to revolutionize agricultural practices, thereby promising a sustainable future for global food production.

While our application has proven to be effective in its current form, there is still ample scope for further enhancements and expansion. Some potential areas for future work include:

Integration of Additional Data Sources: Incorporating real-time weather data, soil moisture levels, and pest surveillance data can enhance the accuracy of crop recommendations and disease detection.

Scaling for different regions and crops: Adapting the application to specific geographical regions and a wider range of crops would increase its applicability and usefulness to farmers nationwide.

Engaging Agricultural Experts: Partnering with agronomists and researchers can offer invaluable perspectives and domain knowledge, guaranteeing that the application remains current with the latest advancements in the field.

References

1. Contribution of Agriculture Sector Towards GDP Agriculture Has Been the Bright Spot in the Economy despite COVID-19. Available online: https://www.pib.gov.in/indexd.aspx (accessed on 29 September 2022).
2. Li, L.; Zhang, S.; Wang, B. Plant Disease Detection and Classification by Deep Learning—A Review. IEEE Access 2021, 9, 56683–56698.
3. Khan, M.A.; Akram, T.; Sharif, M.; Javed, K.; Raza, M.; Saba, T. An automated system for cucumber leaf diseased spot detection and classification using improved saliency method and deep features selection. Multimed. Tools Appl. 2020, 79, 18627–18656. [Google Scholar] [CrossRef]
4. Yun, S.; Xianfeng, W.; Shanwen, Z.; Chuanlei, Z. PNN based crop disease recognition with leaf image features and meteorological data. Int. J. Agric. Biol. Eng. 2015, 8, 60–68. [Google Scholar] [CrossRef]
5. Barbedo, J.G.A. Factors influencing the use of deep learning for plant disease recognition. Biosyst. Eng. 2018, 172, 84–91. [Google Scholar] [CrossRef]
6. Vardhini, P.H.; Asritha, S.; Devi, Y.S. Efficient Disease Detection of Paddy Crop using CNN. In Proceedings of the 2020 International Conference on Smart Technologies in Computing, Electrical and Electronics (ICSTCEE), Bengaluru, India, 9–10 October 2020; pp. 116–119. [Google Scholar]

7. Panigrahi, K.P.; Das, H.; Sahoo, A.K.; Moharana, S.C. Maize leaf disease detection and classification using machine learning algorithms. In Progress in Computing, Analytics and Networking; Springer: Singapore, 2020.
8. Sujatha, R.; Chatterjee, J.M.; Jhanjhi, N.; Brohi, S.N. Performance of deep learning vs machine learning in plant leaf disease detection. Microprocess. Microsyst. 2021, 80, 103615. [Google Scholar] [CrossRef]
9. Aldhyani, T.H.; Alkahtani, H.; Eunice, R.J.; Hemanth, D.J. Leaf Pathology Detection in Potato and Pepper Bell Plant using Convolutional Neural Networks. In Proceedings of the 2022 7th International Conference on Communication and Electronics Systems (ICCES), Coimbatore, India, 22–24 June 2022; pp. 1289–1294.
10. Barbedo, J.G.A. Factors influencing the use of deep learning for plant disease recognition. Biosyst. Eng. 2018, 172, 84–91.
11. Vardhini, P.H.; Asritha, S.; Devi, Y.S. Efficient Disease Detection of Paddy Crop using CNN. In Proceedings of the 2020 International Conference on Smart Technologies in Computing, Electrical and Electronics (ICSTCEE), Bengaluru, India, 9–10 October 2020; pp. 116–119.
12. Mohanty, S.P.; Hughes, D.P.; Salathé, M. Using Deep Learning for Image-Based Plant Disease Detection. Front. Plant Sci. 2016, 7, 1419.
13. Apeksha, R.G.; Swati, S.S. A brief study on the prediction of crop disease using machine learning approaches. In Proceedings of the 2021 International Conference on Computational Intelligence and Computing Applications (ICCICA), Nagpur, India, 18–19 June 2021; pp. 1–6.
14. Kumar, R.; Shukla, N.; Princee. Plant Disease Detection and Crop Recommendation Using CNN and Machine Learning. In Proceedings of the International Mobile and Embedded Technology Conference (MECON), Noida, India, 10–11 March 2022; pp. 168–172.
15. Scientist, D.; Bengaluru, T.M.; Nadu, T. Rice Plant Disease Identification Using Artificial Intelligence. Int. J. Electr. Eng. Technol. 2020, 11, 392–402.

Technologies for Energy, Agriculture, and Healthcare – Shailesh Nikam et al. (eds)
© 2024 Taylor & Francis Group, London, ISBN 978-1-032-98028-7

33

Evaluating Various Learning Algorithms for Crop Disease Detection in Precision Agriculture—A Comparative Study

Deepali Shrikhande[1]
PhD Student, Information Technology,
Pillai College of Engineering,
India

Sushopti Gawade[2]
Professor, Information Technology,
Vidyalankar Institute of Technology,
India

Abstract: The existence of humanity depends heavily on agriculture, and improving agricultural output and quality requires tackling the problem of crop disease detection. Over recent years, machine learning (ML) and deep learning (DL) methods have displayed promising outcomes in identifying crop diseases. This study assesses the performance of numerous cutting-edge machine learning and deep learning models in the context of MultiCrop disease detection. Precision agriculture has emerged as a transformative strategy for optimizing crop production while conserving resources. One pivotal aspect of precision agriculture is early disease detection, which can significantly affect crop yield and quality. The study also emphasizes the significance of selecting appropriate features and employing data augmentation techniques to enhance model performance. These findings can be utilized to create a precise decision support system for MultiCrop disease detection, aiding farmers in making informed choices regarding crop management.

In this investigation, a comparison and evaluation of different techniques for detecting MultiCrop diseases within a precise decision support system were

[1]deepalishrikhande6@gmail.com, [2]sushoptiekrishimitra@gmail.com

DOI: 10.1201/9781003596707-33

conducted. A dataset containing images of various crop diseases was used to train and assess these techniques. The results indicated that deep learning techniques achieved the highest levels of accuracy and speed, whereas machine learning techniques exhibited moderate accuracy and speed.

Keywords: MultiCrop disease detection, Machine learning, Precise agriculture, Decision support system

1. Introduction

Crop diseases pose a significant challenge to the agricultural sector, resulting in substantial losses in yield and economic harm on a global scale. Detecting these diseases early and effectively managing them are paramount to enhancing crop production and ensuring food security. Recent progress in machine learning technologies has opened new avenues for the development of precise decision support systems that aid farmers in real-time crop disease detection and management.

Machine learning techniques involve training a model using labelled images to classify new images, while deep learning methods employ artificial neural networks to automatically learn features, achieving high accuracy and speed. This study evaluates these techniques based on their accuracy, speed, and reliability, making use of a dataset that includes pictures of different crop diseases. The findings of this study will help determine the best method for recognizing various crop diseases inside an accurate decision support system.

A precise decision support system for agriculture is essential for several reasons. Early detection enables farmers to take necessary actions such as targeted spraying, quarantine measures, or disease-resistant crop varieties, minimizing crop losses and reducing the spread of diseases.

A decision support system assists farmers improve resource allocation through providing rapid and accurate information on crop health, soil conditions, and variations in the weather.

A decision support system helps farmers to take informed decisions regarding crop selection, Plant schedules, and harvest strategies. With the help of historical data, weather forecasts, and market trends, the system can recommend suitable crop rotations, predict yield potential, and advise on the optimal time to harvest. This helps optimize production, minimize risks, and improve profitability.

2. Literature Review

Various algorithms are available in machine learning and deep learning for disease detections and prevention. Despite the numerous benefits of precise decision support systems for multi crop multi disease detection, still there are some research gaps in this area that need to be taken care of. Major challenges are basically the lack of standardization in the collection and annotation of image datasets. The accuracy of algorithms heavily relies on the quality and quantity of data used to train them. Hence, it is crucial to have standardized image datasets that cover a wide range of crops and diseases to improve the accuracy and generalizability of these algorithms.

Another research gap is the limited availability of open-source software tools for multi crop multi disease detection. While several commercial software solutions exist, they are often expensive and may not be affordable for small-scale farmers.

Most of the research in this area has focused on the technical aspects of developing machine learning algorithms for disease detection. Farmers need to be educated on the benefits of these systems and trained on how to use them effectively to maximize their impact.

In 2021, Ahmed, A.A., and Reddy, G.H.[6] collected 96, 000 datasets. The developed method utilizes Convolutional Neural Networks-CNN as the fundamental Deep learning engine to identify 38 disease categories. It's essential to take accurate photos of plants because they may have artificial date settings. Additional plant and vegetable varieties can be added.

V. Pallagani, V. Khandelwal, B. Chandra, V. [7] in 2019, A deep convolutional neural network algorithm is used in this paper. The trained model has achieved an accuracy of 99.24% and can identify 14 crop species and 26 diseases. Dataset used consists of only 54, 000 images to improve the efficiency a bigger database of different crops can be used.

Hervé Goëau, Sue Han Lee, Pierre Bonnet, and Alexis Joly[12] in 2020, One of the biggest obstacles to ensuring sustainable agriculture and global food security is the control of plant diseases. A number of recent research have suggested using cutting-edge deep learning-based automatic recognition of images systems to improve current protocols for the early diagnosis of plant diseases. Therefore, this discovery motivates future study to reconsider the crop disease categorization in practice paradigm as it exists.

Gabriele Sottocornola, Sanja Baric, Maximilian Nocker, Fabio Stella, Markus Zanker[15] in 2021, The suggested method might be readily altered and adapted to related areas where the user's dynamic engagement with an image-based knowledge base could be utilized to assist in the identification of disease. The primary unaddressed issue for the system's continued development is how to incorporate structured expert knowledge into DSSApple's reasoning engine.

3. Algorithm Analysis

Machine Learning in Precision Agriculture Machine learning plays a significant role in precision agriculture for disease detection. It helps farmers and agronomists to monitor and manage crop health more effectively. It is a valuable tool for early detection and efficient management of crop diseases, helping to optimize yields and reduce the use of pesticides and other resources. It allows for proactive disease management, reducing the economic and environmental impact of crop diseases. Deep learning algorithms have been increasingly applied in Precision Agriculture to improve crop yield, reduce resource usage, and enhance overall farm efficiency.

These algorithms leverage the power of neural networks to process large volumes of data and make predictions or recommendations.

Here are some popular algorithms and techniques commonly used for multi-crop disease detection in Precision Agriculture:

Convolutional Neural Networks (CNNs): CNNs are widely used for image analysis tasks in Precision Agriculture. They can identify and classify diseases, pests, and crop health from images captured by drones or cameras. Applications include plant disease detection, weed identification, and crop health monitoring.

Recurrent Neural Networks (RNNs): RNNs are suitable for time-series data and can be used to predict crop growth, weather patterns, and soil moisture levels. Long Short-Term Memory (LSTM) networks, a type of RNN, are particularly useful for handling sequences of data.

Transfer Learning: Transfer learning involves using pre-trained CNN models (e.g., ResNet, Inception, VGG) and fine-tuning them on datasets specific to different crops. This approach leverages the knowledge learned from one crop to benefit others.

ResNet: ResNet (Residual Network) is a deep neural network architecture that has shown great results in various computer vision tasks, including image classification. ResNet can be fine-tuned for MultiCrop MultiDisease detection by replacing the final layer with a new layer that predicts the disease class for the target crops.

Inception-v3: Inception-v3 is a deep convolutional neural network architecture that was designed for image classification. It can be fine-tuned for MultiCrop MultiDisease detection by replacing the final layer with a new layer that predicts the disease class for the target crops.

VGG16: VGG16 is a deep convolutional neural network that was originally developed for image classification. It can be fine-tuned for MultiCrop MultiDisease detection by replacing the final layer with a new layer that predicts the disease class for the target crops.

GoogLeNet: GoogLeNet, also known as the Inception network.GoogLeNet has ability to perform well while being relatively computationally efficient compared to deeper networks like VGG or ResNet. GoogLeNet was designed to be computationally efficient and accurate, and it introduced several innovative concepts to improve the performance of deep learning models.

Table 33.1 Publications that use machine learning and deep learning techniques for Disease detection in precise agriculture [1][2][3][4][5][6][7][8][9][10][12][15]

Paper	Year	Crop	Dataset	Algorithm
[1]	2020	Apple	Self Collected dataset	CNN
[2]	2021	Sugarcane	Self Collected dataset	AlexNet, VGG-16, GoogLeNet, and ResNet101
[3]	2022	Tomato (Leaf)	Self Collected dataset	CNN
[4]	2019	Apple, Tomato, Cherry, Potato	Plantvillege dataset	Deep CNN SVM Decision Tree KNN
[5]	2019	Banana Fruit	Self Collected dataset	CNN
[6]	2021	Apple, Grape, Tomato	Github	CNN
[7]	2019	Sugarcane	Kaggle	ResNet50, ResNet34, AlexNet
[8]	2021	Tomato, Potato, Corn, Rice, Apple, Grape	Self Collected dataset	CNN
[9]	2021	Apple	Self Collected dataset	YOLOv4
[10]	2018	Apple, Banana, Cabbage, Grape, Onion, Orange	Self Collected dataset	AlexNet, GOOGLENET, OVERFET, VGG
[12]	2020	Apple, Banana, Corn, Grape, Pepper, Orange	Self Collected dataset	GOOGLENET BN, INCEPTION V3, VGG 16, GoogLeNet
[15]	2021	Apple- Fruit	Self Collected dataset	CNN

Figure 33.1 Publications that used Machine learning and deep learning techniques for Disease detection in precise agriculture.

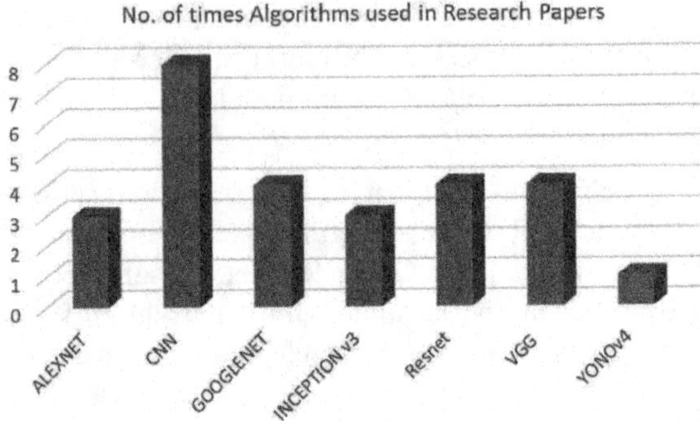

Fig. 33.1 No. of times algorithms used in research papers referred [1][2][3][4][5][6][7][8][9[10][12][15]

Developing a precise decision support system for multi-crop disease detection is a complex task that involves various stages of data collection, preprocessing, feature extraction, model selection, and evaluation. Algorithm analysis is crucial to ensure the system's accuracy, efficiency, and reliability.

4. Conclusion

In this study, on a comprehensive exploration of different machine learning algorithms for the critical task of crop disease detection in precision agriculture. Precision agriculture has gained increasing importance in the modern world, as it allows for more efficient resource utilization, increased crop yields, and reduced environmental impact. Crop disease detection, a pivotal aspect of precision agriculture, holds the potential to revolutionize farming practices by enabling early intervention and informed decision-making. My research involved the evaluation of several learning algorithms, including Support vector machines (SVMs), k-Nearest Neighbors (k-NN), Convolutional Neural Networks (CNNs), Alexnet, and Googlenet are a few examples. We examined each algorithm's strengths and shortcomings in terms of preciseness, computational efficiency, and suitability for realistic agricultural scenarios by a thorough comparison study.

The datasets used for this article of review are from Kaggle as well as self-collected resources.

The findings of this study gives valuable information for stakeholders in agriculture, including farmers, agronomists, and researchers scholars, who seek to implement robust system detection of disease. The selection of algorithm should be tailored to specific needs and available resources.

Ultimately, our research underscores the importance of leveraging advanced machine learning techniques in precision agriculture. By making informed decisions based on accurate disease detection, farmers can mitigate crop losses, reduce the need for chemical treatments, and enhance sustainability. This study contributes to the ongoing effort to optimize crop disease management in agriculture, enhancing the potential for increased food production and environmental conservation.

References

1. Wani, J.A., Sharma, S., Muzamil, M. et al. Machine Learning and Deep Learning Based Computational Techniques in Automatic Agricultural Diseases Detection: Methodologies, Applications, and Challenges. Arch Computat Methods Eng 29, 641–677 (2022). https://doi.org/10.1007/s11831-021-09588-5
2. M. Alencastre-Miranda, R. M. Johnson and H. I. Krebs, "Convolutional Neural Networks and Transfer Learning for Quality Inspection of Different Sugarcane Varieties, " in IEEE Transactions on Industrial Informatics, vol. 17, no. 2, pp. 787–794, Feb. 2021, doi: 10.1109/TII.2020.2992229.
3. Sakkarvarthi, G.; Sathianesan, G.W.; Murugan, V.S.; Reddy, A.J.; Jayagopal, P.; Elsisi, M. Detection and Classification of Tomato Crop Disease Using Convolutional Neural Network. Electronics 2022, 11, 3618. https:// doi.org/10.3390/electronics11213618
4. Geetharamani G., Arun Pandian J., Identification of plant leaf diseases using a nine-layer deep convolutional neural network, Computers & Electrical Engineering, Volume 76, 2019, Pages 323–338, ISSN 0045–7906, https://doi.org/10.1016/j.compeleceng.2019.04.011.
5. Selvaraj, M.G., Vergara, A., Ruiz, H. et al. AI-powered banana diseases and pest detection. Plant Methods 15, 92 (2019). https://doi.org/10.1186/s13007-019-0475-z
6. Ahmed, A.A.; Reddy, G.H. A Mobile-Based System for Detecting Plant Leaf Diseases Using Deep Learning. AgriEngineering 2021, 3, 478–493. https://doi.org/10.3390/agriengineering3030032
7. V. Pallagani, V. Khandelwal, B. Chandra, V. Udutalapally, D. Das and S. P. Mohanty, "dCrop: A Deep-Learning Based Framework for Accurate Prediction of Diseases of Crops in Smart Agriculture, " 2019 IEEE International Symposium on Smart Electronic Systems (iSES) (Formerly iNiS), 2019, pp. 29–33, doi: 10.1109/iSES47678.2019.00020.
8. Kabir, M.M., Ohi, A.Q., Mridha, M.F. (2021). A Multi-Plant Disease Diagnosis Method Using Convolutional Neural Network. In: Uddin, M.S., Bansal, J.C. (eds) Computer Vision and Machine Learning in Agriculture. Algorithms for Intelligent Systems. Springer, Singapore. https://doi.org/10.1007/978-981-33-6424-0_7
9. Roy, Arunabha M. and Bhaduri, Jayabrata, "A Deep Learning Enabled Multi-Class Plant Disease Detection Model Based on Computer Vision"ISSN = 2673–2688
10. Konstantinos P. Ferentinos, Deep learning models for plant disease detection and diagnosis, Computers and Electronics in Agriculture, Volume 145, 2018, Pages 311–318, ISSN 0168-1699, https://doi.org/10.1016/j.compag.2018.01.009.
11. John M. Antle, James W. Jones, Cynthia Rosenzweig, "Next generation agricultural system models and knowledge products: Synthesis and strategy", Agricultural

Systems, Volume 155, 2017, Pages 179–185, ISSN 0308-521X, https://doi.org/10.1016/j.agsy.2017.05.006.

12. Sue Han Lee, Hervé Goëau, Pierre Bonnet, Alexis Joly, "New perspectives on plant disease characterization based on deep learning, Computers and Electronics in Agriculture", Volume 170, 2020, 105220, ISSN 0168–1699, https://doi.org/10.1016/j.compag.2020.105220.
13. Zhaoyu Zhai, José Fernán Martínez, Victoria Beltran, Néstor Lucas Martínez, "Decision support systems for agriculture 4.0: Survey and challenges", Computers and Electronics in Agriculture, Volume 170, 2020, 105256, ISSN 0168–1699, https://doi.org/10.1016/j.compag.2020.105256.
14. Julie Ingram, Damian Maye, Clive Bailye, "What are the priority research questions for digital agriculture?", Volume 114, 2022, 105962, ISSN 0264–8377, https://doi.org/10.1016/j.landusepol.2021.105962.
15. Gabriele Sottocornola, Sanja Baric, Maximilian Nocker, Fabio Stella, Markus Zanker, Picture-based and conversational decision support to diagnose post-harvest apple diseases, Expert Systems with Applications, Volume 189, 2022, 116052, ISSN 0957–4174, https://doi.org/10.1016/j.eswa.2021.116052.
16. Kamran Munir, Mubeen Ghafoor, Mohamed Khafagy, Hisham Ihshaish, AgroSupportAnalytics: A Cloud-based Complaints Management and Decision Support System for Sustainable Farming in Egypt, Egyptian Informatics Journal, Volume 23, Issue 1, 2022, Pages 73–82, ISSN 1110–8665, https://doi.org/10.1016/j.eij.2021.06.002.
17. Meselu Tegenie Mellaku, Ashebir Sidelil Sebsibe, Potential of mathematical model-based decision making to promote sustainable performance of agriculture in developing countries: A review article, Heliyon, Volume 8, Issue 2, 2022, e08968, ISSN 2405–8440, https://doi.org/10.1016/j.heliyon.2022.e08968.
18. Evan D.G. Fraser, Malcolm Campbell, Agriculture 5.0: Reconciling Production with Planetary Health, One Earth, Volume 1, Issue 3, 2019, Pages 278-280, ISSN 2590–3322, https://doi.org/10.1016/j.oneear.2019.10.022.
19. Qiao Jie, Precision and intelligent agricultural decision support system based on big data analysis, Acta Agriculturae Scandinavica, Section B — Soil & Plant Science, 72:1, 401–414, DOI: 10.1080/09064710.2021.2008477
20. Weckesser, F.; Beck, M.; Hülsbergen, K.-J.; Peisl, S. A Digital Advisor Twin for Crop Nitrogen Management. Agriculture 2022, 12, 302. https://doi.org/10.3390/agriculture12020302
21. Kong, J.;Wang, H.; Yang, C.; Jin, X.; Zuo, M.; Zhang, X. A Spatial Feature-Enhanced Attention Neural Network with High-Order Pooling Representation for Application in Pest and Disease Recognition. Agriculture 2022, 12, 500. https://doi.org/10.3390/agriculture12040500
22. Pilvere, I.; Nipers, A.;Krievina, A.; Upite, I.; Kotovs, D. "LASAM Model: An Important Tool in the Decision Support System for Policymakers and Farmers." Agriculture 2022, 12, 705. https://doi.org/10.3390/ agriculture12050705
23. Amiri-Zarandi, M.; Hazrati Fard, M.; Yousefinaghani, S.; Kaviani, M.; Dara, R. A Platform Approach to Smart Farm Information Processing. Agriculture 2022, 12, 838. https://doi.org/10.3390/agriculture12060838

24. Escorcia-Gutierrez, J.; Gamarra, M.; Soto-Diaz, R.; Pérez, M.; Madera, N.; Mansour, R.F. Intelligent Agricultural Modelling of Soil Nutrients and pH Classification Using Ensemble Deep Learning Techniques. Agriculture 2022, 12, 977.https://doi.org/10.3390/agriculture12070977
25. Saiz-Rubio, Veronica, Rovira-Más, Francisco 2020/02/03 "From Smart Farming towards Agriculture 5.0: A Review on Crop Data Management" 10.3390/agronomy10020207 Agronomy
26. Demestichas, K.; Daskalakis, E. Data Lifecycle Management in Precision Agriculture Supported by Information and Communication Technology. Agronomy 2020, 10, 1648. https://doi.org/10.3390/agronomy10111648
27. Anu Jose, S. Nandagopalan, Chandra Mouli Venkata Srinivas Akana. (2021). Artificial Intelligence Techniques for Agriculture Revolution: A Survey. Annals of the Romanian Society for Cell Biology, 2580
28. Shivappa, Himesh. (2018). Digital revolution and Big Data: A new revolution in agriculture. CAB Reviews: Perspectives in Agriculture, Veterinary Science, Nutrition and Natural Resources. 13. 10.1079/PAVSNNR201813021.
29. Dontrey Bourgeois, Suxia Cui, Pamela H. Obiomon, Yonghui Wang, 2015, Development of a Decision Support System for Precision Agriculture, INTERNATIONAL JOURNAL OF ENGINEERING RESEARCH & TECHNOLOGY (IJERT) Volume 04, Issue 10 (October 2015),
30. Ozdogan. B., A. Gacar and H. Aktas, (2017). Digital agriculture practices in the context of agriculture 4.0.. Journal of Economics, Finance and Accounting (JEFA), V.4, Iss.2, p.184–191.

Technologies for Energy, Agriculture, and Healthcare – Shailesh Nikam et al. (eds)
© 2024 Taylor & Francis Group, London, ISBN 978-1-032-98028-7

34

THEORETICAL PERFORMANCE ANALYSIS OF SOLAR PASTEURIZATION VEHICLE

Ishan Upadhyay*

Vidyalankar Institute of Technology,
Mumbai University,
Mumbai, India

Shailesh R. Nikam and Kashianth N Patil

K. J. Somaiya College of Engineering,
Somaiya Vidyavihar University,
Mumbai, India

Abstract: Present work explores the concept of solar milk pasteurization as an environmentally friendly and sustainable solution for ensuring safety and quality of dairy products. By harnessing solar energy, we can address both the global challenges of energy consumption and the need for safe milk processing in regions with limited access to conventional energy sources. The study aims at assessing the performance of a solar milk pasteurization system under various operating conditions. The system consists of a concentrated solar collector, heat exchanger for heating milk, holding tube and heat exchanger for cooling. Various parameters are varied to assess the performance of solar milk pasteurization system. The viability of employing solar energy for mobile pasteurization and evaluating a vehicle equipped with solar technology is verified. Solar pasteurization, utilizing sunlight to heat fluids and eliminate harmful germs aims to offer an eco-friendly solution for on-the-go milk treatment. Assessment of solar collector efficiency, heat transport, and overall system performance is also done. Findings indicate that solar radiation influences milk temperature in the pasteurization system, with air flow rate, ambient temperature, and wind velocity also impacting the process. Valuable insights into mobile solar-powered treatment are also provided.

Keywords: Solar pasteurization, Milk pasteurization, Mobile pasteurization

*Corresponding author: ishan.upadhyay@vit.edu.in

DOI: 10.1201/9781003596707-34

1. Introduction

Milk production and consumption in India are centuries old, and dairy products constitute a significant part of millions of people's daily diets. India produces a large amount of milk every day, which contributes significantly to the economy and meets the nutritional needs of its massive population. Its annual production is 165 million tonnes. Despite this availability, a large amount of milk is lost every day, providing challenges to environmental sustainability (Panchal and Shah 2013). Milk waste can be caused by a variety of factors, including insufficient storage facilities, transportation concerns, and a lack of awareness about proper handling practices.

Milk is perishable, making it susceptible to spoilage, and without adequate infrastructure, a large portion is lost before reaching customers. Unstable storage and maintenance methods lead to the waste of around 5 million tonnes of milk, or more than 3%. A combined study undertaken by MRSS, and the Associated Chambers of Commerce and Industry of India (Assocham) finds that USD $440 billion (about $1,400 per person in the United States) in produce—dairy and vegetables—is wasted. This trash contributes for 40–50% of total production. According to the study, India can only handle 11% of its total perishable produce (Misset). Raw milk has only a two-hour shelf life at room temperature. If milk is not treated properly, it spoils or is wasted.(Ramkumar et al. 2022). Bacteria and harmful pathogens are eliminated when milk is cooked to a specific temperature. Innovative ways are being researched to address this issue and reduce milk waste. One approach is to use current storage and transit technologies like refrigerated vehicles and cold storage facilities. Furthermore, informing farmers and stakeholders about best methods for milk handling and storage can significantly reduce waste.

Solar pasteurisation is a great approach to reduce milk waste. These systems employ solar energy to heat milk to a temperature that eliminates hazardous germs while preserving its nutritious value. Solar pasteurisation, which uses renewable energy sources, not only ensures the safety of milk but also contributes to environmental goals by reducing reliance on traditional energy sources. Solar milk pasteurisation can be a cost-effective alternative to standard procedures in sunlight-rich locations, particularly in off-grid communities. Implementing such solutions is vital for more than just decreasing waste. It increases the dairy industry's economic viability, improves farmer livelihoods, and provides a more sustainable and efficient supply chain. As India deals with the challenges of milk waste, implementing innovative technologies and sustainable practices becomes crucial to ensuring the future of its dairy industry. The current study focuses on a mobile solar milk pasteurisation device, which aims to reduce milk waste. It would also help to reduce milk degradation that occurs during transit to pasteurisation facilities by providing a mobile solution.

The suggested design features a 'mobile pasteurization system' utilizing air for the pasteurization process, chosen for its practicality and widespread accessibility. This system serves the purpose of pasteurizing collected milk during transit, effectively preventing spoilage. An aspect open to debate is the potential use of water instead of air for pasteurization. Nonetheless, challenges related to water storage, maintenance, and the continuous need for freshwater procurement present practical hurdles in implementation.

2. Methods and Materials

2.1 Literature Survey

Milk pasteurisation is an important thermal technique for extending the shelf life of milk. Milk pasteurisation theory covers the process of heating milk to a specific temperature for a set amount of time in order to eliminate or reduce harmful germs while preserving the milk's flavour and nutritional value. The primary purpose of pasteurisation is to make the product safer to consume by reducing microbial burden while preserving the food's taste, nutritional value, and other important characteristics. Solar milk pasteurisation is a method of pasteurising milk that utilises sunshine. Solar milk pasteurisation works by collecting and concentrating solar energy to heat the milk to the required temperature for pasteurisation. Figure 34.1 shows a block diagram of the solar milk pasteurisation system. An overview of the technique can be offered as follows:

1. **Solar Collectors:** Solar collectors collect sunlight and transform it into heat. There are various types of solar collectors, such as flat-plate collectors and concentrating collectors. These collectors are intended to capture as much solar radiation as possible.

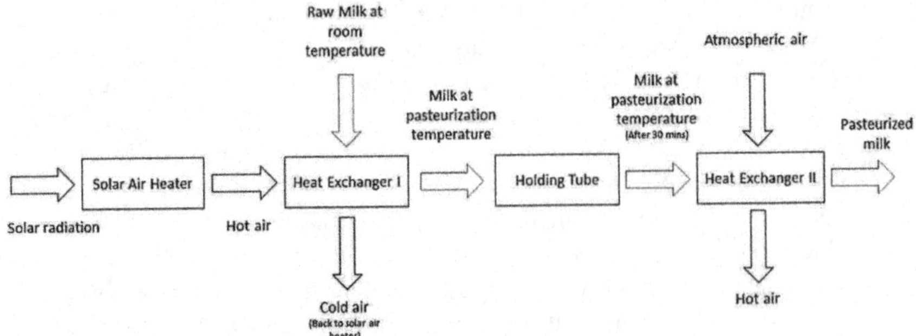

Fig. 34.1 Block diagram of solar milk pasteurization process

2. **Heat Transfer System:** Solar heat is collected and transferred to a heat transfer fluid, which could be air, a liquid like water, or a heat transfer oil. This fluid circulates throughout the solar collector system, carrying the absorbed heat to the next stage of the process.
3. **Pasteurization System:** The heated fluid transfers its heat to the milk via a heat exchanger. The milk is cooked at a temperature that destroys harmful bacteria and pathogens without appreciably altering its flavour or nutritional value. Pasteurisation typically requires heating milk to temperatures ranging from 63°C to 72°C for a predetermined time.
4. **Storage:** After pasteurisation, the milk is immediately cooled to a safe storage temperature. This technique is crucial for limiting the growth of any residual germs and ensuring the quality and safety of the milk.

Solar air heaters are installed to collect solar energy and heat the air passing through them. The parabolic through collector is chosen based on factors such as maximum temperature, concentration ratio, efficiency, sun tracking system, and cost. Based on a standard ambient temperature of 25°C, a wind velocity of 4 m/s, and a milk pasteurisation rate of approximately 35 lit/hr, the sizing obtained to achieve milk pasteurisation temperature at an air flow rate of 0.038 kg/s is 4 m2 collector area and two concentration ratios. This would result in an air exit temperature of 67.37 °C. Hot air from the solar air heater is routed to a heat exchanger, which distributes heat to raw milk. An air-to-liquid finned tube heat exchanger is utilised for this purpose. Raw milk at atmospheric temperature operates as a cold fluid, but air at 67.37 °C is considered hot. The heat exchanger is sized to raise the milk outflow temperature above pasteurisation. A finned tube heat exchanger having 0.3 m2 of frontal area and 0.3 m of height and length. This heat exchanger has an efficiency of 0.98 to one. This would yield a milk outflow temperature of 65.02 °C. This milk should be stored at or above the pasteurisation temperature. This is accomplished within a 0.3 m long and 0.2 m wide holding tube. A 30-minute holding time would provide a heat loss of 55.17 kJ/s. After 30 minutes, the milk temperature exiting the holding tube was 63.62 °C. Following that, the milk is cooled to room temperature via an air-to-liquid finned tube heat exchanger. A comparable heat exchanger is used to chill milk to its regular temperature. In this heat exchanger, the hot fluid will be milk, and the cool fluid will be ambient air. Maintaining an air flow rate of 0.01 kg/s yields a milk output temperature of 23.78 °C.

2.2 Mathematical Modelling

Following equations were used for sizing of the system:

The following sets of equations analyze the performance of solar air dryers.

Useful energy gain can be given as, $Q_u = A_C\left[S - U_L\left(T_{pm} - T_a\right)\right]$ (1)

Holland et. al (Hollands et al. 1976)give the relationship between the Nu and Ra for tilt angle between 0 to 75° as,

$$Nu = 1 + 1.44\left(1 - \frac{1708 * (\sin\sin 1.8\beta)^{1.6}}{Ra * \cos\cos \beta}\right)\left(1 - \frac{1708}{Ra * \cos\cos \beta}\right) + \left(\left(\frac{Ra\beta}{5830}\right)^{\frac{1}{3}} - 1\right) \quad (2)$$

The following sets of equations analyze the performance of heat exchangers (Heat_Exchangers_Selection),

Heat transfer coefficient,

$$G_{fluid} = \frac{\dot{m}_{fluid}}{A_{min}} = \frac{\dot{m}_{fluid}}{\sigma_{fluid} * A_{fr_{fluid}}} \;\&\; Re_{fluid} = \frac{G * D_h}{\mu_{fluid}} \quad (3)$$

For gas side, $j = St.Pr.^{\frac{2}{3}}$ (4)

For liquid side, $f = (1.58\, ln\, ln\, Re - 3.28)^{-2}$ (5)

Figure 34.2 depicts a solar milk pasteurization process with 0.038 kg/s airflow rate through solar collector, 4 m/s wind velocity, 1000 W/m^2 solar radiation, and 25°C ambient temperature.

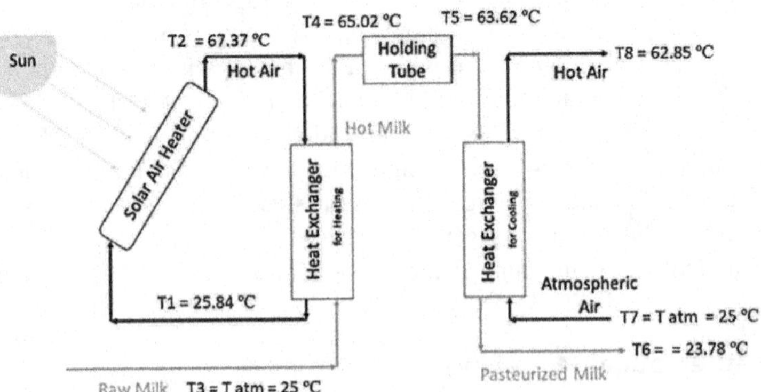

Fig. 34.2 Typical temperatures of solar milk pasteurization system

3. Performance Analysis

A performance study is carried out by creating a mathematical model that evaluates the performance of a solar milk pasteurization system under different conditions of sun insolation, wind velocity, and ambient temperature. A concentrated solar collector, heat exchanger for heating milk, holding tube, and heat exchanger for cooling comprise the solar milk pasteurization system.

Figure 34.3 shows temperature variation at crucial points of the solar milk pasteurization system in relation to changes in solar radiation in W/m^2. Milk temperature rises from 56.80 °C to 73.47 °C when solar radiation rises from 800 W/m^2 to 1250 W/m^2. When sun radiation increases by 56%, the obtained milk temperature rises by 29.32%. The graph clearly shows that as solar radiation grows, so does the temperature of the solar air heater outlet, and hence the temperature of the milk at the heat exchanger's departure. Any sun energy exceeding 930 W/m^2 raises the temperature of milk above the pasteurization point. Changes in sun radiation have no discernible effect on milk temperature after cooling.

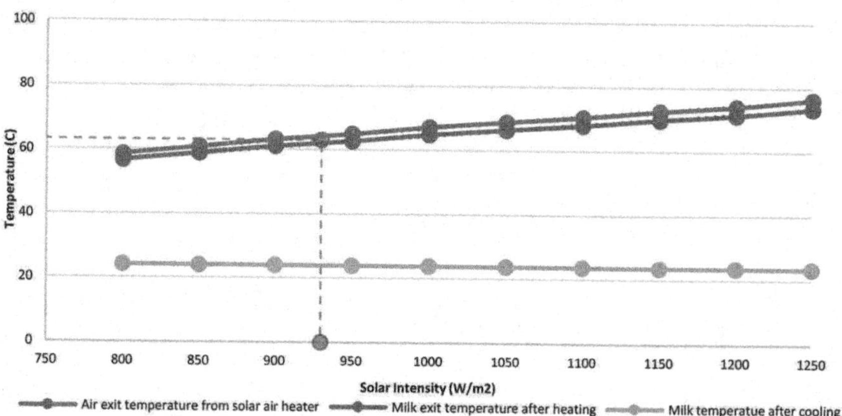

Fig. 34.3 Variation in system temperatures at different solar radiation intensities

Figure 34.4 shows temperature variations at key nodes of the solar milk pasteurization system about changes in air flow rate via the solar air heater in kg/s. When the air flow rate changes from 0.018 kg/s to 0.063 kg/s, the air outlet temperature falls from 81.37 °C to 65.94 °C. Milk temperature at the heat exchanger exit, on the other hand, rises from 50.22 °C to 67.55 °C. The data reveals a threshold at a flow rate of 0.043 kg/s, after which the effectiveness of the heat exchanger hits 1 and the exit temperatures of air and milk remain practically constant. We can conclude from the graph that an air flow rate of more than 0.035 kg/s is required to accomplish milk pasteurization temperature.

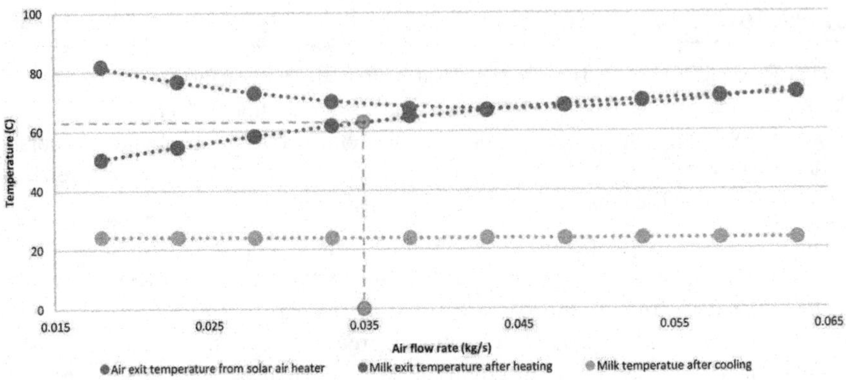

Fig. 34.4 Variation in system temperatures at different air flow rates

Figure 34.5 depicts variation in temperatures at important nodes of the solar milk pasteurization system with respect to change in ambient temperature in °C. When ambient temperature increases from 23 °C to 32 °C, milk outlet temperature increases from 62.47 °C to 70.90 °C. Ambient temperature contributes to heat loss in the form of top, bottom, and side heat loss from solar air heater. Higher ambient temperature drives heat loss and hence when ambient temperature increases, heat loss decreases and milk temperature increases. It is observed that for ambient temperature above 24 °C, obtained milk temperature is above pasteurization temperature.

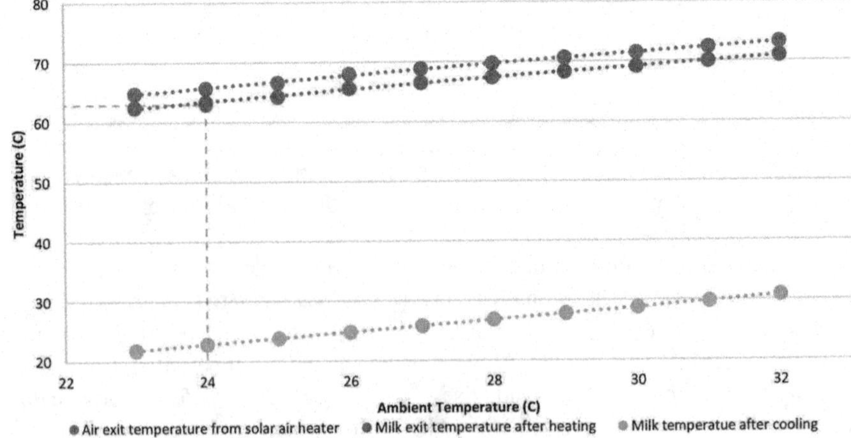

Fig. 34.5 Variation in system temperatures at different ambient temperatures

Figure 34.6 shows the temperature variation at key nodes of the solar milk pasteurization system in relation to changes in wind velocity in m/s. When the wind speed increases from 3 m/s to 7.5 m/s, the milk outlet temperature drops from 67.93 ℃ to 60.24 ℃. Convective heat loss from the top of a solar air heater is aided by wind velocity. Higher wind velocity causes higher heat loss, therefore as wind velocity rises, heat loss rises and milk temperature falls. When the wind velocity is less than 4 m/s, the resultant milk temperature is higher than the pasteurization temperature.

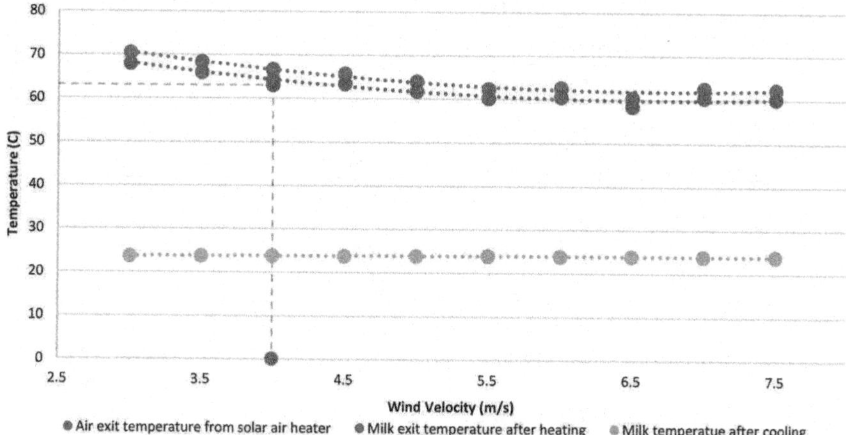

Fig. 34.6 Variation in system temperatures at different wind velocities

4. Summary

The research looks into the potential of employing solar energy for pasteurisation in a mobile context. The study looks at the design and performance of a vehicle equipped with solar pasteurisation technology. Solar pasteurisation is the use of sunlight to heat water or other fluids, killing harmful microorganisms. The analysis takes into account the efficiency of solar collectors, heat transfer systems, and overall system performance. The solar pasteurisation vehicle aims to give a sustainable and environmentally friendly option for on-the-go treatment by utilising renewable energy. The theoretical study evaluates the feasibility, energy efficiency, and potential impact of this unique approach, shedding light on the topic of mobile solar-powered water treatment. It also demonstrates how the entire system operates using a well-defined block diagram. The graph of temperature variation versus change in solar radiation clearly shows that as solar radiation grows, so does the temperature of the solar air heater output, and therefore the temperature of the milk as it exits the heat exchanger. It may also be determined that an air flow rate larger than 0.035 kg/s is required to achieve the milk pasteurisation temperature. When the ambient temperature surpasses 24 °C, milk temperature rises above pasteurisation level. As wind velocity increases, so does heat loss, and milk temperature decreases. When the wind velocity is less than 4 m/s, the milk temperature rises over the pasteurisation temperature.

Table 34.1

Symbol	Parameter
Qu	Useful heat gain in kJ/s
I	Solar Radiation in W/m^2
T_{amb}	Ambient temperature in °C
V_{wind}	Wind Velocity in m/s
m_{air}	Air flow rate through solar air heater in kg/s

References

1. Sadik K., Hongtan L. (2002). Heat Exchangers Selection, CRC Press, 2nd Edition.
2. Hollands, K G T, T E Unny, G D Raithby, L. Konicek. (1976) Free Convective Heat Transfer Across Inclined Air Layers. Journal of Heat Transfer 98 (2): 189–193.
3. Misset, Uitgeverij. Dairy Global. https://Www.Dairyglobal.Net/Dairy/Milking/Indias-Solutions-to-Cut-Milk-Waste/
4. Panchal, Hitesh N., P. K. Shah. (2013) Performance Analysis of Double Basin Solar Still with Evacuated Tubes. Applied Solar Energy 49 (3): 174–179.
5. Ramkumar, G., B. Arthi, S. D. Sundarsingh Jebaseelan, M. Gopila, P. Bhuvaneswari, R. Radhika, Geremew Geidare Kailo. (2022). Implementation of Solar Heat Energy and Adsorption Cooling Mechanism for Milk Pasteurization Application. Adsorption Science and Technology. Hindawi Limited.

Note: All the figures and table in this chapter were made by the authors.

Technologies for Energy, Agriculture, and Healthcare – Shailesh Nikam et al. (eds)
© 2024 Taylor & Francis Group, London, ISBN 978-1-032-98028-7

35

Nurse Bot: Autonomous Remotely Operating Nurse Assistant Robot

Bhavish Sangtani[1],
Tanmaay Dhanawade[2]
Student, Electronics Engineering,
K.J Somaiya College of Engineering,
Somaiya Vidyavihar University,
Vidyavihar, Mumbai

Kaushik Iyer[3]
Student, Computer Engineering,
K.J Somaiya College of Engineering,
Somaiya Vidyavihar University,
Vidyavihar, Mumbai

Sudha Gupta[4]
Professor, Electronics Engineering,
K.J Somaiya College of Engineering,
Somaiya Vidyavihar University,
Vidyavihar, Mumbai

Ayesha Hakim[5]
Assistant Professor, Electronics Engineering,
K.J Somaiya College of Engineering,
Somaiya Vidyavihar University,
Vidyavihar, Mumbai

Abstract: In modern healthcare, the demand for innovative solutions to alleviate the workload of medical staff and enhance patient care is ever-growing. This paper introduces Nursebot, an intelligent robotic system designed for medical purposes within hospital settings. Nursebot integrates a line-following robot platform with a

[1]bhavish.sangtani@somaiya.edu, [2]tanmaay.d@somaiya.edu, [3]kaushik.iyer@somaiya.edu, [4]sudhagupta@somaiya.edu, [5]ayesha.hakim@somaiya.edu

DOI: 10.1201/9781003596707-35

robotic arm controlled by an Arduino Uno and ESP8266 microcontroller. Mounted on the robot are vital sign sensors, including a pulse sensor and a temperature sensor, facilitating real-time patient monitoring. Utilising a mobile application interface, healthcare professionals can remotely access patient data collected by Nursebot, ensuring timely intervention and personalised care. Moreover, Nursebot is equipped with a tray to transport medications, enabling efficient delivery to patients' bedsides. The primary objectives of Nursebot encompass reducing nursing workload, enhancing patient monitoring in isolated environments, and streamlining medication distribution processes.

Keywords: Healthcare, Nurse robot, Telemedicine, Vital signs

1. Introduction

In contemporary healthcare environments, nurses are confronted with an array of responsibilities, including the meticulous monitoring of patient vital signs and the timely administration of medications. These routine tasks often demand substantial time and attention, detracting from nurses' ability to focus on more specialised and critical aspects of patient care. To address this challenge, the development of Nursebot—a novel robotic solution—aims to streamline these essential but time-consuming duties. Nursebot is comprised of a line-following robot integrated with a robotic arm, both controlled by an Arduino Uno and ESP8266, communicating via the I2C protocol. Leveraging WIFI connectivity, Nursebot can be remotely operated through a mobile application interface. Vital sign monitoring is facilitated through the incorporation of a pulse sensor (MAX30105) and a temperature sensor (MLX90614) mounted on a tray, enabling real-time data transmission to the accompanying mobile app. The primary objective of Nursebot is to automate routine tasks such as patient monitoring and medication delivery within hospital wards. By delineating black paths for the line-following robot to navigate, Nursebot can autonomously reach designated patients, assess their vital signs, and relay this information to healthcare professionals via the mobile app. Additionally, Nursebot's equipped tray and robotic arm—with three SG90 servo motors—enable transportation and dispensation of medications to patients. (See Fig. 35.1 for the flowchart depicting the Nursebot system.)

Following this introduction, the subsequent sections will delve into a thorough literature review, detailing existing technologies and research relevant to Nursebot's development. The methodology section will elucidate the design and implementation process, while the results and discussion section will analyse the

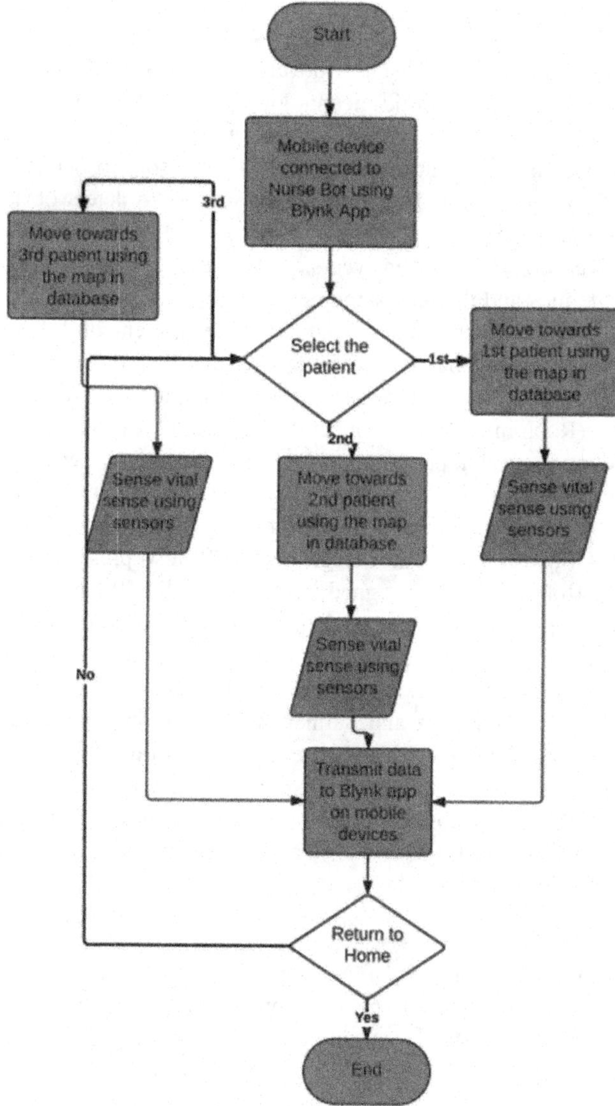

Fig. 35.1 Flow chart

Source: Author

performance and implications of Nursebot in hospital environments. Finally, the paper will conclude with reflections on the significance of Nursebot and avenues for future research

2. Literature Review

(Kumar and Savadatti 2020, 16) discuss a novel approach to healthcare management through the utilisation of robotics technology. Specifically, they describe the development of a line-following robot equipped with a medicine delivery system. The robot utilises light-dependent resistor (LDR) sensors to detect and follow lines on the floor, and infrared (IR) proximity sensors to detect obstacles in its path. (Saddam 2016) discusses that the field of healthcare robotics has seen significant advancements in recent years, particularly in response to challenges posed by pandemics and the increasing demand for efficient healthcare delivery. Two notable projects in this domain are the development of a line-following robot equipped with a medicine delivery system and the creation of Virobot, an artificial nurse robot designed to assist medical personnel and disinfect hospital environments. (Robu.in 2021) introduces MedRobo, a robot designed to deliver medicine and measure vital parameters like heart rate and temperature. It utilises RFID for navigation, line following for path detection, and a vending machine for medication delivery. Data is transmitted to the cloud via IoT, with alerts sent to doctors via a GSM Module. (IoT Design Pro 2019) also presents an Automatic Line Follower Robot (ALFR) designed to assist COVID-19 patients in isolation rooms. ALFR delivers essential medicines and food, maintains emergency wireless communication between doctors and patients, collects and disposes of waste, and sanitises the room using a disinfectant machine. ALFR's performance is highly effective. (Abdelaziz and Salman 2020) introduce a project about an autonomous assistive robot for healthcare, designed to dispense medication to individual patients in a care facility. Utilising infrared sensors for line detection and ultrasonic sensors for obstacle detection, the robot halts when necessary and notifies staff upon successful delivery via a Telegram message.

3. Methodology

The successful implementation of Nursebot involves a meticulous fusion of hardware, software, and communication protocols to create a seamless and efficient healthcare assistance system. The methodology encompasses the following key components:

3.1 System Architecture

Line Following Robot utilises an Arduino R3, L298N motor driver, four 150 RPM motors, and four IR sensors for precise navigation within hospital wards. Robotic Arm comprising a NodeMCU and three SG90 servo motors, the robotic arm facilitates the transport of medications with three degrees of freedom.

Sensor Tray is Equipped with MAX30105 and MLX90614 sensors for real-time monitoring of patient vital signs, including temperature and pulse (see Fig. 35.2 for the Arduino circuit diagram). The printed circuit board (PCB) layout serves as the physical foundation for housing and connecting the electronic components. By carefully designing the PCB layout to optimise signal integrity and minimise electromagnetic interference, engineers can enhance the system's performance and reliability.

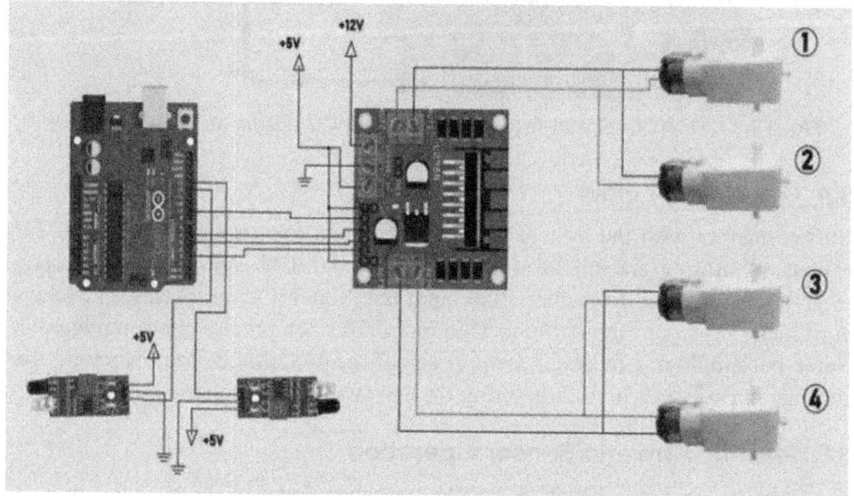

Fig. 35.2 Arduino circuit design (Robu.in,2021)

3.2 Connectivity Protocols

The Nursebot utilises two main connectivity protocols for communication between its components. The UART Protocol enables communication between the Arduino Uno in the line following robot and the NodeMCU in the robotic arm, facilitating coordinated movement and action. The *I2C* Protocol establishes communication between the NodeMCU, MAX30105, and MLX90614 sensors. For a visual representation of the UART connectivity between the Arduino and NodeMCU, refer to Fig. 35.3.

3.3 Control Interface

Nursebot is controlled through a dedicated mobile app built on the Blynk platform. The Blynk app acts as a user-friendly interface, allowing nurses to input patient details, direct the robot to specific locations, and initiate medication delivery.

Fig. 35.3 UART between Arduino and NodeMCU (forum.arduino.cc, 2020)

3.4 Operational Flow

Nurses interact with the system through the Blynk app, issuing commands for patient monitoring and medication delivery. NodeMCU commands relay upon receiving commands from the Blynk app, the NodeMCU processes and relays movement signals to the Arduino Uno via UART for precise robot navigation. Nurse bot employs a feedback loop to signal the NodeMCU upon reaching the designated patient location, triggering the activation of the robotic arm.

3.5 Robotic Arm and Sensor Operation

The robotic arm movement, controlled by the NodeMCU, executes precise movements for medication transport upon reaching the patient. Additionally, the MAX30105 and MLX90614 sensors gather real-time temperature and pulse data from the patient. The NodeMCU orchestrates this process using the *I2C* protocol. For a visual representation of the servo circuit connected to the NodeMCU, refer to Fig. 35.4. The obtained vital sign data is then transmitted via WIFI to the Blynk app, providing healthcare professionals with timely and accurate patient information.

3.6 Path Following Algorithm

The path following algorithm implemented in Nursebot relies on continuous data acquisition from four sensors strategically positioned at the robot's front bottom. These sensors play a pivotal role in facilitating various 4 manoeuvres, including forward motion, subtle adjustments to maintain trajectory, sharp 90-degree turns, and complete U-turns. Positioned at extreme left, middle left, middle right, and extreme right, these sensors enable the robot to perceive its surroundings and navigate effectively. Notably, the algorithm incorporates conditional logic

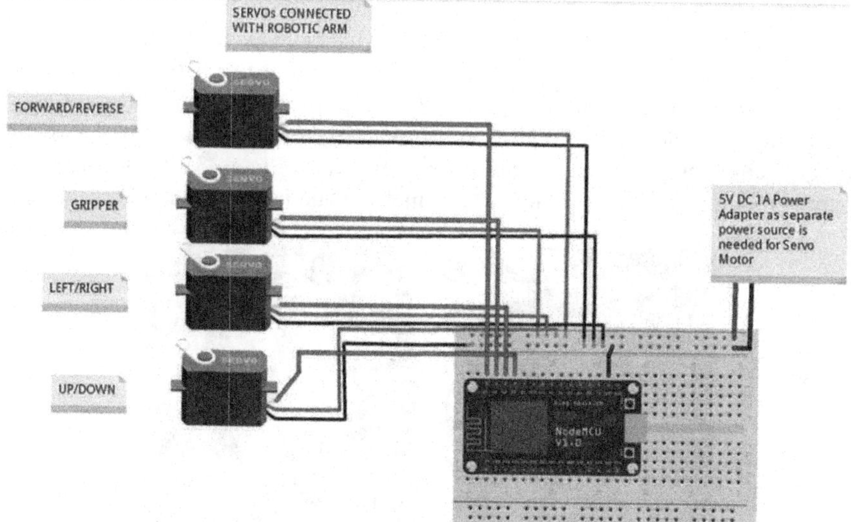

Fig. 35.4 Servo circuit for NodeMCU (IoT Design Pro, 2019)

tailored to different movement scenarios, adapting Nursebot's behaviour based on environmental cues. For instance, when approaching junctions resembling a '+', Nursebot intelligently disregards turns leading away from the target patient bed. This adaptive approach enhances Nursebot's efficiency in navigating complex hospital environments and ensures precise delivery of healthcare services.

4. Result and Discussion

The implementation of Nursebot has yielded promising results. The results are presented across key dimensions of system performance, patient vital sign monitoring, and medication delivery efficiency.

4.1 System Performance

Nursebot demonstrated robust performance in navigating hospital wards using the line-following robot. The integration of four IR sensors ensured accurate path following, allowing the robot to reach specified patient locations with precision. The coordination between the line-following robot and the 3-degree-of-freedom robotic arm proved seamless, enabling efficient and controlled medication transport. The exceptional performance of the Nursebot system was further underscored by the near-instantaneous communication between the ESP8266 and the Blynk app. Whether issuing commands for the robot to navigate to a specific patient bed or transmitting real-time sensor results to the Blynk app, the entire process unfolded in a matter of milliseconds. This rapid exchange of information played a crucial role in enhancing the system's efficiency. The integration of this real-time communication aspect further solidifies Nursebot's position as a technologically advanced solution in automated healthcare assistance, where timely actions and data accessibility are paramount. Figure 35.5 illustrates the hardware setup of Nursebot, showcasing its front and isometric views.

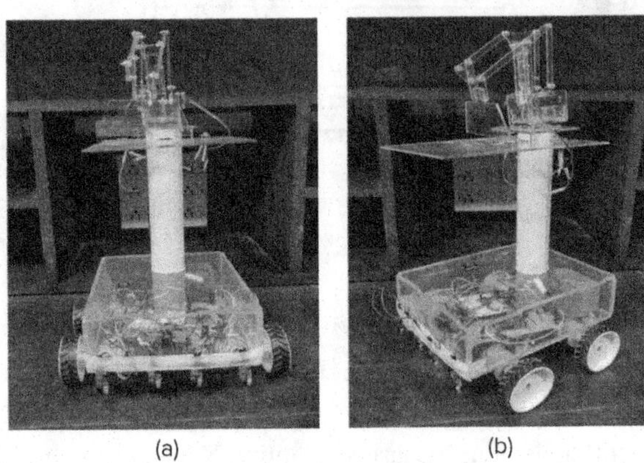

Fig. 35.5 Nurse bot hardware setup (a) Front view (b) Isometric view

Source: Author

4.2 Patient Vital Sign Monitoring

Following the successful line-following phase, Nursebot transitioned into the vital sign monitoring phase. This transition was marked by the Arduino autonomously sending a signal to the ESP8266, initiating the sensing functions. The first step involved activating the temperature sensing module, specifically using the Adafruit MLX90614 infrared thermometer. This sensor, operational for a predetermined duration of 10 seconds, employed infrared technology to non-invasively measure the patient's surface temperature.

Subsequently, the focus shifted to pulse sensing, engaging the MAX30105 pulse oximeter for an additional 10-second interval. This sensor, leveraging its optical characteristics, detected variations in blood volume beneath the skin, thereby enabling the measurement of the pulse rate. The MAX30105's capability to read the infrared (IR) signal was crucial for determining the presence of a finger and, subsequently, for the heartbeat detection.

The pulse rate was calculated by identifying time intervals between successive heartbeats and averaging these over a set period to yield the beats per minute

(BPM). Throughout this phase, the I²C protocol played a pivotal role in ensuring seamless and reliable communication between the Adafruit MLX90614, MAX30105 sensors, and the ESP8266.

4.3 Medication Delivery Efficiency

Nursebot's medication delivery mechanism, orchestrated through the robotic arm, demonstrated efficiency in reaching patients as directed by the Blynk app. The system responded promptly to nurse-initiated commands, showcasing its potential for timely medication administration. The feedback loop, activated upon reaching the patient, ensured that the robotic arm executed precise movements for medication delivery.

4.4 User Interaction and Control

The Blynk app, serving as the central control interface, provided healthcare professionals with a user-friendly platform to interact with Nursebot. Notably, the interface was organised with four distinct buttons, each corresponding to a different patient. Upon the selection of a button for a specific patient, the system automatically disabled the other three buttons, effectively preventing any command override and ensuring a streamlined workflow for each patient interaction. Above these patient-specific buttons, configured in two rows and two columns, were two real-time displays. These displays were dedicated to showcasing the latest temperature and pulse sensing results, which were transmitted from the ESP8266 using Wi-Fi and the Blynk library. For a visual representation of the Blynk app's user interface, refer to Fig. 35.6.

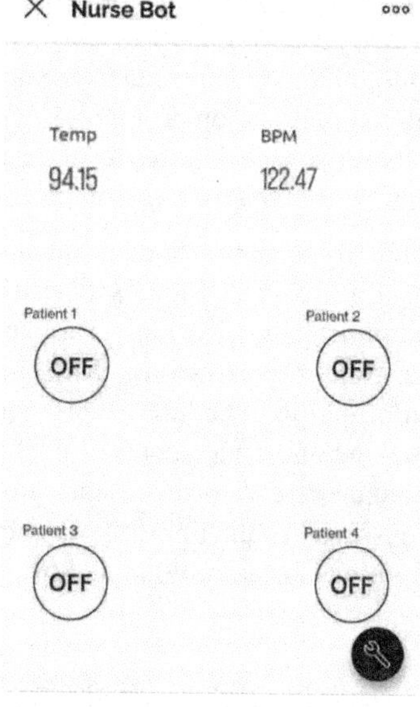

Fig. 35.6 Blynk App UI

Source: Author

4.5 Data Transmission

The intricate communication protocols employed within the system contributed to the reliability and accuracy of information dissemination. Internally, between the ESP8266 and Arduino Uno, data transmission was orchestrated through UART communication protocols. This involved the use of serial printing and receiving, enabling the robust exchange of messages. Specifically, messages communicated the initiation and completion of critical events within a workflow, including but not limited to line following, turning manoeuvres, and data collection phases.

During various operational phases, such as patient monitoring and medication delivery, internal communication instances were prevalent. For instance, the ESP8266 transmitted patient identification numbers to the Arduino Uno, instructing it to navigate to specific patient beds. Conversely, upon successful completion of the line following procedure, the Arduino Uno relayed pertinent data back to the ESP8266, indicating readiness for the next task. Moreover, during the testing phase, additional data transmission scenarios were utilised, such as exchanging information while executing turning and U-turn manoeuvres. Externally, the WiFi-based data transmission from the ESP8266 to the Blynk app showcased a different facet of the communication architecture.

4.6 Discussion

The results affirm the efficacy of Nursebot in navigating hospital environments with precision, responding promptly to nurse-initiated commands, and autonomously returning to its start position after completing tasks. The integration of communication protocols, including UART for intra-robot communication and I2C for sensor data retrieval, has been pivotal in ensuring seamless coordination among the diverse components of Nursebot.

The Blynk app serves as an intuitive and user-friendly interface, empowering healthcare professionals to effortlessly direct the robot, monitor patients, and administer medications. The successful transmission of vital sign data via Wi-Fi further underscores Nursebot's potential to provide timely and accurate information to healthcare providers. This aspect is crucial for informed decision-making and underscores Nursebot's role in facilitating personalised and efficient patient care. However, the project also encountered certain challenges. The sensitivity of the line-following capabilities required specific conditions related to lighting, surface colour, and tilt, with deviations potentially leading to path errors.

Table 35.1 Control conditions

Control	Extreme Left	Left	Right	Extreme Right
Move Forward	1	0	0	1
Turn Left	1	0	1	1
Turn Right	1	1	0	1
Sharp Turn (90 Degree) Left/Right	0	0	0	0

Note: The conditions for Sharp Turn Left and Sharp Turn Right may vary based on the patient bed number due to the presence of plus junctions.

Source: Author

Additionally, navigating complex pathways, particularly at '+' and 'T' junctions, was challenging with the limited sensor array employed in this iteration of Nursebot. The current sensor setup, while effective in straightforward navigation scenarios, struggled to consistently and accurately negotiate these more complex junctions. This observation suggests that augmenting the sensor array or employing more sophisticated navigation algorithms could significantly enhance Nursebot's operational efficacy in diverse hospital environments.

5. Conclusion

The development journey of Nursebot has been marked by various challenges, each of which has been met with dedication and innovation. Building an efficient structure, selecting compatible components, ensuring data accuracy, and navigating paths using IR sensors presented significant hurdles throughout the development process. However, through perseverance and problem solving, these challenges were successfully overcome, resulting in the creation of a versatile and effective solution for patient monitoring and medication delivery in hospital wards. One compelling avenue for future development involves the integration of a camera and speaker, enabling telemedicine capabilities.

As we reflect on Nursebot's journey, it is clear that its impact extends beyond the boundaries of traditional healthcare. In conclusion, Nursebot stands as a testament to the power of technology to transform healthcare delivery, improve patient outcomes, and enhance the overall quality of care.

References

1. Kumar M B, Punith, and D Manikant Amaresh Savadatti (2020). Virobot the Artificial Assistant Nurse for Health Monitoring, Telemedicine and Sterilization through the Internet. International Journal of Wireless and Microwave Technologies 10, no. 6: 16–26. https://doi.org/10.5815/ijwmt.2020.06.03.
2. Saddam. "WIFI Controlled Robot Using Arduino." Arduino Based WiFi Controlled Robot, April 6, 2016. https://circuitdigest.com/microcontroller-projects/arduino-wifi-controlled-robot.
3. How to Make a Line Follower Robot Using Arduino - Connection & Code - Robu.in: Indian Online Store: RC Hobby: Robotics/ Robu.in | Indian Online Store | RC Hobby | Robotics, May 3, 2021. https://robu.in/how-to-make-a-line-follower-robot-using-arduino-connection-code/.
4. IoT Design Pro.Iot Based Robotic Arm Using NodeMCU (2019). https://iotdesignpro.com/projects/iot-based-robotic-arm-using-esp8266.
5. Abdelaziz, Marwa, and Aymen Salman(2020). Mobile Robot Monitoring System Based on IoT. Journal of Xi'an University of Architecture Technology 12 : 5438-5447. https://www.researchgate.net/publication/341626062_Mobile_Robot_Monitoring_System_based_on_IoT

6. Mishra, Shubhanvit(2020). Low Cost IoT Based Remote Health Monitoring System. https://doi.org/10.13140/RG.2.2.19645.61923.
7. Zafar, S., et al(2018). An IoT Based Real-Time Environmental Monitoring System Using Arduino and Cloud Service.Engineering, Technology & Applied Science Research 8, no. 4 : 3238–3242. https://doi.org/10.48084/etasr.2144.
8. Jambotkar, C(2017). Pick and Place Robotic Arm Using Arduino. International Journal of Science, Engineering and Technology Research (IJSETR) 6, no. 12 :2278-7798. https://www.researchgate.net/publication/332565132_Pick_and_Place_Robotic_Arm_Using_Arduino
9. Chaudhari, Jagruti, et al(2019). Line Following Robot Using Arduino for Hospitals. In 2019 2nd International Conference on Intelligent Communication and Computational Techniques (ICCT). https://doi.org/10.1109/icct46177.2019.8969022
10. Holovatyy, Andriy(2021). Development of IoT Weather Monitoring System Based on Arduino and ESP8266 Wi-Fi Module. IOP Conference Series: Materials Science and Engineering,. https://doi.org/10.1088/1757-899X/1016/1/012014.

Technologies for Energy, Agriculture, and Healthcare – Shailesh Nikam et al. (eds)
© 2024 Taylor & Francis Group, London, ISBN 978-1-032-98028-7

36

Comparative Study of Conventional Jaggery Making Process

Ganesh D Katale[1]
Research Scholar,
K J Somaiya College of Engineering Mumbai,
Somaiya Vidyavihar University,
Mumbai

Nandkumar R Gilke[2]
Professor,
K J Somaiya College of Engineering Mumbai,
Somaiya Vidyavihar University,
Mumbai

Kashinath N Patil[3]
Professor,
K J Somaiya College of Engineering Mumbai,
Somaiya Vidyavihar University,
Mumbai

Abstract: Jaggery is traditional sweetener produced by evaporation of water from the sugarcane juice. Sugarcane juice contains around 12% sucrose and 88% moisture (water). The evaporation of water from the sugarcane juice is carried out in open pan juice boiling processes. This process requires significant heat energy. The open pit furnaces are used for sugarcane juice heating, which has lower thermal efficiencies. The present study measures the performances of traditional jaggery making plants are evaluated on the basis of heat supplied by bagasse combustion in the open pit furnaces; heat required for the water evaporation from the juice, heat utilized to produce every kilogram of jaggery production, and instantaneous thermal efficiency of plants. The field survey carried out for

[1]ganesh.katale@somaiya.edu, [2]nandkumar@somaiya.edu, [3]kashinath@somaiya.edu

DOI: 10.1201/9781003596707-36

five jaggery manufacturing plants. Instantaneous efficiency of these plants is evaluated and it is observed that it varies from 35.17% to 46.02%. The amount of jaggery obtained from the sugarcane juice is in the range of 186.9 gm to 208.3 gm per litre of sugarcane juice.

Keywords: Sugarcane, Jaggery, Instantaneous thermal efficiency, Juice evaporation

1. Introduction

In India, sugarcane cultivation in 2020-21 is 52.25 lakh hectares; there is 8 % growth as compared to 48.41 lakh hectares in 2019-20. Almost 80% of the sugarcane is used for white sugar production, around 12% for jaggery production and approximately 8% is used as seed for sugarcane cultivation [Handbook of Processing of Jaggery (2020)]. In India around 1,00,000 jaggery manufacturing plants are working under ministry of micro, small and medium enterprises (MSME) sector. Jaggery (also called as Gur in India) is produced from sugarcane in addition to sugar. Jaggery having composition of Fe-11%, Ca-0.4%, P-0.045%, CH_2OH-10%, $C_6H_{12}O_6$-15%, RCH-0.25% and fat (0.05%) [Yogesh Shankar Kumbhar (2016)]. The jaggery is having various nourishing and favourable benefits for human body in contrast with the white sugar. The consumption of white sugar and its products is growing extremely fast however it has major impact on health issues like diabetes and obesity. The encouragement of jaggery uses might improve the human health and minimize the consumption of white sugar [Abhai Kumar and Smita Singh (2020)]. The jaggery manufacturing is basically farmers driven activity thereby they must be aware of recent trends in sugarcane cultivation and jaggery processing [P. K. Pattnayak and M. K. Misra (2004)].

2. Literature Review

In the traditional jaggery making process, primary fuel used is a dry bagasse to produce heat energy for boiling of sugarcane juice. Recent reports suggest that the jaggery industry used about 30% of the sugarcane produced [PVK Jagannadha Rao, Madhushweta Das and SK Das (2007)]. Jaggery production is basically a small scale/cottage industry located in rural area and owned by farmers. The raw sugarcane juice boiling is carried out with an open pan heating process. In conventional jaggery manufacturing single pan is used for sugarcane juice boiling. In single pan jaggery producing plants requires 3.85 kg of bagasse for every kg of jaggery production and efficiency of plant is around 14.75% [Rakesh Kumar,

Mahesh Kumar (2021)]. The use of multiple pans in jaggery making process with improvement in furnace reduces the bagasse consumption up to 1.44 kg per kg of jaggery production and the efficiency of plants increased up to 55% [Sanober Khattak , Richard Greenough, Vishal Sardeshpande and Neil Brown (2018)]. The energy efficiency of furnace used in jaggery production is significantly improved by using fins around the sugarcane juice boiling pans. The changes in pan structure shows better heat transfer from furnace wall to boiling juice and thereby reduction in bagasse fuel and also minimized heat loss. The bagasse saved from minimizing heat loss is then can be used for paper production or other similar applications. The lower cycle time for sugarcane juice extraction helps in quality improvement of the jaggery production [S.I. Anwar (2010)]. In case of four pan jaggery making process the plant is working for 24 hours a day. Plant capacity varies according to the plant crushing capacity. In case of single pan juice boiling system requires 360 litre of sugarcane juice and produces around 90 kg of jaggery. The every batch requires around 56 minutes for sugarcane juice boiling. The bagasse fuel required is 1.44 kg per kg of jaggery produced; having thermal efficiency of the plant is around 46%. The plant could able to produce 2 tonnes of jaggery in a single day [Kiran Y. Shiralkar, Sravan K. Kancharla, Narendra G. Shah and Sanjay M. Mahajan (2014)]. The CFD simulation tool can be used for analysis of jaggery producing furnace with various improvements in existing furnaces, such as use of fire bricks for furnace construction, the improved pan design for juice heating so as to develop thermal efficient furnaces [Sunildatta N. Kulkarni & Babruvahan P. Ronge (2018)]. The analysis is carried out to evaluate convective heat and mass transfer coefficient with respect to heat supplied. Convective heat and mass transfer coefficient is reliant on the heat supplied. Variation of convective heat and mass transfer coefficient is from 50.65 W/m^2 °C to 345.20 W/m^2 ^{0}C for the heat supplied range from 160 W to 340 W [G. N. Tiwari, Om Prakash and Subodh Kumar (2004)]. The results are analysed from the NCS (jaggery) production technologies in elsewhere. It shows that efficiency of the furnace is highly depends on technological development in the jaggery making. The efficiency of the system is improved by optimizing the energy exchange and well organized implementation of multiple evaporation process [Rodriguez Jader, Velasquez Fabian, Espitia John, Escobar Sebastian, Mendieta Oscar (2018)]. The study shows that for jaggery preparation effectively utilized 45 % of energy to that of total energy supplied and 55 % of heat is loss through furnace wall, hot flue gases and ash. Extensive energy is required to evaporate water from the sugarcane juice is around 39.22% of the total energy utilized in jaggery making process. Preheating of sugarcane juice up-to the boiling temperature requires 2360.44 kJ of energy and 0.2360 kg of bagasse for per kilogram of jaggery production[Lakshmi Pathi Jakkamputi and Mohan Jagadeesh Kumar Mandapati (2016)]. The adequate precautions to be taken in terms of cleanliness and hygiene in jaggery making process such as; boiling of sugarcane juice, settling of jaggery, packaging and storing, etc. To provide hygienic working conditions in jaggery making process is required to implement good practices to reduce microbiological risks. Therefore it is necessary to provide advanced technological knowhow to all labours to reduce impurity risk and nourishment of jaggery [Tatiana Vera-Gutierrez, Maria Cristina Garcia,Angela Maria and Oskar Mendieta (2019)].

The present study discuss the detailed findings of five jaggery producing units in terms of, Heat Supplied, Heat Utilized, Heat Lost through furnace Wall, Heat Required for Evaporation, Heat Required for Jaggery processing, Heat loss through hot flue gases, Instantaneous Thermal Efficiency, Jaggery Fraction, and Water Evaporation in case of jaggery processing.

3. Case Study

The jaggery manufacturing units in Nashik District, North-Western part of Maharashtra state are considered for field survey. To analyse the conventional jaggery manufacturing process the field visits are carried out. Figure 36.1 shows typical layout of four pan jaggery making unit. The jaggery processing employs unskilled labours. Around 8-10 labours are used for small to medium size jaggery processing unit. The jaggery manufacturing process is labour intensive in nature. The least mechanization is observed in the jaggery making unit under consideration for the analysis. The labours are required for various operations like

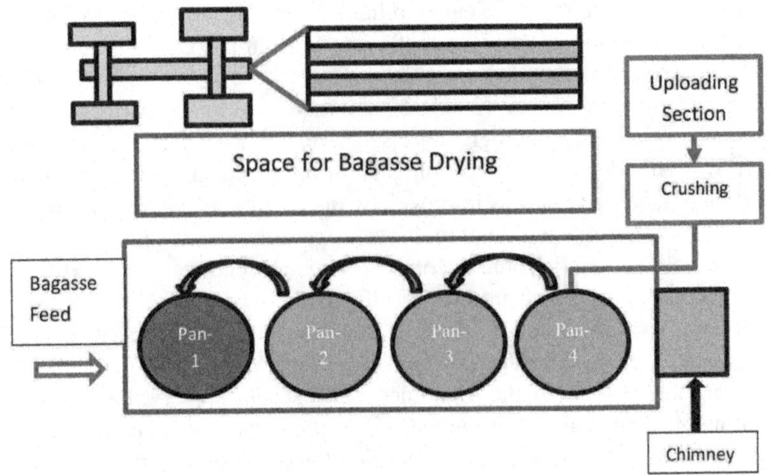

Fig. 36.1 Layout of four pan jaggery processing unit

juice extraction from sugarcane, juice transfer, drying bagasse, bagasse feeding in furnace, juice boiling, settling of jaggery and moulding of jaggery, etc. Usually sugarcane crushing capacity of unit is around 200 tonnes per day produces 23.5 tonne of the jaggery.

Table 36.1 shows the various input parameters for five jaggery making units surveyed carried out near Nashik region of Maharashtra state. The lowest sugarcane crushing capacity capacity is around 15 tonnes to as high as 200 tonne. Generally jaggery production is a batch process operating around 15 to 18 hours per day. For sugarcane juice heating dry bagasse is used as a primary fuel in the conventional open pan heating/concentration process. The survey is carried out for four pan system used for jaggery production. The exhaust flue gas temperature is measured, which is around 380 to 400°C.

Table 36.1 Shows plant capacity with respect to 4-pan jaggery production per day

Sr. No	Description	Unit No-1	Unit No-2	Unit No-3	Unit No-4	Unit No-5
1	Cane Capacity of crushing Unit (tonnes/day)	20	200	15	18	20
2	Working (hours/day)	24	24	15	16	18
3	Sugarcane juice production (litres/day)	12000	120000	9000	11500	10000
4	Jaggery production (kg/day)	2500	23500	1850	2160	2300
5	Fuel used	Bagasse	Bagasse	Bagasse	Bagasse	Bagasse
6	Rate of sugarcane (Rs./tonne)	2200-3000	2200-3000	2200-3000	2200-3000	2200-3000
7	Cost of jaggery (Rs./Kg)	28	24	40	27	26
8	Jaggery production rate (kg of Jaggery/1000 kg of sugarcane)	125 – 180	125 – 180	125 – 180	125 – 180	125 – 180
9	Bagasse consumption (Kg/kg of jaggery)	1.5-2	1.5-2	1.5-2	1.5-2	1.5-2

4. Heat Transfer in Jaggery Making Process

The boiling of sugarcane juice is the most trivial and vital process to remove water from sugarcane juice. The rate of water evaporation dictates the quality of the jaggery produced. The critical rate of heating and accurate temperature control required during the sugarcane juice boiling. This process is completed with combustion of primary fuel bagasse in the open pit furnace. The remaining part of sugar-cane left after juice removal is called bagasse. The bagasse can be used for combustion after open sun drying. Figure 36.2 represents the heat utilized in jaggery making process. The energy required for production of jaggery by vaporising the water present in the sugarcane juice. It is estimated on the basis of water vaporised in the single batch and the final temperature of the product.

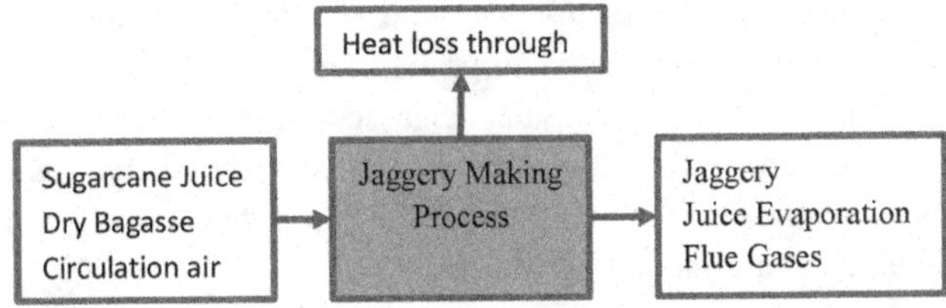

Fig. 36.2 Represents heat utilized in jaggery making process

A sample evaluation for jaggery production unit no. 01 is presented in the following section. The jaggery production unit under consideration uses 12000 litres of sugarcane juice while it produces 2500 kg of jaggery per day. The bagasse required for a batch of 12000 litres of cane juice evaporation is around 3750 kg. The instantaneous thermal efficiency is evaluated for every jaggery producing unit.

Mass of sugarcane juice (m_j) = 12000 liters, Jaggery Produced (jp) = 2500 kg, Required Bagasse (rb) = 3750 kg ,Moisture Removed (mr) = 9500 kg, Calorific value of Bagasse = 18 MJ/kg

$$\text{Heat supplied} = \text{Required bagasse} * CV = 7500 \text{ MJ} \quad (1)$$

Hence, the net heat balance is given as,

$$\text{Heat supplied } (H_s) = H_{ue} + H_{uj} + H_{fg} + H_w \quad (2)$$

Where, Heat supplied (Hs), Heat require to vaporisation (H_{ue}), Heat loss through hot flue gases (H_{fg}) and Heat loss through the furnace walls (H_w).

The heat loss from the furnace wall

$$H_w = h \times A_w \times (T_{skin} - T_{amb}) \quad (3)$$

5. Heat Transfer in Furnace

In traditional jaggery making process boiling of sugarcane juice is carried out with the help of combustion of the bagasse in open pit furnace. Traditionally furnace layout varies according to local manufacturers and trend. The criterion for variation in furnace layout is mainly because of number of pans used for jaggery making process and variable material used for furnace construction. The furnaces are constructed by local artisan. The furnaces are usually made of mud or local soil for wall construction. The imperfect construction of furnace leads to excessive heat loss, resulting in excess fuel consumption. In order to make the furnace more energy efficient, the furnace can be constructed using masonry bricks or refractory bricks. Figure 36.3 shows the furnace wall structure for single pan heating process. Similarly, all four pans are arranged in series for four pan jaggery making process.

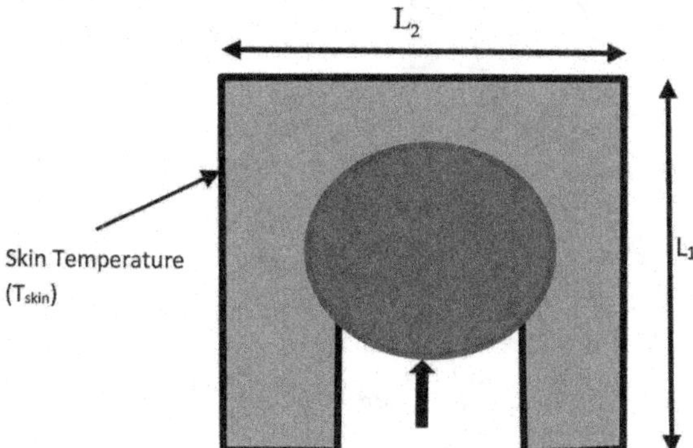

Fig. 36.3 Block layout of jaggery making furnace

Temperature of skin = 180°C and Ambient Temperature = 300°C

$$T_{avg} = \frac{T_{skin} + T_{amb}}{2} = \frac{180+30}{2} = 105°C$$

Table 36.2 Properties of air at average temperature

Temp °C	Density ρ, kg/m³	Specific Heat C_p J/kg-K	Thermal Conductivity k, W/m-K	Thermal Diffusivity α, m²/s	Dynamic viscosity μ, kg/m.s	Kinematic viscosity ν, m²/s	Prandlt Number
105	0.8977	1011	0.03235	3.565×10^{-5}	2.264×10^{-5}	2.522×10^{-5}	0.7073

Where,

g = Gravitational acceleration = 9.81 m/s², β = Coefficient of thermal expansion (β = 1/T) = 1/378 = 2.6455 × 10^{-3} K^{-1}, T_{amb} = Temperature of ambient = 30°C, T_{skin} = Temperature at outer surface = 180°C, Lc = Characteristic length of the geometry = 1.49 m, ν = Kinematic viscosity of the fluid = 2.522 × 10^{-5} m²/s,

Height of furnace = 1.49 m, Length of furnace = 6 m, Width = 1.5 m, σ_b = Stefan – Boltzmann Constant = 5.670×10^{-8} W/m^2.°K,

Grash of Number

$$Gr = \frac{g\,\beta(T_{skin} - T_{amb})Lc\text{^}3}{\nu\text{^}2} \tag{4}$$

$$RaL = Gr\,Pr = 9.61 \times 10^{09} \tag{5}$$

$$Nu_L = 0.59 \times (9.61 \times 10^{09})^{1/4} = 184.72 \tag{6}$$

$$\frac{hL}{k} = Nu_L \tag{7}$$

$$h = 4.01\ W/\,m^2 \cdot k \tag{8}$$

Heat loss through wall considering convective heat transfer coefficient

$$H_w = h \times A_w \times (T_{skin} - T_{amb}) = 43.55\ MJ \tag{9}$$

Heat loss through radiative heat transfer

Area of Furnace = 20.115 m^2

$$Q_{rad} = \sigma b \times A_s \times (T_s^{\,4} - T_{amb}^{\,4}) = 138.29\ MJ \tag{10}$$

Total Heat loss through wall

= Heat loss through wall considering convective heat transfer + Heat loss through radiative heat transfer

= 43.55 + 138.29 = 181.85 MJ

$$\text{Energy lost to flue gases } (H_{fg}) = m_{fg} \times C_p \times (T_{fge} - T_a) \tag{11}$$

Mass of air and fuel = air + fuel = 5.7 kg/kg

Table 36.3 The composition of bagasse for combustion and air required

Sr. No.	Bagasse Composition	% (kg/kg)	O_2 (kg/kg composition)	O_2 (kg/kg bagasse)	Mair (kg/kg bagasse)
1	C	0.46	2.67	1.2	$\frac{1.3132 \times 100}{23}$
2	H	0.06	8	0.48	
3	O	0.42	-	-0.42	
4	N	0.001	-	-	
5	S	0.0008	1	0.0008	
6	A	0.047	-	-	-
				1.3132	5.7

Exit temperature of flue gas as 400°C and initial temperature as 27°C,

Hear loss through hot flue gas = 2126.1 kJ/kg

Bagasse Energy = 15873.9 kJ/kg

$$\text{Required Bagasse} = \frac{\text{Heat per kg of jaggery production}}{\text{Net energy from bagasse}}$$

$$= \frac{10428}{15873.9} = 0.6569 \text{ kg/kg jaggery}$$

Energy from bagasse = 11824.69 kJ/kg

$$\text{Ideal thermal efficiency} = 10428/11824.69 = 88.18\ \% \tag{12}$$

Heat Utilized

$$\begin{aligned} Qt &= \{(m_{we} * C_p * (T_{bw} - T_i) + m_{we} * L_{hw}\}_{water} + \\ &\quad [(mj * Cp * (T_{jf} - T_{ji})]_{jaggery} + H_w + H_{fg} \\ &= 24450.52 \text{ MJ} \end{aligned} \tag{13}$$

$$\text{Efficiency of furnace} = \frac{\text{Heat Utilised (Hu)}}{\text{Heat Supplied of bagasse (Hs)}}$$

$$= \frac{24450.32(Hu)}{67500(Hs)} = 36.22\% \tag{14}$$

Table 36.4 Represents comparative energy analysis for Factory No-1 to factory No -05

Sr. No	Description	Unit No-1	Unit No-2	Unit No-3	Unit No-4	Unit No-5
1	Heat Supplied (MJ)	67500	702000	49950	58050	64800
2	Heat Utilized (MJ)	24450	246898	21665	26715	25450
3	Heat Lost through Wall (MJ)	181	378	235	173	343
4	Heat Required for Evaporation (MJ)	24225	27543	15682	23842	19380
5	Heat Required for Jaggery (MJ)	220	378	250.	189	211
6	Heat loss through hot flue gases (MJ)	20092	200070	14763	17769	19836
7	Instantaneous Thermal Efficiency (%)	36.22	35.17	43.37	46.02	39.27
8	Jaggery Fraction	0.2083	0.2166	0.2055	0.1869	0.24
9	Water Evaporation	0.7916	0.7833	0.7944	0.8130	0.76

6. Discussion and Results

The performance of traditional jaggery manufacturing units are evaluated on the basis of Heat supplied, Heat utilized, Heat loss through furnace wall, Heat loss through flue gases and Instantaneous thermal efficiency. It is observed that unit no 2 is having lowest instantaneous thermal efficiency which is around 35.17% as compared with remaining four jaggery processing units. The heat loss through furnace wall and hot flue gases are significant which is around 28.5% thereby affects the instantaneous thermal efficiency. Thus it results in excess bagasse consumption by the said unit. By appropriate measures such as sufficient air supply for combustion and reduction in heat loss through the furnace lining could improve the thermal efficiency. The unit no 1 having instantaneous thermal efficiency is around 36.22% slightly greater than that of unit no 2. The jaggery fraction for unit no 2 is 0.2166 which is higher than the jaggery fraction of unit no 1. The heat loss in unit no 1 is 1.5% higher than that of the unit no 2. The unit no 4 is operating with best thermal conditions with highest instantaneous thermal efficiency of 46.02%. The unit no 4 has an instantaneous thermal efficiency is greater than unit no 1 by 9.8%, and to that of for unit no 2 by 10.85%, for unit no 3 by 2.67% and unit no 5 by 6.95%. This is evident from the fraction of water evaporation from the sugarcane juice is 0.8130. It is highest amount of water evaporated by any other unit under consideration. The quality of the jaggery produced in this unit is superior as compared to all other units under consideration. The jaggery fraction for unit no 4 is 0.1869 which is less than 0.9% of unit no 1, 0.7% of unit no 2, 0.9% of unit no 3 and 0.8% of unit no 5. The highest jaggery fraction of 0.24 is in unit no 5 as compared with other four jaggery processing units. This may be due to better sucrose content in the sugarcane used for jaggery processing. It is observed that furnaces could be operated at much higher instantaneous efficiencies and amount of bagasse required could be reduced by improving the furnace designs of all the jaggery processing units.

7. Conclusions

The results showed that the instantaneous efficiency of the jaggery manufacturing units varies from 35.17% to 46.02%. The field survey and comparative analysis of all jaggery manufacturing units suggest that grater scope in energy efficiency improvement of jaggery manufacturing units by minimizing heat losses through the furnace. It is observed that significant losses losses occurred in these jaggery processing units are, heat loss through furnace wall lining and exhaust flue gas.

It is observed that the jaggery processing is very energy intensive process and an open pit furnace has significant lower thermal efficiencies. With detailed survey carried out, states that energy required for every kg of jaggery production is much higher as compared to the white sugar production.

Nomenclature:

H_u = Heat utilised

H_s = Heat Supplied

m = mass

CV = Calorific value

m_{we} = Mass of evaporated water

C_p = Specific heat of water

T_{bw} = Boiling temperature of water

T_i = Initial temperature of juice

L_{hw} = Latent heat of eater

m_j = mass of juice

T_{jf} = Final juice temperature

T_{ji} = Initial temperature of juice

H_{ue} = Heat utilised in evaporation

H_{fg} = Heat loss through flue gases

m_{fg} = Mass of flue gases

H_w = Heat loss through walls

T_w = Outer surface temperate of walls

T_a = Ambient air temperature

A_w = total area of walls

ha = Convective heat transfer

T_{fge} = Final temperature of flue gases at exit

References

1. Abhai Kumar and Smita Singh "The benefit of Indian jaggery over sugar on human health" Dietary Sugar, Salt and Fat in Human Health Elsevier (2020): 347–359.
2. G. N. Tiwari, Om Prakash and Subodh Kumar "Evaluation of convective heat and mass transfer for pool boiling of sugarcane juice". Energy Conversion and Management 45 (2004): 171–179.
3. Indian Institute of Food Processing Technology "Handbook Of Processing Of Jaggery". Ministry of Food Processing Industries, Government of India Pudukkottai Road, Thanjavur Tamil Nadu. (2020–21)
4. Kiran Y. Shiralkar, Sravan K. Kancharla, Narendra G. Shah and Sanjay M. Mahajani "Energy improvements in jaggery making process" Energy for Sustainable Development Elsevier 18(2014) : 36–48.

5. Lakshmi Pathi Jakkamputi, Mohan Jagadeesh Kumar Mandapati "Improving the performance of jaggery making unit using solar energy" Perspective in Science Elsevier (8) (2016) :146–150.
6. Oscar "Thermal performance evaluation of production technologies for non-centrifuged sugar for improvement in energy utilization". Accepted Manuscript, Energy (2018), doi: .1016/j.energy.2018.
7. P. K. Pattnayak, M. K. Misra "Energetic and economics of traditional gur preparation:a case study in Ganjam district of Orissa, India" Elsevier Biomass and Bioenergy 26 (2004) :79–88.
8. PVK Jagannadha Rao, Madhushweta Das and SK Das "Jaggery – A traditional Sweetener" Indian Journal of Traditional Knowledge. Vol.6(1),January (2007) : 95–102
9. Rakesh Kumar, Mahesh Kumar "Performance evaluation of improved and traditional two pan jaggery making plants: A comparative study" Sustainable Energy Technologies and Assessments Elsevier 47 (2021) : 104162
10. Sanober Khattak, Richard Greenough, Vishal Sardeshpande and Neil Brown "Exergy analysis of a four pan jaggery making process" Energy Reports Elsevier 4(2018) : 470–477
11. S.I. Anwar "Fuel and energy saving in open pan furnace used in jaggery making through modified juice boiling/concentrating pans" Energy Conversion and Management Elsevier 51(2010) : 360–364
12. Sunildatta N. Kulkarni & Babruvahan P. Ronge "CFD Simulation and Field Data Assessment Of Open Pan Jaggery Making Furnace". Vol.8(3),June (2018) : 647–652.
13. Rodriguez Jader, Velasquez Fabian, Espitia John, Escobar Sebastian, Mendieta
14. Tatiana Vera-Gutierrez, Maria Cristina Garcia,Angela Maria and Oskar Mendieta "Effect of processing technology and sugarcane varieties on the quality properties of unrefined non-centrifugal sugar". Elsevier Heliyon 5(2019): e02667
15. Yogesh Shankar Kumbhar "Study On Gur (Jaggery) Industry In Kolhapur" International Research Journal of Engineering and Technology (IRJET). Volume: 03 Issue: 02- Feb-2016, : 590–594.

Note: All the figures and tables in this chapter were made by the authors.

Technologies for Energy, Agriculture, and Healthcare – Shailesh Nikam et al. (eds)
© 2024 Taylor & Francis Group, London, ISBN 978-1-032-98028-7

37

Heat Loss Analysis of Solar Cavity Receiver

Muzammil Ahmad Shaikh[1]
Technical Officer,
Somaiya Institute for Research & Consultancy (SIRAC),
Vidyavihar, Mumbai, India

Kashinath Nimba Patil[2]
Professor,
Department of Mechanical Engineering,
K. J. Somaiya College of Engineering,
Vidyavihar, Mumbai, (India)

Abstract: In the realm of solar concentrator systems, cavity receivers emerge as vital components for extracting heat. Enduring high temperatures and dynamic positional adjustments throughout the day, these receivers manifest significant variations in heat losses. This paper undertakes an in-depth exploration of convection and radiation losses from a helical coil-type solar cavity receiver, featuring specific dimensions (coil diameter: 304.88mm, height: 800 mm, tube diameter: 6.35 mm). An experimental investigation is conducted to scrutinize convective and radiative losses for low-temperature fluids at cavity inclinations of 0°, 30°, 45°, 60°, and 90°. Results unveil that lower inclinations and higher mean fluid temperatures escalate convective losses, while radiative losses surge with an increase in receiver inclination and mean fluid temperature. The study incorporates two working fluids, water, and ethylene glycol, with varying concentrations (10% & 20% ethylene glycol mixed with pure water). It is observed that convective heat loss decreases with an ethylene glycol mixture, whereas radiation heat loss increases compared to pure water. Numerical correlations proposed by Prakash et al. are employed for the analysis, serving as a validation mechanism for the experimental results. This research contributes valuable insights into the intricate

[1]muzammilahmad.s@somaiya.edu, [2]kashinath@somaiya.edu

DOI: 10.1201/9781003596707-37

dynamics of solar cavity receivers, offering implications for enhanced efficiency and fluid selection in solar energy applications.

Keywords: Solar cavity receiver, Experimental investigation, Nusselt number, Convection heat loss, Effect of ethylene glycol

1. Introduction

Cavity receiver is a kind of heat exchanger mainly used for the solar concentrator. The efficiency of the solar concentrator system is intricately linked to how well the cavity receiver performs. Hence, it is crucial to enhance the effectiveness of the cavity receiver to minimize losses in the solar concentrator system. Harris and Lenz [1] conducted a thorough analysis of heat loss in various standard geometries, estimating both system efficiency and the heat power profile within the cavity. Their findings underscored that a well-designed cavity receiver has the potential to retain as much as 88% of the energy entering through its aperture. T. Taumoefolau et al. [2] explored the dynamics of an electrically heated cylindrical cavity, investigating free convection across various cavity inclinations within the temperature spane of 450 °C to 650 °C. Trials involving varied aperture diameters (AD) to cavity diameter (CD) were made resulting in a higher convection loss for ratios closer to 1 and an angle of -45° and minimum for the same angle with AD to CD ratio of 0.5. U. Leibfried & J. Ortjohann [3] performed a practical exploration on convective heat losses occurring in both elevated and inverted facing cavity receivers within a solar concentrator. Their research involved scrutinizing the effects of various opening ratios and operating temperatures on convection heat losses, focusing on both globular and semi-globular shaped cavity geometries. Stine & Mc Donald [4] conducted empirical studies on convection heat losses, specifically exploring the influence of cavity orientation, operating temperature, and variations in aperture dimensions. Additionally, they established a relationship that takes into account the effects of incline, geometry, and thermal conditions on the observed convection heat losses. M. Prakash [5] investigated 3 varied cavity receivers, each with an identical heat transfer area. The investigation employed CFD software to explore the impact of diverse aperture ratio at varying working temperatures of the cavity receiver. The outcome of this study led to the creation of a universal correlation applicable to all cavity configurations for the assessment of convection losses. Clausing, A.M. [6] delved into the study of cavity receivers, specifically focusing on convection losses through the presentation of an analytical model. This model emphasized the significance of the cavity receiver's ability to heat air and the influence of air circulation on convective losses. Validation of this framework was achieved by another researcher with a similar focus [7] through a practical experimentation involving a quadrilateral aperture of the cavity receiver. Koenig & Marvin [8] dedicated their research to convection, specifically addressing the high temperatures encountered by cavity receivers ranging from 550 to 900 degree celsius. Their endeavors yielded an phenomenological model. Quere Le, P. et al. [9] delved into the intricacies of a uniform temperature cubical enclosure, dissecting natural convection with an aperture ratio of 1 and inclinations spanning from -90 to +90. The findings unveiled a significant fluctuation in Nusselt numbers derived from experimental data. R.D. Jilte et al. [10] traversed the numerical terrain, investigating the combination of losses from both natural convection and radiation across a spectrum of diverse receiver configuration. Wu, S.Y. et al. [11] ventured into experimental territories, examining diverse thermal losses in a fully exposed cylindrical opening under varying boundary constraints. Kim, M. et al. [12] presented a bidimensional numerical model, unraveling the intricacies of natural convection within a quadrilateral enclosure. Kim, H.N. et al. [13] performed tests on three distinct solar receivers, focusing on thermal performance through the lens of a tubular receiver concept. Clausing, A.M. et al. [14] embarked on a investigation centered around natural convection from a cuboid cavity featuring a Lateral aperture. Kumar, N.S. & Reddy [15] orchestrated a computational simulation, deciphering convection losses in a customized cavity receiver tailored for a solar concentrator. Hess, C.F. & Henze [16] empirically investigated natural convection losses, specifically targeting horizontal extended cavities at various inclination angles. Their findings suggested a potential 10% reduction through strategic implementation of limiting flow at the aperture plane. Abbas et al. [17] delved into the impact of concentration ratio on the thermal prowess of a cone-shaped cavity tube receiver within a solar parabolic dish concentrator system. Azzouzi D et al. [18] conducted a comprehensive investigation, merging experimental and analytical approaches to dissect a cylinder-shaped cavity receiver designed for a solar dish. Wang, F. et al. [19] steered towards an optical efficiency analysis, scrutinizing a cylinder-shaped cavity receiver featuring a convex base surface. Pawar, J.B. et al. [20] engaged in an experimental exploration, unraveling the intricacies of a spirally wound cone-shaped cavity receiver paired with a Scheffler dish concentrator, with a focal point on energy and exergy performance. Martín-Alcántara, A. et al. [21] took a modeling approach, outlining the external flow dynamics of a novel Horseshoe receiver and evaluating its thermal prowess. Reddy, K.S. and Kumar [22] employed a 2-D model to estimate convection losses from a modified cavity receiver, operating under the assumption of a uniform and peak solar flux distribution. They further delved into a numerical study of combined natural convection and radiation using CFD software. Ma, R.Y. [23] explored the influence of wind on convection heat loss from the cavity receiver of a solar concentrator, noting an escalation in convection losses with higher wind speeds. M. Prakash et al. [24, 25] scrutinized heat losses emanating from a cylindrical cavity receiver, featuring an aperture diameter surpassing the cavity diameter. Their investigation extended to the impact of wind on total and convective losses, particularly for low-temperature applications. Numerical analysis played a pivotal role in identifying stagnation and convective zones through CFD software. Shewale, V.C. et al. [26,27] orchestrated a series of experiments and computational assessments focusing on spherical cavity receivers, with a keen eye on different inclination angles of the cavity receiver across varied temperatures.

In this study, a copper tube helical coil cavity receiver with a diameter of 0.385 m is fabricated. The aim is to explore how variations in receiver inclination, mass flow rate of the working fluid, type of fluid, and inlet fluid temperature affect the heat losses from the cavity receiver.

2. Empirical Analysis and Research Methodology

2.1 Specifications

The fabricated cavity receiver (Fig. 37.1) boasts 52 turns with precise spacing (0.001 to 0.003 m) and a length of 0.44 m. Glass wool insulation minimizes outer surface heat loss, clad with an aluminum sheet for comprehensive insulation.

(a) (b)

Fig. 37.1 (a) Isometric view (b) bottom views of fabricated coil

2.2 Experimental Arrangement

Figure 37.2 displays the schematic representation of the experimental setup. It consists of the top tank having heating element to the working fluid under trail. The 4 kW immersion heaters are used to heat the fluid. The hot fluid storage capacity is 78 liters. The test is carried out when the temperature of the fluid is in

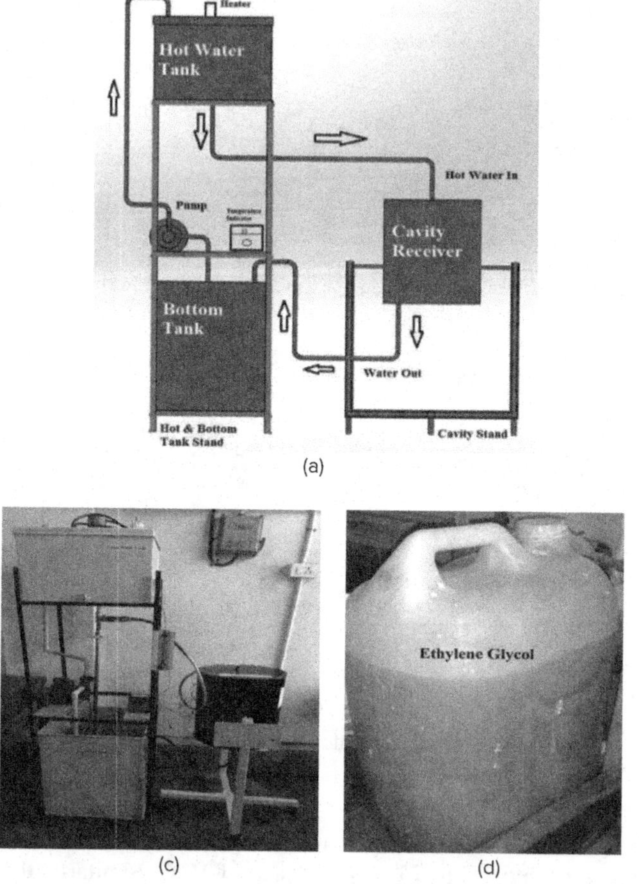

Fig. 37.2 (a) CAD Model and (b) Actual experimental set-up (C) Ethylene glycol

equilibrium. The pump of 0.25 HP capacity is used for circulation. The flow rate was monitored with the help of a rotameter. To monitor the temperature at various locations, Fifteen K-type thermocouples are employed. Alongside monitoring the temperature within the cavity receiver, ambient temperature is also measured. The wind speed was also measured while conducting the experiment. The bottom collection tank collects the fluid coming out of the cavity receiver. The cold water is then pumped to the top hot tank where equilibrium temperature is assured by on-off control of the immersion heater. Figure 37.3 shows the actual photograph of the experimental setup used in the study.

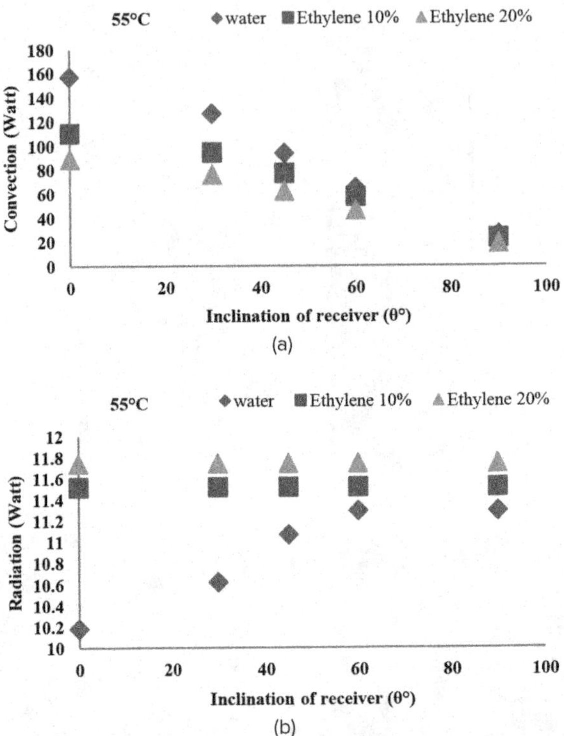

Fig. 37.3 Comparison of (a) Convection variance & (b) Radiative heat loss with inclination at 55°C

The hot water used for the heat loss trials. Initially the temperature of hot water is in the range from 55°C to 75°C with a step of 10°C. Similarly, the study was carried out for the 10% and 20% ethylene glycol solution in the same temperature range. The experiments maintain a mass flow rate of the hot fluid at 0.0025 kg/s and 0.0033 kg/s. The cavity receiver is oriented from 0°, 30°, 45°, 60° and 90° vertically. Figure 37.3(a) shows the variation in the convective and radiation losses at 55°C for various inclinations of cavity receiver. It has been noted that the convective loss for water is the highest, followed by the loss for a 10% ethylene glycol solution, with the least loss observed for the 20% ethylene glycol solution. Highest convective heat loss observed at 0° cavity orientation while least convective heat loss observed at 90° cavity inclination. In case of radiation heat loss exactly opposite observations are seen. Figure 37.3(b) shows the radiation heat loss at various cavity orientations at 55°C.

3. Heat Loss Estimation

3.1 Overall Thermal Dissipation (Experimentally)

$$Q_{total/\theta} = m \cdot C_p(T_{fi} - T_{fo}) \tag{1}$$

3.2 Overall Thermal Dissipation (Theoretically)

The energy equilibrium of a solar cavity receiver involves two predominant modes of heat loss: convection and radiation. Both convection and radiation losses occur through the opening of the cavity receiver. The cavity receiver is enveloped by substantial insulation, and it is presumed that conductive heat loss is minimal (negligible) compared to convection and radiation. Therefore, the energy balance equation is utilized to compute the heat loss within the cavity receiver.

$$Q_{total/\theta\, theoretically} = Q_{rad/\theta} + Q_{conve/\theta} \tag{2}$$

Where, $Q_{(total/\theta\, theoretically)}$, $Q_{(rad/\theta)}$, $Q_{(conve/\theta)}$, are the total, radiation, and convection heat losses for cavity inclination angle θ respectively.

3.3 Radiative Heat Loss

The Radiant energy loss from the cavity aperture is computed by utilizing the accompanying condition at θ at edge of cavity receiver.

$$Q_{rad/\theta\, theoretically} = \epsilon_{eff} A_{ap} \sigma\left(T_m^4 - T_a^4\right) \tag{3}$$

Where,

$$\varepsilon_{eff} = 1/\left[1 - (1 - \varepsilon_w)(1 - A_{ap}/A_w)\right] \tag{4}$$

3.4 Convective Heat Loss

Theoretical determination of convective heat dissipation through the cavity receiver's aperture is performed utilizing the subsequent equation corresponding to the inclination angle θ.

$$Q_{convec/\theta} = h * A_{ap} * (T_m - T_a) \tag{5}$$

The Rayleigh number is computed using the cavity diameter (D) as a reference.

$$R_{aD} = \frac{g\beta \Delta T D^3}{\gamma\alpha} \tag{6}$$

To determine the convective losses under no wind conditions, the Nusselt number is evaluated based on the work by M. Prakesh et al [5]. Fluid inlet temperatures ranging from 55 to 75 degrees Celsius are employed to compute the Nusselt number for all inclination angles of the cavity receiver.

$$N_u = 0.013 * R_a^{\left(\frac{1}{3}\right)} * (1+\cos\theta)^{2.72} * \left(\frac{d}{D}\right)^{0.72} \tag{7}$$

Know,

$$Q_{convec/\theta} = h * A_{ap} * (T_m - T_a) \tag{8}$$

4. Results and Discussion

The total and convective heat losses are assessed at operating temperatures of 55, 65, and 75 degrees Celsius within the cavity receiver under no wind conditions, considering inclination angles of 0°, 30°, 45°, 60°, and 90°. The total and convective heat transfer is greatest at cavity orientation of 0°, while it is least at 90°. An observation reveals that as the inclination of the cavity increases from 0° to 90°, the total and convective heat transfer decreases linearly. In the case of 0° cavity receiver the air movement is perpendicular to the base. This leads to an increase in air residence time inside the cavity receiver. Thus, it increases the convective heat transfer. However, at 90° receiver position the mouth surface of the cavity receiver parallel to the air movement. Thereby, air residence/interaction time is least which reduces the convective heat losses. Further, it is observed that as the cavity orientation increases from 0° to 90°, the radiative heat loss slightly increases. The radiation heat loss is least at cavity orientation at 0°. The radiative heat loss is highest at cavity orientation of 90° for all the temperatures as well as for the different working fluids used for trial. The trails are carried out at a lower temperature therefore it shows linear change in heat transfer however at higher temperature it will vary with fourth power with temperature change. The localized high temperature zone near the mouth of the cavity receiver would lead to an increase in heat transfer at 90° cavity receiver position.

Figure 37.3 shows that with increase in inclination, convection heat loss decreases (Fig. 37.3(a)) and radiation heat loss increases (Fig. 37.3(b)). Three different liquids are used for the analysis such as water, water with 10% ethylene glycol and 20 % ethylene glycol. The mass flow rate (0.0025 kg/s) is kept constant during the heat transfer analysis. The boiling point of solution improves by adding ethylene glycol into water. Thereby, the water boiling will start later. This will facilitate to study the radiation losses at lower temperature. The trend with increasing angle is also reversed in radiation as compared to convection i.e. radiation is maximum for 90° inclination and least for 0° inclination. For same mass flow rate, the convection heat loss due to pure water is maximum followed by water with 10% ethylene glycol and water with 20% ethylene glycol.

Likewise, in Fig. 37.4(a) and 37.5(a), convective heat losses are analyzed for cavity receiver operating temperatures of 65°C and 75°C. These losses peak at a cavity orientation of 0° and diminish at 90°, while also decreasing with a higher percentage of glycol in water.

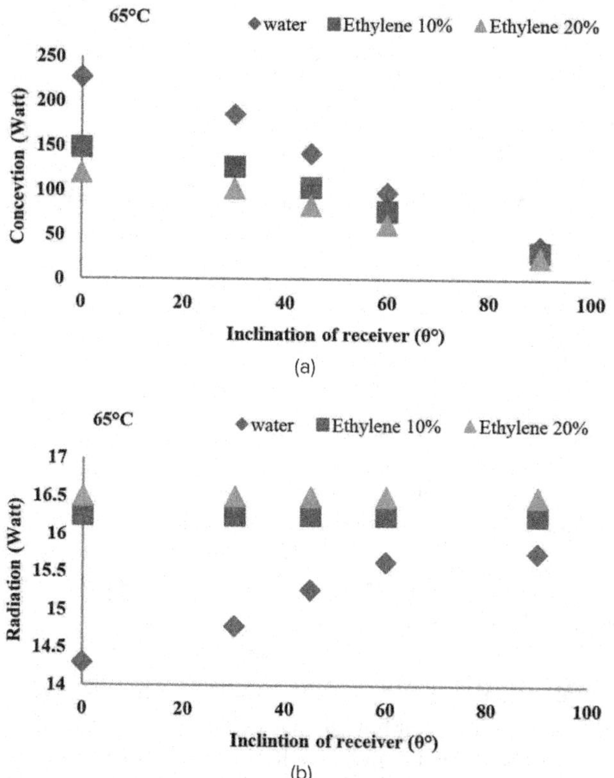

Fig. 37.4 Comparison of (a) Convection variance & (b) Radiative heat loss with inclination at 65°C

In radiation, the maximum heat is lost with 20% ethylene glycol water solution then with 10% ethylene glycol water solution and least heat loss is with pure water as shown in Fig. 37.4(b) and 37.5(b). It is also found that use of ethylene glycol with water as a heat transfer fluid is very useful to improve the thermal efficiency of the parabolic concentrator receiver system and decrease the temperature drop at

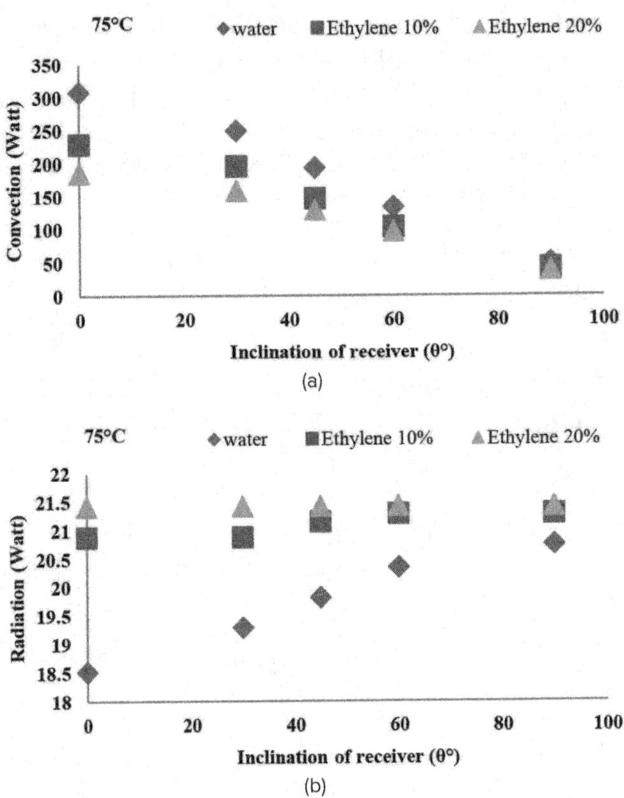

Fig. 37.5 Comparison of (a) Convection variance & (b) Radiative heat loss with inclination at 75°C

low temperature system. In the case of 10% and 20% ethylene glycol solution the change in radiative heat loss in both the cases is not significant. However, water and ethylene glycol solution the radiative heat transfer is significantly changes.

Figure 37.6 (a) and 37.6 (b) depict the overall heat losses of hot water, employed as the working fluid, at operating temperatures of 55, 65, and 75 degrees Celsius within the cavity receiver under no wind conditions. These losses are examined across various inclination angles (0°, 30°, 45°, 60°, and 90°) of the cavity, while maintaining a constant mass flux of 0.0025 kg/s. A notable trend is observed wherein total heat losses diminish as the cavity tilt angle varies from 0° to 90°. Moreover, it was noted that while convection losses do not reach zero, contrary to certain theoretical predictions, they reach a minimum at a 90° inclination angle.

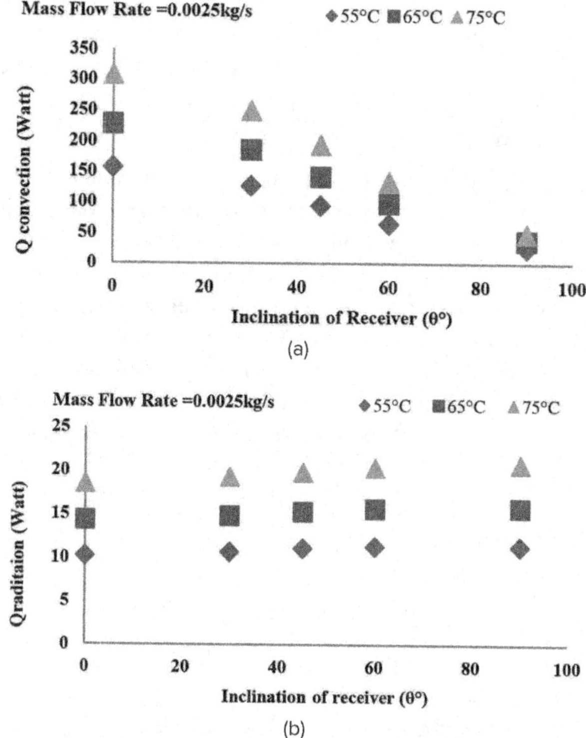

Fig. 37.6 (a) Convection variance & (b) Radiative heat loss with inclination

5. Conclusion

Analyzing the analytical results alongside experimental findings, this research provides a comprehensive understanding of the heat loss dynamics in solar cavity receivers. The observed trends in convective and radiative losses align with theoretical predictions, validating the analytical models used for the study.

The significance of cavity orientation in influencing heat transfer is highlighted, with a clear trend of decreasing convective losses and increasing radiation losses as the inclination angle rises. The experimental and analytical data consistently demonstrate that, at the same temperature, convective heat loss is maximum at 0° orientation, while radiation loss is least. This underscores the critical role of cavity orientation in determining overall heat losses. The experimental results show significant agreement with the numerical findings across all temperature ranges, with a maximum discrepancy of 10-15% observed across all angles of inclination of the cavity receiver.

Furthermore, the incorporation of ethylene glycol-water mixtures unveils intriguing insights. While these mixtures lead to a reduction in convective heat loss, they concurrently contribute to an increase in radiation heat loss compared to pure water. This nuanced understanding aids in making informed decisions regarding fluid selection for optimal performance in solar energy applications.

In summary, the analytical results validate the experimental outcomes and collectively contribute valuable knowledge to the realm of solar cavity receivers. This research not only enhances our understanding of the intricate dynamics involved but also offers practical implications for improving efficiency and selecting appropriate working fluids in solar power systems. The limitation of the current research is testing the system for higher temperatures such as greater than 600 °C. However, in this context need to use molten salts and different flow arrangement. As currently concentrator systems mainly focus on higher temperature applications it will be more relevant to test the system for similar environments.

6. Nomenclature

A_{op}	Opening area of cavity receiver [m^2]
C_p	Specific heat of working fluid [kJ/kg-K]
d	Opening diameter of cavity receiver [m]
D	Diameter of cavity receiver [m]
R_{aD}	Rayleigh Number
h_{op}	Heat transfer coefficient based on opening area of cavity [W$m^{-2}k^{-1}$]
K	Thermal conductivity [W$m^{-1}k^{-1}$]
N_u	Nusselt number [-]
$Q_{total/\theta}$	Total heat loss at cavity receiver angle θ [Watt]
$Q_{convec/\theta}$	Convection heat loss at cavity receiver angle θ [Watt]
$Q_{rad/\theta}$	Radiation heat loss at cavity receiver angle θ [Watt]
T_m	Mean temperature of working fluid [°C]
T_{fi}	Inlet temperature of working fluid entering the receiver [°C]
T_{fo}	Outlet temperature of working fluid leaving the receiver [°C]
T_a	Atmospheric temperature [°C]

Greek letters

ρ	Density [kgm-3]
σ	Stefan-Boltzmann constant [W$m^{-2}K^{-4}$]
ε	Emissivity of cavity surface [-]
θ	Inclination angle of cavity receiver [Degree]

References

1. Harris, J.A. and Lenz, T.G., 1985. Thermal performance of solar concentrator/cavity receiver systems. Solar energy, 34(2), pp.135–142.
2. Taumoefolau, T., Paitoonsurikarn, S., Hughes, G. and Lovegrove, K., 2004. Experimental investigation of natural convection heat loss from a model solar concentrator cavity receiver. J. Sol. Energy Eng., 126(2), pp.801–807.
3. Leibfried, U. and Ortjohann, J., 1995. Convective heat loss from upward and downward-facing cavity solar receivers: measurements and calculations.
4. Stine, W.B. and McDonald, C.G., 1989. Cavity receiver heat loss measurements, Proceeding of Inter. Solar Energy Society World Congress, Kobe, Japan, 1318–1322.
5. Prakash, M., Kedare, S.B. and Nayak, J.K., 2012. Numerical study of natural convection loss from open cavities. International Journal of Thermal Sciences, 51, pp.23–30.
6. Clausing, A.M., 1981. An analysis of convective losses from cavity solar central receivers. Solar Energy, 27(4), pp.295–300.
7. Clausing, A.M., 1983. Convective losses from cavity solar receivers- comparisons between analytical predictions and experimental results, J. of Solar Energy Engineering, 105 29–33.
8. Koenig, A.A. and Marvin, M., 1981. Convection heat loss sensitivity in open cavity solar receivers. Final report, DOE contract No. EG77-C-04-3985, Department of Energy, Oak Ridge, Tennessee.
9. Le Quere, P., Penot, F. and Mirenayat, M., 1981, Experimental study of heat loss through natural convection from an isothermal cubic open cavity, Sandia Laboratory Report SAND81-8014, 165–174.
10. Jilte, R.D., Kedare, S.B. and Nayak, J.K., 2013. Natural convection and radiation heat loss from open cavities of different shapes and sizes used with dish concentrator. Mechanical Engineering Research, 3(1), p.25–43.
11. Wu, S.Y., Guan, J.Y., Xiao, L., Shen, Z.G. and Xu, L.H., 2013. Experimental investigation on heat loss of a fully open cylindrical cavity with different boundary conditions. Experimental Thermal and Fluid Science, 45, pp.92–101.
12. Kim, M., Doo, J.H., Park, Y.G., Yoon, H.S. and Ha, M.Y., 2014. Natural convection in a square enclosure with a circular cylinder according to the bottom wall temperature variation. Journal of Mechanical Science and Technology, 28, pp. 5013–5025.
13. Kim, H.N., Lee, H.J., Lee, S.N., Kim, J.K., Chai, K.K., Yoon, H.K., Kang, Y.H. and Cho, H.S., 2014. Experimental evaluation of the performance of solar receivers for compressed air. Journal of Mechanical Science and Technology, 28, pp.4789–4795.
14. Clausing, A.M., Waldvogel, J.M. and Lister, L.D., 1987. Natural convection from isothermal cubical cavities with a variety of side facing apertures, J. of Heat Transfer, 109 407–412.
15. Kumar, N.S. and Reddy, K.S., 2007. Numerical investigation of natural convection heat loss in modified cavity receiver for fuzzy focal solar dish concentrator. Solar Energy, 81(7), pp.846–855.
16. Hess, C.F. and Henze, R.H., 1984. Experimental investigations of natural convection losses from open cavities, J. of Heat Transfer, 106, 333–338.

17. Abbas, S., Yuan, Y., Hassan, A., Zhou, J., Ahmed, A., Yang, L. and Bisengimana, E., 2023. Effect of the concentration ratio on the thermal performance of a conical cavity tube receiver for a solar parabolic dish concentrator system. Applied Thermal Engineering, 227, p.120403.
18. Azzouzi, D., Boumeddane, B. and Abene, A., 2017. Experimental and analytical thermal analysis of cylindrical cavity receiver for solar dish. Renewable Energy, 106, pp.111–121.
19. Wang, F., Lin, R., Liu, B., Tan, H. and Shuai, Y., 2013. Optical efficiency analysis of cylindrical cavity receiver with bottom surface convex. Solar Energy, 90, pp.195–204.
20. Pawar, J.B. and Tungikar, V.B., 2022. An experimental examination of a helically coiled conical cavity receiver with Scheffler dish concentrator in terms of energy and exergy performance. Sustainable Energy Technologies and Assessments, 52, p.102221.
21. Martín-Alcántara, A., Serrano-Aguilera, J.J. and Parras, L., 2022. Modeling the external flow of a novel HorseShoe receiver and the evaluation of thermal performance. Applied Thermal Engineering, 215, p.118949.
22. Reddy, K.S. and Kumar, N.S., 2009. An improved model for natural convection heat loss from modified cavity receiver of solar dish concentrator. Solar Energy, 83(10), pp.1884–1892.
23. Ma, R.Y., 1993. Wind effects on convective heat loss from a cavity receiver for a parabolic concentrating solar collector (No. SAND-92-7293). Sandia National Lab. (SNL-NM), Albuquerque, NM (United States); California State Polytechnic Univ., Pomona, CA (United States). Dept. of Mechanical Engineering.
24. Prakash, M., Kedare, S.B. and Nayak, J.K., 2009. Investigations on heat losses from a solar cavity receiver. Solar Energy, 83(2), pp.157–170.
25. Prakash, M., Kedare, S.B. and Nayak, J.K., 2010. Determination of stagnation and convective zones in a solar cavity receiver. International Journal of Thermal Sciences, 49(4), pp.680–691.
26. Shewale, V.C., Dongarwar, P.R. and Gawande, R.R., 2016. Heat loss investigation from spherical cavity receiver of solar concentrator. Journal of Mechanical Science and Technology, 30, pp.5233–5238.
27. Shewale, V.C., Dongarwar, P.R. and Gawande, R.R., 2017. Experimental and numerical analysis of convective heat losses from spherical cavity receiver of solar concentrator. Thermal Science, 21(3), pp.1321–1334.

Note: All the figures in this chapter were made by the authors.

Technologies for Energy, Agriculture, and Healthcare – Shailesh Nikam et al. (eds)
© 2024 Taylor & Francis Group, London, ISBN 978-1-032-98028-7

38

Predicting Solar Panel Output Performance with Advanced Machine Learning Techniques

Deep Khatri[1], Kevin Gladstone[2], Umang Patel[3]
K J Somaiya College of Engineering,
Somaiya Vidyavihar University

Abstract: Solar energy utilization is very important for long term electricity generation. Solar panel performance is affected through several parameters such as weather and solar radiation, making forecast difficult. The goal of this study is to create machine learning models which can accurately estimate solar panel power production. The rapid growth of renewable energy sources, particularly solar power, has revolutionized the global energy landscape. This proposed method focuses on harnessing the potential of machine learning techniques to predict power output from solar panels, thereby enhancing energy generation and utilization. In this research, historical data of solar panel power generation, including environmental variables such as irradiance, temperature, and time of day, are collected and pre-processed.

Keywords: Solar power, Solar panel, Power prediction, Machine learning

1. Introduction

Power prediction using machine learning is a fascinating subject that combines the power of data analysis and algorithms to forecast future power consumption or generation. It has various applications in energy management and optimization. Machine learning models can take into account various factors like weather

[1]deep.khatri@somaiya.edu, [2]kevin.gladstone@somaiya.edu, [3]umang@somaiya.edu

DOI: 10.1201/9781003596707-38

conditions, time of day, historical usage patterns, and even events or holidays that might impact power consumption. By considering these variables, the models can make more accurate predictions and assist in making informed decisions. These predictions can be incredibly useful in various areas [1],[2],[3] such as energy management, grid optimization, and renewable energy integration. For example, utility companies can use power prediction models to optimize their energy generation and distribution, ensuring a more efficient and reliable power supply.

The study focuses on a predictive model that increases the precision and reliability of solar power forecasts. Using the capabilities of machine learning [4],[5] this initiative aims to optimize the performance of solar panels, making them more adaptable to the dynamic and unpredictable nature of sunlight. As the demand for renewable energy escalates and the world transitions towards a more sustainable future, the ability to accurately predict solar power output becomes a critical factor in achieving energy efficiency [6]. Traditional forecasting methods [7],[8],[9] often fall short in capturing the intricacies of solar radiation patterns, necessitating the exploration of innovative approaches. By collecting historical solar power data, weather patterns, and other relevant variables, the machine learning model is designed to learn and adapt, providing real-time predictions of solar power output with unprecedented accuracy. The contributions to this proposed method are as follows:

1. This study outperforms earlier research by using a variety of regression models, offering a larger examination of modelling methodologies, and so increasing the adaptability and usefulness of our methodology.
2. In addition to standard approaches, we present an easy text-based prediction tool that enables users to estimate particular characteristics such as AC power. This user-centric feature increases our project's practical value
3. Unlike datasets of 20 points [7], our approach uses a large dataset of 3100 unique values. This significant increase allows for more sophisticated and representative investigation of underlying trends.

The subsequent sections of this paper are carefully crafted to provide a comprehensive understanding of our contributions. In Section 2, we look into prior studies on solar panel power prediction, forecasting and data collection researched by the co-author. In Section 3, we look into the objective, motivation and analyze study period of the data. The methodology formulated by the primary author, model specifications, experimental setup, are discussed in Section 4. The results of our analysis are detailed in Section 5, providing further insight into the effectiveness of our method. Finally, in Section 6, we conclude with our findings see the main takeaways and outline suggestions for future research in this area.

2. Literature Review

In the paper by Garg et al. [1] machine learning emerges as a transformative force. The study employs diverse algorithms, surpassing traditional methods in accuracy and adaptability. Addressing challenges like weather variability, machine learning enables real-time forecasting, enhancing grid stability and resource allocation. Comparative analyses underscore its superiority, marking a paradigm shift in solar power prediction. This work contributes a vital foundation for future research emphasizing the potential of data-driven approaches in advancing renewable energy systems. Bamisile et al. [2] investigates the use of artificial neural networks (ANNs) in predicting solar radiation by utilizing various backpropagation techniques and meteorological data. The research demonstrates the adaptability of ANNs in capturing complex interactions within solar radiation patterns. The study digs into optimization tactics by applying multiple backpropagation algorithms, with the goal of improving prediction accuracy. The incorporation of meteorological data refines the model further, emphasizing the need of complete input variables.

This review of the literature underlines the importance of ANNs in enhancing solar radiation prediction approaches for better renewable energy planning. It specifically focuses on the integration of photovoltaic power prediction into optimal control strategies for solar micro-grids. Examining seminal works, including Kallio et al. [3], the research underscores the crucial role of accurate PV power forecasting in optimizing micro-grid performance. Various prediction models and methodologies are discussed, highlighting advancements in real-time forecasting techniques. The study emphasizes the importance of integrating predictive analytics into micro-grid control strategies to enhance the efficiency and reliability of solar energy systems in off-grid or distributed settings.

Solar power generation prediction is based on deep learning stresses the revolutionary influence of deep learning models by examining major contributions in the area, such as Chang et al. [4] In effectively estimating solar power generation, neural networks, particularly recurrent and convolutional designs, play a critical role. The research emphasizes the benefits of deep learning, such as its ability to handle complicated temporal and geographical correlations in solar data. Deep Learning Enhanced Solar Energy Forecasting with AI-Driven Internet of Things investigates the integration of deep learning and AI-driven IoT for improved solar energy forecasting. The research emphasizes the synergy between deep learning algorithms and IoT technologies by examining seminal works in this field, such as Zhou et al. [5]. The study demonstrates the benefits of using real-time data from IoT devices to make accurate solar energy predictions. The review emphasizes this integrated approach's transformative potential in improving the precision and reliability of solar energy forecasting for optimal grid integration and sustainable energy management.

Solar photovoltaic power prediction using different machine learning looks at the landscape of solar photovoltaic power prediction, with a focus on various machine learning methods. The study highlights the versatility of machine learning algorithms in forecasting solar power output by examining notable contributions such as Zazoum et al. [7]. The effectiveness of various methods, such as support vector machines, neural networks, and decision trees, is investigated. The importance of accurate prediction models for optimizing solar energy utilization, grid integration, and resource planning is emphasized in the literature. This in-depth examination offers insights into the evolving methodologies and advancements in machine learning applications for solar photovoltaic power prediction.

3. Motivation

3.1 Motivation

Solar panels are becoming more prevalent, and accurately predicting their power output can greatly enhance their efficiency and integration into the power grid. By using machine learning techniques, we can develop models that taking into account various factors like weather conditions, panel orientation, shading, and historical data to forecast the power output of solar panels. This can be incredibly valuable for solar energy system operators, as it allows them to optimize energy production, plan maintenance schedules, and better manage energy flow into the grid. Additionally, accurate power prediction can also benefit consumers who have installed solar panels on their homes or businesses. It can help them monitor and optimize their energy usage, make informed decisions about energy storage or selling excess power back to the grid, and ultimately maximize the return on their investment in solar energy. The study was done keeping in mind the development of a more sustainable transition towards a more sustainable future, for accurate prediction of solar power output for achieving energy efficiency.

3.2 Block Diagram

The block diagram formulated by the co-author involves two parameter sets: independent (solar irradiance, ambient temperature, and module temperature) and dependent (DC power, total yield, and daily yield). Both datasets undergo preprocessing before entering regression models: linear regression (LR) [8],[9], random forest (RF) [9],[10], decision tree (DT) [11], extreme gradient boosting (XGB) [12], light gradient boost machine (LGBM), and k-nearest neighbor regression (k-NN) [13],[14]. These models analyze preprocessed data to uncover trends. Combined outputs, processed in the Prediction block, yield an overview of expected values for DC power, total yield, and daily yield based on input parameters. Preprocessed data undergoes feature selection and normalization to enhance model robustness. Chosen models collectively analyze interrelationships to capture patterns. Synthesized outputs in the Prediction block provide a comprehensive forecast, enhancing understanding of the system's behavior.

Parameter Selection: Independent and dependent parameters were selected by the authors for this study. They are classified based on the level of control which the tester has on them during the real time analysis. Selection of these parameters increase the potential of gaining near accurate results thereby showing efficiency and smooth operation.

Dependent Parameter: These are parameters which can be controlled by the user, these parameters have a direct effect on the results since these are variables that can be altered by the user.

Independent Parameter: These parameters are inherent in nature and cannot be controlled by the user. These parameters have an indirect effect on the obtained results and changes cannot be controlled by the user.

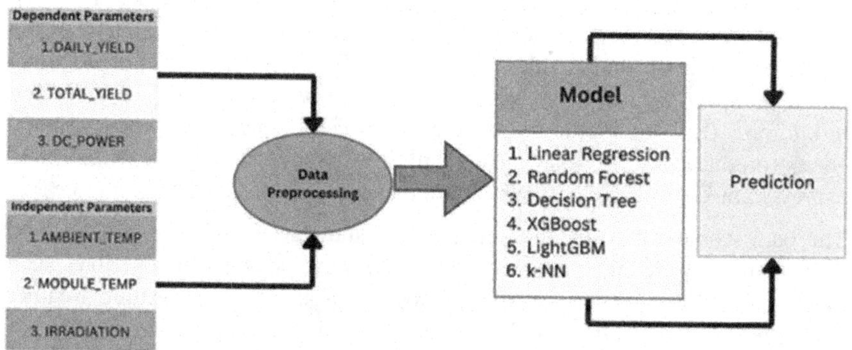

Fig. 38.1 Block diagram of the proposed method

Source: Made by the Author

4. Methodology and Model Specification

4.1 Methodology

The primary objective of this study is to predict solar panel power output using machine learning algorithms. Models will be trained on historical data from a Kaggle dataset, including solar panel power and meteorological data such as temperature, humidity, and solar radiation. The primary author obtained the solar panel dataset from Kaggle, ensuring it includes relevant independent parameters (solar irradiance, module temperature, ambient temperature) and dependent parameters (dc power, ac power, daily yield, total yield). The dataset underwent meticulous cleaning processes to address missing values and outliers, ensuring data integrity. He split the dataset into training and testing sets. We then train each regression model using the training dataset. Imputation techniques were employed where necessary to enhance completeness. To optimize model performance, he conducted feature engineering, creating time-based features to capture diurnal patterns. Categorical variables underwent encoding, and numerical features were normalized. Model assessment utilized metrics such as mean squared error (MSE) [15],[16], r-squared (R^2) [15],[16], and bias/variance. Training and testing scores were recorded to evaluate the model's performance on training and unseen data, respectively. The comprehensive approach taken in data preparation and model evaluation contributes to the robustness of the study, providing valuable insights into the predictive capabilities of machine learning algorithms for solar panel power output [17],[18].

4.2 Dataset

The dataset consists of two sets of information: power and yield generated & alongside weather conditions recorded by sensors. Comprehensive data has been collected from both solar panels and weather sensors. We have referred to the dataset available on [19] kaggle. Power Output refers to the amount of electrical power produced by the solar panels over a given period. It might be measured in kilowatts (kW) or another appropriate unit. Yield typically refers to the total energy produced by the solar panels within a specified timeframe. It could be expressed in kilowatt-hours (kWh) or another applicable energy unit.

The data would likely be time stamped to show when the power and yield measurements were taken. This enables analysis over specific time periods, such as hourly. The weather aspect gives information about the prevailing weather conditions at the location of the solar panels. It may encompass a range of meteorological parameters, such as solar irradiance, temperature and humidity. Solar irradiance indicates the intensity of solar radiation reaching the Earth's surface. It's a crucial factor influencing solar panel performance. Ambient temperature can affect the efficiency of solar panels. Higher temperatures, for instance, can lead to reduced efficiency. Humidity levels in the air may impact solar panel performance indirectly by influencing temperature and atmospheric conditions

4.3 Experimental Setup

For the hardware setup the experimental setup is n1-highmem-2 instance with a 2vCPU @ 2.2GHz, a 13GB RAM along with 100GB Free Space. It will consist of an idle cut-off 90 minutes and maximum cutoff of 12 hours

4.4 Model Specifications

We explored regression algorithms for solar panel power prediction: LR [8],[9] for simplicity, RF Regression [9],[10] for non-linearity, and DT Regression [11] for single-tree assessment. XGB [12], LGBM, and k-NN [13],[14] were also used. Models underwent dataset training with hyperparameter tuning for optimization and validation on the testing dataset.

5. Results and Discussion

The emphasis is on the installation and assessment of six different regression models for AC power forecasting. A thorough study was undertaken, focusing on important performance metrics such as training and testing scores, as well as the R^2 score and MSE, to offer a full assessment of each model's predictive capabilities. This section seeks to provide a clear and informative review of the results obtained from empirical testing of various models, providing light on their unique strengths and limitations in the context of AC power forecast.

Table 38.1 Training, testing and R^2 score along with MSE

Sr. No	Model	Training Score	Testing Score	R^2 Score	MSE	
					Training	Testing
1.	LR	0.9999	0.9999	100	0.8634	0.7460
2.	RF	0.9999	0.9999	100	0.0315	0.1009
3.	DT	1	0.9999	100	3.15E-29	0.1852
4.	XGB	0.9999	0.9999	100	0.9590	2.1443
5.	LGBM	0.9999	0.9999	100	1.66194	2.0525
6.	k-NN	0.9465	0.9076	89.47	7021.6257	11998.4227

Source: Adapted Kaggle Dataset [19], Calculations done by the Author

It is clear that the training score, testing score, and R^2 score all show little inaccuracy across the models, showing strong performance in AC power prediction. Notably, the DT and RF models outperform the others in terms of accuracy and precision. Specifically, the DT model demonstrates superior predictive skills with high scores and a robust R^2 score, reinforcing its efficacy in projecting AC power with minimal errors. The RF model, known for its ensemble approach, also performs admirably, showcasing its suitability for precise AC power forecasting. In contrast, the k-NN model attains significantly lower accuracy ratings than its counterparts. It is crucial to highlight its distinctive feature of higher precision, evidenced by a bias of 76058.1672 and a variance of 76058.1672. This indicates that, although the k-NN model may struggle with overall accuracy, it excels at maintaining precision in its predictions.

6. Conclusion

This study has tested various regression models for optimum solar panel power efficiency. The dataset has been pre-processed and every model has been run through keeping in mind several performance parameters. The study clearly indicates the model which provides power prediction most efficiently and least efficiently. The methodology followed is rigorous and therefore provides accurate results.

6.1 Main Takeaway

1. The study uses six regression models for estimating AC power, with a user-friendly interface that improves accessibility and usefulness across all models.
2. Assessment measures like as training and testing scores, R^2, MSE, bias and variance give a thorough assessment of model accuracy and generalization abilities.
3. The DT model shines out with high scores and a great R^2, demonstrating accurate AC power forecasts despite worries about overfitting.
4. In comparison, the k-NN Regression model has the poorest overall performance, highlighting accuracy and generalization issues, emphasizing the significance of rigorous model selection and validation.
5. Despite the DT model's success, the conclusion emphasizes the importance of careful model selection and understanding each model's differences and limitations in the field of AC power forecast.

6.2 Future Scope

The study will experience development, including dedicated gear for extracting data from solar panels and generating a full dataset. Integrated environmental sensors and IoT devices will give real-time weather and sun exposure data, improving the machine-learning model's ability to estimate AC power. Concurrently, the proposed method investigates sophisticated machine learning approaches, such as deep learning and ensemble methods, with deployment decisions depending on continuing research and developments.

In addition, a user-friendly interface will be created, ensuring accessibility for end users and stakeholders. This comprehensive method, which combines hardware innovation, algorithm optimization, and the creation of a simple interface, seeks to provide a more robust, flexible, and applicable solution for estimating solar panel power.

References

1. U. Garg, D.K. Chohan, and D.C. Dobhal, "The prediction of power in solar panel using machine learning," In Proc. International Conference on Computational Performance Evaluation (ComPE), December 2021, pp. 354–358. IEEE.
2. O. Bamisile, D. Cai, A. Oluwasanmi, C. Ejiyi, C.C. Ukwuoma, O. Ojo, M. Mukhtar, and Q. Huang, "Comprehensive assessment, review, and comparison of AI models for solar irradiance prediction based on different time/estimation intervals," Scientific Reports, vol. 12, no. 1, 2022, Art. 9644.
3. S. Kallio and M. Siroux, "Photovoltaic power prediction for solar micro-grid optimal control," Energy Reports, vol. 9, 2023, pp. 594–601.
4. R. Chang, L. Bai, and C.-H. Hsu, "Solar power generation prediction based on deep learning," Sustainable Energy Technologies and Assessments, vol. 47, 2021, Art. 101354.
5. H. Zhou, Q. Liu, K. Yan, and Y. Du, "Deep learning enhanced solar energy forecasting with AI-driven IoT," Wireless Communications and Mobile Computing, vol. 2021, pp. 1–11.
6. E.F. L. Fernández and F.M. Almonacid, "Optimum cleaning schedule of photovoltaic systems based on levelised cost of energy and case study in central Mexico," Solar Energy, vol. 209, 2020, pp. 11–20.
7. B. Zazoum, "Solar photovoltaic power prediction using different machine learning methods," Energy Reports, vol. 8, 2022, pp. 19–25.
8. A. Geetha, J. Santhakumar, K.M. Sundaram, S. Usha, T.T. Thentral, C.S. Boopathi, R. Ramya, and R. Sathyamurthy, "Prediction of hourly solar radiation in Tamil Nadu using ANN model with different learning algorithms," Energy Reports, vol. 8, 2022, pp. 664–671.
9. K. Anuradha, D. Erlapally, G. Karuna, V. Srilakshmi, and K. Adilakshmi, "Analysis of solar power generation forecasting using machine learning techniques," In E3S Web of Conferences, vol. 309, 2021, Art. 01163. EDP Sciences.
10. P.E. Rubini, G.D. Reddy, N.J. Chandan, K. Dikshith, and G. Ajay, "An Efficient Energy and Water Management in Agricultural Land using Sensors and Machine Learning Algorithm," Journal of Green Engineering, vol. 10, no. 7, 2020, pp. 3350–3360.
11. S.N. Shorabeh, N.N. Samany, F. Minaei, H.K. Firozjaei, M. Homaee, and A.D. Boloorani, "A decision model based on decision tree and particle swarm optimization algorithms to identify optimal locations for solar power plants construction in Iran," Renewable Energy, vol. 187, 2022, pp. 56–67.
12. S.H. Choi and J. Hur, "Optimized-XG boost learner based bagging model for photovoltaic power forecasting," Transactions of the Korean Institute of Electrical Engineers, vol. 69, no. 7, 2020, pp. 978–984.
13. M.E. Celebi, H.A. Kingrayi, and P.A. Vela, "A comparative study of efficient initialization methods for the k-means clustering algorithm," Expert Systems with Applications, vol. 40, no. 1, 2013, pp. 200–210.

14. P.S. Bradley and U.M. Fayyad, "Refining Initial Points for k-Means Clustering," Proceedings of the Fifteenth International Conference on Machine Learning, 1998.
15. C. Vennila, A. Titus, T.S. Sudha, U. Sreenivasulu, N.P.R. Reddy, K. Jamal, D. Lakshmaiah, P. Jagadeesh, and A. Belay, "Forecasting solar energy production using machine learning," International Journal of Photoenergy, 2022, pp. 1–7.
16. E. Izgi, A. Oztopal, B. Yerli, and A.D. Şahin, "Short-midterm solar power prediction by using artificial neural networks," Solar Energy, vol. 86, no. 2, 2012, pp. 725–733.
17. M. Zhang, H. Beaudin, H. Zareipour, and D. Wood, "Forecasting Solar Photovoltaic power production at the aggregated system level," in Proceedings of the North American Power Symposium, 2014, pp. 1–6.
18. A. Botchkarev, "Evaluating performance of regression machine learning models using multiple error metrics in Azure Machine Learning Studio," SSRN 3177507, 2018.
19. https://www.kaggle.com/datasets/pythonafroz/solar-power

Technologies for Energy, Agriculture, and Healthcare – Shailesh Nikam et al. (eds)
© 2024 Taylor & Francis Group, London, ISBN 978-1-032-98028-7

39

Charcoal Production using an Experimental Vertical Kiln

Doti Baqe[1]
Department of Agricultural Engineering,
Egerton University, Kenya

Jane G. Nyaanga[2]
Department of Crops Horticulture & Soils,
Egerton University, Kenya

Samuel Nyakach[3],
Wilberforce Okwara[4], **Daudi Nyaanga**[5]
Department of Agricultural Engineering,
Egerton University, Kenya

Abstract: The overreliance on fossil fuels due to increase in energy demand by ever growing population has resulted into depletion of these products and environmental degradation through air pollution. An alternative source of energy in form of biomass material can remarkably reduce the dependence on fossil fuels and eliminate its threat to the environment. The current study used twigs of acacia, black wattle and eucalyptus in producing charcoal to be used as source of energy through slow pyrolysis process using an experimental kiln. The result showed that an average of 37.18% of charcoal can be produced from slow pyrolysis of forest wastes with acacia generating the highest (39.46%). The acacia charcoal characteristics (moisture content, volatile matter, ash content and fixed carbon) were of a higher quality compared to those from black wattle and eucalyptus for cooking energy with less smoke thus contribute to achieving the sustainable development goal seven (SDG7) on affordable and clean energy.

Keywords: Energy, Biomass, Slow pyrolysis, Charcoal characteristics, Experimental kiln

[1]baqedoti95@gmail.com, [2]jgnyaanga@egerton.ac.ke, [3]samwel.nyakach@egerton.ac.ke, [4]okwaramayabi@gmail.com, [5]dmnyaanga@egerton.ac.ke

DOI: 10.1201/9781003596707-39

1. Introduction

Global energy demand is on the rise in response to population growth and technological advancement. An increase in population has resulted into high demand for energy production and this has been estimated to be rising by about 1.7% annually to 2030 (Mahood et al., 2015). To meet an increasing energy demand, fossil fuel has been a dominant source being approximately 80% while renewables only contributes just about 11% of required gross (Mahood et al., 2015). Overreliance on the fossil fuel has resulted into an increase in carbon footprint, hence need for a consistent use of renewable energy sources to reduce greenhouse gas emissions (Kapoor et al., 2020) including pyrolysis products of organic waste.

Biomass material which is a source of renewable energy can be easily converted to clean fuel through thermochemical process (Hou et al., 2018) including slow pyrolysis (through inert or partial inert thermal decomposition) of organic waste into a high quality char for briquettes or charcoal (Qin et al., 2020). Climate smart processes that convert biomass into char for briquette with less GHG emissions coupled with circularity will lead to carbon offsets and increase in environmental protection by reduction in fossil fuel usage and efficient increase the reuse of waste biomass (Thibanyane et al., 2019). Other associated products from pyrolysis including biochar and wood vinegar also termed pyroligneous acid which are chemical fertilizer and pesticide substitutes will further advance reduction in the carbon footprint (Doti et al., 2023).

In the production of charcoal, the common method used is by cutting down the trees and burning them directly. This technique results to desertification hence increasing earth's temperature while huge amounts of agroforestry wastes and their management is a challenge (Rabiu & Zakaria, 2017). The traditional charcoal production methods of direct burning of these wastes and crop residue often leads to low recovery rate and hence energy translation efficiency often accompanied with low calorific value (Thibanyane et al., 2019) due to poor design and control of the pyrolysis process. The production of high-quality charcoal from agroforestry wastes as source of energy will not only reduce the greenhouse gas emissions into the atmosphere but also aid in achieving the SDG7.

This paper presents the thermal decomposition using an experimental (vermiculite insulated medium sized metallic) kiln (Doti et al., 2023) of twigs cum trimmings from acacia, black wattle and eucalyptus (blue gum) trees into biochar (for agronomic use), char (for briquette production) and charcoal and their quality attributes.

2. Materials and Methods

2.1 Material Preparation

The kiln feedstock materials used for the study (acacia, black wattle and eucalyptus) were collected from the university in form of wastes during tree harvesting as form of forestry management. The materials were selected and loaded based on sizes of 2 cm to 5 cm diameters (usual firewood pieces) and height of 0.7 m that would fit in the experimental kiln.

The moisture content of crushed portions of the feedstock was determined using the air oven method at a temperature of 105°C for a period of 2 hours as in line with Igalavithana et al. (2017) protocol to less than 20% (Yaashikaa et al., 2019). This was after freshly harvested agroforestry waste (twigs) were sun-dried for an average of 6 hours/day for 14 days.

2.2 Charcoal Production

The different agro/forestry feedstock materials (acacia, black wattle and eucalyptus) wastes (branches and twigs) all at a moisture between 12% and 20% was weighed to a mass of 100 kg (for each species) using a weighing balance and fed into the experimental carbonization kiln shown in Fig. 39.1 below.

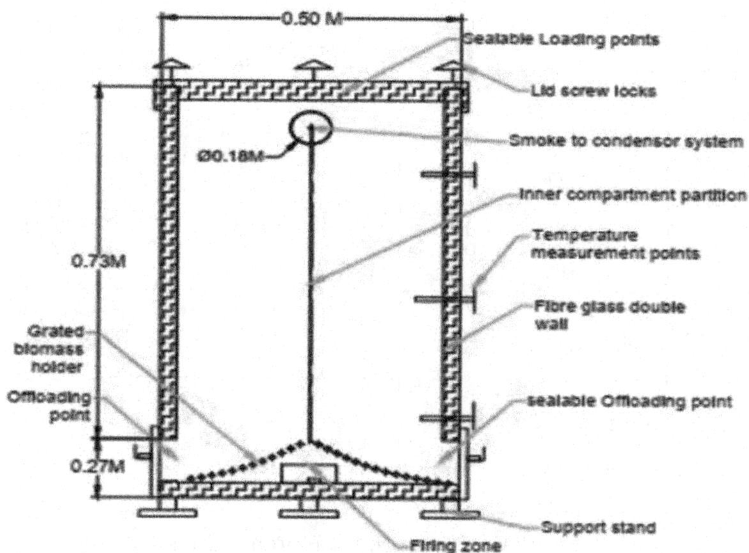

Fig. 39.1 Experimental carbonization kiln [1]

Each sample material was carbonized for a period of 180 minutes at a slow pyrolysis temperatures ranging between 240°C and 400°C (averaging 320°C). After the process, the system was allowed to cool to ambient temperature and charcoal produced offloaded from the kiln and weighed using the weighing balance. The charcoal production rate from each batch was calculated using Equation 1 below:

$$C_p = \frac{M_c}{M_f} \times 100 \tag{1}$$

where; C_p = charcoal production rate (%)
M_c = weight of charcoal (kg)
M_f = weight of raw kiln feedstock (kg)

The procedure was repeated three times for each material and the data analyzed and discussed.

2.3 Charcoal Characteristics

Three samples of each charcoal batch from the three feedstock types were analyzed for quality, that is, their moisture content, volatile matter, ash content and fixed carbon using the standard formulae as given below (Table 39.1).

Table 39.1 Formulae for computation of charcoal characteristics [1]

Quality	Formula	Eqn. No.	Parameter Description
Moisture Content (%)	$MC = \left[\frac{m_1 - m_2}{m_1}\right] \times 100$	2	m_1 = weight of air-dried raw sample (g) m_2 = weight of oven-dried sample at 105°C for 2 hours (g)
Volatile Matter (%)	$VM = \left[\frac{m_2 - m_3}{m_2}\right] \times 100$	3	m_3 = weight of muffle furnace dried sample at 950°C for 6 minutes (g)
Ash Content (%)	$Ash = \left[\frac{m_4}{m_2}\right] \times 100$	4	m_4 = weight of muffle furnace dried sample at 750°C for 6 hours
Fixed Carbon (%)	$FC = \left[\frac{m_2 - m_3 - m_4}{m_2}\right] \times 100$		All the parameters remain as defined above

The mass of the samples were determined using ASTM procedures before and after drying at 110 ± 5 °C for 1 hour and read to nearest 1 mg. The charcoals' calorific value (CV) and smoke emission levels (in terms of CO2) were estimated using available literature and relating them to common Kenyan charcoal cookstove and cooking environments.

3. Results and Discussion

3.1 Charcoal Production Rate and Density

The amount of charcoal produced as a percent of the raw material fed into the kiln for each of the three forest wastes is given in Table 39.2.

Table 39.2 Charcoal production from specific forest wastes

Feedstock material	Charcoal Production Rate (%)	Charcoal Density ($kg.m^{-3}$)
Acacia	39.46[a]	800[a]
Black wattle	37.11[b]	740[b]
Eucalyptus	34.97[c]	700[c]
LSD	0.74	35.46
Average	37.18	746.7

Means followed by the same letter(s)in the same column are not significantly different at $\alpha = 0.05$

Source: Authors

The charcoal production from the three feedstocks (averaging 37.18%) was slightly higher than the 35% reported by Wu et al. (2019) while that of eucalyptus was same to the whole number. The other two feedstocks (acacia and black wattle yielded 39.46% and 37.11%, respectively), were significantly different from that of eucalyptus at $\alpha = 0.05$. This difference can be attributed to difference in the genetic makeup of the forestry wastes since the pyrolysis temperature (320°C) used was same for both experiments. The charcoal produced from the current study is as shown in Fig. 39.2.

Acacia Black Wattle Eucalyptus

Fig. 39.2 Charcoal produced from specific forest waste

Source: Authors

The study further determined the density of each feedstock material which was obtained as 800 $kg.m^{-3}$, 740 $kg.m^{-3}$ and 700 $kg.m^{-3}$ for acacia, black wattle and eucalyptus, respectively. The differences in the amount of charcoal produced (also termed as recovery rate) can be associated with the different feedstocks and relative densities as in agreement with Rodrigues and Braghini (2019).

The calorific value of briquettes was estimated to be 13,500 kJ/kg and CO2 emissions as 240.6 ppm as reported by Otieno et al. (2022). This agrees with Nyaanga et al. (2023) and Kabango et al. (2023) who have reported a reduction of greenhouse gases by use of briquettes in rather than firewood.

3.2 Charcoal Characteristics and Quality

The quality characteristics of the charcoal produced through carbonization of acacia, black wattle and eucalyptus branches and twigs (forestry waste material) in the experimental kiln {in terms of moisture content (MC), volatile matter (VM), ash content (AC) and fixed carbon (FC)} are presented in Table 39.3 below.

Table 39.3 Charcoal characteristics and quality

Feedstock material	Proximate analysis (Wt.%, air-dried basis)			
	MC (%)	VM (%)	AC (%)	FC (%)
Acacia	8.2^{b}	28.2^{b}	6.8^{a}	56.8^{a}
Black wattle	9.6^{a}	28.9^{a}	7.0^{a}	54.5^{b}
Eucalyptus	9.5^{a}	29.2^{a}	7.0^{a}	54.3^{b}
LSD	*0.59*	*0.72*	*0.27*	*0.83*
Average	*9.1*	*28.8*	*6.9*	*55.2*

Means followed by the same letter(s) in the same column are not significantly different at $\alpha = 0.05$.

Source: Authors

The average values of moisture content 9.1%, volatile matter 28.8% and ash content 6.9% were higher than 3.6% 27.0% and 1.5%, respectively from cypress sawdust mixed with olive mill waste while fixed carbon of 55.2% was lower than 68.0% reported by Haddad et al. (2021). This difference could be associated with the use of olive mill waste in the previous research. Generally, the charcoal compares will with the properties reported for the charcoal from acacia and eucalyptus forestry branches and twigs. These current results differ from those of Haddad et al. (2021) due to difference in feedstock material, pyrolysis temperature and residence time by the two studies.

4. Conclusions and Recommendation

Forestry waste from acacias and eucalyptus can be converted into charcoal using a vermiculate insulated metallic experimental kiln that allow slow pyrolysis of the biomass operating at an average of 320 °C (240-400 °C) over a period of about 3 hours when the feedstock had been air-dried to below 20% moisture content. The average conversion (recovery) rate charcoal was 37 % with a density of 747 $kg.m^{-3}$ with average moisture content, volatile matter, ash content an fixed carbon of 9.1%, 28.8%, 6.9% and 55.5%, respectively. Acacia waste had a lower moisture content but higher of the other attributes that were significantly different at 0.05 than black wattle and eucalyptus. The charcoal had lower emissions and higher calorific value hence recommended that direct use of the forestry waste biomass as a source of energy cooking in rural Kenya.

Acknowledgements

The study was funded by Kenya Climate Smart Agriculture Project (KCSAP) while the writing of the manuscript was supported by Center of Excellence in Sustainable Agriculture and Agribusiness Management (CESAAM).

References

1. Doti, B. S., Daudi, M. N., & Nyakach, S. (2023). Designing, Developing and Testing of a Pyrolysis System: A Case Study of Biochar and Pyroligneous Acid. *Journal of Energy, Environmental & Chemical Engineering*, *8*(1), 1–9. https://doi.org/10.11648/j.jeece.20230801.11
2. Haddad, K., Jeguirim, M., Jellali, S., Thevenin, N., Ruidavets, L., & Limousy, L. (2021). Biochar production from Cypress sawdust and olive mill wastewater: Agronomic approach. *Science of The Total Environment*, *752*, 1–11. https://doi.org/10.1016/j.scitotenv.2020.141713
3. Hou, X., Qiu, L., Luo, S., Kang, K., Zhu, M., & Yao, Y. (2018). Chemical constituents and antimicrobial activity of wood vinegars at different pyrolysis temperature ranges obtained from *Eucommia ulmoides* Olivers branches. *RSC Advances*, *8*(71), 40941–40949. https://doi.org/10.1039/C8RA07491G
4. Igalavithana, A. D., Mandal, S., Niazi, N. K., Vithanage, M., Parikh, S. J., Mukome, F. N. D., Rizwan, M., Oleszczuk, P., Al-Wabel, M., Bolan, N., Tsang, D. C. W., Kim, K. H., & Ok, Y. S. (2017). Advances and future directions of biochar characterization methods and applications. *Critical Reviews in Environmental Science and Technology*, *47*(23), 2275–2330. https://doi.org/10.1080/10643389.2017.1421844
5. Kapoor, L., Bose, D., & Mekala, A. (2020). Biomass pyrolysis in a twin-screw reactor to produce green fuels. *Biofuels*, *11*(1), 101–107. https://doi.org/10.1080/17597269.2017.1345360

6. Mahood, H. B., Campbell, A. N., Thorpe, R. B., & Sharif, A. O. (2015). Heat transfer efficiency and capital cost evaluation of a three-phase direct contact heat exchanger for the utilisation of low-grade energy sources. *Energy Conversion and Management*, *106*, 101–109. https://doi.org/10.1016/j.enconman.2015.09.023
7. Kabango K., F G D Thulu, T. Mlowa, C. Chisembe, C.C. Kaonga. (2023). Effect of carbonisation on combustion characteristics of faecal sludge and sawdust blended briquettes. *Environmental Sustainability*. DOI: 10.1007/s42398-023-00269-6
8. Nyaanga, D.M., Okwara, W., Nyaanga J.G. (2023). Performance of forestry-waste-based briquettes in a ceramic cook stove. *Kenya Climate Smart Agricultural Project (KCSAP) Bioenergy Report. Interim Report Egerton University.*
9. Otieno A.O, Home, P.G.,Raude, J.M. Murunga, S.I. and Gachanja, A. (2022). Heating and emission characteristics from combustion of charcoal and co-combustion of charcoal with faecal char-sawdust char briquettes in a ceramic cook stove. Heliyon. 8(8): e10272. Published online 2022 Aug 18. https://doi:10.1016/j.heliyon.2022.e10272.
10. Qin, L., Shao, Y., Hou, Z., & Jiang, E. (2020). Effect of temperature on the physicochemical characteristics of pine nut shell pyrolysis products in a screw reactor. *Energy Sources, Part A: Recovery, Utilization, and Environmental Effects*, *42*(22), 2831–2843. https://doi.org/10.1080/15567036.2019.1618993.
11. Rabiu, Z., & Zakaria, Z. A. (2017). Pyrolignous Acid Production from Palm Kernel Shell Biomass. *Journal of Applied Environmental and Biological Sciences, 7*(2s), 59–62.
12. Rodrigues, T., & Braghini, J. A. (2019). Charcoal: A discussion on carbonization kilns. *Journal of Analytical and Applied Pyrolysis*, *143*, 1–46. https://doi.org/10.1016/j.jaap.2019.104670.
13. Thibanyane, N., Agachi, P., & Danha, G. (2019). Effects of biomass/coal copyrolysis parameters on the product yield: A review. *Procedia Manufacturing*, *35*, 477–487. https://doi.org/10.1016/j.promfg.2019.07.007.
14. Wu, X. F., Zhang, J. J., Huang, Y. H., Li, M. F., Bian, J., & Peng, F. (2019). Comparative investigation on bio-oil production from eucalyptus via liquefaction in subcritical water and supercritical ethanol. *Industrial Crops and Products*, *140*, 1–7. https://doi.org/10.1016/j.indcrop.2019.111695.
15. Yaashikaa, P. R., Senthil K. P., Varjani, S. J., & Saravanan, A. (2019). Advances in production and application of biochar from lignocellulosic feedstocks for remediation of environmental pollutants. *Bioresource Technology*, *292*, 1–11. https://doi.org/10.1016/j.biortech.2019.122030.

Technologies for Energy, Agriculture, and Healthcare – Shailesh Nikam et al. (eds)
© 2024 Taylor & Francis Group, London, ISBN 978-1-032-98028-7

40

Comprehensive Evaluation of Automated Stroke Prediction: Harnessing Machine Learning

Nitin Pal[1],
Girish Gidaye[2], Uma Jaishankar[3]
Department of Electronics and Computer Science,
Vidyalankar Institute of Technology,
Mumbai, India

Ravindra Sangle[4]
Department of Computer Engineering,
Vidyalankar Institute of Technology,
Mumbai, India

Jagannath Nirmal[5]
Department of Electronics Engineering,
K J Somaiya College of Engineering,
Mumbai, India

Abstract: Stroke is a global threat, causing early deaths and disability. Stroke cannot be a big problem if it is detected early and to detect strokes early, accurate and efficient prediction methods are needed. Nowadays, various researchers and doctors delve into artificial intelligence by using clinical data and reliable machine-learning techniques to predict strokes. However, challenges include handling imbalanced datasets, employing complex algorithms with limited interpretability, and model performance variations exposing limitations in Explainable AI (XAI). To address these, our proposed methodology uses oversampling and normalization, selects relevant features, and uses seven well-known classifiers for enhanced predictions. The results are superior, with Random Forest and Gradient Boost models outperforming other classifiers by achieving nearly 99% accuracy.

[1]nitin.pal@vit.edu.in., [2]girish.gidaye@vit.edu.in, [3]uma.jaishankar@vit.edu.in, [4]ravindra.sangale@vit.edu.in, [5]jhnirmal@somaiya.edu

DOI: 10.1201/9781003596707-40

The higher area under the curve score emphasizes the model's efficiency and reliability. The suggested approach can help in decision-making and improve stroke treatment choices to prevent this devastating disease.

Keywords: Comparative analysis, Healthcare informatics, Machine learning, Predictive models, Prognosis, Stroke

1. Introduction

Stroke, defined by the World Health Organization (WHO) in 1970, is a sudden disruption of brain function lasting at least a day, affecting specific or broad areas, with no apparent cause except vascular issues, resulting in millions of lives lost annually and leaving countless individuals battling disability (Warlow, 1998).

A stroke is a sudden brain attack that cuts off blood flow, causing potentially fatal damage. It has long-term impacts, including impaired speech, cognitive issues, and mobility loss. Recovery rates are low, with only 20% regaining full upper limb function (Jaafar et al., 2021). Artificial intelligence (AI) is revolutionizing stroke risk assessment by utilizing machine learning algorithms. These algorithms are demonstrating remarkable accuracy in predicting stroke risk, particularly for men aged 45-75, a demographic with a high risk. Stroke is a major cause of death, with 5.5 million people affected annually. The high morbidity and mortality rates result in up to 50% of survivors being disabled (Reeves et al., 2008). Machine learning techniques have shown promising accuracies in stroke prediction, enabling clinicians to identify and prioritize high-risk patients for potential stroke mitigation. These models reveal individual risk factors, allowing for precise treatment plans. Stroke impacts societies through reduced quality of life and overwhelming healthcare systems. (Rajpurkar et al., 2022; Hügle et al., 2020).

The objective of this research is to propose a machine learning-based brain stroke detection system and to find out the most discriminative parameter using a parameter ranking algorithm.

The rest of the paper is structured as follows. Section 2 reviews the extant literature. Section 3 explains the research methodology. Section 4 presents the experimental results along with a discussion of the findings. Section 5 summarises the paper.

2. Literature Review

Machine learning can predict strokes by identifying high-risk individuals, improving prevention, and enhancing diagnosis and treatment outcomes. The study of Arslan et al. (2016) shows the support vector machine model's superiority in ischemic stroke prediction, with an accuracy of 97.89% and an AUC of 97.83%. To make models more reliable, Islam et al. (2022) applied explainable AI (XAI) in healthcare, using gradient boosting and interpretability tools such as Explain Like I'm 5 (Eli5) and Local Interpretable Model-agnostic Explanations (LIME), achieved 80% accuracy in rapid patient diagnosis. Dritsas et al. (2022) investigated early stroke identification using the Linear Regression (LR) model. Employing a stacking approach yielded 98.9% AUC and 80% accuracy in their study. Choi et al. (2020) achieved 98.1% accuracy in stroke prediction using decision tree ID3 (Iterative Dichotomiser 3), Classification, and Regression Tree algorithms.

Dev et al. (2022) used Neural Networks (NN) describing the combination of a perceptron neural network and four essential features yielded 77% of accuracy among all models. A hybrid machine learning approach by Liu et al. (2019) successfully predicted cerebral strokes using imbalanced data from 43,400 subjects, focusing on 783 stroke patients, with a 71.6% accuracy rate and 19.1% false-negative rate using a Deep Neural Network (DNN). Tazin et al. (2021) addressed imbalanced data with the Synthetic minority oversampling technique (SMOTE), the Random Forest model yielded the best accuracy of approximately 96% for stroke prediction. The study by Raju et al. (2020) found that Min-Max Scaler, a supervised classification model, showed a 5-10% accuracy improvement through normalization techniques. Various machine learning models like support vector machine, gradient boosting, linear regression, decision trees, and random forest, have shown promising results in stroke prediction, with accuracies ranging from 80% to 98%. Techniques such as explainable AI, data balancing, and normalization contribute to enhancing model reliability and better performance.

3. Proposed Methodology

The study uses a stroke prediction dataset from the Kaggle community, which has 4981 rows and 12 columns. Figure 40.1 depicts the proposed machine learning methodology for pre-processing the raw datasets to eliminate noise, handle outliers, and ensure numerical compatibility. It removes outliers related to age and BMI and deletes missing values.

The pre-processed data is then divided into training and testing sets for robust model evaluation. During the training stage, a 10-fold cross-validated classification model was developed using training recordings. In the testing phase, unseen records are tested on a cross-validated trained model. This mitigates the risk of overfitting and allows for reliable assessment of model performance on unseen data. To address potential class imbalance, the training data undergoes Synthetic Minority Over-sampling Technique (SMOTE) and min-max normalization. A

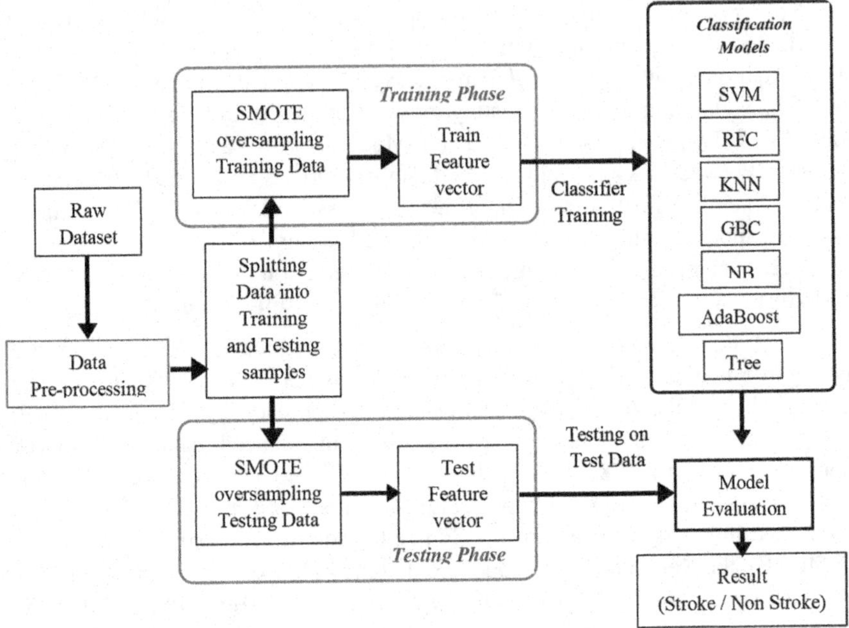

Fig. 40.1 The workflow of proposed methodology

Source: Author's compilation

diverse set of classifiers, including Support Vector Machine (SVM), Random Forest Classifier (RFC), Gradient Boosting Classifier (GBC), Naive Bayes (NB), K-Nearest Neighbor (KNN), AdaBoost, and Decision Tree (Tree) are trained on the pre-processed data. Feature vectors are extracted from the training data for each classifier. The testing phase uses consistent SMOTE oversampling by connecting data to trained classifiers to test data on the trained models.

Chi-square test-based feature selection is employed for enhanced evaluation reliability. This selection of Top 3, Top 5, and All 10 features are considered from the dataset parameters namely gender, age, hypertension, heart disease, ever married, work type, Residence type, avg_glucose_level, BMI, and smoking status. The performance of each classifier is evaluated to assess its accuracy in classifying instances as either stroke or non-stroke.

The evaluation highlights the effectiveness of a machine learning methodology for stroke prediction, which uses advanced data pre-processing and diverse classifiers for reliable results. This structured workflow serves as a valuable framework for future healthcare analytics and predictive modeling.

4. Result and Discussion

The features extracted are utilized for classification by seven well-known classifiers namely SVM, RFC, GBC, KNN, along with NB, Tree, and AdaBoost. The SMOTE technique was utilized for both training and testing data to evaluate the performance of classifiers. The model is evaluated by considering the AUC score, precision, recall, and F1 score. This critical step gauges the models' ability to generalize to unseen data and accurately predict instances of stroke or non-stroke.

The scatter plot depicted in Fig. 40.2 unveils the intricate correlations between variables such as BMI, age, and work type, thereby enhancing our comprehension of stroke occurrence. The segregation of data points into stroke and non-stroke categories highlights the distinct influences of these variables on stroke susceptibility. Particularly noteworthy is the prevalence of higher BMI and advanced age among stroke-afflicted individuals, underscoring their potential as predictive features in stroke classification. Moreover, the plot's utility extends beyond visualization, fostering hypothesis generation and informed decision-making in model development. By offering a comprehensive view of the data landscape, it aids in outlier identification, facilitates the discernment of complex interactions, and optimizes model performance.

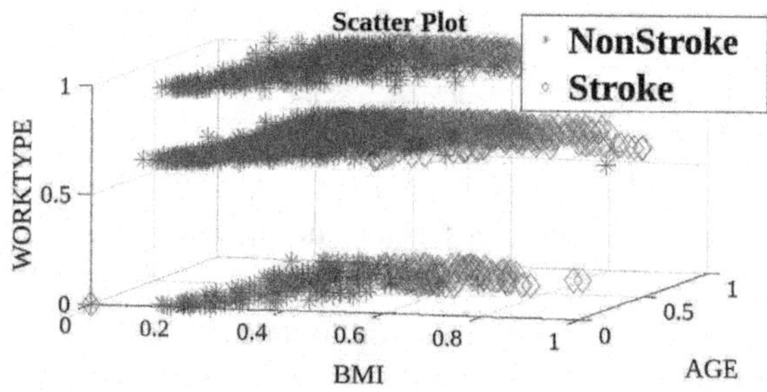

Fig. 40.2 3D scatter plot for top 3 features

Source: Author's compilation

The Chi-square test results depicted in Table 40.1 provide valuable insights of the relationships between categorical variables, enabling us to make informed judgments about which features were most relevant.

Feature selection was tailored for three subsets: Top 3, Top 5, and All 10 features. For the Top 3, chi-square focused on BMI, work type, and age, streamlining the analysis to reduce computational costs and model complexity. The Top 5 features included ever married and avg_glucose_level, adding a bit more complexity while

Table 40.1 Chi-square feature selection-based model performance

Metrics	No. of features	Classifier Model						
		RFC	SVM	KNN	GBC	NB	Ada-Boost	Tree
AUC	Top 3	0.999	0.685	0.995	1.000	0.997	0.995	0.969
	Top 5	0.999	0.800	0.996	1.000	0.997	0.995	0.969
	All 10	0.999	0.904	0.982	1.000	0.996	0.995	0.969
CA (%)	Top 3	99.40	63.00	99.10	99.50	97.80	99.40	96.50
	Top 5	99.30	66.40	99.00	99.50	96.50	99.40	96.50
	All 10	98.90	81.90	94.10	99.50	96.90	99.20	96.50
F1 Score	Top 3	0.994	0.628	0.991	0.995	0.978	0.994	0.965
	Top 5	0.993	0.664	0.990	0.995	0.965	0.994	0.965
	All 10	0.989	0.819	0.941	0.995	0.969	0.992	0.965
Precision (%)	Top 3	99.40	63.20	99.20	99.50	97.80	99.40	96.50
	Top 5	99.30	66.50	99.00	99.50	96.50	99.40	96.50
	All 10	98.90	81.90	94.40	99.50	96.90	99.20	96.50
Recall (%)	Top 3	99.40	63.00	99.10	99.50	97.80	99.40	96.50
	Top 5	99.30	66.40	99.00	99.50	96.50	99.40	96.50
	All 10	98.90	81.90	94.10	99.50	96.90	99.20	96.50

Source: Author's compilation

maintaining efficiency. The comprehensive set of All 10 features aimed to capture a detailed understanding of numerous factors, despite the increased computational load.

While all models achieved satisfactory performance, the RFC emerged as the most proficient model, consistently outperforming its counterparts. Higher chi-square test statistics indicated stronger feature associations across all selections, contributing to improved model predictions. The structured approach not only optimized computational resources but also effectively balanced model complexity and performance.

Table 40.2 presents a comparison of existing and proposed methods in stroke prediction. From Table 40.2 it was observed that the proposed methodology consistently outperforms existing methods, with higher accuracy and AUC values. The SVM algorithm in Arslan et al.'s study achieved 97.89% accuracy, while the proposed methodology using KNN achieved 81.9% accuracy and an AUC of 0.904. The stacking method in Dritsas et al.'s study yielded 80.0% accuracy and

Table 40.2 Comparison with existing approaches

Reference	Algorithms	Accuracy (%)	AUC	Proposed Methodology	
				Accuracy (%)	AUC
(Arsalan et al., 2016)	SVM	97.89	0.978	81.90	0.904
(Dritsas et al., 2022)	Stacking, LR, KNN, NB	80.00	0.989	KNN = 99.10, NB = 97.80	KNN = 0.996, NB = 0.997
(Liu et al., 2019)	DNN	71.60	–	–	–
(Choi et al., 2020)	Tree	98.10	–	96.50	0.969
(Tazin et al., 2021)	RFC	96.00	–	99.40	0.999
(Dev et al., 2022)	NN	77.00	–	–	–

Source: Author's compilation

an AUC of 0.989. The proposed methodology achieved 99.1% accuracy with an AUC of 0.996 for KNN and 97.8% with an AUC of 0.997 for NB. Similarly, Tazin et al.'s study got an accuracy of 96% whereas the proposed methodology achieved 99.4% accuracy for RFC. Table 40.2 serves as a valuable benchmark for evaluating the potential of the proposed methodology in stroke prediction.

5. Conclusion

Clinical decision support systems that use AI to predict stroke using models can be beneficial potential to be valuable tools for doctors and clinicians providing a second opinion to make diagnosis faster and more reliable. However, these AI-driven tools lack some transparency in their practical adoption. Transparency in the approach is essential for clinicians to understand the model's reasoning and relevance which enables them to integrate for improvement of deciding for patient care. To overcome these shortcomings, we demonstrated research towards introducing a new machine learning framework by harnessing the suitable hyper-parameters in the models for the prediction of stroke that prioritizes both accuracy and explainability. We were able to achieve an impressive accuracy of more than 95% using Random Forest, Gradient Boosting, Naïve Bayes, k-nearest neighbor, and AdaBoost Classifiers and extremely good AUC score exceeding 0.96 for respective highest achieving classifiers, but the true innovation lies in the model's transparency. This study offers a promising glimpse into the future of personalized informed decisions to deliver the best possible care to their patients. The study highlights the potential of explainable machine learning models in stroke prediction but emphasizes the need for comprehensive evaluation metrics. Future research should focus on refining models by incorporating interpretability and efficient evaluation metrics, thereby enhancing accuracy and transparency in stroke prediction systems.

References

1. Arslan, A.K., Colak, C. and Sarihan, M.E. (2016). "Different medical data mining approaches based prediction of ischemic stroke". Computer Methods and Programs in Biomedicine. 130:87–92. DOI: 10.1016/j.cmpb.2016.03.022.
2. Choi, Y. and Choi, J.W. (2020). "Stroke prediction using machine learning based on artificial intelligence". Int. J. Adv. Trends Comput. Sci. Eng. 9(4):1–6.
3. Dev, S., Wang, H., Nwosu, C.S., Jain, N., Veeravalli, B. and John D. (2022). "A predictive analytics approach for stroke prediction using machine learning and neural networks". Healthcare Analytics 2. p.100032. DOI: 10.1016/j.health.2022.100032.
4. Dritsas, E. and Trigka, M. (2022). "Stroke risk prediction with machine learning techniques". Sensors. 22(13):4670. DOI: 10.3390/s22134670.
5. Hügle, M., Omoumi, P., Van Laar, J.M., Boedecker, J. and Hügle, T. (2020). "Applied machine learning and artificial intelligence in rheumatology". Rheumatology Advances in Practice. 4(1):rkaa005. DOI: 10.1093/rap/rkaa005.
6. Islam, M.S., Hussain, I., Rahman, M.M., Park, S.J. and Hossain, M.A. (2022). "Explainable artificial intelligence model for stroke prediction using EEG signal". Sensors. 22(24):9859. DOI: 10.3390/s22249859.
7. Jaafar, N., Daud, A.Z.C., Roslan, N.F.A. and Mansor, W. (2021). "Mirror Therapy Rehabilitation in Stroke: A Scoping Review of Upper Limb Recovery and Brain Activities". Rehabilitation Research and Practice. p.9487319 (12 pages). DOI: 10.1155/2021/9487319.
8. Liu, T., Fan, W. and Wu, C. (2019). "A hybrid machine learning approach to cerebral stroke prediction based on imbalanced medical dataset". Artificial Intelligence in Medicine 101. p.101723. DOI: 10.1016/j.artmed.2019.101723.
9. Rajpurkar, P., Chen, E., Banerjee, O. and Topol, E. (2022). "AI in health and medicine". Nature Medicine. 28:31–38. DOI: 10.1038/s41591-021-01614-0.
10. Raju, V.N.G., Lakshmi, K.P., Jain, V.M., Kalidindi, A. and Padma, V. (2020). "Study the Influence of Normalization/Transformation process on the Accuracy of Supervised Classification", 2020 Third International Conference on Smart Systems and Inventive Technology (ICSSIT), Tirunelveli, India. 729–735. DOI: 10.1109/ICSSIT48917.2020.9214160.
11. Reeves, M.J., Bushnell, C.D., Howard, G., et al. (2008). "Sex differences in stroke: Epidemiology, clinical presentation, medical care, and outcomes". Lancet Neurology. 7:915–926. DOI: 10.1016/S1474-4422(08)70193-5.
12. Tazin, T., Alam, M.N., Dola, N.N., Bari, M.S., Bourouis, S. and Khan, M.M. (2021). "Stroke disease detection and prediction using robust learning approaches". Journal of Healthcare Engineering. 2021:7633381. DOI: 10.1155/2021/7633381.
13. Warlow, C.P. (1998). "Epidemiology of Stroke". The Lancet. 352:1–4. DOI: 10.1016/S0140-6736(98)90086-1.

Technologies for Energy, Agriculture, and Healthcare – Shailesh Nikam et al. (eds)
© 2024 Taylor & Francis Group, London, ISBN 978-1-032-98028-7

41

Modeling Attack Tree for CAN and ECU Vulnerabilities in Electric Vehicles to Drain its Battery

Irfan Siddavatam[1], Rahil Wahedna[2], Mohammed Haaris Shaikh[3], Kashmira Sonar[4]
Department of Information Technology,
K. J. Somaiya College of Engineering,
Somaiya Vidyavihar University,
Mumbai, India

Abstract: Electric Vehicles (EVs) represent a promising advancement in transportation technology, offering numerous benefits such as reduced emissions and lower operating costs. However, as EVs become increasingly integrated with complex digital systems, they are also becoming more susceptible to cyber threats. The presented work delves into the critical examination of vulnerabilities inherent in Electric Vehicles (EVs), specifically focusing on the Controller Area Network (CAN) and Electronic Control Units (ECUs). The primary goal is to develop and analyze an attack tree model to understand potential threats that could lead to the malicious draining of an EV's battery. Through meticulous analysis of existing literature, vulnerabilities in the CAN network and ECUs are found, encompassing compromise methods such as physical access, remote intrusion, and malware injection. Critical vulnerabilities in communication, software, and battery management systems are revealed, underscoring the multifaceted nature of cyber threats. Additionally, risks associated with overcharging, over-discharging, and battery drain are explored, as these vulnerabilities can be exploited through remote access, physical tampering, and social engineering, posing significant risks. The future scope involves implementing strategies to mitigate identified vulnerabilities, thereby enhancing the security and resilience of EV systems against cyber threats.

[1]irfansiddavatam@somiya.edu, [2]rahil.w@somaiya.edu, [3]mohammedhaaris.s@somaiya.edu, [4]kashmira.sonar@somaiya.edu

DOI: 10.1201/9781003596707-41

Keywords: Electric vehicles (EVs), Controller area network (CAN), Electronic control units (ECUs), Cyberattacks, Battery management systems, Battery draining

1. Introduction

Electric vehicles (EVs) are increasingly reliant on sophisticated software and interconnected systems, making vulnerabilities within their Controller Area Network (CAN) and Electronic Control Units (ECUs) prime targets for cyber threats. This paper conducts a comprehensive examination of existing research on vulnerabilities within CAN and ECU systems in EVs, aiming to shed light on this critical yet underexplored domain.

The CAN bus serves as the central nervous system of an EV, facilitating the transmission of crucial data among various sensors, actuators, and control units. However, it was originally designed without robust security features. This lack of security mechanisms, combined with the constant flow of information, presents an attractive target for malicious actors. Unauthorized access to the CAN bus could enable attackers to manipulate critical functions, compromise safety features, and even steal sensitive data. Similarly, vulnerabilities within ECUs, responsible for managing various EV subsystems, could grant attackers control over crucial aspects such as braking, steering, and battery management, potentially leading to catastrophic consequences.

The growing complexity of EV technology has created a fertile ground for cyber threats. Malicious actors can exploit vulnerabilities in CAN and ECU systems to manipulate critical vehicle functions, compromise sensitive data, and even cause physical harm. This investigation draws upon a comprehensive analysis of twenty research papers sourced from reputable academic databases and publications, including IEEE and Springer, to identify key vulnerabilities and potential attack goals specifically related to draining the battery of EVs.

2. Literature Review

The following literature review provides a comprehensive analysis of vulnerabilities in Electric Vehicles (EVs) and explores potential battery draining attacks along with mitigation strategies and implications for EV security and sustainability.

CAN Vulnerabilities: Vulnerabilities in the CAN protocol pose significant risks to EV security, particularly concerning battery draining attacks. These vulnerabilities encompass authentication and encryption shortcomings, broadcast-based communication, priority-based message scheduling, physical access vulnerabilities, remote access vulnerabilities, limited security features, and challenges in implementing security updates [1],[3],[4],[6],[9],[10],[15],[16],[17] ,[20]. While these references shed light on various aspects of CAN vulnerabilities, they may not comprehensively cover emerging threats or specific attack vectors relevant to battery draining attacks.

ECU Vulnerabilities: ECUs introduce additional vulnerabilities to EVs, increasing the risk of battery draining attacks. These vulnerabilities encompass software vulnerabilities, physical tampering, supply chain attacks, access control inadequacies, insecure communication protocols, and challenges in patching vulnerabilities [2],[3],[10]. Some studies might focus on specific ECU vulnerabilities, potentially neglecting broader systemic issues or emerging threats relevant to battery draining attacks.

Attack Vectors and Consequences: Various attack vectors, including remote attacks, physical tampering, social engineering, and denial-of-service attacks, pose significant threats to EV battery security and operational integrity [1],[3],[5],[9], [11],[14],[15],[17],[20]. The discussed attack vectors may not comprehensively cover all threats related to battery draining attacks, and the consequences mentioned might not fully illustrate the severity of damages or risks associated with such attacks.

Mitigation Strategies: Effective mitigation strategies are crucial for safeguarding EV batteries against draining attacks. These strategies include hardware-based solutions like tamper-proof hardware, software-based measures such as secure coding practices, network-based defenses like segmentation, and other tactics like regular software updates and user awareness training [1],[6],[7],[8],[9],[11],[12], [17]-[20]. The effectiveness of mitigation strategies may vary based on implementation challenges, resource constraints, and evolving threat landscapes, which are not extensively discussed in some references.

Implications: Battery draining attacks on EVs can compromise vehicle functionality and pose safety risks. Additionally, such attacks can lead to environmental concerns due to inefficient charging cycles and increased carbon footprint. Ensuring the cybersecurity of EV batteries is paramount for building trust in EV technology and ensuring a sustainable future [8],[10]-[14],[18],[19]. While the implications of battery draining attacks are discussed, some references may not fully explore the specific challenges in addressing these issues or regulatory considerations relevant to EV battery cybersecurity.

3. Attack Tree

Using a scenario-driven methodology, an attack tree was constructed to assess potential Attack Paths that threat actors may employ. Six Attack Goals were identified within the context of Draining Electric Vehicle Battery:

- Direct Battery Manipulation
- Indirect Battery Drain Through Component Control
- Denial of Service (DoS) Attacks Targeting Battery
- Social Engineering Attacks to Gain Physical Access
- Malware Injection to Manipulate Battery Management System
- Physical Destruction or Tampering with Battery Components

Additionally, 14 Attack Paths were identified as possible initial steps for attackers to execute these goals. The attack tree is categorized into four sections based on the level of risk: very high, high, medium, and low:

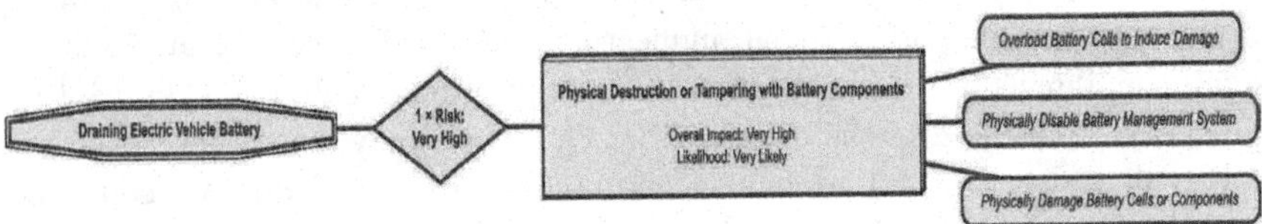

Fig. 41.1 SEQ figure * ARABIC 1: Attack tree - very high-risk attack paths

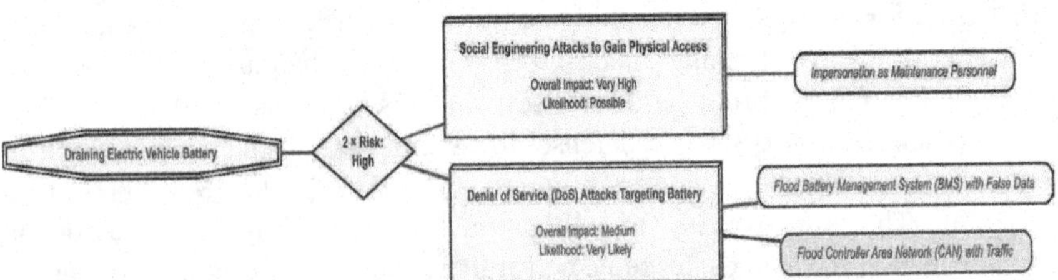

Fig. 41.2 Attack tree - high risk attack paths

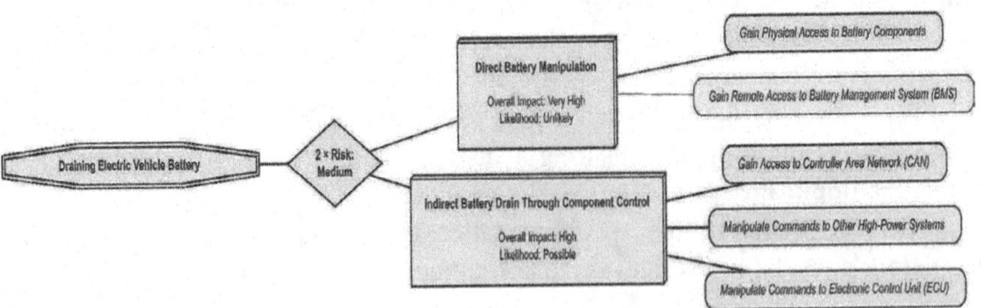

Fig. 41.3 Attack tree - medium risk attack paths

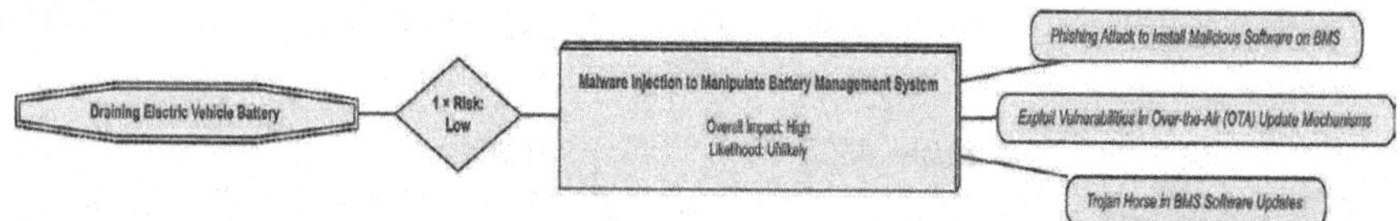

Fig. 41.4 Attack tree - low risk attack paths

3.1 Threat Actors

The following Threat Actors were considered as part of the analysis:

Script Kiddie: A novice attacker who uses pre-written scripts or tools to exploit vulnerabilities in electric vehicles and drain their batteries.

Social Engineer: An attacker who manipulates individuals or employees to gain access to EVs and initiate battery-draining attacks through persuasive or deceptive techniques.

Compromised Employee: A legitimate employee whose access credentials or privileges have been compromised, allowing them to perform battery-draining attacks on EVs.

Evil Admin: A malicious system administrator with privileged access who could exploit their position to execute unauthorized actions, including draining the battery of EVs, for nefarious purposes.

Attacker group outside: A coordinated group of malicious actors external to the organization or system, using advanced techniques and resources to infiltrate EV systems remotely and execute sophisticated battery drain attacks.

Table 41.1 provides a structured overview of the occurrence values assigned to each threat actor based on defined criteria such as expertise level, resources needed, historical data, current trends, and accessibility.

Table 41.1 Occurrence of threat

Threat Actor	Expertise Level	Resources Needed	Accessibility	Occurrence
Script Kiddie	Low	Low	High	High
Attacker Group Outside	High	High	High	Very High
Social Engineer	Moderate	Moderate	Moderate	Medium
Compromised Employee	Moderate	Moderate	Moderate	Medium
Evil Admin	High	High	Low	Low

Table 41.2 outlines the criteria used to determine the occurrence values for threat actors.

Table 41.2 Basis of occurrence

Criteria	Description
Expertise Level	The level of technical expertise required to execute attacks.
Resources Needed	The resources, including time, money, and equipment, required to carry out attacks.
Accessibility	Accessibility of attack tools and techniques, considering factors like ease of use and availability on the dark web.

4. Results and Discussion

The following discussion evaluates the six identified Attack Goals, assessing their potential impact and risk levels. By analyzing each goal in detail, a better understanding of the vulnerabilities and threats posed to electric vehicle (EV) security can be gained.

4.1 Physical Destruction or Tampering with Battery Components (Fig. 41.1)

Attackers aim to physically damage or tamper with the components of the battery in an electric vehicle. Following attack paths are identified:

Overload Battery Cells to induce damage: Attackers may deliberately exceed the recommended charging or discharging limits of the battery, leading to overheating, swelling, or catastrophic failure. By overloading the battery cells, attackers can cause irreversible damage, compromise safety, or trigger hazardous conditions such as fire or explosion.

Physically Disable Battery Management System: Attackers may target the BMS components directly, such as cutting power connections or removing critical components. By disabling the BMS, attackers can undermine the monitoring, control, and protection functions of the battery, potentially leading to reduced performance, or permanent damage to the battery cells.

Physically Damage Battery Cells or Components: Attackers may employ various methods, such as puncturing battery cells, cutting wiring connections, or applying excessive force to critical components. By causing physical damage to the battery, attackers can disrupt EV operations or make the vehicle unusable.

4.2 Social Engineering Attacks to Gain Physical Access (Fig. 41.2)

Social engineering attacks are aimed at manipulating individuals to gain unauthorized physical access to an electric vehicle. Following attack paths are identified:

Impersonation as Maintenance Personnel: Attackers may impersonate maintenance personnel by assuming the guise of technicians or service workers employed by legitimate maintenance companies. Once inside, attackers may exploit the opportunity to engage in nefarious activities, such as tampering with battery components, installing unauthorized hardware or software.

4.3 Denial of Service (DoS) Attacks Targeting Battery (Fig. 41.3)

This goal involves launching denial of service (DoS) attacks specifically targeting the battery of an electric vehicle. Following attack paths are identified:

Flood Battery Management System (BMS) with False Data: Attackers may deploy automated tools or scripts to flood the BMS with a high volume of fabricated requests, exceeding its processing capacity and causing it to become unresponsive or malfunction. This can disrupt normal battery operation, potentially leading to incorrect charging/discharging decisions, reduced battery efficiency, or even system crashes.

Flood Controller Area Network (CAN) with traffic: Attackers may deploy techniques such as packet flooding or broadcast storms to overwhelm the CAN bus with a large number of messages or commands. This flood of traffic can congest the network, disrupt communication between vehicle components, and potentially cause delays or failures in battery management functions.

4.4 Direct Battery Manipulation (Fig. 41.4)

This goal involves directly accessing and manipulating the battery of an electric vehicle (EV) to alter its state, such as draining its power or causing it to malfunction. Following attack paths are identified:

Gain Physical Access to Battery Components: Attackers may directly tamper with the battery management system (BMS), wiring, or individual battery cells to manipulate their behavior. By gaining physical access, attackers can perform actions such as altering charging parameters, initiating overcharging or over-discharging, or modifying the battery's health parameters.

Gain Remote Access to Battery Management System: Attackers exploit vulnerabilities in software or communication protocols to gain unauthorized access to the BMS. Once access is obtained, attackers can send false commands to the BMS, such as initiating charging or discharging processes, altering charging rates, or modifying battery health parameters.

4.5 Indirect Battery Drain Through Component Control (Fig. 41.5)

In this scenario, attackers aim to indirectly drain the battery of an EV by gaining control over other components or systems within the vehicle. Following attack paths are identified:

Gain Access to Controller Area Network (CAN): By gaining unauthorized access to the CAN network, an attacker can intercept, modify, or inject messages exchanged between various electronic control units (ECUs) within the vehicle. This access provides the attacker with the ability to manipulate critical vehicle functions and systems, potentially leading to battery drain. For example, the attacker could manipulate commands sent to the AC system, causing it to operate more frequently or at higher power levels than necessary.

Manipulate Commands to Other High-Power Systems: This attack targets other high-power systems within the electric vehicle that are not directly controlled by the ECU. By manipulating commands to these systems, such as the lighting, entertainment, or auxiliary power systems, the attacker can increase power consumption, putting additional strain on the vehicle's battery and potentially leading to accelerated battery drain.

Manipulate Commands to Electronic Control Unit (ECU): By tampering with the commands sent to the ECU, the attacker can induce actions that increase the power consumption of components such as the traction motor or engage continuous regenerative braking. These actions can lead to accelerated battery drain and compromise the overall performance of the vehicle.

4.6 Malware Injection to Manipulate Battery Management System (Fig. 41.6)

This goal involves injecting malware into the battery management system (BMS) of the EV, either remotely or through physical access. Following attack paths are identified:

Phishing Attack to install Malicious Software on BMS: In this scenario, attackers use phishing techniques to trick authorized personnel into installing malicious software onto the BMS. The phishing attack could involve deceptive emails or messages containing links or attachments that, when clicked or opened, install the malware onto the BMS. Once installed, the malware grants attackers unauthorized access to the BMS, enabling them to manipulate the charging and discharging processes of the battery.

Exploit Vulnerabilities in Over-The-Air (OTA) Update Mechanisms: Attackers exploit vulnerabilities in the OTA update mechanisms used to update the BMS software remotely. By exploiting these vulnerabilities, attackers can inject malicious code into the OTA update process. Once the malicious code is deployed to the BMS during an update, it grants attackers unauthorized access to manipulate the BMS functions.

Trojan Horse in BMS Software Updates: This involves injecting a malicious code disguised as a legitimate software update for the Battery Management System (BMS). Once the update is installed, the malicious code gains unauthorized access to the BMS, allowing attackers to manipulate its functions.

Table 41.3 provides a summary of risk level, impact, and the most likely threat actor associated with each attack goal.

Table 41.3 Summary of attack goals

Attack Goal	Risk	Impact	Threat Actor
Physical Destruction or Tampering with Battery Components	Very high risk due to potential irreversible damage	Very high impact as any tampering can lead to safety hazards or complete system failure.	Script Kiddie, due to lower expertise and resources for physical attacks.
Social Engineering Attacks to Gain Physical Access	High risk due to likelihood of successful manipulation	Very high impact as it can result in various malicious activities.	Social Engineer, for moderate expertise and effectiveness in gaining physical access to EVs.
Denial of Service (DoS) Attacks Targeting Battery	High risk as DoS attacks can disrupt normal operation.	Medium impact as it can disrupt battery operation but may not lead to complete failure.	Script Kiddie, due to simplicity in executing DoS attacks compared to other methods.
Direct Battery Manipulation	Medium risk requiring specific expertise and resources.	Very high impact as it can lead to safety hazards and malfunctions in EV operation.	Evil Admin, for higher expertise and access rights to execute complex attacks.
Indirect Battery Drain Through Component Control	Medium risk requiring moderate expertise and resources.	High impact as it can accelerate battery drain and affect EV performance.	Attacker Group Outside, for high expertise and resources for coordinated attacks.
Malware Injection to Manipulate Battery Management System	Low risk due to specific conditions required for execution.	High impact as compromising BMS can lead to unauthorized control over battery functions.	Compromised Employee, for insider access and potential involvement in installing malware.

5. Conclusion

In conclusion, this study has comprehensively examined the vulnerabilities inherent in Electric Vehicles (EVs), particularly focusing on the Controller Area Network (CAN) and Electronic Control Units (ECUs). Through the development and analysis of an attack tree model, the research has identified six critical Attack Goals, each with its associated attack paths and potential impact levels. The findings highlight the diverse range of threats facing EVs, including physical tampering, social engineering attacks, denial of service (DoS) attacks, direct battery manipulation, indirect battery drain through component control, and malware injection. Understanding these vulnerabilities enables stakeholders to implement targeted security measures to mitigate risks and enhance the resilience of EV systems against cyber threats. Actionable recommendations may include securing CAN bus communications, implementing intrusion detection systems, and conducting regular security audits. While this study provides valuable insights, it is important to acknowledge its limitations and the need for continued research in this area. Further investigation into emerging threats, evolving attack vectors, and advanced defense mechanisms will be essential for ensuring the safety, reliability, and sustainability of electric transportation technology.

References

1. S. V. Kumar, G. A. A. Mary, P. Suresh, and R. Uthirasamy, "Investigation On Cyber-Attacks Against In-Vehicle Network," IEEE Xplore, Feb. 01, 2021. https://ieeexplore.ieee.org/abstract/document/9383720
2. J. Edwards, A. Kashani, and G. Iyer, "Evaluation of Software Vulnerabilities in Vehicle Electronic Control Units," Sep. 2017, doi: https://doi.org/10.1109/secdev.2017.26.
3. K. Iehira, H. Inoue, and K. Ishida, "Spoofing attack using bus-off attacks against a specific ECU of the CAN bus," 2018 15th IEEE Annual Consumer Communications & Networking Conference (CCNC), Jan. 2018, doi: https://doi.org/10.1109/ccnc.2018.8319180.
4. S. Hounsinou, M. Stidd, U. Ezeobi, H. Olufowobi, M. Nasri, and G. Bloom, "Vulnerability of Controller Area Network to Schedule-Based Attacks," *IEEE Xplore*, Dec. 01, 2021.
5. Y. Kim, Saqib Hakak, and A. Ghorbani, "DDoS Attack Dataset (CICEV2023) against EV Authentication in Charging Infrastructure," Aug. 2023, doi: https://doi.org/10.1109/pst58708.2023.10320202.
6. Asma Alfardus and D. B. Rawat, "Evaluation of CAN Bus Security Vulnerabilities and Potential Solutions," Mar. 2023, doi: https://doi.org/10.1109/wids-psu57071.2023.00030.
7. Giordano Reyes Euceda, Aditya Akundi, and S. Luna, "Cybersecurity Challenges in Electric Vehicles: An initial literature review and research agenda," Apr. 2023, doi: https://doi.org/10.1109/syscon53073.2023.10131069.

8. J. Ye et al., "Cyber–Physical Security of Powertrain Systems in Modern Electric Vehicles: Vulnerabilities, Challenges, and Future Visions," IEEE Journal of Emerging and Selected Topics in Power Electronics, vol. 9, no. 4, pp. 4639–4657, Aug. 2021, doi: https://doi.org/10.1109/jestpe.2020.3045667.
9. Paul Carsten, Todd R. Andel, Mark Yampolskiy, and Jeffrey T. McDonald. 2015. In-Vehicle Networks: Attacks, Vulnerabilities, and Proposed Solutions. In Proceedings of the 10th Annual Cyber and Information Security Research Conference (CISR '15). Association for Computing Machinery, New York, NY, USA, Article 1, 1–8. https://doi.org/10.1145/2746266.2746267.
10. M. Mar, J. Noel, and J. Eric Dietz, "Cyber-Physical Review of a Battery Electric Vehicle Power Train: Vulnerabilities and Challenges," Sep. 2021, doi: https://doi.org/10.1109/isgtlatinamerica52371.2021.9543036.
11. Z. Tao, R. Fu, H. Shen, Y. Gu, and L. Qi, "Voltage Regulation Defense Strategy for Electric Vehicle Charging System under DoS Attacks," May 2023, doi: https://doi.org/10.1109/ccdc58219.2023.10327356.
12. K Dhananjay Rao, Manasa Taddi, Tharun Sriramula, Dilip Kumar Baliga, Akhil Simhadri, and Parth Sarathi Panigrahy, "Detection of Cyber Attacks on Wireless BMS of Electric Vehicles using Long Short-Term Memory Networks," Nov. 2023, doi: https://doi.org/10.1109/csitss60515.2023.10334240.
13. D. Reeh, F. Cruz Tapia, Y.-W. Chung, B. Khaki, C. Chu, and R. Gadh, "Vulnerability Analysis and Risk Assessment of EV Charging System under Cyber-Physical Threats," IEEE Xplore, Jun. 01, 2019. https://ieeexplore.ieee.org/document/8790593
14. S. I. Jeong and D.-H. Choi, "Electric Vehicle User Data-Induced Cyber Attack on Electric Vehicle Charging Station," IEEE Access, vol. 10, pp. 55856–55867, 2022, doi: https://doi.org/10.1109/access.2022.3177842.
15. D. Oladimeji, A. Rasheed, C. Varol, M. Baza, H. Alshahrani, and A. Baz, "CANAttack: Assessing Vulnerabilities within Controller Area Network," Sensors, vol. 23, no. 19, p. 8223, Jan. 2023, doi: https://doi.org/10.3390/s23198223.
16. M. Bozdal, M. Samie, S. Aslam, and I. Jennions, "Evaluation of CAN Bus Security Challenges," Sensors, vol. 20, no. 8, p. 2364, Apr. 2020, doi: https://doi.org/10.3390/s20082364.
17. Abdulmalik Humayed and B. Luo, "Using ID-Hopping to Defend Against Targeted DoS on CAN," Apr. 2017, doi: https://doi.org/10.1145/3055378.3055382.
18. S. Acharya, Y. Dvorkin, H. Pandzic, and R. Karri, "Cybersecurity of Smart Electric Vehicle Charging: A Power Grid Perspective," IEEE Access, vol. 8, pp. 214434–214453, 2020, doi: https://doi.org/10.1109/access.2020.3041074.
19. R. Metere, Zoya Pourmirza, S. E. Walker, and M. Neaimeh, "An Overview of Cyber Security and Privacy on the Electric Vehicle Charging Infrastructure," arXiv (Cornell University), Sep. 2022, doi: https://doi.org/10.48550/arxiv.2209.07842.
20. M. Bozdal, M. Samie, and I. Jennions, "A Survey on CAN Bus Protocol: Attacks, Challenges, and Potential Solutions," 2018 International Conference on Computing, Electronics & Communications Engineering (iCCECE), Aug. 2018, doi: https://doi.org/10.1109/iccecome.2018.8658720.

Note: All the figures and tables in this chapter were made by the authors.

Technologies for Energy, Agriculture, and Healthcare – Shailesh Nikam et al. (eds)
© 2024 Taylor & Francis Group, London, ISBN 978-1-032-98028-7

42

Detection of Birds from Natural Images Using an Improved Version of Swin Transformer

Nupur Pathrikar[1]
CSE Dept., MGM University, Chh. Sambhajinagar

Deepa Deshpande[2]
CSE Dept., MGM University, Chh. Sambhajinagar

Abstract: Real-world object identification is an essential topic in machine vision because various real-world applications necessitate object detection for distant objects. It is a challenging problem due to the appearance of small objects in images being noisy, blurry, and less informative. This work proposes an innovative technique for recognizing birds in aerial and natural photos, which is crucial for farmers and other real-world applications. To enhance feature representation capabilities, the technique employs an upgraded Swin Transformer as the core feature extraction network of Fast RCNN, which is further augmented with channel attention and an improved binary self-attention mechanism. The proposed approach is tested on an annotated and classified COCO Dataset.

Keywords: Computer vision, Object detection, RCNN, Deep learning

1. Introduction

Agriculture plays a critical part in feeding the world's population. Increased environmental consciousness has resulted in an improvement in the natural environment and an increase in the bird population in recent years. However,

[1]vhanmante@mgmu.ac.in, [2]ddeshpande@mgmu.ac.in

DOI: 10.1201/9781003596707-42

due to their eating and breeding habits, birds can seriously threaten crops in farm fields. Birds may wreak havoc on crops and food production. Bird-related incidents, according to research, account for more than 32% of total crop loss, trailing only lightning and external causes.

Farmers had to rely on manual bird detection, which was time-consuming, labor-intensive, and frequently unreliable. As technology progresses, CCTV has emerged as a viable solution for agricultural inspection, offering a safer and more efficient type of inspection. Aerial images of birds can be collected by CCTV outfitted with high-resolution cameras, allowing preventive steps to be taken. The precise detection of birds from other objects using CCTV photographs is a crucial challenge in intelligent framing and maintenance. However, the detecting technique is challenging due to factors such as intricate backdrops and fluctuations in image quality, which further complicate the task.

In recent years, researchers have become interested in the automatic recognition of birds in CCTV images. Several studies employing various deep-learning architectures and image-processing technologies have developed object detection systems for autonomously finding birds in agricultural field photographs. However, there is still room for improvement in terms of accuracy and robustness. Based on an improved Swin Transformer, we present a new bird detection technique for crops and farm field pictures in this work. The technique improves the feature maps and creates a new object detection model by utilizing a channel attention mechanism and a self-attention mechanism. In addition, we construct, annotate, and classify a picture dataset of farm field birds. Experimental findings reveal that the suggested method is more accurate and robust in spotting birds than existing strategies.

2. Literature Review

Earlier computer research includes template matching and invariant position methods, in addition to recent machine learning and deep learning algorithms such as Haar cascade [1], support vector machine (SVM) [2], You Only Look Once (YOLO) [3], and Faster RCNN [4], are discussed. A brief introduction to visual transformers, like a deep learning architecture used for various computer visualization tasks, is also included in this section. Recently, bird detection has been a research emphasis, and several approaches have been advanced for this purpose [5]. This section will go through both classic and current bird detection techniques. The goal of bird recognition is to recognize and pinpoint faulty targets in patrol photos. Image-based bird recognition research is divided into classical approaches and computer vision detection algorithms [6]. Visual examination and manual detection are two traditional procedures. Visual inspection entails human professionals inspecting the farm field, which is time-consuming and labor-intensive. Manual detection entails examining photographs of crops from various angles to assess the presence of birds over the crops or in the entire agricultural field.

Machine learning-based approaches for detecting birds have grown in popularity in recent years. These methods extract visual features and train a classifier for bird detection using feature descriptors such as SIFT [7]. They proposed a deep Convolutional Neural Network & Scale-invariant Features Transformation (SIFT) to overcome the drawbacks of complex object backgrounds, dense situations, and similarities among different objects. Bird image classification has emerged through study in Machine Learning & Deep Learning, covering background variations, lighting conditions in bird images, and different bird postures. YOLOv5 helps an object detection algorithm identify various birds. To address the issues of multiple identification and classification of birds from other objects, the suggested method combines an improved version of the YOLOv5 algorithm with position limitations [8]. A lightweight attention mechanism convolution module is built by combining the Ghost Net Module with ECA, increasing detection capabilities for small objects and minimizing scale differences. The identification procedure focuses especially on CCTV patrol panoramas, and the birds' hazard levels are classified based on their position limits. A deep learning-based system for automatic bird detection in CCTV pictures termed ROI mining Fast-RCNN [4]. Xavier et al. [10] proposed the approach using K-means clustering to improve anchor generation, focal loss to balance training samples, and an ROI mining module to solve classification imbalance and problematic samples in terms of mean average precision (mAP) and F1 score. Cai, Z., Vasconcelos [9] proposed a multistage object detection framework, the cascade R-CNN, for the design of high-quality object detectors to avoid the problems of overfitting during training and quality mismatch during interference.

3. Methodology

This work proposes a new object detection technique for identifying birds in agricultural farm fields and over photographs of crops, which is critical for crop protection. The suggested technique replaces Fast RCNN's backbone feature extraction network with an enhanced Swin Transformer, resulting in the construction of a new object identification model. Figure 42.1 depicts the model's broad framework. By adding a channel attention mechanism into the traditional Swin Transformer blocks, the algorithm hopes to increase detection accuracy.

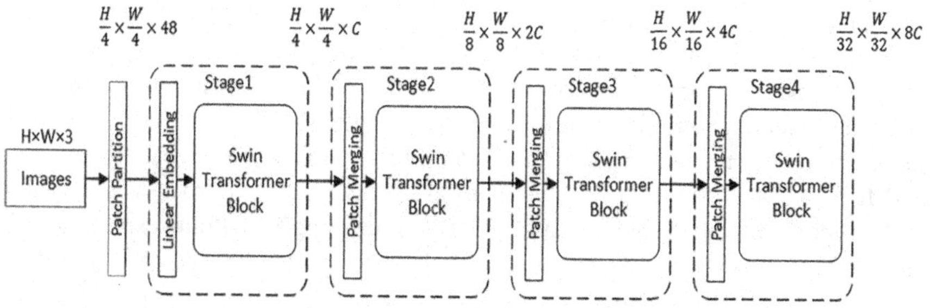

Fig. 42.1 Swin transformer architecture

Source: Adapted from IEEE proceedings Swin transformer: hierarchical vision transformer using shifted windows

3.1 Swin Transform

Figure 42.1 depicts the smaller version (SwinT), which delivers an overview of the Swin Transformer architecture [11]. A patch-splitting module, such as ViT, first divides an input image provided in RGB form into non-overlapping patches. Each patch is handled as a "token," with its feature set as a concatenation of the raw RGB values of the pixels. At the implementation time, we utilize a patch size of 4 × 4, hence the feature dimension of each patch is 4 × 4 × 3 = 48. On this raw-valued feature, a linear embedding layer is used to project it to an arbitrary dimension (denoted as C). On these patch tokens, many Transformer blocks with modified self-attention computation (Swin Transformer blocks) are applied. The Transformer blocks keep the token count (H/4 × W/4) and are referred to as "Stage 1" coupled with the linear embedding.

To generate a hierarchical representation, the network employs patch-merging layers to reduce the number of tokens as it progresses deeper. In the initial patch merging layer, the features of each group of neighbouring 2 × 2 patches

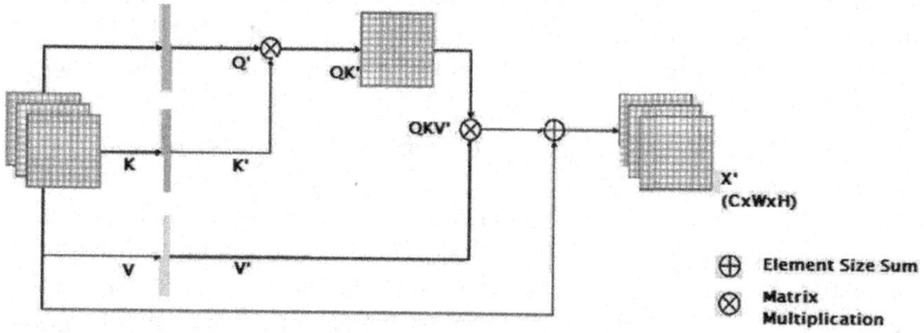

Fig. 42.2 Self-attention block

Source: Author

are concatenated, and a linear layer is applied to the resulting 4C-dimensional concatenated features. This down-samples the tokens by a factor of 2 × 2 = 4 (resulting in a 2 × decrease in resolution), and the output dimension is adjusted to 2C. Subsequently, Swin Transformer blocks are applied for feature transformation, maintaining the resolution at H/8 × W/8. This initial block of patch merging and feature transformation is referred to as "Stage 2". This process is iterated twice more, termed "Stage 3" and "Stage 4", resulting in output resolutions of H/16 × W/16 and H/32 × W/32, respectively. These stages collectively produce a hierarchical representation, with feature map resolutions akin to typical convolution networks, allowing the suggested architecture to conveniently replace backbone networks in existing methods for various vision tasks.

Additionally, modifications are made to the Swin Transformer's self-attention mechanism to enhance feature representation capability. Object proposals are generated using the Region Proposal Network once the feature map is derived from the Swin Transformer. Following this, classification and bounding box regression are performed using ROI pooling and two fully connected layers. Each input image is partitioned into a series of non-overlapping patches before being input into a sequence of transformer blocks. The input image is resized to a fixed size (e.g., 224 × 224 pixels) before undergoing processing by a Convolution Network to extract low-level information.

The image is subsequently divided into a predetermined number of non-overlapping patches, which are flattened and passed through a series of transformer blocks. Each transformer block utilizes a multi-head self-attention mechanism to learn patch relationships, followed by a feed forward network to generate feature representations.

To improve feature representation ability even further, the proposed algorithm incorporates a channel attention mechanism into the output of each transformer block, which adaptively recalibrates the relevance of distinct feature channels. In addition, the proposed approach includes a feature map attention module to modify the Swin Transformer's self-attention mechanism. Specifically, the output of the self-attention mechanism is multiplied with the original feature map to improve the input feature's feature representation capacity.

Each transformer block is made up of numerous attention and feedforward layers that learn to attend to and refine relevant picture patches. A self-attention mechanism is utilized within each attention layer to generate a weighted sum of the input patch embeddings, with the weights learned based on their pairwise similarities. The attention output is then subjected to nonlinear modifications by the feedforward layers, which further refine the patch embeddings. The patch embeddings are initially processed by a stage, which is a group of transformer blocks with shared weights, at each step of the network.

Each stage's output is then down-sampled by a set factor with a convolutional layer with a stride of 2. This decreases the features' spatial resolution while increasing their receptive field. Finally, the final stage output is sent to a series of detecting heads, which estimate the existence, category, and position of objects within the image. The detection heads are made up of a combination of fully connected layers and convolutional layers trained to provide high-quality object proposals and reliably categorize them. As the ultimate result of the suggested method, a set of bounding boxes with matching confidence scores is generated.

The cross-entropy loss is utilized for classification and the smooth L1 loss is used for bounding box regression to improve detection performance during training.

3.2 Self-Attention Block

we provide a more detailed explanation of the modified binary self-attention mechanism for feature extraction in object detection. The primary goal of binary self-attention is to filter out coarse features and extract more refined features, thereby improving the accuracy and robustness of object detection algorithms.

A linear layer is utilized to transform the coarse features into three matrices: Q (query), K (key), and V (value). These matrices capture essential information for subsequent operations.

In this case, the three vectors can be formulated as follows:

$$Q = L(CF) \tag{1}$$

$$K = L(CF) \tag{2}$$

$$V = L(CF) \tag{3}$$

Where L represents the linear layer,
CF represents the Coarse feature,
Q matrix is multiplied by the K matrix, which establishes the relationships between different features.

The resulting matrix is then normalized to ensure the attention weights fall within a suitable range. A SoftMax operation is applied to obtain the attention distribution, highlighting the importance of various features within the map. Next, the attention activation map is multiplied element-wise with the matrix V, enabling the mechanism to focus on relevant features while suppressing irrelevant ones. This multiplication operation generates an updated representation of the features, emphasizing those that contribute more to the final detection results. The whole process can be formulated as follows

$$\text{Refined} - \text{feature} = CF * \text{Bin}(S(QK)K) \tag{4}$$

where CF denotes the Coarse-feature, Bin denotes Binary, and S denotes SoftMax.

Overall, the binary self-attention mechanism offers a refined approach to feature extraction in object detection. It introduces a series of operations, including linear transformation, attention computation, and binary combination, to filter and refine the feature map, resulting in improved detection performance Transformer Block The transformer block serves as a vital element of the feature extraction process, designed to capture intricate features. As illustrated in Fig. 42.3, the transformer block consists of three steps. The first and third steps are similar to those of standard Swin Transformer blocks, with the self-attention mechanism enhanced by the binary attention map. addition, the SW-MBSA is performed based on W-MBSA, with W-MBSA and WS-MBSA alternately present, and W-MBSA always preceding WS-MBSA.

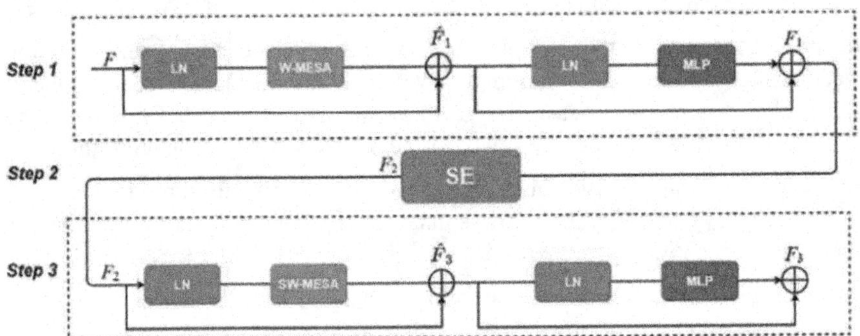

Fig. 42.3 Transformer block diagram

Source: Author

To capture the interaction information between channels, the proposed algorithm incorporates a squeeze-and-excitation (SE) module in the second step.

The SE module is designed to learn the channel-wise dependencies in the feature map and adaptively recalibrate the channel features by considering their relevance to different object categories. These relationships can be represented with the following equations:

$$F1 = W - MBSA(LN(F)) + F \tag{5}$$

$$F2 = SE(MLP(LN(F1)) + F1) \tag{6}$$

$$F3 = SW - MBSA(LN(F2)) + F2 \tag{7}$$

$$F4 = MLP(LN(F3)) + F3 \tag{8}$$

3.3 Contributed Block

It learns channel attention by aligning channel-wise feature maps and was first introduced by Hu et al. in 2017 for object classification. In this technique, the shape of the feature map, denoted as (H, W, C), is first globally pooled to a shape of (1, 1, C). This pooling operation can be mathematically expressed as follows:

$$Z_C = \frac{1}{H \times W} \sum_{i=1}^{H} \sum_{j=1}^{W} fc(i, j) \tag{9}$$

where *fc* represents the 2-dimensional matrix of the input feature map f for channel c, and H and W represent its height and width, respectively. i and j denote the row and column indices of fc, respectively.

It is aimed at enhancing the interaction information between channels. It models the relationship between different feature channels and assigns importance to each channel based on its correlation with other channels. To model the correlation between channels and assign their corresponding weights, the SE approach utilizes a fully connected neural network with a RELU activation function, followed by a sigmoid function. To generate the final output, the normalized weights obtained from the excitation operation are applied to the original feature maps through a scale operation, which can be expressed as follows:

$$F_{ex}(Z, W) = Sigmoid(W_2 \times ReLU(W_1 \times Z)) \tag{10}$$

3.4 Improved Version of Swin Transformer

The proposed technique uses the swin transformer method for bird object identification. It includes the following steps:

Feature Enhancement – It involved the enhancement of feature representation capabilities by modifying to self-attention mechanism within the swin transformer blocks.

Channel Attention Mechanism – By integrating channel attention mechanism into the swin transformer to recalibrate the relevance of different feature channels in focusing on more informative features over less relevant once.

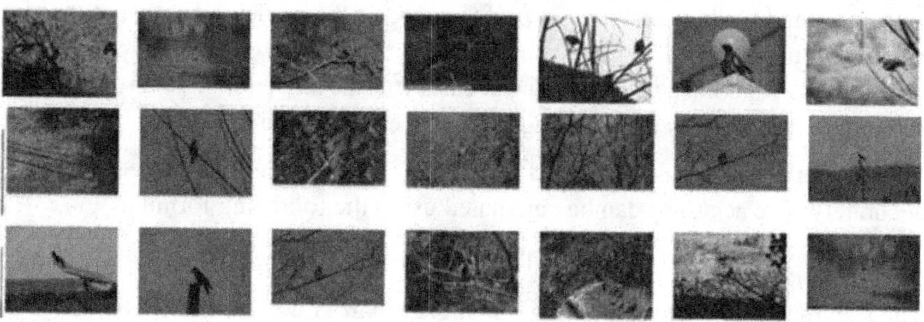

Fig. 42.4 Input bird images

Source: Author

Modified Binary Self Attention Mechanism – By incorporating a binary attention map with the channel attention mechanism swin transformer to filter out course features and extract more refined features enhancing the accuracy and robustness of object detection algorithms .

Object Detection Model Construction – The improved swin transformer has the feature extraction network for developing a new object detection model. To achieve better performance in detecting birds in the environment, the model uses a combination of channel attention and a modified self-attention mechanism.

5. Experiment

5.1 Dataset

The natural bird images taken of various sizes and taken from various sources like digital cameras and CCTV are considered to perform the experimental analysis The collections include high-resolution bird photos as well as annotation files. We considered 127 photos for training purposes and 100 photographs for testing purposes. Some sample images are shown below

5.2 Hyper-parameters

Our proposed algorithms use the following sets of hyper-parameters. The following table shows the hyperparameters used and their associated values considered Image Size - 224 × 224, Batch Size – 256, Learning rate - 0.001, Decay - 0.001, Momentum – 0.02, Epoch-20, embed-dim-768, dim_head -64.

To effectively compare the application effects of different target detection technologies, it is essential to use both visual inspection and objective evaluation metrics to evaluate the validity of the detection results. To assess the performance of the object detection model, various standard indicators are used in the field, including mAP, AP, and recall rate. To classify the detection results into four categories and further evaluate the performance of the model, a confusion matrix is often constructed. The four categories are true positive (TP), false positive, true negative, and false negative. The accuracy of the detection model reflects the proportion of TP predictions among all positive predictions.

The model's ability is measured based on recognizing the target category accurately. The accuracy can be calculated using the following formula:

$$P = TP/(TP + FP) \tag{11}$$

The accuracy of the proposed model is higher than YOLOv4, Faster RCNN, and ViT Transformer algorithms.

Fig. 42.5 Output images

Source: Author

6. Conclusion

The research proposes a novel method for recognizing birds in photos to protect farm crops. Unlike existing algorithms, the suggested method extracts features using an upgraded Swin Transformer and combines a channel attention mechanism and a modified self-attention mechanism to improve feature representation. The research also includes a dataset of captured bird image dataset with bird annotations and classifications. When compared to previous methods, experimental results show that the suggested algorithm detects birds with more accuracy and robustness. Future studies might look into how the suggested method performs in real-world scenarios and compare it to other cutting-edge object detection systems.

References

1. Sander Soo: Object detection using the Haar-cascade classifier. Institute of Computer Science, University of Tartu 2(3), 1–12 (2014)
2. Bazi Y., Melgani F.: Convolutional SVM Networks for Object Detection in UAV Imagery. Transactions on Geoscience and Remote Sensing (February 2018). https://doi.org10.1109/TGRS.2018.2790926.
3. Redmon, J.; Divvala, S.; Girshick, R.; Farhadi, A. You only look once: Unified, Real-Time Object Detection. In Proceedings of the IEEE Conference on Computer Vision and Pattern Recognition, Las Vegas, NV, USA, 27–30 June 2016; pp. 779–788. [Google Scholar]
4. Ren, S.; He, K.; Girshick, R.; Sun, J. Faster RCNN: Towards Real-Time Object Detection with Region Proposal Networks. Adv. Neural Inf. Process. Syst. **2015**
5. Krizhevsky, A.; Sutskever, I.; Hinton, G.E. ImageNet Classification with Deep Convolutional Neural Networks. Commun. ACM **2017**, 60, 84–90. [Google Scholar] [CrossRef]

6. Chauhan, R.; Ghanshala, K.K.; Joshi, R. Convolutional neural network (CNN) for image detection and recognition. In Proceedings of the 2018 First International Conference on Secure Cyber Computing and Communication (ICSCCC), Jalandhar, India, 15–17 December 2018; pp. 278–282.
7. Rashid, M., et al.: Object Detection and Classification: A Joint Selection and Fusion Strategy of Deep Convolutional Neural Network and SIFT Point Features. Multimedia Tools and Applications 78(12), 15751–15777 (2019). https://doi.org/10.1007/s11042-018-7031-0
8. Hoang-Tu Vo, Nhon Nguyen Thien, Kheo Chau Mui : Bird Detection and Species Classification: using YOLOv5 and Deep Transfer Learning Model, IJACSA-International Journal of Advanced Computer Science and Applications Vol-14 No.7, 2023.
9. Cai, Z., Vasconcelos, N.: Cascade r-cnn: delving into high-quality object detection. In: Proceedings of the IEEE Conference on Computer Vision and Pattern Recognition, pp. 6154–6162 (2018).
10. Xavier, A.I.; Villavicencio, C.; Macrohon, J.J.; Jeng, J.H.; Hsieh, J.G. Object Detection via Gradient-Based Mask R-CNN Using Machine Learning Algorithms. Machines 2022, 10, 340. [Google Scholar]
11. Liu, Ze, et al.: Swin transformer: hierarchical vision transformer using shifted windows. In: Proceedings of the IEEE International Conference on Computer Vision, pp. 10012–10022 (2021).

Technologies for Energy, Agriculture, and Healthcare – Shailesh Nikam et al. (eds)
© 2024 Taylor & Francis Group, London, ISBN 978-1-032-98028-7

43

Classification and Stress Analysis Using Artificial Neural Network

Parth Matalia[1]
Student,
K J Somaiya College of Engineering,
Somaiya Vidyavihar University

Amod Zack[2]
Student,
K J Somaiya College of Engineering,
Somaiya Vidyavihar University

Dinesh Chawde[3]
Assistant Professor,
Department of Mechanical Engineering,
K J Somaiya College of Engineering,
Somaiya Vidyavihar University

Abstract: Pressure vessels, vital in the Energy Sector, are required to go through rigorous structural safety assessments. Traditional stress analysis methods are effective but integrating artificial intelligence (AI), particularly deep learning with neural networks like Convolutional Neural Networks (CNN), enhances precision. This work explores Artificial Intelligence & Machine Learning (AI & ML) application in revolutionizing pressure vessel's stress analysis, encompassing dataset examination, AI model development, robust evaluation metrics, and real-world user interactions. Results indicate the successful model functioning, paving the way for further exploration in this transformative concept.

Keywords: Pressure vessels, Stress analysis, Neural networks

[1]p.matalia@somaiya.edu, [2]amod.zack@somaiya.edu, [3]dineshc@somaiya.edu

DOI: 10.1201/9781003596707-43

1. Introduction

The energy sector heavily relies on pressure vessels for various purposes, such as containing harmful gasses in oil refineries and metal-works. Nuclear power plants utilize specialized pressure vessels called Reactor Pressure Vessels (RPVs), which are crucial in withstanding high temperatures, pressures, and neutron irradiation. To avoid the failure of these pressure vessels, it is crucial to perform stress analysis accurately. Finite Element Analysis (FEA), a software-based method widely adopted in engineering, utilizes calculations, models, and simulations to predict an object's behavior under varying conditions. The finite element method (FEM) is employed, breaking down the structure into elements connected at nodes. This numerical technique generates algebraic equations for engineers to analyze stress, thermal behavior, seismic calculations, etc. [10]. Leading companies like Autodesk, Ansys, and Dassault Systems provide FEA software for comprehensive engineering solutions. Artificial Neural Network (ANN), a Deep Learning (DL) method inspired by the human nervous system, employs interconnected nodes to transfer data between layers as depicted in Fig. 43.1 [9]. Widely used in deep learning for image classification, speech/character/image recognition, natural language processing, and computer vision, ANN functions as a sophisticated data processing system. The fusion of FEA and ANN represents a novel research avenue, exploring the potential of combining established engineering analysis with advanced artificial intelligence. This paper delves into this uncharted territory, presenting the initial steps towards a groundbreaking concept. The discussion explores insights and findings, laying the groundwork for a potential paradigm shift in engineering practices.

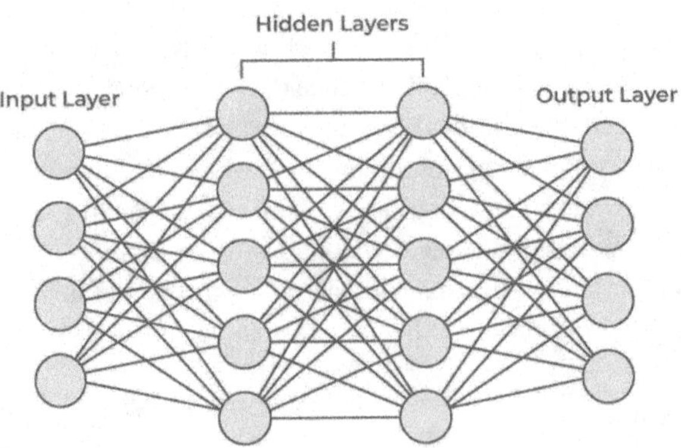

Fig. 43.1 A generic artificial neural network [9]

2. Literature Review

This section emphasizes literature surveys done in the three major areas - Machine Learning & Pressure Vessel Safety Assessment, Neural Networks in Structural Damage Recognition, Custom Imageset Creation Techniques.

2.1 Machine Learning and Pressure Vessel Safety Assessment

There have been many recent contributions to the field of pressure vessel safety and structural integrity assessment through ML techniques. DL is the foremost refinement of ML and various Neural Networks are being trained as of this moment. The focus is on various studies that leverage ML techniques for safety assessment, risk evaluation, stress prediction, burst pressure estimation, damage recognition in pressure vessels, pipelines and how to employ Neural Nets for the same. *Shah et al.* [1] explore the application of machine learning for design-by-analysis of pressure equipment, showcasing how ML can supplement or replace manual procedures in finite element analysis (FEA). Another perspective on risk assessment for pipelines with active defects is provided by *Anghel* [2], emphasizing the use of artificial intelligence methods to evaluate risks in in-service pressure piping containing defects. From the Analytical Perspective, *Chawde and Bhandakkar* [11], addressed semi analytical method to solve mixed boundary value problems, contributing to the understanding of structural behavior of functionally graded pressure vessel. Khadke and Chawde [12], designed a thin cylindrical pressure vessel subjected to multiple loading and performed finite element analysis using ANSYS. Therefore, Artificial Intelligence is pivotal in manufacturing and detecting errors in the Pressure Vessel industry.

2.2 Neural Networks in Structural Damage Recognition

A safety assessment approach to pressure vessels based on machine learning by *Zhang et al.* [3] investigates the elastoplastic 3D J-integral of a crack tip, providing insights into safety assessment methodologies using ANN. *Zolfaghari and Izadi* [4] propose a burst pressure prediction model for cylindrical vessels using artificial neural networks (ANNs), incorporating input parameters such as internal and outer diameter, thickness, ultimate and yield strength. *Johnson et al.* [5] demonstrate the use of ANN for predicting burst pressure of both thick and thin-walled pressure vessels, addressing various strain hardening carbon steels. *Fan and Hu* [6] present a paper on pressure vessel nozzle local stress prediction software, incorporating ABAQUS and ML techniques to obtain accurate stress values. *Modarres et al.* [7] introduce a CNN-based approach for automated damage recognition and identification of structural damage types in pressure vessels. The above section illustrates that Deep Learning can be used to further improve the efficiency of ML Techniques used in Stress Analysis and Detection of Pressure Vessels.

2.3 Custom Imageset Creation Techniques

Utilizing learnable convolutional filters, *Bajpai and He* [8] detail a methodology for creating custom datasets and utilizing CNN models for image processing operations in Google's T-Rex game, showcasing the versatility of ML in various applications. Thus, the creation of custom datasets and application of CNN models in image processing is easier with Machine Learning techniques.

3. Problem Definition, Methodology and Model Used

3.1 Problem Definition

This study proposes a method that combines ANN and stress analysis to provide an accurate and efficient model for classification of pressure vessels and prediction of stresses. Eventually, this model aims to provide an approach to enhance safety management and operational efficiencies in the industrial sector. For the imageset, illustrations are created using images of pressure vessels that can be found online. The images are stored in Google Drive and are used for the process that is followed in this model.

3.2 Methodology

Our innovative approach involves the integration of ANN and Stress Analysis for classifying pressure vessels and predicting stress. A diverse dataset, encompassing various pressure vessel types, material parameters, and results, underwent meticulous preprocessing for ANN training. Multiple ANN structures were explored, optimizing through regression methods and hyperparameter tuning to achieve the desired performance. Trained models underwent rigorous testing for accuracy and precision.

3.3 Model Used

The architecture is divided into three flow structures, with the primary focus on classifying pressure vessels. A homogeneous image dataset was crucial for this, leading to the first structure.

- **Imageset Creation**
- **Classification of Pressure Vessels**
- **Stress Prediction**

Imageset Creation

The following Fig. 43.2 depicted the steps in imageset creation as:

- **Data Preprocessing**

  Google Drive integration in Colab facilitated dataset access. Libraries like os, PIL.Image, and Sklearn were imported for preprocessing.

- **Image Processing**

  Images underwent processing, including format conversion, resizing, and detail filtering. Dataset organization into training and testing sets followed.

- **Dataset Creation**

  Directories were defined, existing ones removed, and new ones created. The dataset was split into training and testing sets using train_test_split. This laid the foundation for neural network training.

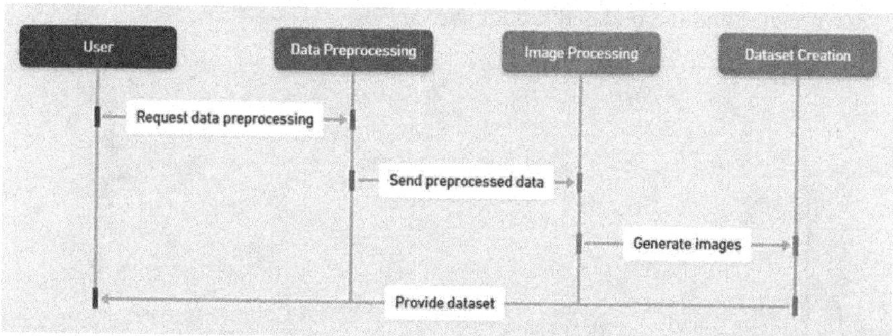

Fig. 43.2 The flow and interconnectivity of the process to obtain the training and testing Imageset

Source: Author

Classification of Pressure Vessels

The steps involved in classifying the pressure vessel consist of following steps and are depicted in Fig. 43.3.

- **Setup and Data Preprocessing**

  Google Drive was mounted, TensorFlow and relevant modules imported. Image data generators with augmentation techniques were configured.

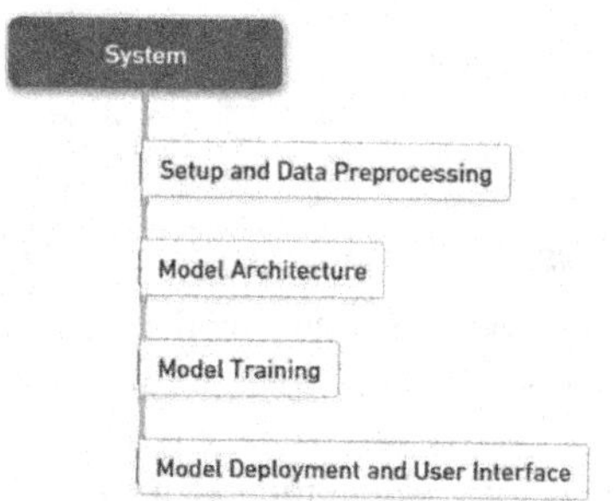

Fig. 43.3 The linear flow of the intermediate process to classify the obtained Imageset

Source: Author

- **Model Architecture**
  A Convolutional Neural Network (CNN) with three convolutional layers, max-pooling, and dense layers was designed for binary classification as shown in Fig. 43.4. The model was compiled with Adam optimizer and categorical cross-entropy loss.
- **Model Training**
  The model was fitted to the augmented training dataset using Model Checkpoint and Early Stopping callbacks. Training involved 20 epochs.
- **Model Deployment and User Interface**
  Saved models were loaded, providing a user interface for predictions. Users could correct predictions through feedback buttons
- **User Feedback Loop**
  User feedback influenced model re-evaluation. Corrected labels were added to images, and the updated model was saved.

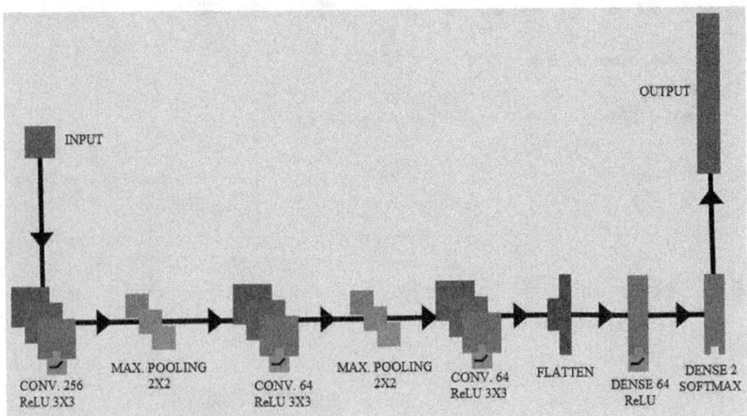

Fig. 43.4 Architecture of the convolution model

Source: Author

Stress Prediction

This stress prediction model, as shown in Fig. 43.5, uses a feedforward neural network for stress prediction, incorporating input, hidden, and output layers. K-fold cross-validation ensured robust evaluation. The model, compiled with Adam optimizer, utilized mean squared error as the loss function and mean absolute error as a metric. A test dataset containing the parameters temperature (T), thickness (t), radius (R), inner radius (r), ellipticity (a), and pressure (P) of the vessel as well as the recorded stress values at such conditions are used in this model.

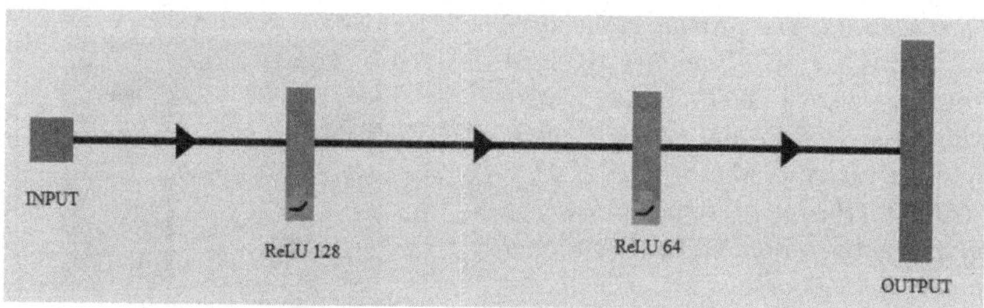

Fig. 43.5 Stress prediction model

Source: Author

4. Results and Discussion

Vertical pressure vessels are large structures widely used in the oil and gas industry to store oil and other liquids. They have several different trays designed to mix rising air and falling water. The container looks like a horizontal drum with two mold heads, one at the top and one at the bottom. It is supported by a coat welded to the lower head. The following two test cases were considered to test the classification model designed by us.

TEST CASE 1: The classification model was provided with a pre-processed image of a used 225-gallons (GAL) Ventech Carbon Steel Tank; Rated 232 psig @ 752 F; Dished top and bottom; Leg mounted Vertical Pressure Vessel with an estimated weight of 2,000 lbs; Built in 2015 [13]. *The model predicted incorrectly as shown in Fig. 43.6, thus the feedback loop was established, and the model learned the correct category of the image as shown in Fig. 43.7.*

Fig. 43.6 Vertical pressure vessel

Source: Author

Fig. 43.7 Predicted vertical pressure vessel

Source: Author

TEST CASE 2: The classification model was provided with another Vertical Pressure, LOB's WTP heated vessel with conical outlet and heated support brackets vessel as shown in Fig. 43.8. LOB's WTP-heating system serves as an alternative to the half-tube coil and double-walled jacket [14]. *The model predicted correctly; thus, the stress calculation model was called upon. This model, which was provided with the parameters and dataset predicted the stress with an error of 0.265 MPa as shown in Fig. 43.9.*

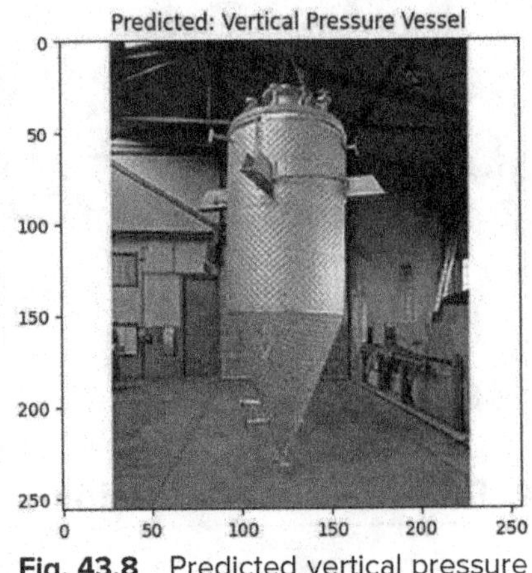

Fig. 43.8 Predicted vertical pressure vessel

Source: Author

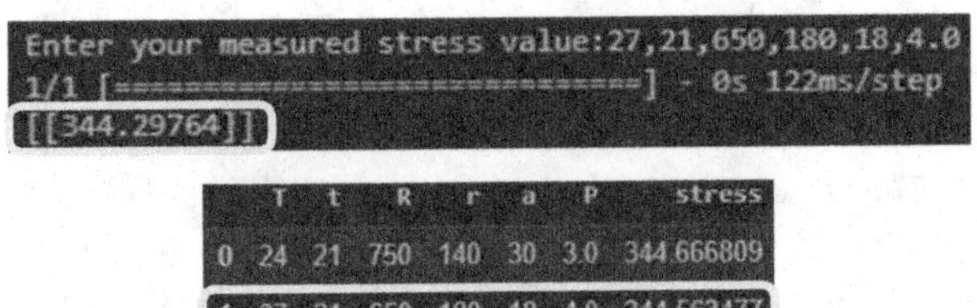

Fig. 43.9 User–Input of parameter and obtained result

Source: Author

5. Conclusion

The completed study demonstrates a reliable and efficient Convolutional Neural Network method for classifying images and analyzing stresses of pressure vessels. Constructed using TensorFlow's Keras API, the CNN architecture exhibits a deliberate design intended to extract complex patterns from input images to produce precise predictions. Three convolutional layers make up the model. To minimize categorical cross-entropy loss during the training phase, the Adam optimizer is employed. A feedforward neural network for stress prediction is incorporated by the model. Also, a test data set containing various vessel parameters such as temperature, pressure, thickness, etc. is used for stress prediction. Furthermore, the study has a comprehensive user interface for real-time prediction and feedback. With the help of this interactive feature, end users can offer constructive insights, which helps the model go through an iterative improvement cycle. With this study as a foundation, designing, monitoring, as well as maintenance of Pressure Vessels can be made highly efficient. The real-time monitoring of various parameters of a working pressure vessel is highly useful, especially in the Energy Sector where the margin of error is extremely low.

References

1. Shah, Chandnani, Raykar, Mavinkurve, "Application of Machine Learning for Design-by-Analysis of Pressure Equipment" (2019), DOI:10.1109/ICNTE44896.2019.8945858
2. Anghel, "Risk assessment for pipelines with active defects based on artificial intelligence methods" (2009), DOI: 10.1016/j.ijpvp.2009.01.009
3. Zhang, Hu, Shi, Liang, Xu, Cao, "A safety assessment approach to pressure vessels based on machine learning" (2022), DOI:10.3389/fmats.2022.1051890
4. Zolfaghari, Izadi, "Burst Pressure Prediction of Cylindrical Vessels Using Artificial Neural Network"(2020), DOI:10.1115/1.4045729
5. Johnson, Zhu, Sindelar, Wiersma, "Artificial Neural Networks for Predicting Burst Strength of Thick and Thin-Walled Pressure Vessels"(2023), DOI:10.1115/PVP2023-106471
6. Fan, Hu, "Pressure vessel nozzle local stress prediction software based on ABAQUS-machine learning" (2023), DOI: 10.1016/j.softx.2023.101550
7. Modarres, Astorga, Lopez-Droguett, Meruane, "Convolutional neural networks for automated damage recognition and damage type identification"(2018), DOI:10.1002/stc.2230
8. Bajpai, He, "Custom Dataset Creation with Tensorflow Framework and Image Processing for Google T-Rex" (2020), DOI: 10.1109/CICN49253.2020.9242565
9. R.M. Baphana, S. Gupta and S. Alegavi, "Fundamentals of Neural Networks." Deep Learning (2023)
10. Seshu, P., "Finite-element analysis", ISBN:9788120323155, 8120323157, Publisher: PHI Learning, (2004)
11. Chawde, Dinesh P., and Tanmay K. Bhandakkar, "Mixed boundary value problems in power-law functionally graded circular annulus." International Journal of Pressure Vessels and Piping 192 (2021), DOI: 10.1016/j.ijpvp.2021.104402
12. Kunal. Khadke and Dinesh Chawde.: Design & Finite Element Analysis of Pressure Vessels. International Journal for Research in Applied Science & Engineering Technology, vol. 10, no. 7, pp. 4741–4748 (2022). DOI: 10.22214/ijraset.2022.46076
13. Test Case 1: "225 GAL Ventech Carbon Steel Tank", S/N 8000-032-V591, NB # 64
14. Test Case 2: "LOB's WTP heated vessel with conical outlet and heated support brackets", SG100332-Kopie-2.jpg

Technologies for Energy, Agriculture, and Healthcare – Shailesh Nikam et al. (eds)
© 2024 Taylor & Francis Group, London, ISBN 978-1-032-98028-7

44

Design and Development of Drip Monitoring of IV System

R. Nandita Srivatsan[1],
Niyati A. Rolia[2], Gurveer Singh Bharj[3],
Sandeep Sainkar[4], Sudha Gupta[5], Kavita Kelkar[6]
Somaiya Vidyavihar University,
Mumbai, India

Abstract: Intravenous (IV) therapy is crucial in healthcare, efficiently delivering substances into the bloodstream for swift absorption. Despite widespread use, challenges like drip rate errors persist. Technological advancements, including IoT integration, aim to enhance IV therapy safety. IV fluid drip systems face potential inaccuracies, posing risks to patient health, with manual regulation attributing to life-threatening errors. This paper addresses such challenges by developing an IoT-based drip monitoring system, emphasising the critical need for self-sufficient, non-contact systems, especially in the context of the Covid-19 pandemic. The paper illustrates the objective of alerting healthcare professionals, reducing the risk of manual errors and ensuring timely interventions for enhanced patient care through the utilization of sensors and IoT technology.

Keywords: IV drip monitoring, Intravenous drip, Ultrasonic sensor HC-sr04, IR sensor, Arduino UNO R3, Design and development, Healthcare

1. Introduction

In modern healthcare, Intravenous (IV) therapy plays a crucial role in direct administration of fluids, medications, and nutrients into the bloodstream for

[1]r.srivatsan@somaiya.edu, [2]niyati.rolia@somaiya.edu, [3]gurveer.b@somaiya.edu, [4]sandeepsainkar@somaiya.edu, [5]sudhagupta@somaiya.edu, [6]kavitakelkar@somaiya.edu

DOI: 10.1201/9781003596707-44

rapid absorption. Despite its widespread use, IV fluid drip systems encounter challenges, particularly in ensuring the accuracy of drip rate. Variations in flow rates, stemming from factors such as improper setup, equipment malfunctions, human error, absence of healthcare faculties, etc., pose significant risks to patient health. These risks include overhydration, dehydration, reverse blood flow and other complications related to medication overdosage and underdosage [10]. Hence, observing and regulating the drip rate of intravenous fluid infusion becomes vital, as even a minute change in drip rate might cause severe side effects [11]. The demanding nature of healthcare, witness during the Covid-19 pandemic, emphasizes the need for self-sufficient, non-contact systems.

[1] proposes a method using pressure sensors and RF technology, for developing a system that effectively monitors and regulates the flow of the IV solution, while also minimizing the need for healthcare professionals to be in immediate vicinity of the patient. An upgraded system was built by the authors of literature [2], where an automatic IV fluid feed system that automatically controlled administration of intravenous fluids and or drugs to patients to eliminate the need for continuous supervision was designed.

Hospitals use a wide variety of fluids and drug combinations for intravenous infusion. *S. Gayathri et al. (2017)* in [3] developed a system that indicated the level or amount of glucose liquid present in the trip bottles in hospitals using a load cell. A common trend observed in multiple papers that aimed for an alerting feature, was the utilisation of GSM modules [3]-[5]. Such automation technologies that facilitate SMS transmissions help reduce manual labour to an extent in the healthcare industry. One system deployed the GSM module along with a solenoid valve for regulating the drip via signals sent through an actuator [4], whereas another system deployed the similar SMS facility along with a UART module [5] for lower power consumption.

The authors of literature [8], *S. Joseph et al. (2019)* suggested an alternative use of IoT along with a mobile application for remote access to constantly monitor the patient's condition. An enhanced system was implemented [6] as a low-cost automated fluid control device with the use of smartphones to aid the regular life of an affected person. [7] suggested that intravenous fluid level indicator has various advantages such as low cost, smaller size, high accuracy, and easy handling.

A simple prototype using LDR was developed [9], for building a control mechanism that can alert the nurses or doctors if the fluid levels in the IV infusion bag drops down a certain level to prevent air embolism and avoid reverse flow of blood. The authors of this paper also consulted with medical practitioners to learn and understand medical expectations in terms of creating an assistive technology for IV drip, thereby aiming to develop a feasible working system that fulfils the requirements from a medical point of view. Out of multiple highlighted points, designing a device that monitors the drips and displays the data remotely with an additional warning alarm was focused upon as the objective.

On identifying the challenges faced in the various existing systems, the possible solution to overcome limitations like use of GSM module, lack of physical alert, speed of data transmission using RF transmitter or other data transmission and processing methods, etc., is by designing and developing a system equipped with IR sensors and an ultrasonic sensor to ensure reliable drop detection and fluid level monitoring. Beyond technical considerations, the device integrates IoT for system control, data transfer, and visualisation. Significantly, it activates alarms promptly in response to any deviations in the fluid flow rate, aiming to enhance the accuracy, reliability, and safety of IV fluid drip systems.

This paper initially outlines the methodology and calculations used in the development of the system. Further, it describes the hardware specifications along with a block diagram to visualise the connections designed in the model. A detailed explanation follows, regarding the integration of components to provide alerting and IoT features. The results and accuracies of the model were observed and analysed across multiple test cases, each of which is discussed along with conclusions.

2. Proposed System

In hospitals with a substantial number of patients, healthcare professionals and nurses find it laborious to constantly monitor and keep a check on each patient's IV drip status. The proposed system was designed to track the fluid levels present in the IV fluid bottle, which is in turn essential in measuring the drip rate of the fluid being infused into the patient. The data for fluid level and drip rate were graphically represented over a webpage, with each patient displayed on a separate channel for easy navigation. Furthermore, in case of IV fluids decreasing to critical levels, an alert was triggered to warn the caretaker or nurses for immediate intervention in the treatment procedure.

An ultrasonic sensor was placed on the extreme end of the bottle to determine the fluid levels in the container. The initial fluid capacity was measured using the height and radius of the bottle, as parameters of the volumetric equation of a cylinder,

$$V = \pi r^2 h \tag{1}$$

In this case, the amount of fluid is equivalent to the "fluid level" that nurses can view. The distance between the top surface of the bottle and the fluid inside,

Table 44.1 Comparative analysis of existing methods [1]–[11]

Method	Technology Used	Key Features	Advantages	Limitations
[1]	Pressure sensors, RF technology	Monitors and regulates IV flow effectively	Minimizes the need for immediate healthcare supervision by ensuring precise fluid regulation	Limited to pressure-based monitoring, may not capture all aspects of IV administration
[2]	Automatic IV fluid feed system	Automatically controls IV administration	Eliminates the need for continuous supervision, enhancing efficiency and reducing human error	May require a sophisticated setup and ongoing maintenance for optimal performance
[3]	Load cell, GSM module	Indicates glucose levels in IV bottles accurately	Offers a cost-effective solution with easy handling and SMS alerts for timely interventions	Limited to monitoring glucose levels, not suitable for other IV fluids or drugs
[4]	GSM module, solenoid valve	Regulates IV drip via SMS signals	Reduces manual labour and enables remote control, enhancing flexibility in IV fluid management	Dependence on GSM network availability might pose connectivity issues in certain areas
[5]	UART module, GSM module	Provides SMS alerts with lower power consumption	Facilitates efficient data transmission and remote monitoring while conserving energy resources	Dependence on GSM network availability might limit functionality in areas with poor coverage
[6]	IoT, mobile application	Enables remote monitoring of patient's condition in real-time	Offers a cost-effective solution for continuous patient monitoring, improving patient care	Reliability dependent on network connectivity, may require periodic updates for stability
[7]	Intravenous fluid level indicator	Offers a low-cost, compact solution with high accuracy	Enhances safety by providing easy handling and accurate fluid level indication	Limited to monitoring fluid levels, may not provide comprehensive data on IV administration
[8]	IoT, mobile application	Facilitates constant monitoring via a mobile app	Allows for remote access and real-time updates, improving healthcare delivery	Reliability dependent on network connectivity, potential issues with data security and privacy
[9]	LDR, consultation with medical practitioners	Alerts for low fluid levels to prevent complications	Enhances safety by preventing air embolism and reverse flow, following medical best practices	Limited to monitoring fluid levels, may require calibration for different IV setups

which posed as a reference point for determining the fluid's height, was measured using the HC-SRF05 Ultrasonic sensor. The fluid height (fh), was determined by subtracting this distance from the height of the bottle. The volumetric formula becomes,

$$fl = \pi r^2(fh) \tag{2}$$

From formula (3), the total volume of fluid was then utilised to measure the "Flow Rate" or the "Infusion Rate". This was calculated to determine the amount of fluid to be administered for a duration of time and is measured in mL per hour. It was calculated by dividing the total amount of liquid by total time in hours:

$$\text{Infusion Rate(mL/hrs)} = \frac{\text{Total Volume (mL)}}{\text{Total Time(hrs)}} \tag{3}$$

This value was then used to evaluate the required parameter "Drip Rate (gtt/min)". The Drip Rate helps to calculate the number of drips trickling per minute to ensure that the fluid infusion is balanced, continuous and precise. This was calculated by multiplying the infusion rate from formula (4) with a term known as "Drop Factor", which refers to the number of drops per mL. This factor is usually mentioned on saline bags, depending on the tubing of the drip chamber (Micro or Macro tubing).

$$\text{Drip Rate(gtt/min)} = \text{Infusion Rate} \times \text{Drop Factor} \tag{4}$$

An IR sensor with transmitter and receiver was placed at the drip chamber for detecting the drip during the process. The sensor data was then sent to the microcontroller and the drip was observed real-time with negligible delay on the web interface via the interfaced ESP8266. Additionally, the system was programmed to have a threshold for fluid level, which was set a few units over the critical level, so that an alert via a buzzer or LED was triggered as a warning, to prevent fatal situations, at the same time allowing the medical practitioners to take necessary actions on time and at ease.

To indicate the level of fluid, an HC-SRF05 Ultrasonic sensor was placed at the top of the saline bottle as seen in the block diagram. Based on the SONAR principle, an ultrasonic sensor sends soundwaves, which are reflected and detected to measure the output (Fluid level). The IR sensor was inserted inside the drip chamber and tuned such that the drip was detected at least distance. An LED or Buzzer is linked to an Arduino such that a physical alarm is triggered in case of critical situations. The NodeMCU connects the prototype to the cloud webpage for the seamless integration of Internet of Things applications into the system. All these hardware components were linked via the microcontroller board, Arduino UNO R3.

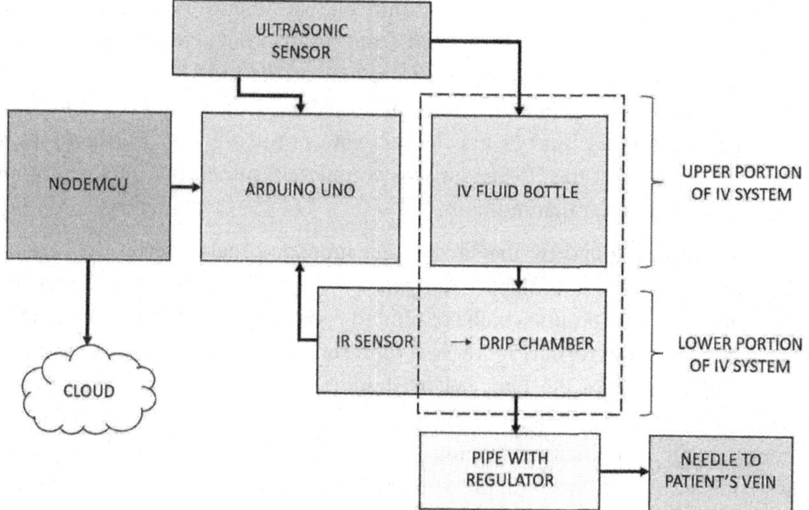

Fig. 44.1 Block diagram

Source: Author

3. Result and Analysis

Figure 44.2 shows the output obtained for three different cases. Figure 44.2(a) depicted the first case when no drip was detected, and the fluid level was over the threshold. It was observed that the LED was turned off, indicating that the distance measured by the Ultrasonic Sensor was above the set threshold. The IR Sensor, unable to detect any drips at that point in time, did not light up. Figure 44.2(b) depicted the second case when no drip was detected, and the fluid level was under the threshold. It was observed that the LED was turned on, indicating that the fluid

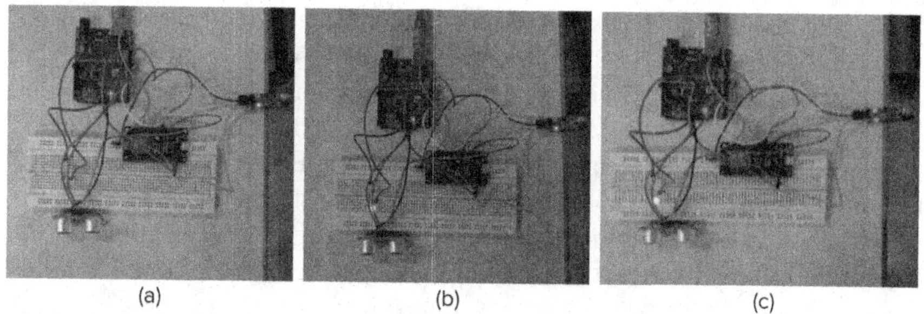

Fig. 44.2 Observed output

Source: Obtained real-time by authors

in the IV bag was below the threshold limit. The IR Sensor was partially turned on to display only VCC, which resulted from no drip detection. Figure 44.2(c) depicted the third case when a drip was detected, and the fluid level was under the threshold. It was observed that the LED was turned on, indicating that the adjusted threshold levels for the fluid in the IV bag were not being met. Furthermore, it was also observed that the IR Sensor was completely turned on, thus indicating the presence of a drip at that moment.

The data values obtained by the sensors get uploaded real-time on the channel of the cloud platform, and display the pattern of events. Figure 44.3 shows the representation of data points which are plotted every 15 seconds on the respective chart in the cloud platform. Figure 3(a) represents the distance over time chart, Fig. 44.3(b) represents the changes in fluid level over time and Fig. 44.3(c) represents the drip rate changes over time. After analysis of the plotted charts, it is observed that as the distance increases, the fluid level decreases. The drip rate shows constant reading per minute which is a positive sign. Many irregularities in the drip rate can prove to have fatal consequences and hence, the nurse can be alerted without having to physically monitor the patient at regular intervals.

4. Conclusion

IV fluid drip systems often face potential irregularities, related to manual regulation, attributing to life-threatening errors. The prototype proposed in this paper addresses these concerns by displaying the monitored data values on cloud platforms. Such a system would allow free mobility for healthcare professionals in dire situations of mass patient numbers like the Covid-19 pandemic. The alarm and warning mechanism allows the nurses in charge to intervene in the drip process as and when needed to avoid fatal conditions. Furthermore, Automated regulators, which are made using potentiometers, can also be used to make the system more feasible without being in contact. To enhance cost-effectiveness, optimizing PCBs with minimal materials is advisable. Moreover, utilizing advanced sensors with adjustable sensitivities can significantly improve drip detection efficiency. Additionally, integrating a dedicated cloud application tailored specifically for healthcare facilities can bolster the security of medical data further. To ensure the proposed system is both scalable and feasible, it's preferable to develop features that don't necessitate alterations at the manufacturing stage. Rather, an extended system that can seamlessly integrate or detach from the existing IV design as required would offer greater compatibility and flexibility.

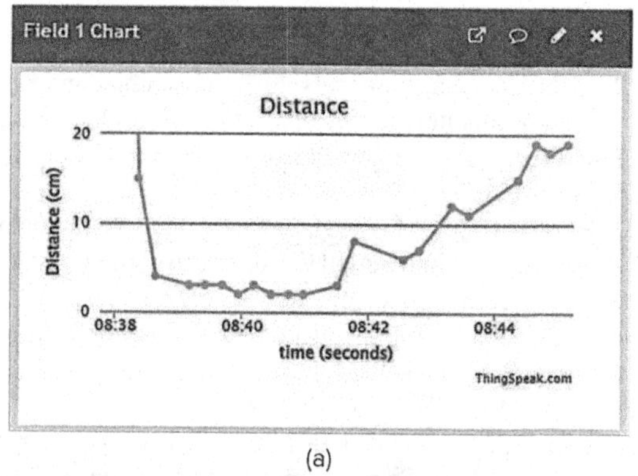

(a)

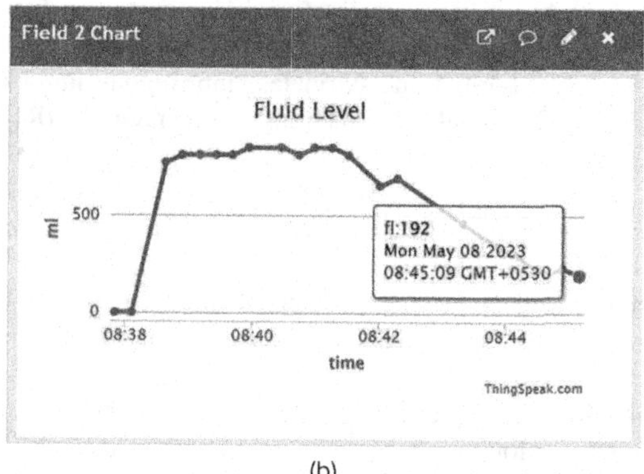

(b)

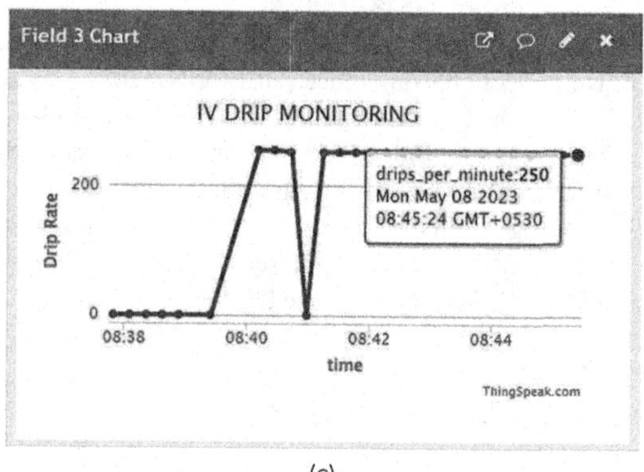

(c)

Fig. 44.3 Representation of data in cloud

Source: Obtained real-time by authors

References

1. K. Ramisha Rani, N Shabana, P Tanmayee, S Loganathan and Velmathi Guruviah, "Smart drip infusion monitoring system for instant alert-through nRF24L01", International Conference on Nextgen Electronic Technologies: Silicon to Software (ICNETS2), pp. 452–455, 2017
2. M. Farhan, M. Rashik Mojid, M. Kifayath Chowdhury, M. Farhan and M. Farhan, "Design and Fabrication of Automatic IV Fluid Feed System", International Journal of Scientific and Engineering Research, pp.989–997, 2017
3. S. Gayathri and C.S. Sundar Ganesh, "Automatic indication system of glucose level in glucose trip bottle", International Journal of Multidisciplinary Research and Modern Education (IJMRME), pp.148–151, 2017
4. H. Rashid, S. Shekha, S. Muhammad Taslim Reza, I. Uddin Ahmed, Q. Newaz and M. Rasheduzzaman, "A Low-Cost Automated Fluid Control Device using Smart Phone for Medical Application", International Conference on Electrical, Computer and Communication Engineering (ECCE), pp. 809–814, 2017
5. A. Jora, D. Laveena, Earlina and Nirmala, "Intravenous Fluid Level Indicator", International Research Journal of Engineering and Technology (IRJET), pp. 525–529, 2018
6. M. Anand, Pradeep MM, S. Manoj, M. Arockia Raj and P. Thamaraikani, "Intravenous Drip Monitoring System", Indo-Iranian Journal of Scientific Research (IIJSR), pp. 106–113, 2018
7. P. Sardana, M. Kalra and A. Sardana, "Design, Fabrication, and Testing of an Internet Connected Intravenous Drip Monitoring Device", Journal of Sensor and Actuator Networks, pp. 1–20, 2018
8. Ms. S. Joseph, Ms. N. Francis, Ms. A. John, Ms. B. Farha and Mrs. A. Baby, "Intravenous Drip Monitoring System for Smart Hospital using IOT", 2nd International Conference on Intelligent Computing, Instrumentation and Control Technologies (ICICICT), pp. 835–839, 2019
9. R. Maniktalia, S. Tanwar, R. Billa and Deepa K, "IoT Based Drip Infusion Monitoring System", IEEE Delhi Section Conference (DELCON), 2022
10. Snijder, Roland A., Konings, Maurits K., Lucas, Peter, Egberts, Toine C. and Timmerman, Annemoon D. "Flow variability and its physical causes in infusion technology: a systematic review of in vitro measurement and modeling studies" Biomedical Engineering / Biomedizinische Technik, vol. 60, no. 4, 2015, pp. 277–300.
11. International Journal of Engineering Research & Technology (IJERT) ISSN: 2278–0181 IJERTV9IS090485 Vol. 9 Issue 09, September-2020

Technologies for Energy, Agriculture, and Healthcare – Shailesh Nikam et al. (eds)
© 2024 Taylor & Francis Group, London, *ISBN 978-1-032-98028-7*

45

Computational Analysis of Ballistic Material against High Energy Impact

M J Pawar[1], Mayuresh Pathak[2]
Mechanical Engineering Department,
K J Somaiya College of Engineering,
Mumbai, India

Vikash Gautam[3]
Mechanical Engineering Department,
Swami Keshvanand Institute of Technology,
Jaipur, India

Vikash Kukshal[4]
Mechanical Engineering Department,
National Institute of Technology,
Srinagar, Uttarakhand, India

Abstract: High strength steel plates are majorly used in protection against high energy impact for variety of threat levels. Vehicular Light Armour demand has grown exponentially over past few years. Material properties of Aluminium like least density, better strength as well as least prone to corrosion make them a viable option for defence and aerospace operations.

Physical testing of various materials against high energy impact isn't all-time feasible in initial stages of development. The experimental setup for high energy impact also needs to pass the international testing standards and it can be costly. To deal with this, finite element analysis helps in understanding failure of models well in advance of the physical testing of materials against high energy impact. Computational analysis for performance of armour material is presented in this article. Various constitutive models have been proposed but only few are practically

[1]manoj.jp@somaiya.edu, [2]mayuresh.phatak@somaiya.edu, [3]gautam.mnitj@gmail.com, [4]vikaskukshal@nituk.ac.in

DOI: 10.1201/9781003596707-45

usable. Johnson-Cook strength and failure model is employed for projectile and target material. The aluminium alloy AA2014 target plate of 500 mm x 500 mm with thickness of 15 mm. The projectile is spherical shape with diameter 10 mm. Two different materials were considered for projectile, viz; deformable soft iron and rigid/non-deformable projectile. The target plate is clamped at rectangular surface along thickness. Impact of projectile with high energy considered for this work is normal to surface of target plate at centre. FEA code using ANSYS was used for computational analysis. Ballistic performance of target plate was compared for six different velocities between 750 m/s to 1300 m/s. Residual velocity of projectile, total deformation and back-face deformation was predicted as output.

In-case of soft iron projectile, the target plate was not perforated for all the velocities under consideration. However, for rigid projectile the target plate was non-perforated up to impact velocity of 869 m/s. At impact velocity 995 m/sec and above the target plate was completely perforated and residual velocity is observed increasing with initial velocity of projectile. For deformable projectile the deformation at the backside of aluminium plate was observed as 1.52 mm at projectile velocity of 761 m/s and 8.72 mm at projectile velocity of 1264 m/s. However, for rigid projectile the values of back-face deformation were 2.11 mm at projectile velocity of 803 m/sec and 9.12 mm at projectile velocity of 1247 m/s.

Keywords: Computational analysis, Ballistic velocity, Residual velocity

1. Introduction

Civil and defence vehicles need strong and durable armour for protection against high energy ballistics. New and new material for armour is being developed using metal-metal alloys or metal-non-metal alloy, such as, aluminium alloys, high strength steel and using foam for covering of these plates for vests. Experimental setup required for testing of these plates is many times not feasible to material unavailability and costs incurred for this setup. If one manages to setup an experiment for testing, it should be able to comply the requirements of testing requirements and standards globally. Finite Element Analysis is one of the best solutions for analysing high energy impact. Aydin et. al (2020) evaluated performance of Al6061-SiC sandwich plate against high energy impact, with increasing layer quantity and volume proportion. Test specimen were fabricated by high temperature press method of powder stacking.

Zhang et. al (2018) illustrated the shield properties of aluminium based sandwich with honeycomb structure. Four different thicknesses were considered for face sheets. Surface sheets and core were bonded together by means of polymeric adhesive and were kept pressed for 24 hours at atmospheric conditions. Jena et. al (2010) demonstrated ballistic impact of metallic armour materials against 7.62 mm flexible projectiles. Projectile on hitting formed a round disc because of resistance by armour plate. Sharma et. al (2018) have discussed the ballistic impact of soft iron and commercially available ball bearings of diameter 10mm with aluminium alloy(AA2014 T652) target plate using experimental setup as well as the software results of the same. Bore propellant gun with 30mm bore diameter is used to fire the projectile. Shasthri et. al (2020) explored effect of ballistic impact on titanium alloy and carbon steel bi-layer armour plate. Projectile was incident on two angles, i.e., 45° and 90°. Maximum deformation was observed for edge impact rather than centre impact. Pawar et. al (2016) evaluated high energy impact behaviour of Al_2O_3 and AlN using AUTODYN hydrocode. Target plates were glued to circular metal plates with a polyurethane adhesive. The backing plate deformation is directly proportional to chemical composition of the ceramic material. Pawar et. al (2021) presented simulation work relevant to high energy impact behaviour of Aluminium-Alloy sheets of varying thickness. Also, thickness of sheets for completely stop projectile is estimated numerically. Bagwe et. al (2021) performed studies relevant to FEA of Armour Steel and Aluminium-alloy impacted by projectile 7.62 mm with high energy.

In this study computational analysis for performance of armour material is presented. Johnson-Cook strength and failure model is used for both the counterpart of high energy impact. Two different materials were considered for projectile, viz; deformable soft iron and rigid/non-deformable projectile. The target plate is clamped at rectangular surface along thickness. Impact of projectile with high energy considered for this work is normal to surface of target plate at centre. FEA code using ANSYS was used for computational analysis. Residual velocity of projectile, total deformation and back-face deformation of target plate was predicted as output.

2. Materials and Methods

Model parameters from Sharma et. al. [4] was taken as input variables for geometry of target plate and experimental setup model. Target plate and projectiles were modelled and loaded into Ansys software. Initial conditions were set such as fixed support and velocity to projectile. Output of this analysis was set as total deformation, von-Mises stresses, and residual velocity of projectile and back-face deformation of target plate. Analysed values were then compared with the experimental values for each run. Ballistic impact for soft iron projectiles and ball-bearing projectiles was done for a set of 5 different velocities. At the end of each run, residual velocity and back-face deformation was noted. Figure 45.1 is illustrated geometry of high energy impact configuration. Also the boundary conditions for target plate and projectile is indicated.

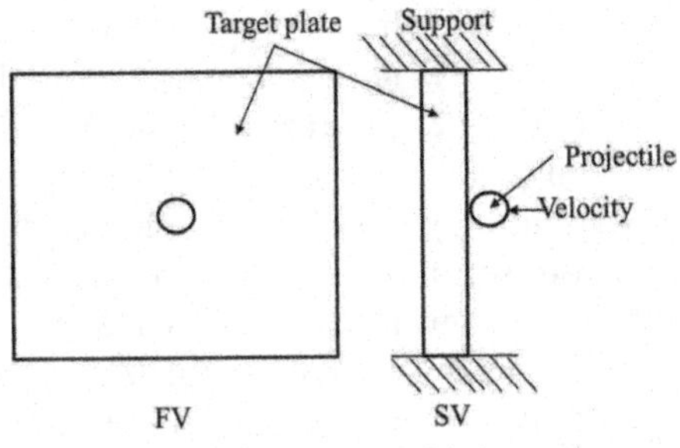

Fig. 45.1 Geometry with boundary condition

Target plate material is selected as aluminium alloy AA2014 T652 and thickness of plate is 15mm. For projectiles, one deformable material and other rigid material is selected. Geometry was generated in Design Modeller of Ansys in the Explicit Dynamics analysis systems. In engineering data, Johnson-Cook model parameters were considered for target plate, soft iron projectile and rigid projectile. Size of target plate is 500mm × 500mm × 15mm (Fig. 45.2 (a)). Projectile is incident on the target plate at angle of 90° with respect to plane of target plate. To reduce computational time and reduce nodes and elements to be analysed, quarter geometry is selected (Fig. 45.2 (c)). Figure 45.2 (d) shows the exploded view of quarter geometry. Meshing was done using 2 techniques; first meshing was for overall geometry and second was done using edge sizing technique – sphere of influence (Fig. 45.3). In sphere of influence, the point of contact of projectile and target plate is selected as centre of sphere and radius of 16 mm was given to that sphere so as to get a fine mesh for contact region.

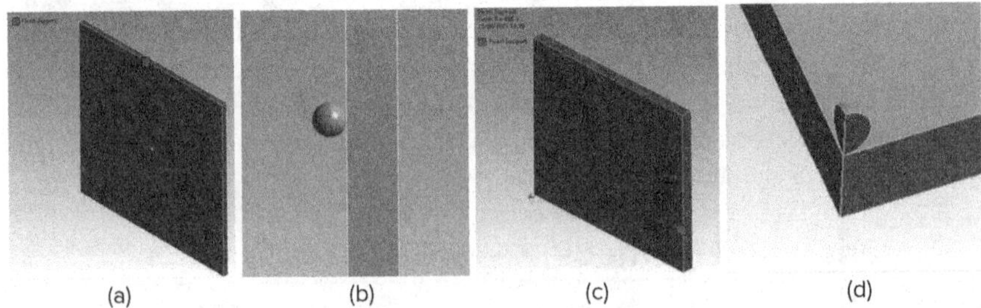

Fig. 45.2 (a) Complete geometry, (b) Side View of Complete Geometry, (c) Quarter geometry, (d) Exploded view of quarter geometry

There are two main boundary conditions in this setup. First is the fixed support (Fig. 45.3(a)) provided to target plate on all 4 sides and second is the velocity of the projectile (Fig. 45.3(b)). To help in analysing deformation of target plate due to initial and residual velocity of projectile post impact, probes were added. Velocity probe is fixed on the other end of the projectile which is not in contact

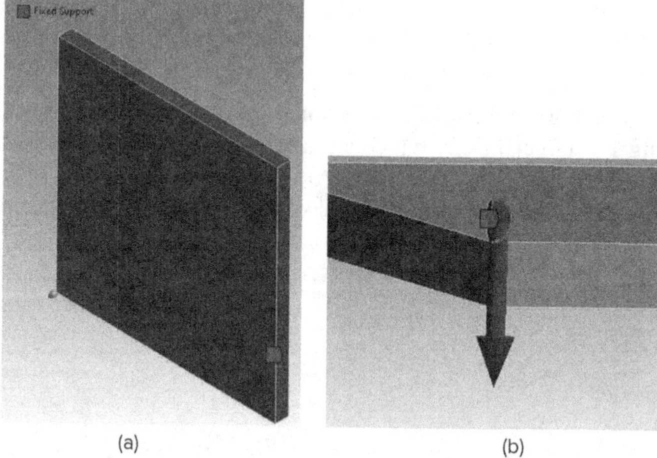

(a) (b)

Fig. 45.3 Boundary conditions (a) Fixed support and (b) Velocity of projectile

with the target plate. The deformation probe is fixed to back-face of target plate which bulges on impact with projectile. After solving the mathematical model in Workbench, von-Mises stresses, total deformation, back-face deformation and residual velocity for each velocity of both projectiles were recorded.

3. Results and Discussion

In present study the influence of increase in initial velocities were analysed computationally for two different projectile materials. Figure 45.4 showed behaviour of target plate when impacted by soft iron projectile with 1264 m/s

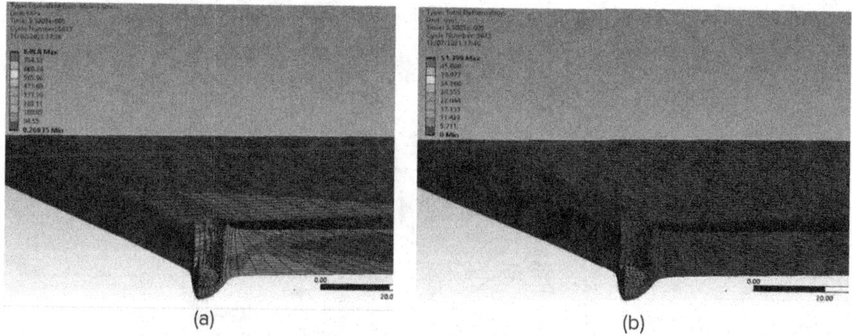

(a) (b)

Fig. 45.4 Target plate subjected to impact of soft iron core with velocity of 1264 m/sec a) stress developed and b) back-face deformation velocity. Figure 45.4a illustrated the stress developed in target plate and projectile. It was observed the maximum value of stress developed at the point of perforation on the target plate. Back-face deformation of target plate is illustrated in Fig. 45.4b.

Figure 45.5 and Fig. 45.6 depicted the reduction of velocity with respect to time after hitting the target plate. Immediate after impact the velocity is observed decreasing rapidly for 15 microseconds. After that rate of reduction of velocity is slow down for both the projectile. It was also observed that the time taken to arrest the projectile is more in case ball bearing balls as compared to soft iron balls.

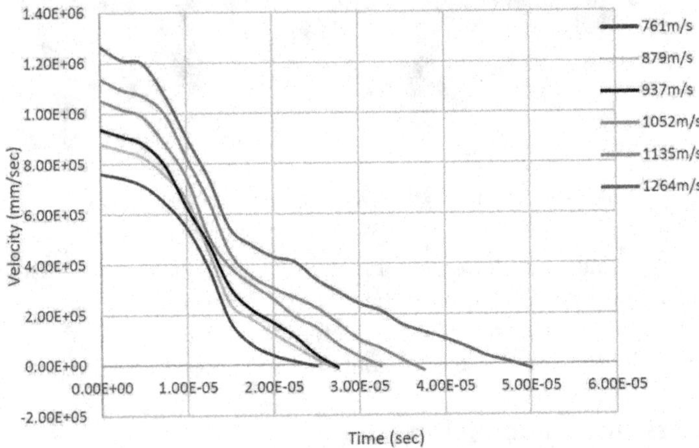

Fig. 45.5 Velocity-time graph of high energy impact for soft iron projectiles

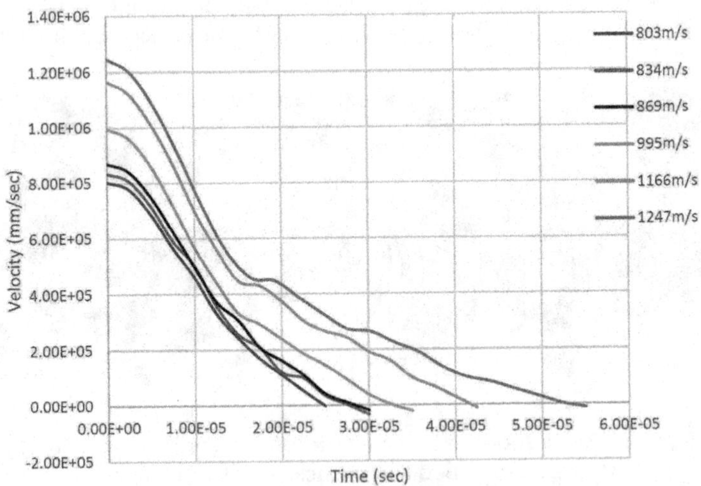

Fig. 45.6 Velocity-time graph of high energy impact for ball bearing projectiles

Table 45.1 illustrated residual velocity and back-face deformation of target plate for high energy impact with soft iron core. It was observed from computation results of this study that the target plate is remaining un-perforated for the velocity of impact from 761 m/s to 1264 m/s. In other report [4] it was experimentally evident that the target plate was got perforated for impact of soft iron core for velocities 1052 m/s and above. The maximum back-face deformation of 8.72 mm was observed for impact of deformable projectile with 1264 m/s velocity. Table 45.2 illustrated that the target plate remains un-perforated for the high energy impact with ball bearing projectile up to velocity of 869 m/s. However, it was perceived that target plate got perforated for high energy impact with velocity of 995 m/s and above. In another report [4] it was experimentally observed that the target plate got perforated for impact of ball bearing projectile with velocity of 834 m/s and above. The residual velocity is also mentioned in Table 45.2. The variation in the experimental results [4] and predicted results for complete perforation and residual velocity is because of assumption made in strength and failure model of material.

Table 45.1 Back-face deformation for soft iron projectiles

Sr. No.	Initial Velocity (m/s)	Residual Velocity (m/s)		Back face deformation (mm)
		Experimental [4]	Numerical	(Observed data)
1	761	NP	NP	1.52
2	879	NP	NP	2.37
3	937	NP	NP	2.95
4	1052	192	NP	4.25
5	1135	269	NP	5.53
6	1264	423	NP	8.72

Table 45.2 Back-face deformation for Ball Bearing projectiles

Sr. No.	Initial Velocity (m/s)	Residual Velocity (m/s)		Back-face deformation (mm)
		Experimental [4]	Numerical	(Observed data)
1	803	NP	NP	2.11
2	834	57	NP	2.33
3	869	229	NP	2.64
4	995	421	324	3.96
5	1166	661	578	6.79
6	1247	786	692	9.12

4. Conclusion

The ballistic performance of Al 2024-T652 against rigid and deformable particles is tested. Computational analysis of all high energy impacts were conducted using the well proven Johnson-Cook models. Computational analysis predicts the back-face deformation of target sheet and remaining velocity of projectile after impact. Immediate after impact the velocity is observed decreasing rapidly for 15 microseconds. After that rate of reduction of velocity is slow down for both the projectile.

It was observed from computation results of this study that the target plate is remaining un-perforated for the velocity of impact from 761 m/s to 1264 m/s. However, it was perceived that target plate got perforated for high energy impact with velocity of 995 m/s and above. The variation in the experimental results [4] and predicted results for complete perforation and residual velocity is because of assumption made in strength and failure model of material.

References

1. Aydin, M., Apalak, M.K., and Apalak Z.G.,(2020). Experimental study on structure optimization of functionally graded sandwich plates under ballistic impact. J. Compos. Mater. 54(26):3967–3980.
2. Zhang, Q.N., Zhang, X.W., Lu, G.X., and Ruan, D., (2018). Ballistic impact behaviors of aluminum alloy sandwich panels with honeycomb cores: An experimental study. J. Sandw. Struct. Mater. 20(7):861–884.
3. Jena, P.K., Mishra, B., Siva Kumar, K., and Bhat, T.B., (2010). An experimental study on the ballistic impact behavior of some metallic armour materials against 7.62mm deformable projectile. Mater. Des. 31(7):3308–3316.
4. Sharma, P., Chandel, P., Bhardwaj, V., Singh, M., and Mahajan, P.,(2018). Ballistic impact response of high strength aluminium alloy 2014-T652 subjected to rigid and deformable projectiles. Thin-Walled Struct. 126:205–219.
5. Shasthri, S. and Kausalyah, V., (2020). Effect of ballistic impact on Ti6Al-4V titanium alloy and 1070 carbon steel bi-layer armour panel. Int. J. Struct. Integr. 11(4):557–565.
6. Pawar, M.J., Patnaik, A., Biswas, S.K., Pandel, U., Bhat, I.K., Chatterjee, S., Mukhopadhyay, A.K., Banerjee, R., Babu, B.P., (2016). Comparison of ballistic performances of Al2O3 and AlN ceramics. Int. J. Impact Eng. 98:42–51.
7. Pawar, M.J., Gautam, V., Kumar, A., Kukshal, V., (2021). Computational analysis of aluminium alloy plates against conical-nose steel projectile. Materials Today: Proceedings. 46(15):6552-6557.
8. Bagwe, S., Thale, S., Sawant, P., Pawar, M.J., (2021). Finite element analysis of armor steel and aluminium alloy under the impact of 7.62 mm projectile. Materials Today: Proceedings. 44(6):4086–4091.

Note: All the figures and tables in this chapter were made by the authors.

Technologies for Energy, Agriculture, and Healthcare – Shailesh Nikam et al. (eds)
© 2024 Taylor & Francis Group, London, ISBN 978-1-032-98028-7

46

Interference Modeling Over Terahertz Band Wireless Communication Network

Surendra Sutar[1]

Department of Electronics and Telecommunication Engineering,
Rajiv Gandhi Institute of Technology, Mumbai University,
Mumbai, India

Sanjay Deshmukh[2]

Department of Electronics and Telecommunication Engineering,
Rajiv Gandhi Institute of Technology, Mumbai University,
Mumbai, India

Abstract: Nowadays, the fifth generation (5G) mobile network is being stationed worldwide, and various researchers are even trying to investigate very promising 6G terahertz wireless technology. Compared to microwave radio communication, the transmission range of terahertz radio is limited. Frequency reuse becomes much more adaptable and even feasible in the vicinity of close transceiver pairs in the terahertz band. Notwithstanding, the recurrence reuse can likewise bring about co-channel severe obstruction and low value Signal to Interference Noise Ratio (SINR) in the terahertz band, which at last decreases signal discovery and network reliability. This is because of the coordination and spatial arrangement of the frequency reuse. This work's proposed composite channel model can be used to model co-channel interference. The proposed channel model accurately captures THz communication's key characteristics, including dynamic shadowing, spreading loss, and absorption loss. These characteristics are significantly more complex than the low-frequency band. In addition, the mutual information maximization problem and the interference model characterization have been considered.

Keywords: Mutual information, Nano networks, THz band, TS-OOK (time-spread on-off keying)

[1]surendra.sutar@mctrgit.ac.in, [2]sanjay.deshmukh@mctrgit.ac.in

DOI: 10.1201/9781003596707-46

1. Introduction

Answer to increase in demand for data transfer at extremely high speeds in wireless networks, researchers have proposed a number of methods for implementing terabit-per-second (Tbps) links, such as terahertz (THz) communication and nanotechnology. Due to the lack of bandwidth in the microwave frequency range for conventional wireless communication systems, higher frequencies are being pursued for use in delay-sensitive wireless applications. The THz band has proven capable of meeting the growing demand for wireless IoT devices at 100 Gb/s and even 1 Tb/s with unprecedented bandwidth (Akyildiz et al., 2022).

Terahertz-based nano-networks are emerging as cutting-edge technology with the potential to transform future medical applications. These networks can accurately quantify important numbers, like as a patient's viral load, and even anticipate septic shock and heart attacks before they occur. THz spectroscopy has great potential for the qualitative and quantitative identification of key substances in the biomedical field, such as early cancer diagnosis, accurate boundary determination of pathological tissue, and non-destructive detection of superficial tissue, due to its non-ionizing, non-invasive, high penetration, high resolution, and spectral fingerprinting features.

For coverage probability analysis, developed outdoor terahertz network model, taking into consideration blocking, directional antennas, and molecule absorption. The suggested model's accuracy is evaluated using Monte Carlo simulations. Studied how node density, transmission distance, and surrounding conditions affect coverage and bandwidth. Result shows that coverage probability and bandwidth is lower denser network, long distances, humid condition and low latitude. (Chen et al.,2021).

Researchers have also focused on using nanomaterials like graphene to implement THz in a variety of military, environmental, and biomedical fields owing to the capability of THz communication technologies. Channel capacity modeling and THz wave propagation channel modeling with various power allocation schemes were investigated for wireless communications. Using a variety of power allocation schemes and channel molecular compositions, a brand-new THz propagation model is utilized to carry out a numerical evaluation of the THz band channel capacity. However, in order to achieve ultra-high THz transmission rates over wireless, it is necessary to overcome a number of technical obstacles because of the great complexity required to perfectly represent THz band and interference models between nodes (Zhang et al., 2020).

In order to effectively address the aforementioned issues, trying to make an effort to maximize mutual information and create an interference model in the terahertz band. Especially, create models for terahertz band wireless communication considering spreading loss, molecular absorption, shadowing effect. The addition of simulation results validates and evaluates our proposal.

2. Terahertz Wireless Communication Network System Model

There is continuous increase in the demand for high date rate. To meet this requirement, researchers are proposing various techniques including terahertz communication. Terahertz communication has potential applications in biomedical, environmental, and military fields. There are many challenges in high frequency band such as accurately characterizing high frequency band wireless communication and interference over it. This research work will propose interference model, test its performance, and will also be maximizing mutual information in terahertz band.

$$m = \text{radius of covered area}$$

$$n = \text{radius of dead zone}$$

Probability Density Function (PDF) of distance between transmitter and receiver is:

$$\begin{aligned} f_D(r_k) &= \frac{2r_k}{m^2 - n^2} \quad \text{for } n < r_k < m \\ &= 0 \quad \text{otherwise} \end{aligned} \tag{1}$$

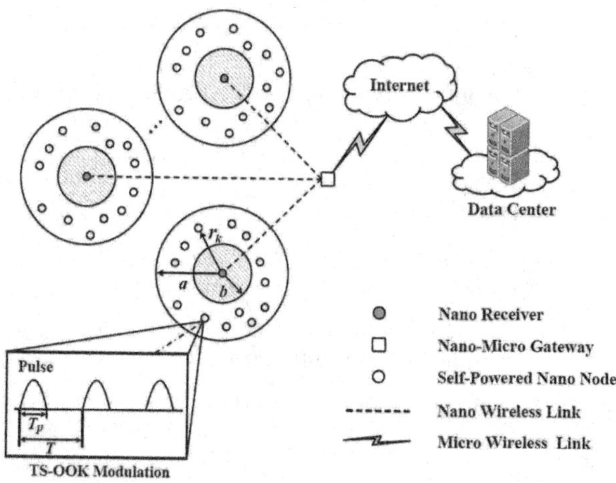

Fig. 46.1 System block diagram [4]

2.1 Path Loss Model

Total path loss, $L_k(f, r_k)$ is sum of the spreading and molecular absorption loss.

$$L_k(f,r_k) = L_{k,\text{spr}}(f,r_k)L_{k,\text{abs}}(f,r_k) \tag{2}$$

Where,
$$L_{k,\text{spr}}(f,r_k) = \left(\frac{4\pi f r_k}{c}\right)^2 \tag{3}$$

and
$$L_{k,\text{abs}}(f,r_k) = e^{\alpha_{\text{abs}} r_k} \tag{4}$$

α_{abs} = molecular absorption coefficient

Molecules present in the channel absorbs radiated electromagnetic energy and to different extent. The effects of molecular absorption loss on graphene-enabled wireless communication (GWC) scenarios are minor. Then, by taking into account only the spreading loss, the total attenuation is obtained. As can see from Fig. 46.2 that spreading loss increases with distance and square of frequency.

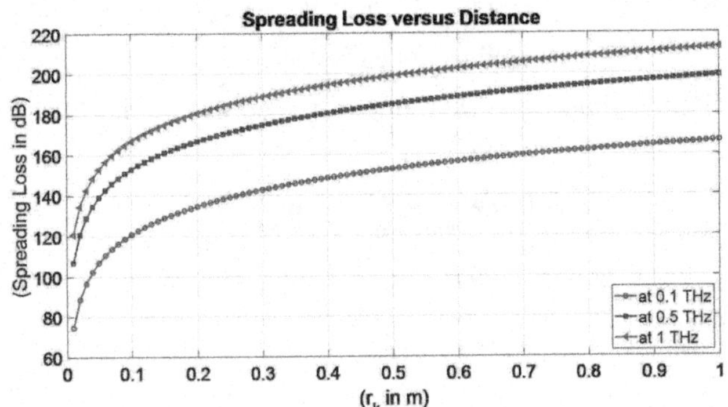

Fig. 46.2 Spreading loss vs distance at 200 GHz and 500GHz

Source: Author

The power of the received pulse, considering spreading loss, molecular absorption loss and shadowing effect is

$$P(r_k) = \int_B (r_k)^{-\eta} G 10^{\frac{\xi_k}{10}} S(f) \left(\frac{c}{4\pi f r_k}\right)^2 e^{-\alpha_{\text{abs}} r_k df} \tag{5}$$

where, $S(f)$ = psd of the transmitted symbol, B = channel bandwidth,

G = system gain, ξ_k = random variable representing shadowing characteristics

2.2 Noise Model

In terahertz band the molecular absorption noise is the primary source of noise. (Zhang et al., 2020)

$$N_{total}(r_k) = N_{back} + N_{self}(r_k) \tag{6}$$

where

$$N_{back} = \int_B B(T_0, f)\left(\frac{c}{\sqrt{4\pi f_0}}\right)^2 df \text{ background noise} \tag{7}$$

$$N_{self}(r_k) = \int_B S(f)\left(1 - e^{-\alpha_{abs} r_k}\right)\left(\frac{c}{4\pi f r_k}\right)^2 df \text{ self-induced noise} \tag{8}$$

$B(T_0, f)$ = Planck's function

2.3 Modulation Scheme Model

Energy is main constraint in terahertz communication, thereby cannot use carrier based modulation scheme. Proposed work is using TS-OOK (time spread on-off keying) modulation based on traffic of femtosecond pulses between devices.

For '0' transmission we use silence and we consider only background noise

$$N_0 = \int_B B(T_0, f)\left(\frac{c}{\sqrt{4\pi f_0}}\right)^2 df \tag{9}$$

For '1' transmission we use femtosecond long pulses and will consider both background noise and self-induced noise.

$$N_1 = \int_B B(T_0, f)\left(\frac{c}{\sqrt{4\pi f_0}}\right)^2 df + S(f)\left(1 - e^{-\alpha_{abs} r_k}\right) \tag{10}$$

3. Analysis of Interference

The SINR (signal-to-interference-plus-noise ratio) function denoted by $\gamma_k(r)$ can be calculated using

$$\gamma(r) = \frac{\text{Signal power, } P(r)}{\text{Interference, } N_I(r) + \text{Noise, } N_{total}(r)} \tag{11}$$

3.1 Mean

The mean of total interference is

$$\mathrm{E}_r\left[I_k(r)\right] = \frac{2\lambda\pi m^2 C p_X(x_1)\left(\frac{2}{\alpha_{abs}}\right)^{1-\frac{\eta}{2}}}{(m^2 - n^2)\beta}\left[\gamma\left(1 - \frac{\eta}{2}, \frac{\alpha_{abs} m}{2}\right) - \gamma\left(1 - \frac{\eta}{2}, \frac{\alpha_{abs} n}{2}\right)\right] \tag{12}$$

3.2 Variance

The variance of the interference can be calculated using

$$N_{I_k}(r) = \frac{p_X(x_1)}{\beta}\left\{\begin{array}{l}\dfrac{2\lambda\pi m^2 \widetilde{C}(\alpha_{\text{abs}})^{-\eta}}{m^2-n^2}\left[\gamma(-\eta,\alpha_{\text{abs}}m)-\gamma(-\eta,\alpha_{\text{abs}}n)\right]+\sum\limits_{i=1,i\neq k}^{K+1} N_{i,b} \\ +\dfrac{2\lambda\pi a^2\widetilde{C}}{m^2-n^2}\left[\log\left(\dfrac{m}{n}\right)-\text{E}_i\left(-\alpha_{\text{abs}}m\right)+\text{E}_i\left(-\alpha_{\text{abs}}n\right)\right]\end{array}\right\}$$

$$+\left(\frac{p_X(x_1)}{\beta}\right)^2\left\{\frac{4\lambda\pi m^2\widetilde{C}(\alpha_{\text{abs}})^{-\eta}}{m^2-n^2}\left[\gamma(-\eta,\alpha_{\text{abs}}m)-\gamma(-\eta,\alpha_{\text{abs}}n)\right]\right\}$$

$$-\left(\frac{2\lambda\pi m^2 C p_X(x_1)}{\left(m^2-n^2\right)\beta}\right)^2\left(\frac{2}{\alpha_{abs}}\right)^{2-\eta}\left[\gamma\left(1-\frac{\eta}{2},\frac{\alpha_{abs}m}{2}\right)-\gamma\left(1-\frac{\eta}{2},\frac{\alpha_{abs}n}{2}\right)\right]^2 \quad (13)$$

3.3 Maximization of Mutual Information

Mutual information between two random variables, spreading facor β $\left(\beta \triangleq \dfrac{T_s\,(\text{symbol duration})}{T_p\,(\text{pulse duration})} \gg 1\right)$ and source probability is

$$I\left(\beta, p_Y(y_1)\right) = -\sum_{a=0}^{1} p_Y(y_a)\log_2 p_Y(y_a)$$

$$-\int\sum_{a=0}^{1}\left[\begin{array}{l}\dfrac{1}{\sqrt{2\pi\left(N_{I_k}(r)+N_a\right)}}\exp\left\{-\dfrac{\left(y-\text{E}\left|I_k(r)\right|-m_a\right)}{2\left(N_{I_k}(r)+N_a\right)}\right\} \\ \times p_Y(y_m)\log_2\left\{\sum\limits_{K=0}^{1}\begin{array}{l}\dfrac{p_Y(y_K)}{p_Y(y_a)}\sqrt{\dfrac{N_{I_k}(r)+N_a}{N_{I_k}(r)+N_K}} \\ \times\exp\left\{-\dfrac{\left(y-\text{E}\left|I_k(r)\right|\right)^2}{2\left(N_{I_k}(r)+N_0\right)}+\dfrac{\left(y-\text{E}\left|I_k(r)\right|-m_a\right)^2}{2\left(N_{I_k}(r)+N_1\right)}\right\}\end{array}\right\}\end{array}\right]dy \quad (14)$$

Formulation: Problem: $\underset{\{\beta, p_Y(y_1)\}}{\arg\max}\; I\left(\beta, p_Y(y_1)\right)$ subject to: $\gamma_k(r) \geq \gamma_{\text{th}}$

For proper signal reception, $\gamma_k(r) \geq \gamma_{th}$ (threshold value of signal-to-interference-plus-noise ratio is set to 10 dB)

4. System Analysis

Validated and evaluated suggested model through MATLAB simulations.

For proposed model following parameters are set:

Bandwidth, B = 1 *THz*, Reference Temperature, T_0 = 296 K, Radius of covered region, m = 1 m, Radius of dead zone n = 5 mm,

Nano node density λ = 0 to 500 nodes per square mm

SINR γ_{TH} = 10 dB, Spreading factor, β = 200 to 1200, Total signal energy = 500 pJ

Figure 46.3 shows total interference power as a function of frequency and node density for proposed model. As the nodes density increases, the aggregate interference initially rises before settling at a constant level. This shows that molecular absorption noise has negligible effect and compared to interference. Also, for given same node density interference decreases with increase in frequency which implies spreading loss is directly proportional to square of frequency and values of molecular absorption coefficients are bigger at higher frequencies in the THz band.

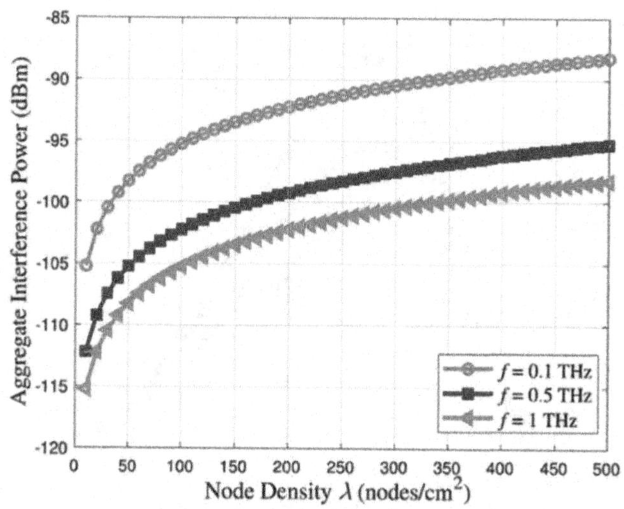

Fig. 46.3 Total interference vs λ (node density)

Source: Author

Figure 46.4a shows plot of spreading factor β and mutual information. As can be observed, as the spreading factor grows, so does the mutual information, which reduces as the distance increases. This demonstrates that a smaller transmission distance results in lower path loss and higher mutual information. Figure 46.4b shows that with increasing transmission distance, the mutual information

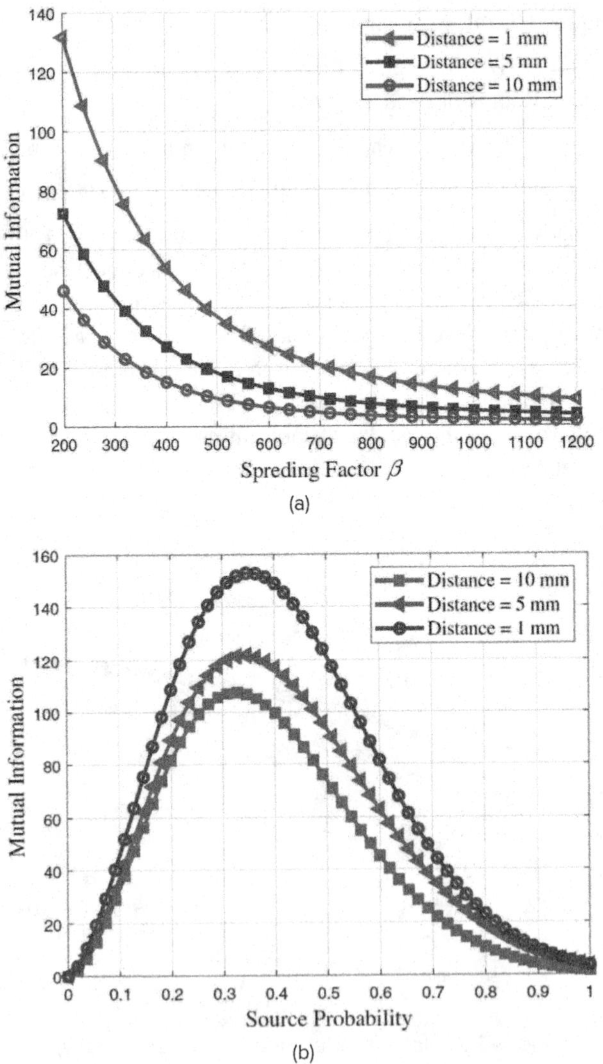

Fig. 46.4 I (Mutual information) vs β (spreading factor) and probability of Source

Source: Author

decreases. This indicates that a lower path loss results from a shorter transmission distance, resulting in a higher mutual information value. Also there exist optimum source probability which is less than 0.5, that maximizes mutually information

5. Conclusion

This work proposed an analytical model for interference, signal to interference and noise ratio, and mutual information in dense randomly distributed nano network. Model is validated over range of input metrics using MATLAB simulations. Total aggregate interference depends upon transmission distance, molecular composition between transmitter and receiver, and node density of nano nodes. Work also demonstrate dependance of mutual information on transmission distance, source probability and spreading factor. Even though it has a limited range, wireless communication in the THz band can transmit extremely quickly and safely. Modulation speed and depth of modulation are two main challenges for the THz amplitude modulator also there is difficulty in finding proper material and structure which can effectively and quickly responds to the THz wave. Because of lack of proper instruments available at THz frequencies the research progress is been quite slow.

References

1. Akyildiz, I., Han, C., Hu, Z., Nie, S., and Jornet, J. (2022). Terahertz Band Communication: An Old Problem Revisited and Research Directions for the Next Decade. IEEE Transactions on Communications vol. 70, no. 6, pp. 4250–4285.
2. Ye, J., Dang, S., Shihada, B., and Alouini, M. (2021). Modeling Co-Channel Interference in the THz Band. IEEE Transactions on Vehicular Technology, vol. 70, no. 7, pp. 6319–6334.
3. Chen, W., Li, L., Chen, Z., and Quek, T. (2021). Coverage Modeling and Analysis for Outdoor THz Networks With Blockage and Molecular Absorption. IEEE Wireless Communications Letters, vol. 10, no. 5, pp. 1028–1031.
4. Zhang, X., Wang, J., and Poor, H. (2020). Interference Modeling and Mutual Information Maximization Over 6G THz Wireless Ad-Hoc Nano-Networks. IEEE Global Communications Conference, Taipei, Taiwan, pp. 1–6.
5. Rappaport, T., Xing, Y., Kanhere, O., Ju, S., Madanayake, A., Mandal, S., Alkhateeb, A., and Trichopoulos, G. (2019). Wireless Communications and Applications Above 100 GHz: Opportunities and Challenges for 6G and Beyond. IEEE Access, vol. 7, pp. 78729–78757.
6. Juttula, H., Kokkoniemi, J., Lehtomaki, J., Makynen, A., and Juntti, M. (2019). Rain Induced Co-Channel Interference at 60 GHz and 300 GHz Frequencies. IEEE International Conference on Communications Workshops (ICC Workshops), Shanghai, China, pp. 1–5.

Technologies for Energy, Agriculture, and Healthcare – Shailesh Nikam et al. (eds)
© 2024 Taylor & Francis Group, London, ISBN 978-1-032-98028-7

47

Smart IoT-Based Healthcare Monitoring System

Mrunmayee Utekar[1], Tanisha Jadhav[2], Chaitanyaa Shinde[3], J.H Nirmal[4], Makarand Kulkarni[5], Madhura Pednekar[6]

K.J. Somaiya College of Engineering, Vidyavihar, Mumbai

Abstract: Introducing our Smart Healthcare System, a ground-breaking application of the Internet of Things (IoT) framework. Combining NodeMCU, Electrocardiogram (ECG), Pulse sensor, Buzzer, Web interface, and OLED, this system offers a user-centric approach to continuous health monitoring. Focusing on key parameters like ECG signals and pulse rates, our innovation provides comprehensive insights into heart and vascular health. The versatile NodeMCU ensures seamless real-time data transmission to a centralized Blynk Cloud-based server for storage and analysis. In critical conditions, a buzzer triggers an immediate alert, notifying patients that their doctor has received vital information via email. The web server interface grants remote access to real-time health metrics for individuals, facilitating convenient self-monitoring. Simultaneously, healthcare professionals can analyze this data remotely, enabling prompt interventions. Displaying alerts and patients' medical data, this interface enhances communication between patients and healthcare providers. Our project not only demonstrates the successful integration of IoT components but also paves the way for further research in refining and expanding the capabilities of such systems, contributing to the realm of proactive healthcare management.

Keywords: Buzzer, ECG, Led, NodeMCU, OLED, Pulse sensor

[1]mrunmayee.u@somaiya.edu, [2]tanisha.jadhav@somaiya.edu, [3]chaitanyaa.s@somaiya.edu, [4]jhnirmal@somaiya.edu, [5]makarandkulkarni@somaiya.edu, [6]madhura.pednekar@somaiya.edu

DOI: 10.1201/9781003596707-47

1. Introduction

Smart Healthcare System uses (IoT) to connect various devices like ECG (P. Chavan et al. 2016) and Pulse sensors to the internet (Valsalan P et al. 2018). This system allows continuous heart health monitoring by collecting and transmitting real-time data. With features like alerts and an OLED display (Aghenta et al.2019), users get instant feedback on their health. We've integrated a buzzer for additional alerts and created a user-friendly webpage interface (Aziz D. A.2017) for easy access to health data. This innovation aims to empower individuals to actively manage their well-being while providing healthcare providers with remote monitoring capabilities. Our focus is on contributing to a shift in healthcare towards personalized and proactive monitoring (Udit S et al. 2017) and ultimately improving the quality of life. This IoT-based system, using NodeMCU technology, redefines healthcare technology by making comprehensive health data easily accessible and promoting a seamless connection between individuals and their health information (Ruhani Ab. Rahman et al. 2017).

2. Proposed System

The primary goal of this project is to develop and deploy an efficient Smart IoT system in the healthcare industry to expedite patient treatment. In the proposed system (refer to Fig. 47.3), two crucial sensors are employed on the patient's body. ECG sensors capture the electrical heart rate waves, while pulse sensors ascertain the accurate pulse rate. These Pulse readings are displayed on the OLED display and ECG readings are transmitted to the Blynk cloud server for real-time monitoring. Additionally, the system generates reports sent via email to the doctor. For communication between the doctor and patient, a buzzer and a dedicated web server are integrated. The dedicated web server facilitates secure and private interactions, allowing the doctor to remotely access real-time ECG data.

2.1 Sensors

The ECG sensor and easy pulse sensor are connected to the Analog pin of the NodeMCU. These sensors provide analog readings, allowing the creation of a graph and analysis of value variations. The ECG sensor captures a signal with distinct peaks and vital biological features. The AD8232 ECG sensor measures heartbeat and heart rate variability (HRV). HRV is calculated using a window of 10 seconds, with HR derived from RR peaks in that interval. The electrical signals exhibit voltage signals that is amplified and filtered by the AD8232 sensor before being outputted. The Pulse sensor is interfaced with OLED, displaying readings. Optical heart-rate sensors, like the Pulse sensor, operate by shining green light (~550nm) on the finger and measuring reflected light with a photo sensor, a technique known as Photoplethysmogram. This method is similar to observing your heartbeat pulsing when shining a flashlight through your fingers.

2.2 IoT Server

Table 47.1 Data

Sensor Data	Data Type
ECG Sensor	Float
Easy Pulse Sensor	Float

Source: Author

In our proposed IoT system, data from the ECG and easy pulse sensors smoothly goes to the Blynk cloud server using the analog pin of the NodeMCU. This cloud setup helps us watch the data from anywhere, making it easy to access in real-time. The Blynk cloud acts like a central hub where we carefully look at all the collected data. While we keep an eye on things, the system automatically sends notifications, like emails and SMS, to a specific healthcare professional, usually the attending doctor. These notifications quickly update the doctor about the patient's vital signs and any unusual findings in the readings. Once the doctor gets the automated message, a chain of communication starts. The doctor tells the receptionist, who plays a crucial role in passing important information to the patient. This communication process happens through a dedicated web server using HTML. The receptionist, armed with the necessary details, talks to the patient using various methods. For quick and effective communication, our system includes a buzzer connected to the product. The receptionist can make the buzzer vibrate to alert the patient about the received message or provide essential information. This multi-layered approach ensures that patients get timely updates about their health, allowing a quick response to any emerging medical concerns. The interconnected web servers and cloud-based communication systems create a high efficiency.

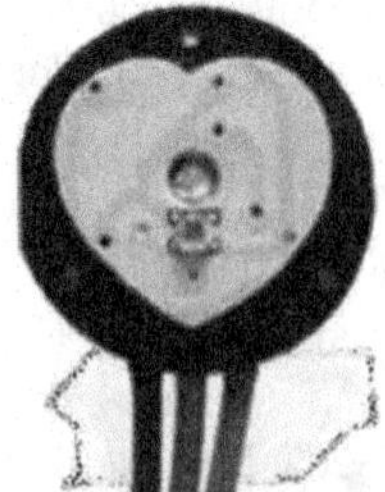

Fig. 47.1 Pulse sensor

Source: Adapted from https://shorturl.at/WV1U9

Fig. 47.2 AD8232 ECG sensor

Source: Adapted from https://shorturl.at/g0mgj)

3. Experimental Setup

Table 47.2 Pulse values

Pulse Rate	State
0-60 Bpm	Low
60-100 Bpm	Normal
100- above Bpm	High

Source: Author

The experimental setup is configured to display heart rate and pulse sensor readings on an OLED screen, while the ECG sensor readings are visualized on the serial plotter. Simultaneously, the ECG data is promptly transmitted to the Blynk cloud server, following the workflow depicted in Fig. 47.3. The system continuously monitors these outputs and, upon detection of critical conditions, automatically sends an email notification to the designated doctor. In response to the received email, a buzzer is activated through a separate web server managed by the doctor's receptionist. Table 47.2 determines the actual values of pulse sensors where the unit is correctly measured in Beats per minute. For proper determination of values, the ECG sensors should be positioned according to the guidelines outlined illustrating the significance of various features and intervals derived from the ECG signal shapes. The electrical signals exhibit voltage variations ranging from approximately 300+ to -200, with an average output of around 500 which is considered a normal reading. Based on

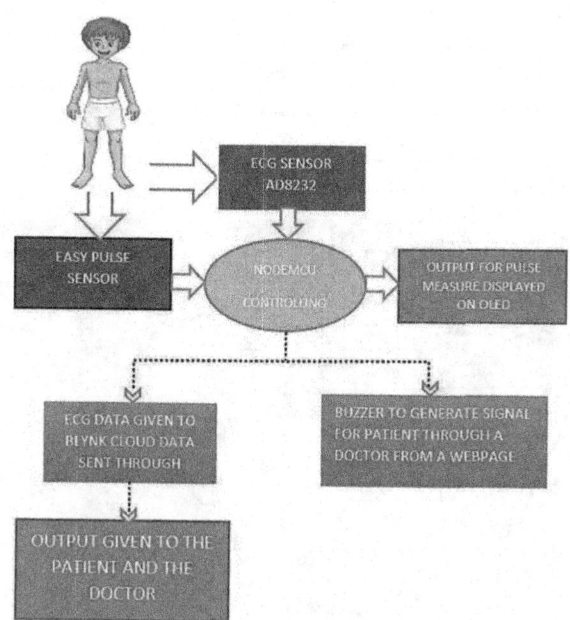

Fig. 47.3 Workflow

Source: Author

these characteristics and their respective intervals, a multitude of cardiac diseases and irregularities can be diagnosed. Some examples include:

1. *Irregular heartbeat or absence of P-wave:* Indicative of Atrial Fibrillation.
2. *Resting Heart Rate exceeding 100:* Suggestive of Tachyarrhythmia.
3. *Presence of Tachyarrhythmia and delta wave:* A possible indication of Wolf- Parkinson-White (WPW) syndrome.
4. *Presence of Tachyarrhythmia and delta wave:* A possible indication of Wolf- Parkinson-White (WPW) syndrome

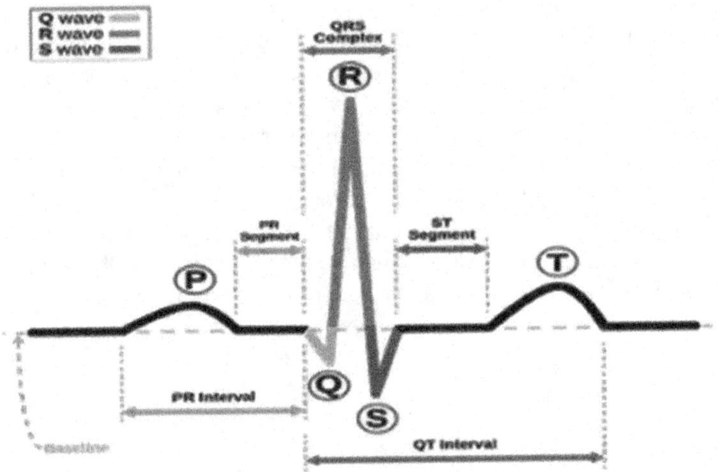

Fig. 47.4 ECG peaks

Source: Adapted from https://rb.gy/n22z4g

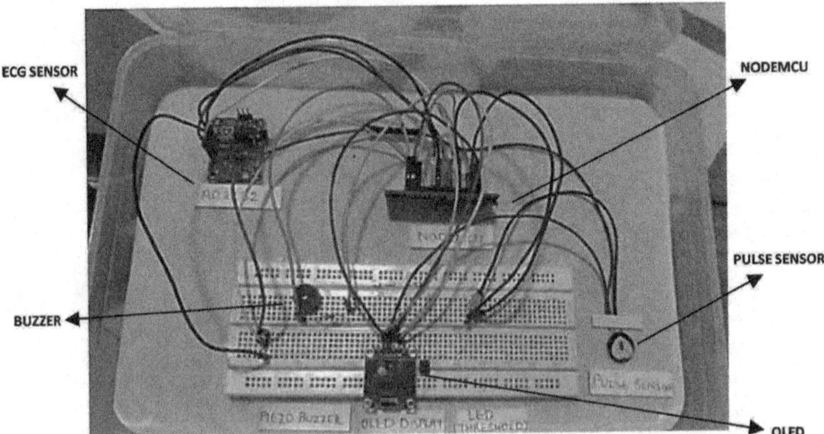

Fig. 47.5 Proposed system setup of the IoT-based healthcare

Source: Author

5. *Sawtooth P wave:* Associated with Atrial flutter.
6. *Depression of the ST-segment:* Potential sign of Ischemia.
7. *Levation of the ST-segment:* May indicate myocardial Infarction.

BLYNK cloud is interfaced with the data which gives appropriate monitoring reads and emails of these reports are being sent to the doctor with a delay according to the situation. HTML webserver is interfaced with a receiver which will be used by the Hospital.

4. Experimental Results and Discussion

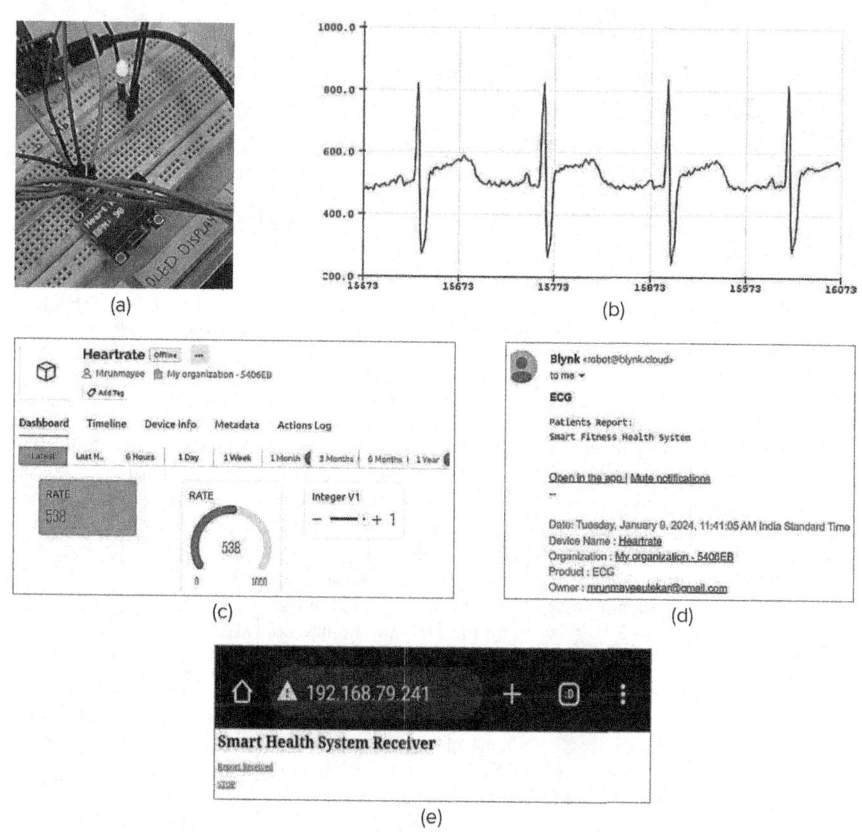

Fig. 47.6 (a) Pulse reading displayed on OLED, (b) ECG readings monitored on serial plotter X axis- time Y axis - Voltage, (c) ECG monitored on Blynk app, (d) Email of the report sent to the doctor, (e)Webpage operated by the medical faculty.

Source: Author

The IoT-based smart healthcare system implementation demonstrates the feasibility and effectiveness of integrating ECG and pulse monitoring with a responsive alert system and a user-friendly web server interface. It allows for the integration of additional sensors and is demonstrated with other healthcare platforms. This healthcare system can potentially improve patient outcomes by adding features like notifying the patient if the ECG or the pulse value becomes abnormal. As we get the output, the ECG value reaches the threshold value or the danger value, we get a mail from the Blynk app notifying us that "Patients Report: Smart Fitness Health System". Also, the doctor or hospital receptionist will receive the ECG data of the patient, and the website is created for the buzzer switch so that the receptionist will press the buzzer which means that the ECG data has been received by the doctor.

Table 47.3 Novelty research of proposed system

Sr. No	References with Year	Report and Technique Used	Novelty	Remark
1	Rahman, R. A. A., Aziz, N. A., Kassim, M., & Yusof, M. I. (2017)	This system Monitors diabetic level with breath using a gas sensor with the integration of Arduino and Esp8266	Potentially lacks comprehensive communication features, hindering direct interaction between patients and healthcare providers. May face challenges in integrating with existing healthcare systems or platforms.	The paper targets diabetic patients using glucose and activity sensors, its limitations include a narrow focus, lack of robust communication features, and potential integration challenges.
2	Sidhartha, G., Dwara-kanath, M. L., Vineeth, K. C., & Sukruth, R. (2021)	It aims at respiratory and circulatory health for people using pulse oximeter with Arduino and Blynk server.	Limited focus on respiratory and circulatory health may look broader cardiovascular concerns. Also Lack of real-time feedback mechanisms and direct interaction channels for patients and healthcare providers.	The paper concentrates on respiratory and circulatory health using pulse and oxygen saturation sensors, its limitations include a narrow focus and absence of real-time feedback mechanisms.
3	Proposed System (2024)	This system is reliable where Nodemcu with ECG and pulse sensor. The Blynk server is used for monitoring and sending mail while the HTML webpage is used for feedback	E-mail communication between patient or app user and blynk app. 2-way communication between the doctor and the patient or the app user.	The paper provides a healthcare monitoring solution with broader sensor use, real-time feedback features like OLED display and buzzer, and enhanced communication channels through the Blynk app and webpage for feedback from the healthcare clinic or hospital.

Source: Author

5. Conclusion and Future Scope

In conclusion, the implementation of a smart healthcare system integrating ECG and pulse sensors, which gives the Pulse rate in BPM and the ECG value along with a buzzer and a webpage, this system leverages technology to provide timely, personalized, and accessible healthcare. It not only gives continuous and accurate data or tracking of vital signs but also gives immediate alerts to avoid any worst circumstances. It allows doctors to interact with patients' health data by notifying the patient about their abnormal ECG or pulse value. Our cost-effective healthcare solution is invaluable for severe heart patients, offering crucial support in urban, rural, and remote areas to make a healthy world. With a focus on efficiency, our product ensures timely care, operating 24/7 to meet the diverse healthcare needs of individuals across different regions. The future of smart healthcare systems lies in continuous innovation, leveraging emerging technologies, and addressing evolving healthcare challenges by adding extra features when the patient gets the email notification of a doctor appointment, we will include that when the time of taking medicine the Blynk app will also remind you to take medicine through the email notification by using the clock module We will include a temperature sensor so that we can check whether the patient has a fever or not. A GPS will be also included so that if the patient is in an emergency condition, the hospital ward boy or the nurse will be notified by receiving the patient's location through email so that the medical service would be quick and faster to avoid any casualties. As we have made a prototype, we will integrate all the components and convert them into a wearable device like a smartwatch by designing a PCB board, we will be notified through the wearable device or smartwatch rather than checking it on a different device. Due to the short duration, we were unable to validate our work with direct doctor supervision. However, we will prioritize validation in future research efforts. Additionally, the pulse measurements were accurate and validated by a qualified medical professional.

Acknowledgment

The research reported in this presented work was supported by K J Somaiya College of Engineering, A constituent College of Somaiya Vidyavihar University.

References

1. Chavan, P., More, P., Thorat, N., Yewale, S., & Dhade, P. (2016). ECG- Remote patient monitoring using cloud computing. Imperial Journal of Interdisciplinary Research, vol. 2.
2. Rahman, R. A. A., Aziz, N. A., Kassim, M., & Yusof, M. I. (2017). IoT-based Personal Health Care Monitoring Device for Diabetic Patients. IEEE AIMS Electronics and Electrical Engineering, 4(1), 57–86.

3. Surendran, V. P., & Surendran, P. (2018). IoT-based breath sensor for mycobacterium tuberculosis. Journal of Advanced Research in Dynamical and Control Systems.
4. Sidhartha, G., Dwarakanath, M. L., Vineeth, K. C., & Sukruth, R. (2021). IoT-Based Pulse Oximeter. International Journal for Research in Applied Science & Engineering Technology (IJRASET), vol. 9.
5. Hasan, T. S. I., Hasan, M. R. M., & Hossain, M. S. H. (2019). IoT-based Portable ECG Monitoring System for Smart Healthcare. IEEE 2019 1st International Conference on Advances in Science, Engineering and Robotics Technology (ICASERT).
6. Bansal, M., & Gandhi, B. (2017). IoT-based smart healthcare system using CNT electrodes (for continuous ECG monitoring). IEEE 2017 International Conference on Computing, Communication and Automation (ICCCA).
7. Aghenta, L. O., & Iqbal, T. (2019). Design and implementation of a low-cost, open-source IoT-based SCADA system using ESP32 with OLED, ThingsBoard and MQTT protocol. AIMS Electronics and Electrical Engineering, 4(1), 57–86.
8. Satija, U., Ramkumar, B., & Manikandan, M. S. (2017). Real-Time Signal Quality-Aware ECG Telemetry System for IoT-Based Health Care Monitoring. IEEE Internet of Things Journal.
9. Vasalan, P., Baomar, T. A. B., & Baomar, T. A. B. (2020). IoT-Based Health Monitoring System. ResearchGate.
10. Krishna, C. S., & Sampath, N. (2017). Healthcare Monitoring System Based on IoT. IEEE Bengaluru, India.
11. Aziz, D. A. (2018). Webserver-based smart monitoring system using ESP8266 node MCU module. Int. J. Sci. Eng. Res, 9(6), 801–808.

Technologies for Energy, Agriculture, and Healthcare – Shailesh Nikam et al. (eds)
© 2024 Taylor & Francis Group, London, ISBN 978-1-032-98028-7

48

DESIGNING A LARGE-SCALE FACE AUTHENTICATION SYSTEM

Akhil Nagar[1]
Student, Computer Engineering,
K.J. Somaiya College of Engineering,
Somaiya Vidyavihar University

Hetvi Gudka[2]
Student, Computer Engineering,
K.J. Somaiya College of Engineering,
Somaiya Vidyavihar University

Manish Potey[3]
Professor, Computer Engineering,
K.J. Somaiya College of Engineering,
Somaiya Vidyavihar University

Sangeeta Nagpure[4]
Assistant Professor, Information Technology,
K.J. Somaiya College of Engineering,
Somaiya Vidyavihar University

Abstract: The primary focus of this paper is to develop a software architecture that ensures secure and scalable user authentication across various scenarios from traditional security to modern services like flight bookings and event ticketing. The motivation behind this application stems from the realisation of the critical role that secure and scalable user authentication play in different industries. Current authentication techniques, like physical identity cards, not only provide operational difficulties but are also vulnerable to security breaches. The goal is to bring about a paradigm shift by creating a large-scale face authentication system which offers a safe one stop solution for user verification. The scope of the project

[1]akhil.nagar@somaiya.edu, [2]hetvi.gudka@somaiya.edu, [3]manishpotey@somaiya.edu, [4]sangeetanagpure@somaiya.edu

DOI: 10.1201/9781003596707-48

includes the development of a Large-Scale Authentication System based on face recognition technology, adopting a microservices approach for architectural design.

Keywords: Authentication, Face recognition, Load testing, Microservices, Mongo, Nosql

1. Introduction

In an increasingly interconnected and digital world, the need for secure and efficient user authentication systems has never been more critical. Traditional authentication methods employed in various scenarios are often vulnerable to security breaches and operational inconveniences. They often suffer from prolonged compute times and high processing power requirements, limiting their applicability to large-scale scenarios. With the rapid expansion of digital services and the growing need for secure and seamless user authentication, there is a clear gap in the market for a face recognition system that can scale efficiently to accommodate millions of users. Real-time user verification is a pivotal feature of the system. By implementing face recognition APIs, the system can instantly authenticate users, eliminating the need for traditional, time-consuming authentication methods. This not only enhances security but also offers a user-friendly experience, aligning with the project's broader goal of providing a convenient and efficient means of accessing services. The project is centred around the development of a state-of-the-art Large Scale Authentication System using face recognition technology. This ambitious initiative aims to transcend the limitations of current face recognition tools by focusing on scalability, efficiency, and real-time user verification, authentication methods.

2. Literature Survey

2.1 Need and Adoption of Face Recognition System in India

The unprecedented growth in Indian travel, with over 1.73 billion domestic visits and 6.43 million foreign arrivals (Invest India, n.d.), highlights the urgency to streamline airport processes. Current manual verification and paper-based documentation create inefficiencies, errors, and inconvenience. To address these challenges, the Indian government has envisioned a transformative shift to a large-scale biometric digital identification system for air travel, detailed in

"From Paperwork to Biometrics: Accessing the Digitization of Air Travel in India through Digi Yatra" (George et al., 2023). This initiative aims to enhance the efficiency and reduce errors by eliminating the repetitive need for identity and travel documents at every stage, ensuring a smoother and more user-friendly air travel experience. Similarly, the need for secure and efficient digital identification is paramount in other sectors, including metro stations, hotel bookings, etc.

2.2 Microservices Architecture for Scalable Systems

The rise of microservices architecture stems from the increasing complexity and scalability demands of modern software. The paper "Monolith vs. Microservice Architecture: A Performance and Scalability Evaluation" defines *"A Microservices architecture involves breaking down a business domain into small, consistently bounded contexts implemented by autonomous, self-contained, loosely coupled, and independently deployable services."* (Blinowski, 2023, p. 20357). A scalable architecture for deploying a deep learning based facial recognition system using distributed and microservices is proposed (Timur et al., 2019). It involves encapsulating the system into Docker containers, ensuring efficient handling of data volumes, and flexible adaptation. This architecture strikes a compromise between security, adaptability, and scalability, making it a good choice for real world scenarios.

3. Methodology

Important Design Decisions to be taken during the process include:

1. Estimation of the load and requirements for different use cases which provide a baseline for designing a system for large scale data handling.
2. Choose an appropriate face recognition algorithm which will serve the purpose well.
3. Determine the different data elements, storage requirements. Choosing an appropriate database system, creating a database schema which can be efficiently queried through.
4. Specify system functionalities and design a microservices architecture to ensure scalability.
5. Focus on the non-functional requirements such as scalability, maintainability to guide system architecture decisions.
6. Develop a novel efficient architecture that aligns with the requirements.
7. Conduct thorough testing to validate the functionalities and accuracy.

4. Back of the Envelope Calculation

This technology holds great potential for transformative applications in various real-world applications such as:

1. **Hotel Booking:** Guests can experience a seamless and contactless check-in process by scanning their faces upon arrival, eliminating the need for the physical identification documents.
2. **Airports and Railway Stations:** It enables biometric boarding of passenger's identities and issues boarding passes electronically. This will seamlessly verify passenger's identities and issue boarding passes electronically. Immigration and security officers estimated a reduction of average passenger processing time by 8-12 minutes during peak hours (George et al., 2023). According to data from the Airports Authority of India (2023), airports like Indira Gandhi International Airport in Delhi can handle over 50 million passengers annually. Assuming an even distribution, this would be approximately 137,000 passengers per day. If we consider that 25% of these passengers use face recognition for check-in or security purposes, that's roughly 34,250 people per day (Goyal, 2023), which is approximately 0.4 people per second. Delhi metros carry 2.6 million passengers daily. If 25% of the passengers use the application, that's 650,000 people per day which is approximately 7.5 people per second.
3. **MyGate application:** Residents and authorised visitors can get seamless access to the premises by scanning their faces at the gate, streamlining the verification process. It increases security by preventing unauthorised access. This app is used by 3.5 million homes across 25,000 societies (Wikipedia, n.d.) as of 2023. It includes more than 3 lakh security guards and residents and operates in around 11 major cities in India.
4. **Private Events:** The entry process of attendees can be accelerated by scanning their faces at the entry gates, which reduces the wait time, eliminates the use of physical ticket checks and prevents the use of counterfeit tickets. Taking into consideration numbers from the music festival Lollapalooza in Mumbai, 2023, total fans were 60,000. Assuming all fans used the face identification for entry, it estimated 41.67 people per minute.

5. Face Recognition

The proof-of-concept hinges on the assumption that our face recognition system is state-of-the-art. The current project scope excludes the optimization of existing face recognition algorithms; our goal is solely to leverage existing technologies on a large scale. We are bound by the accuracy and precision of the models used, but we firmly believe that improvements will come over time. Specifically, we utilise the Dlib Face Recognition library, developed by Davis E. King (2023). This library operates through a combination of advanced computer vision and Convolutional Neural Networks (CNNs) to achieve precise facial recognition. Key features include facial landmarks detection and feature extraction. The model has an accuracy of 99.38% on the Labeled Faces in the Wild benchmark (King, 2023).

6. Data Modeling

6.1 Data Requirements

The first step is to understand the data requirements of the system. Based on a rough estimate we should be able to approximate the number of users and clients on-board. We must determine the type of data stored and the frequency of usage. We will require one database for all user data such as name, phone number, username, password, and the face biometric stored as an embedding. One database for clients where client and event code are auto allotted.

We also need to consider the read/write capabilities of different databases; it must possess a high level of atomicity. We have used two different datasets for faces namely Labelled Faces in the Wild (Huand, 2007), which include 13000 faces from the web and Microsoft DigiFace1M which includes 1 million synthetic faces. We have artificially generated 10,000 users with random data. Each user has two images, one with which he signs up, second on which we will test.

6.2 Database Schema

Ensuring scalability for millions of users and accommodating diverse data structures, NoSQL's schemaless data models fit well for varied client requirements, such as hotel bookings and airport details. Our choice of MongoDB stems from its SaaS offering called Atlas. We initially created the following databases with their respective collections as shown below.

1. **Actors:** a) Users: Stores the personal information of the users, their login credentials along with all the events they are registered in. b) Client: Stores the information of the client, their login credentials and maintains a list of all the events created by the client.
2. **Hotel Chain *(client_code)*:** a) Event: In this case the event will be a particular location and will contain a manifest of all the bookings. b) Event-date: A list of all guests residing or checking in on that date. This acts as a feeder to the caching mechanism which will be setup in the later stage.

Similarly, there will be a different database for other clients like Mumbai Airport, Cinemas, etc.

6.3 Functional Requirements

In a microservices architecture, APIs serve as communication channels between individual services. Each microservice exposes an API, facilitating the exchange of information among services. For our tech stack, we have selected Flask, a lightweight Python framework, driven by its simplicity, quick deployment, and our team's expertise, enabling efficient delivery of scalable APIs for our project.

6.4 API Endpoints

1. **Auth API:** Handles user authentication and authorization for the face recognition system. Provides endpoints for Sign up, KYC, Login, and Logout.
2. **Booking API:** Allows users to manage event bookings. Provides endpoints for creating, viewing, and cancelling bookings.
3. **Verify API:** Enables clients to scan user faces and verify event attendance. Provides endpoints for face scanning and verification results. Clients capture the faces using an imaging device and compare with the database of registered users and hence verify them.

The APIs collaborate to offer a seamless event booking and verification experience.

6.5 System Flow

The system process has three main phases: Pre-Event, Event Day, and Post-Event. During the Pre-Event phase, clients upload event details like name, date, time, and location. The system automatically generates a unique ID and prepares for the upcoming event. On the event day, registered attendees undergo authentication at the registration area, where facial recognition cameras capture and process their images. Facial features are observed, converted into face embeddings, and compared with the database for authentication. Successfully authenticated individuals gain access, and alternative verification methods are available in case of mismatches. The system logs authentication events, including timestamps and attendee information. In the post-event phase, the system generates detailed reports on attendance and authentication, providing valuable insights. It is illustrated in Fig. 48.1.

6.6 Non-Functional Requirements

Martin Klepmann (2017) defines the 3 pillars supporting a Data Intensive Application as:

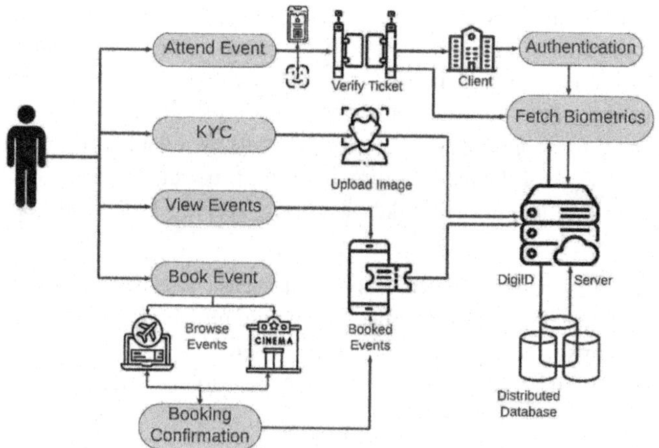

Fig. 48.1 System flow

Source: Author

1. **Scalability:** To handle high volumes and prevent bottlenecks, we deploy multiple flask instances running in parallel on servers. Gunicorn, a Python web server gateway interface (WSGI), manages worker nodes for parallel request handling. Using multiple flask apps on different ports with a Load Balancer like Nginx ensures effective load distribution. AutoScaling enables efficient utilisation of hardware resources
2. **Reliability:** For reliability, systems should work correctly despite adversity, such as hardware or software faults. Strategies include global replication, disaster recovery sites, and Content Delivery Networks (CDNs) for static content. Logging is crucial for learning from failures and improving software.
3. **Maintainability:** Maintainable code is modular, easy to read, and follows standard coding practices. Version control tools like Git aid in rollback to working versions. Testing is essential to avoid single points of failure, and separating development and production environments is recommended. DevOps practices, with tools like Jenkins and Azure DevOps, streamline feature integration in the software development cycle.

7. Architecture

The novelty of our research lies in the design decisions taken in order to make this face-recognition algorithm feasible for large scale purposes. Without important software components, database structure and system design. Once we gained intricate knowledge behind the working of distributed systems, we were able to come up with a suitable architecture which we felt would suit the requirements of this project. Our objective from the beginning was to introduce the idea of Microservices architecture to build this system. The rise of microservices architecture stems from the increasing complexity and scalability demands of modern software. Benefits of this microservices architecture include enhanced fault tolerance, horizontal scaling and simplicity in resource augmentation. We present an overview of the architecture diagram. Illustrated in Fig. 48.2, the entire architecture is divided into four key sections: Actors, Functions, Services and Databases.

1. **Actors:** The primary actors for this use case consist of the users and the clients. The users are the public, who would utilise our product to streamline and optimise their time at locations where our product is deployed and enhance their experience. Clients are the organisations who would like

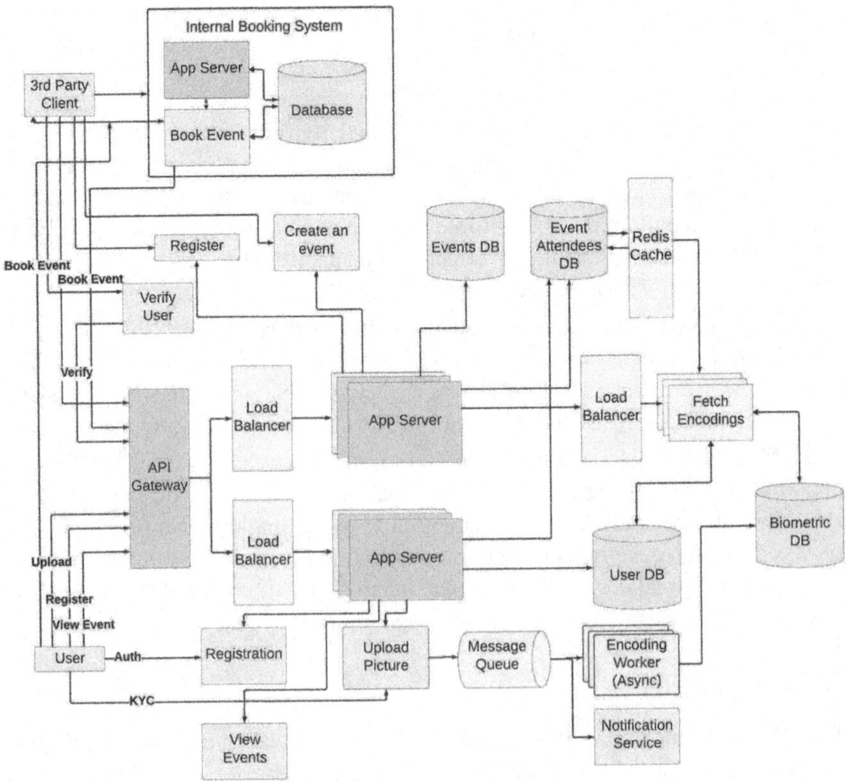

Fig. 48.2 Proposed system architecture

Source: Author

to avail services offered by us. They are predominantly online booking agencies who can speed up their ticketing process.

2. **Functions:** This section is further categorised into two parts: User Functions and Client Functions.

   User Functions: A user on our platform must register with a username and password, complete KYC by uploading a picture, and provide a unique phone number serving as their digital id. Through their phone number, users can easily book events on all partner websites. On our app, they can view all booked events, and manage all vendors that have access to their personal biometric information.

   Client Functions: Clients undergo onboarding, integrating their systems with ours and receive unique credentials. They can create events via their admin panel, with dedicated databases allocated for each client. As a prerequisite they are required to possess the necessary hardware which will facilitate the face recognition system. Once they are logged into our system, they can verify if an individual has purchased a ticket for a specific event.

3. **Services:** This section includes components which get rid of performance bottlenecks that may occur at scale.

   Nginx: Nginx serves as a load balancer, distributing traffic among three instances of front-end services. The Round-robin algorithm ensures even load distribution, preventing server overload. We are running 3 instances of User, Client and Verify on 6 different ports.

   Redis: Redis acts as a server-side cache for biometric data, reducing response time for database reads. Cached data speeds up processing for subsequent requests, enhancing system performance.

   Docker: Docker containerizes all services, facilitating independent deployment, scaling, and management. The Docker compose file orchestrates communication between services.

This architecture ensures efficiency and responsiveness in handling large-scale biometric identification processes.

8. Testing

As mentioned before, testing allows developers to find points of failure. For the scope of this project, we shall be Load Testing a sequence of critical operations. We have 3 Major Flows tested on the following hardware configuration:

System: Intel(R) Core(TM) i5-10210U CPU @ 1.60GHz 2.11 GHz 16GB x64-based processor

8.1 Testing the Face Recognition Algorithm

Testing on a Random Sample

Table 48.1 Confusion matrix

	Actual Positive	Actual Negative
Predicted Positive	763	31
Predicted Negative	38	756

Accuracy: 95.4% Precision: 96.2% Recall: 95.2%

Source: Author

8.2 Testing API Endpoints

1. **Registration:**
    a) *Sign Up:* Create a username, password
    b) *Login:* Login with above credentials
    c) KYC Upload a satisfactory image along with personal information
2. **Hotel Booking:**
    a) *Login:* Login as a user
    b) *Book:* Choose your hotel, check-in and check-out date
3. **Verify:**
    a) *Login:* Client must login with client credentials to access their database
    b) *Verify:* Verify if the user has a booking for today.

9. Results

Traditional Methods of Verification:

Table 48.2 Observational analysis

Scenario	Time	Automation Level	Steps Involved
Airports	45 sec (Entry) + 6 min (Check-in)	Manual Semi-automated (Digi Yatra & QR Scan)	Wait in line, Pull out ID Card, Ticket. Perform routine twice
Hotels	10 mins (Check-in) 7 mins (Check-out)	Manual (Receptionist)	Procure Reservation Details, ID Card. Wait till scanned and processed.
Events	2 mins	Semi-Automated (Reception & RFID)	Verify name with guest list

Source: Author

Fully Automated Face Recognition System:

Table 48.3 Postman results

Report	Throughput (requests/sec)	Average Response Time (ms)	P-90 Time (ms)
1. Registration	4.01	533	765
2. Booking	12.26	197	497
3. Verification	2.50	3568	6810

Source: Author

Booking API

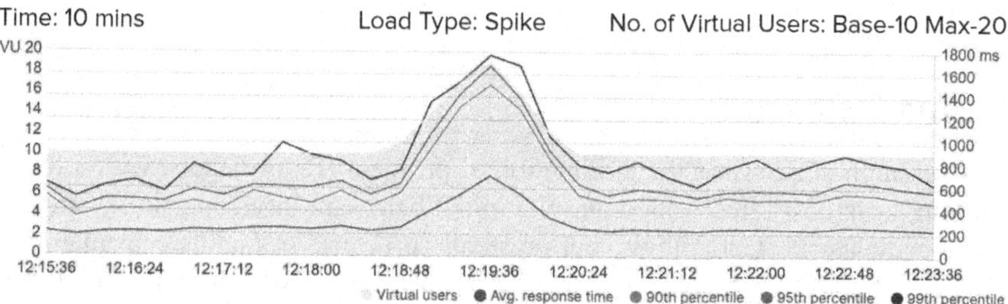

Fig. 48.3 Response time for booking API

Source: Made by the author, from the postman reports

Verification API

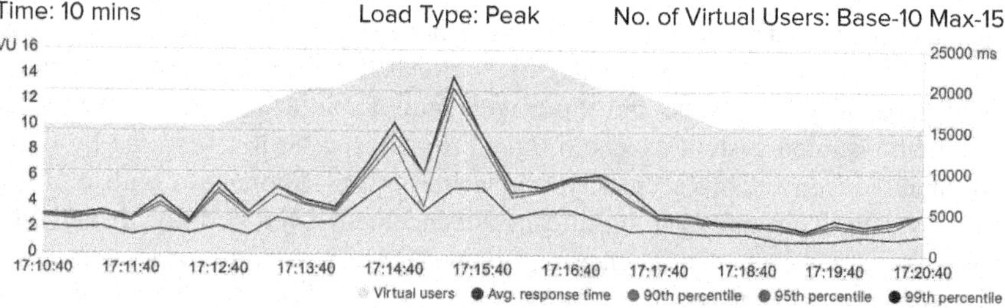

Fig. 48.4 Response time for verification API

Source: Made by the author, from the postman reports

10. Analysis

Scenario: Let us consider a hypothetical yet realistic scenario as seen in Table 48.3 where-in at an Event, an average of 1 attendee arrives every 60 seconds. In the traditional approach of Manual Approach, it would take a Guard, 50 seconds to verify tickets of everyone. In a semi-automated system, it would take 30 seconds to cross-verify with an ID. In the fully automated system designed, our tests show an average verification time of 10 seconds.

Table 48.4 Queue theory

	Avg. no. of customers in queue	**Avg. Waiting Time in System (sec)**	**Expected Waiting time in the queue (sec)**
Manual (Guest List)	4.16	300	250
Semi-Automated (Paper Tickets)	0.5	30	15
Fully Automated (Face Recognition)	0.033	12	2

Source: Author

11. Discussions

Our research emphasises the elimination of physical IDs or tickets, with a focus on long-term cost-effectiveness despite initial hardware investments. Leveraging advancements in technology, our systems prioritise reliability, availability, and scalability. We faced hardware limitations but have optimised hardware requirements to efficiently manage heavy loads. It's important to note that while this guide provides valuable insights, it is not an exhaustive list of steps.

12. Conclusion

In conclusion, our proposed architecture showcases a robust large-scale face recognition system capable of verifying National IDs, facilitating ticket booking at airports, metros, events, and more with remarkable accuracy and reliability. The demonstrated system excels in three critical aspects: it saves time by 60%, eliminates inconveniences entirely, and achieves full automation with a 100% success rate. This marks a significant advancement over traditional methods, heralding a new era of efficiency and convenience in identity verification and access control systems.

Acknowledgments

The authors are grateful to Avish Rodrigues and Kaushik Iyer for their valuable assistance in this research and project. They thank Ishika De and Yashvi Donga for many helpful comments and support.

References

1. Blinowski, Grzegorz, Anna Ojdowska, and Adam Przybyłek. 2022. "Monolithic vs. Microservice Architecture: A Performance and Scalability Evaluation." IEEE Access 10: 20357–74. doi: 10.1109/ACCESS.2022.3152803.
2. DigiFace1M. Microsoft. 2022. http://github.com/microsoft/DigiFace1M.
3. Edgar F. Codd. 1970. "A Relational Model of data for Large Shared Data Banks." Communications of the ACM, volume 13. doi:10.1145/362384.362685
4. George, A., Sagayarajan, S., Baskar, T., and Pandey, D. 2023. "From Paperwork to Biometrics: Accessing the Digitization of Air Travel in India through Digi Yatra." In Partners Universal International Innovation Journal (PUIIJ), 01, no. 04: 110. doi: https://doi.org/10.5281/zenodo.8265983
5. Invest India. 2024. "Tourism & Hospitality." https://www.investindia.gov.in/sector/tourism-hospitality. (accessed January 23, 2024)
6. "Investors Annual Reports." Airports Authority of India. https://www.aai.aero/en/business-opportunities/investors-annual-reports. (accessed January 24, 2024)
7. J. Lewis and M. Fowler. March 2014. Microservices: A Definition of This New Architectural Term. https://www.Martinfowler.com/articles/microservices.html.
8. Kleppmann, Martin. 2017. Designing Data-Intensive Applications: The Big Ideas Behind Reliable, Scalable, and Maintainable Systems. 1st ed. Sebastopol: O'Reilly Media.
9. King, Davis E. 2009. "Dlib-ml: A Machine Learning Toolkit". Journal of Machine Learning Research 10: 1755–58. https://jmlr.csail.mit.edu/papers/volume10/king09a/king09a.pdf
10. Labelled Faces in the Wild (LFW). 2007. University of Massachusetts Amherst. http://vis-www.cs.umass.edu/lfw/.
11. Layog, Vishal. "Guesstimate: Estimating the Number of Delhi Metro Commuters." Medium, https://medium.com/@vishallayog/guesstimate-estimating-the-number-of-delhi-metro-commuters-3e1883c3ad13.
12. "MyGate." Wikipedia. https://en.wikipedia.org/wiki/MyGate. (accessed January 24, 2024)
13. Rahnamoun, Rashin. 2023. "Robustness of Microservice Architecture Design, A Complex Network Approach." In the 7th Iranian Conference on Advances in Enterprise Architecture (ICAEA 2023), Institute of Electrical and Electronics Engineers (IEEE). doi: 10.1109/ICAEA60387.2023.10414440.
14. Timur Tahta D., Ketut Eddy, P., and Subpoena Mardi, S. 2019. "Deploying Scalable Face Recognition Pipeline using Distributed Microservices." International Conference of Computer Engineering, Network, and Intelligent Multimedia (CENIM). 1–6. Institute of Electrical and Electronics Engineers (IEEE). doi: 10.1109/CENIM48368.2019.8973287.
15. Zhang, D., Li, J. and Shan, Z. (2020). "Implementation of DLib Deep Learning Face Recognition Technology." In 2020 International Conference on Robots & Intelligent Systems (ICRIS), 142–47. Institute of Electrical and Electronics Engineers (IEEE). doi: 10.1109/ICRIS52159.2020.00030.

Technologies for Energy, Agriculture, and Healthcare – Shailesh Nikam et al. (eds)
© 2024 Taylor & Francis Group, London, ISBN 978-1-032-98028-7

49

Self-Adaptive Control System for Hydroponics

Kirti Sawlani[1], Deepti Patole[2]
Assistant Professor,
KJSCE, Somaiya Vidyavihar University

Vedant Kelkar[3]
Student,
KJSCE, Somaiya Vidyavihar University

Manas Pange[4]
Student,
KJSCE, Somaiya Vidyavihar University

Kiran Kardile[5]
Lab Assistant,
KJSCE, Somaiya Vidyavihar University

Abstract: Soil-based agriculture faces lots of challenges in urban zones, as it requires lot of water and large land areas for cultivation. The crops grown in soil involve use of chemical fertilizers leading to loss of nutrition value of both crop as well as soil. The concept of urban farming is gaining momentum recently, as there are multitudes of advantages for urban population. Hydroponics is a solution for soilless farming. Hydroponics is a technique that is used for cultivating plants without using soil and directly providing the required nutrients by using water-based nutrient solutions to the plants. A common problem faced in the hydroponics technique is that it requires continuous monitoring and frequent maintenance to keep the temperature, light intensity, humidity, pH, and water circulation under control. Failing to do this may hinder plant growth to some extent. This paper focuses on the automation of the system and continuous monitoring of all the parameters of the plants. Special measures have been taken to maintain the vital

[1]kirttisawlani@somaiya.edu, [2]deeptipatole@somaiya.edu, [3]vedant.kelkar@somaiya.edu, [4]manas.pange@somaiya.edu, [5]kvkardile@somaiya.edu

DOI: 10.1201/9781003596707-49

parameters of the plants like pH, light intensity, and Electrical Conductivity. To achieve this, we have built a low-cost system that can efficiently integrate traditional hydroponics techniques and modern Internet of Things (IoT) technology. The system is a blend of Embedded Systems and IoT ensuring complete automation and minimal human intervention.

Keywords: Hydroponics, Automation, Nutrients, Agriculture, IoT, Embedded systems, Urban farming

1. Introduction

The idea of traditional agriculture roots from the usage of soil for the cultivation of plants. Soil is considered to be a readily available growing medium for plants as it is a source of providing nutrients and water to the plants. However, due to large-scale urbanisation, the availability of cultivable land is scarce and traditional agricultural methods also have a negative impact on the environment. In addition to that, as we move from place to place and the geographical and topographical conditions change, the type and quality of soil also vary. This restricts the growth of certain plants in specific regions since the requirements for the type of soil are different for every plant. Moreover, soil-based farming techniques also require large amounts of land for cultivation, significant amount of labour, and plenty of water supplies.

Unlike traditional farming techniques, plants grown using hydroponics utilises nutrient solutions instead of soil. This method also prevents soil-borne diseases which eventually ensure good yields and better growth of the plants. Another major advantage of using hydroponics is that there are no weeds and the nutrients are fully absorbed by only the plants.

As soil is not used in this technique, the nutrients are provided separately and the user has full control over the amount of nutrients to be added. To hold the plants by the roots, coco peats or coco coir are used as growing mediums. Based on the structure and the method of water supply, hydroponics is further divided into different techniques like aquaponics, aeroponics, deep water culture, drip irrigation, nutrient Film technique, ebb and flow techniques etc. Figure 49.1 shows some of those techniques. Due to the nutrients being added directly into the water, deep water culture is considered to be a better technique that eventually reduces water usage too. Instead of sunlight, hydroponics techniques use artificial lights especially Red, Green, and Blue (RGB) LEDs to illuminate the system with any colour within the visible spectrum of light. Every plant has different colour

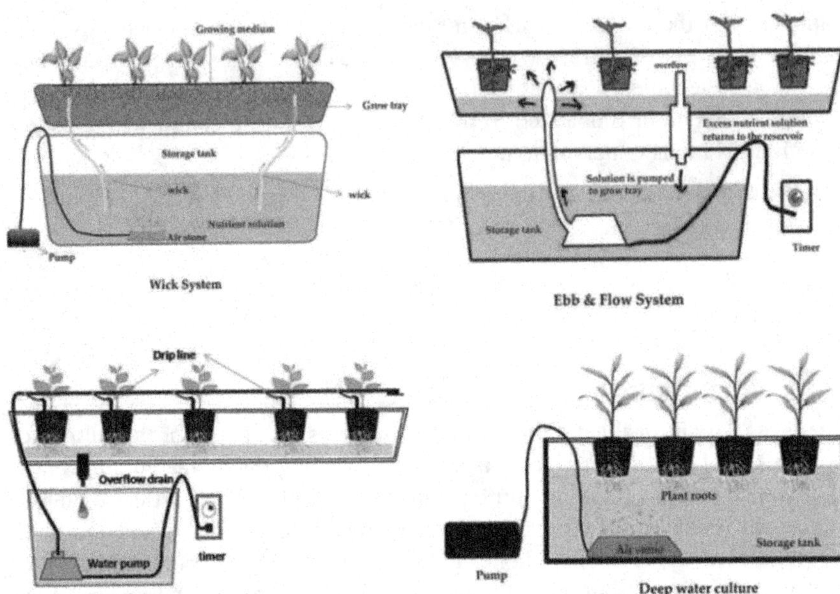

Fig. 49.1 Different types of hydroponics methods. Wick system, Ebb and Flow system, storage tank system, deep water culture. (Sharma N, 2019)

requirements and the colour combinations of the RGB LEDs can be set accordingly. Though the hydroponics technique is more efficient than the conventional farming techniques, it also requires continuous monitoring, handling, and frequent maintenance and hence the method may be tedious. Thus, there is a need for a system that can automate daily tasks and reduce human intervention. Integrating the latest technologies like automation and the Internet of Things (IoT) with these conventional hydroponics can thus prove advantageous. Moreover, the continuous monitoring of data proves useful for in-depth analysis of crops and providing solutions that are useful for excellent plant growth.

2. Literature Review

Dr. D. Saraswathi et al (Saraswathi, 2018) has designed an automated system for maintenance and monitoring of a Greenhouse farm. It aims to eradicate the limitations of a traditional greenhouse environment. These limitations include maintaining the temperature, pressure, humidity, pH, and Electrical Conductivity (EC) value at a particular level. A Raspberry Pi 3 has been used as the controlling unit of the system. Python has been used to program the Raspberry Pi. The microcontroller is interfaced with different sensors for the monitoring of data. A DHT sensor has been used for recording temperature and humidity, a pH sensor for the pH value, and a pressure sensor for monitoring pressure. These values have been published to the cloud platform named Thing speak. And an android application is developed for the user to keep track of the process.

R B Hari Krishna et al (Harikrishna, 2021) has designed a closed loop hydroponics system that provides artificial luminance and also monitors the various parameters that are essential for the overall growth of the plants. The system also included IoT integration making all these parameters easily accessible through a dashboard made using the UBIDOTS cloud platform. The system included hardware parts like ESP32 for IoT integration and controlling, a water flow sensor that regulates the amount of water that flows through the valve, and various other sensors like LDR, MQ135, BMP 180, and the DHT22. The system uses an Ultrasonic sensor HCSR04 for measuring water level and a solenoid valve for supplying water to the system. The paper focuses on growing winter crops since the temperature of the system supports the growth of these crops. The paper focuses on maintaining the nutritional value of the water. The nutrients added in the nutrient tank were Nitrogen (N), Phosphorus (P), and Potassium (K) and were in the weightage of 70ml, 90ml, and 50 ml respectively in 20 litres of water.

Dania Eridani et al (Eridani, 2017) have designed a nutrition feeding automation system for hydroponics. The system uses Arduino Uno, servo motor, proximity sensor, etc to control and monitor the various parameters. The system also included a Total Dissolved Solids (TDS) sensor to detect the electrical conductivity of the nutrient solution. The development stage included the calibration of the TDS sensor by comparing the values with the readings of a TDS meter. The Arduino program fetched the values of various sensors and turned on the Servo motor to add the nutrients into the water. After taking several measurements, an equation was derived and finally included in the implementation of the system. Thus, this hydroponics system met all the functional requirements as planned with very good accuracy of 97.8

Saaid et al (Saaid, 2015) developed a system to support plant growth in a hydroponics method named Deep Water Culture (DWC). It is based on the Arduino Mega 2560 microcontroller. The paper focuses on the importance of maintaining and monitoring pH value in the water. It uses an Atlas sensor and a pH-0091 sensor for measuring pH. It has an automated pH controller system which comprises of pH adjustor solutions to be added to the water, servo motors to help add these solutions, and a stirrer to mix them. It was observed that the pH of the water drops over time because of the reaction of the fertiliser with the water. On day 16 the pH level dramatically dropped, thus the pH adjustor was used after day 16. 2.6 ml of the adjustor solution was added to 10L of water and it was enough to bring the pH to an ideal value.

Nisha Sharma et al (Sharma N. 2019) discuss the various hydroponics techniques and focuses on the limitations and benefits of hydroponics systems. Hydroponics proves beneficial as the water requirements are less because irrigation is not needed. Another benefit of the hydroponics technique is the absence of weeds which in turn removes the necessity of spraying pesticides. Since the number of plants per unit is greater than that of traditional farming techniques, the yields obtained are also higher. However, the hydroponics techniques have a few limitations. It requires a higher initial cost and good technical knowledge. Moreover, maintaining pH and EC values is a must. Failing to do so can result in loss of crops. The paper discusses various Hydroponic structures like the wick system, ebb and flow system, storage tank, deep Water culture, and nutrient film technique. Thus, hydroponics can prove advantageous in areas with limitations in soil and water and scarcity of fertile land. Another paper that talks in detail about different methods of hydroponics was written by Ms.Mamta et al (Sardare, 2013). The different techniques mentioned are Continuous flow solution culture, Nutrient

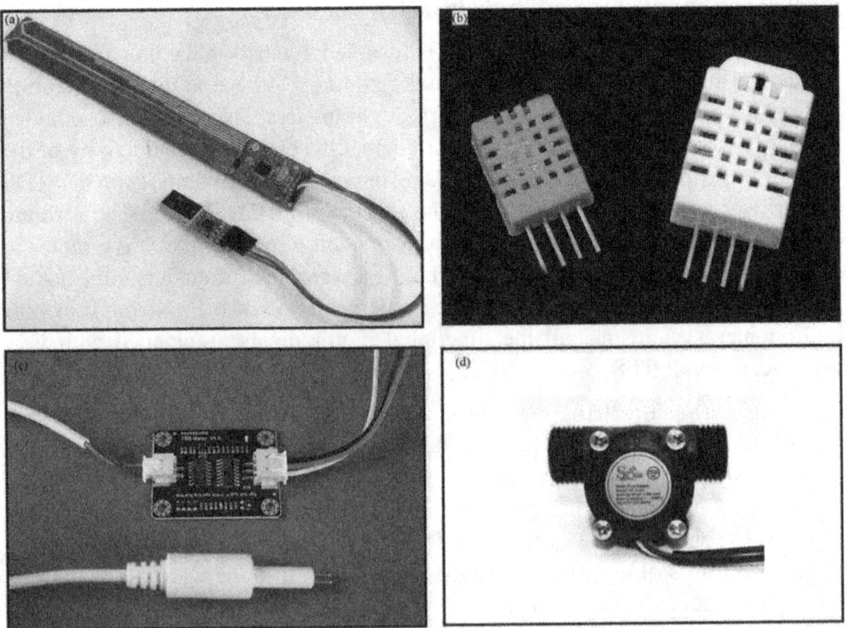

Fig. 49.2 The sensors used for the system. (a) pH sensor. (b) DHT sensor. (c) TDS sensor. (d) flowrate sensor

film technique (NFT), and Deep Flow Technique (DFT). The research also talks about the different factors considered for the selection of one particular technique. It mentions a comparison of the average yield per acre in a conventional farm to that of in a Hydroponic system. Thus concluding that hydroponics uses 1/5th of the space and 1/20th of the water to cultivate the same amount of crop compared to conventional agricultural methods. It also helps in faster growth because the nutrients are fed directly to the roots. Hydroponics helps to achieve increase in yield per acre as well as quality produce by using higher density planting.

Articles (Saputra, 2017)(. M. K. Hafizi, 2020) discuss about an automation system for controlling nutrients in a hydroponics system. The paper (Saputra, 2017) focuses on a hydroponics system that measures the parameters needed by the plants, calculates nutrient needs and takes necessary actions according to the values obtained. Calibration was performed for various sensors by comparing the readings of the sensor to the readings of a standard meter. The average of the difference between the desired value and the actual value is calculated and the calibration factor is adjusted accordingly.

Luechai Promratrak et al (Promratrak, 2015) focuses on the benefits of using LED lights over sunlight for the cultivation of plants. It mentions that light of 430-460nm or 630- 660nm is ideal for the photosynthesis of plants. The benefit of blue light (Wavelength 430-460nm) is that it helps in Chlorophyll production. A red light (Wave length 630-660nm) is helpful in making plants flower and produce fruit. The paper includes a detailed comparison of the growth of plants in sunlight and LED light. The growth of plants in the LED light was found significantly faster. The paper concludes by saying that a mixture of red and blue in a ratio of 3:1 is the most effective for the growth of plants.

3. Methodology and Model Specifications

3.1 Hardware

The hardware components of the proposed system include pH sensor, temperature sensor, humidity sensor, flowrate sensor, Total Dissolved Solids(TDS) sensor, peristaltic motor, RGB LEDs, water level sensor, Switch Mode Power Supply (SMPS), OLED display, ambient light sensor and ESP8266 development board. Figure 49.2 contain the images of all these electronic components. The plants are placed in the slotted sections using special net pots filled with clay balls made for hydroponic purposes. These net pots are placed on the exterior housing of the hydroponics system, such that more than half a portion of the net pots is dipped inside the water. The system consists of a container that stores the water in it that can be used for a long duration of time. A pump and an inlet valve system is connected to the left side of the container cascaded with a flow rate sensor. An outlet valve is connected to the right side of the container for drainage. Two pillars embedded with Red, Green, and Blue(RGB) LEDs are placed on either side of the system such that the light emitted from these LEDs would fall straight on the plants placed in the net pots. Two containers namely pH-UP and pH-DOWN containing basic and acidic solutions respectively are placed just beside the main tank. Three bottles having Nitrogen (N), Phosphorous (P), and Potassium (K) are attached to a wall of the tank. A peristaltic pump is connected to these bottles. The electrodes of the pH sensor and water level sensor are immersed in the water, whereas the temperature sensor and humidity sensor, and ambient light are placed just above the water. The TDS sensor is dipped in the water along one corner of the container.

Table 49.1 pH sensor test (Aliac 2018)

Samples	ph 7	ph 10	ph 4
Sample 1	6.80	9.94	3.87
Sample 2	6.81	9.90	3.82
Sample 3	6.70	9.91	3.80
Sample 4	6.68	9.92	3.78
Sample 5	6.77	9.91	3.77
Sample 6	6.76	9.92	3.79
Sample 7	6.75	9.90	3.85
Sample 8	6.78	9.93	3.77
Sample 9	6.79	9.91	3.77
Sample 10	6.78	9.91	3.76
Average	6.762	9.915	3.798
Percent Error	3%	1%	5%
Margin of Error	± 0.238	± 0.085	± 0.202

The OLED display is present on the front side of the container. All these sensors and peripherals are connected to a control circuit which is completely isolated and controlled by the ESP-8266 development board. The PCB for the controlling unit was designed using EasyEDA. Figure 49.3(a) shows the system design along with its peripherals. Figure 49.5 (b) shows the completed structure of the system.

The system is connected to a WiFi network by using the credentials stored in the memory. The RGB LEDs are then turned on based on the type of plants being grown. The system then establishes a connection with the Adafruit IO Dashboard.

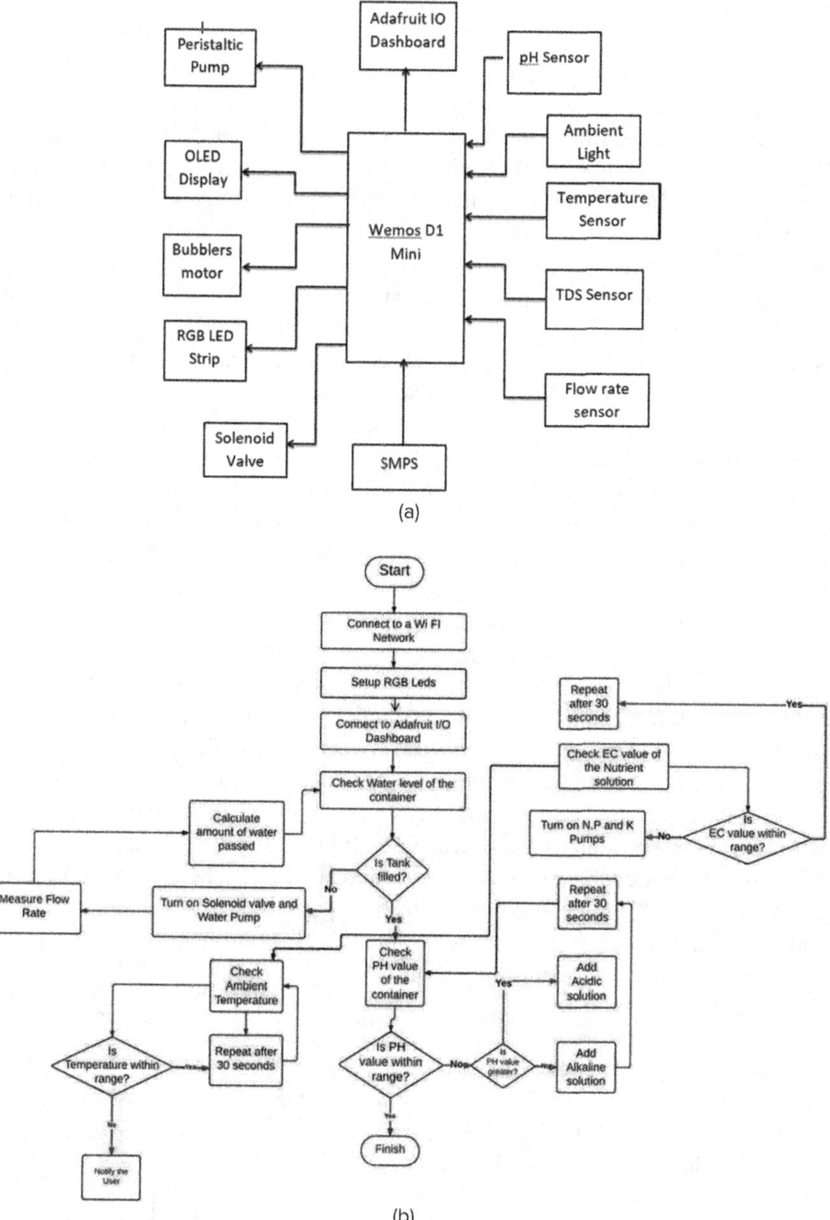

Fig. 49.3 (a) Block diagram of the closed loop hydroponics system, (b) Flow chart of the process

Source: Author

After all these one-time conditions are done, the main loop is executed that performs the following functions periodically. The water level readings are measured and if the tank is not full, the Water pump and valve will turn on until the water level is completely full. Temperature is also measured periodically and if it is not within the range, the system will notify the user through the Adafruit IO Dashboard. The EC value is also continuously monitored and if it goes beyond the threshold values, water in the tank is replenished and the N, P, and K pumps are turned ON to provides the necessary nutrients to the solution. Finally, the pH is also measured repetitively. If the pH value is above the threshold pH value that is required by the plants, the pH is lowered down by adding an acidic solution. Similarly, when the pH value is below the threshold value, the pH value is increased by adding an alkaline solution. All these readings are periodically monitored and sent to the Adafruit IO Dashboard.

3.2 Firmware

The firmware of the proposed system was written in the Arduino IDE. The ESP8266 was programmed to perform various tasks like monitoring and controlling. The ESP8266 module Fig. 49.2(a) is used to connect the system to the internet by using any nearby router. The Wi-Fi credentials are dynamic and can be changed if the current WiFi router is not available or has no internet access. If the network connection is not present, the ESP8266 will act as a WiFi hotspot. The user can then connect to this hotspot and then enter the WiFi credentials of the new WiFi router. These new credentials are then stored in the memory of ESP8266 and from now, the ESP8266 will connect to these new WiFi router credentials. The dashboard for this IoT-based system was developed using the Adafruit IO platform, which uses the MQTT protocol to publish and sub- scribe to messages in order to fetch sensor data and display it on an interactive dashboard. When in Automatic mode, no user intervention is needed. The closed loop system can fetch the vital parameters of the system and can perform necessary actions if the values exceed the pre-defined thresholds. When the system is in Manual mode, the user can turn ON or turn OFF the air pump and peristaltic pumps. Similarly, the inlet and outlet valves of the system can also be turned ON or OFF manually using the interactive dashboard. The dashboard also has a provision for multiple sliders in order to manually adjust the intensity of Red, Green, and Blue colors to generate any

Table 49.2 A table depicting the percentages of N, P and K for specific pH values (Oberoi, 2017)

pH value	N	P	K
4.5	30%	23%	33%
5	53%	34%	52%
5.5	77%	48%	77%
6	89%	52%	100%
7	100%	100%	100%

color within the visible spectrum of lights. The system also has provision for Over-the-air (OTA) updates to upgrade the system to the latest firmware using the internet without the use of any external programmer.

3.3 Working

The proposed system has three major sections: data acquisition section, the controlling section, and the monitoring section.

The vital parameters of the system are fetched by the data acquisition system. These parameters include sensor readings like temperature, humidity, ambient light, and pH value. The water level sensor works on a capacitive principle. The electrodes are submerged in the water, and as the water level increases the capacitance increases. For accurate readings of humidity and temperature, a DHT sensor is used. The pH sensor also has two electrodes and the pH is calculated by the difference in the voltage between the internal electrode and the reference electrode. The TDS sensor outputs an analog value that corresponds to a particular Electrical Conductivity (EC) value. This EC value represents the quality of nutrients in the Water. The TDS sensor measurements are expressed in parts per million. The TDS value is directly proportional to the quality of water. The lower the value of TDS, the purer the water. Using this TDS value, the value of Electrical Conductivity (EC) is calculated which is expressed in micro Siemens/centimetre. This quantity decides the quality of water used in the system.

All these values are published by the ESP8266 using the MQTT protocol over WiFi. These published values are received and displayed on the interactive dashboard made using Adafruit IO. The user is able to monitor these values remotely from anywhere regardless of the location. The vital details like pH and temperature are also physically displayed on the OLED display. This comes under the monitoring section.

Further the controlling system has two options namely, automatic mode and manual mode. Automatic mode, these values are used to automate the functioning of the system. Firstly, the EC value of the system takes time to stabilise, thus the readings are only fetched twice a day. The TDS sensor was calibrated by comparing the readings of the TDS sensor to that of the TDS meter. Figure 49.6(b) plots the readings of the same. If the EC value measured by the system goes beyond a certain threshold it is not ideal for the efficient growth of plants. This situation is indicative that the water in the container needs to be replenished. In this case, the outlet valve is turned on for the water already present in the container to be drained out. Then the input valve and the water pump are switched on to pump fresh water back into the system. The pumping of the water continues until the water level sensor does not indicate that the water is filled to its full capacity. After replenishing the water peristaltic pumps are used to pump the N,P and K nutrients into the water from their respective bottles for the EC to reach its ideal value. Secondly, each plant has a specific pH requirement that needs to be maintained. A pH value that is too low liberates aluminium which in turn stunts root growth and interferes with the plant's uptake of nutrients. In this scenario the water needs to be made less acidic, this is done by adding an alkaline solution. This solution is added to the water in the main container by switching on a peristaltic pump. Whereas a pH value above the ideal range makes the micronutrients less soluble and unavailable for absorption by the plant roots. In this case, an acidic solution is added to the water using the similar method mentioned above. Since these newly added solutions take around two hours to blend well with the water the pH readings are taken every two hours for better accuracy.

The Fig. 49.4 (a) shows the calibration of the pH sensor using the readings of a pH meter. In addition to this, if the light intensity is too high or too low, it will result in reduced plant photosynthesis [5]. According to these values, the brightness of the RGB LEDs in the system is varied to maintain an ideal light intensity for the growth of the plants.

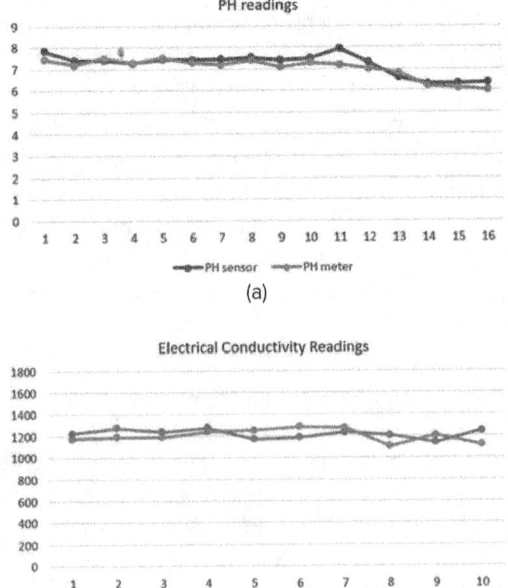

Fig. 49.4 (a) pH readings from the pH sensor and the pH meter from Day 1 to Day 16, (b) A comparison of EC meter readings to the EC sensor readings for calibration

Source: Author

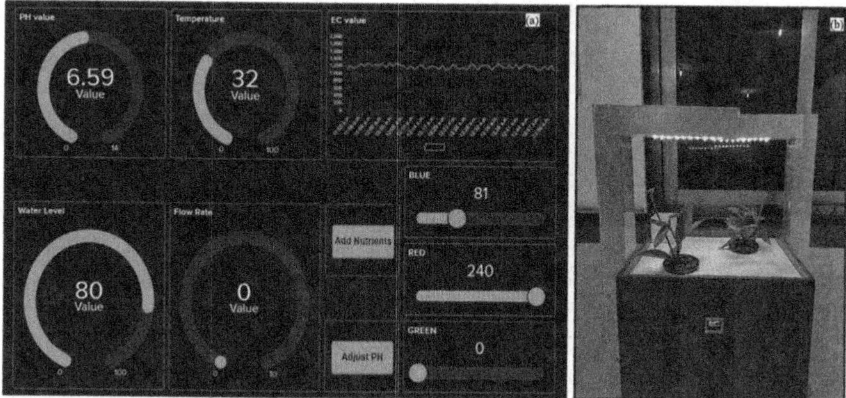

Fig. 49.5 (a) A screenshot of the Adafruit dashboard that depicts all the current parameters of the system and enables the user to control it remotely, (b) A photo of the complete ready and working system displayed in college premises

Source: Author

4. Results and Discussions

Once the system is turned on, the Esp8266 is powered and connected to a WiFi network; it establishes a connection with the Adafruit IO Dashboard. Figure 49.5(a) shows the User Interface of the dashboard. The LEDs were controlled by setting the R, G, and B slider values present on the dashboard. The nutrients were also manually supplied to the system by pressing the 'Add nutrients' button on the dashboard. The 'Adjust PH' switch adjusted the pH of the solution by measuring the current pH value and adding the acidic or alkaline solution accordingly. The water level gauge indicated the amount of water present in the container and the flow rate gauge displayed the rate at which the water flows into the container. The temperature gauge and the pH gauge displayed the real-time values of temperature and pH respectively. In order to analyse the pattern of electrical conductivity, we plotted the EC values in a graph. The electrical conductivity chart displayed the current and past values of Electrical conductivity. This can be seen in Fig. 49.5(a).

As stated in Table 49.1. The pH sensor was calibrated by using known solutions with the pH value of 4.0 and 7.0 and 10.0 as shown in Table 49.1. Since the pH sensor module converts the pH value into a certain voltage, the voltage corresponding to the pH value of the known solutions was calculated. By calculating the corresponding voltage and comparing the values with the instruction manual, the calibration factor for the pH sensor was calculated. The addition of this calibration factor improved the precision of the pH sensor.

The Fig. 49.4(a) plots the pH readings from the pH sensor and the pH meter from day 1 to day 16 of the experiment. The pH value of the water consistently and gradually decreases over time because of its reaction with the nutrients present. It was seen that there was a significant drop in the pH readings after day 14. For calculating the Electrical Conductivity of the solution, a TDS Sensor was used. The TDS sensor indicates the Total Dissolved Solids present in the solution. This value that is measured in ppm is then converted to Electrical conductivity by multiplying the TDS value by a certain calibration factor that also depends on the value of temperature. Thus, based on the readings of temperature and TDS, the final value of Electrical Conductivity is calculated in uS/cm. For calibration of Electrical conductivity, solutions having know EC value of 1413us/cm and 12.88us are used. The calibration factor is updated whenever needed, and stored in memory. Once the calibration stage was completed, the EC values were compared with a standard EC meter and the readings were recorded and printed on a graph as shown in Fig. 49.4 (b). The graph indicates that the calibration factor improves the precision of the TDS sensor and the readings are within the range of expected values of Electrical conductivity.

The thresholds of the system for its automation were set in reference to these values suggested in article (Fallovo, et al., 2009). All these values are predefined in the memory of the controller. While setting up the system a specific plant is chosen and thresholds are set accordingly. For e.g. if lettuce is chosen to be grown the range for pH values will be from 6 to 7 whereas the EC range will be 1.2 to 1.8.

Since hydroponics requires less space and less water, it is immune from soil-borne infections and infestations and ensures the proper supply of nutrients directly to the plant, it proves much more effective than traditional farming. The article by Fallovo (Fallovo, et al., 2009) justifies the same. It compares the yields of different crops grown by hydroponic methods and the ones grown by traditional agricultural methods. It can be observed from the literature that better results are obtained from hydroponic method as compared to traditional agricultural methods.

Our hydroponics system requires a one-time investment but is cheaper in the long run compared to the traditional methods which require recurring maintenance expenses. Our system supports the cultivation of a much more number of plants in a given area than a traditional agricultural system. Thus it proves to be cost-effective. If there is any damage to the hardware, all the components used in our system are easily available and thus easily replaceable. This plug-and-play system can be set up in less than 10 minutes. Once set up, this closed-loop system does not require any monitoring. Since the system does not require sunlight or soil and is very compact it can be employed anywhere with- out any constraints.

5. Conclusion

The study has tested an automation setup in hydroponics system. With the help of literature support and results, it can be concluded that the lab tested assembly developed for an IoT-based hydroponics system could monitor and automate the hydroponic agricultural process. It was observed that the monitoring and control mechanism of hydroponic system helped in continuous observation of the crop using various sensors. The sensors, actuators along with a control unit led to a small scale observation system for the regulated growth of the crop. The outcome was a scalable hydroponics system that has low-cost maintenance, controlled environment and ease of use. A scaled up automated hydroponic system can be used for further studies in urban farming.

Acknowledgement

Authors would like to acknowledge the management of Somaiya Vidyavihar University and K J Somaiya College of Engineering, for supporting this Research Work.

References

1. Aliac, C.J.G., Maravillas, E.: Iot hydroponics management system. In: 2018 IEEE 10th International Conference on Humanoid, Nanotechnology, Information Technology, Communication and Control, Environment and Management (HNICEM), pp. 1–5 (2018). DOI 10.1109/HNICEM.2018.8666372
2. Eridani, D., Wardhani, O., Widianto, and E.D.: Designing and implementing the arduino-based nutrition feeding automation system of a prototype scaled nutrient film technique (nft) hydroponics using total dissolved solids (tds) sensor. In: 2017 4th International Conference on Information Technology, Computer, and Electrical Engineering (ICITACEE), pp. 170–175 (2017). DOI 10.1109/ICITACEE.2017.8257697
3. Harikrishna, R.B., R, S., N, P.P., Anand Kumar A, A., Pandiaraj, S.: Greenhouse automation using internet of things in hydroponics. In: 2021 3rd International Conference on Signal Processing and Communication (ICPSC), pp. 397–401 (2021). DOI 10.1109/ICSPC51351.2021.9451668
4. Oberoi, A., Basavaraju, S., Lekshmi, S.: Effective im- plementation of automated fertilization unit using ana- log ph sensor and arduino. In: 2017 IEEE International Conference on Computational Intelligence and Computing Research (ICCIC), pp. 1–5 (2017). DOI 10.1109/ICCIC.2017.8524170
5. Promratrak, L.: The effect of using led lighting in the growth of crops hydroponics.

Int. J. Smart Grid Clean Energy **6**(2), 133–140 (2017)

6. Saaid, M.F., Sanuddin, A., Ali, M., Yassin, M.S.A.I.M.: Automated ph controller system for hydroponic cultivation. In: 2015 IEEE Symposium on Computer Applications Industrial Electronics (ISCAIE), pp. 186–190 (2015). DOI 10.1109/ISCAIE.2015.7298353
7. Saputra, R.E., Irawan, B., Nugraha, Y.E.: System de- sign and implementation automation system of expert system on hydroponics nutrients control using forward chaining method. In: 2017 IEEE Asia Pacific Conference on Wireless and Mobile (APWiMob), pp. 41–46 (2017). DOI 10.1109/APWiMob.2017.8284002
8. Saraswathi, D., Manibharathy, P., Gokulnath, R., Sureshkumar, E., Karthikeyan, K.: Automation of hydroponics green house farming using iot. In: 2018 IEEE International Conference on System, Computation, Automation and Networking (ICSCA), pp. 1–4 (2018). DOI 10.1109/ICSCAN.2018.8541251
9. Sardare, M.: A review on plant without soil - hydroponics. International Journal of Research in Engi- neering and Technology **02**, 299–304 (2013). DOI 10.15623/ijret.2013.0203013
10. Sharma, N., Acharya, S., Kumar, K., Singh, N., Chaurasia, O.: Hydroponics as an advanced technique for vegetable production: An overview. Journal of Soil and Water Conservation **17**, 364–371 (2019). DOI 10.5958/2455- 7145.2018.00056.5
11. M. K. Hafizi Rahimi, M. H. Md Saad, A. H. Mad Juhari, M. K. Azhar Mat Sulaiman and A. Hussain, "A Secure Cloud Enabled Indoor Hydroponic System Via ThingsSentral IoT Platform," 2020 IEEE 8th Conference on Systems, Process and Control (ICSPC), Melaka, Malaysia, 2020, pp. 214-219, doi: 10.1109/ICSPC50992.2020.9305792. keywords: {Sensors; Temperature sensors; Security; Authentication; Humidity; Monitoring; Agriculture; IoT; 2FA; Cloud Based System; Indoor Hydroponics System}, Fallovo, C., Rouphael, Y., Rea, E., Battistelli, A. and Colla, G. (2009), Nutrient solution concentration and growing season affect yield and quality of *Lactuca sativa* L. var. *acephala* in floating raft culture. J. Sci. Food Agric., 89: 1682–1689. https://doi.org/10.1002/jsfa.3641

Technologies for Energy, Agriculture, and Healthcare – Shailesh Nikam et al. (eds)
© 2024 Taylor & Francis Group, London, ISBN 978-1-032-98028-7

50

Healthcare Transformation—A Comparative Analysis of the Emergence of AI in the Healthcare System

Meghna Vyas[1]
Assistant Prof.
Dept of Mathematics and Statistics,
Somaiya Vidyavihar University

Shaikh Mohammad Bilal Naseem[2]
Assistant Prof.
Dept of Computer Science/IT,
Somaiya Vidyavihar University

Shah Asif Mohd Saleem[3]
Department of Information Technology,
Somaiya Vidyavihar University

Abstract: This research explores how Artificial Intelligence (AI) has changed healthcare services and its operations. This study aims at focusing on healthcare before and after AI, focusing on how things changed, the problems, and the good parts of using AI. We have carefully checked existing studies and analyzed healthcare situations before and after AI to understand the connection between healthcare and AI better. We study how AI affects healthcare systems, especially how it evolves, what issues come up, and the benefits it brings. This is done by looking at information in books and articles and closely examining situations in healthcare before and after AI was introduced. Our goal is to give useful information about how AI and healthcare work together. The primary objective is to understand the impact of AI on healthcare by looking at the past and present. A review of what we already know from other studies and study how healthcare was

[1]meghna.vyas@somaiya.edu, [2]mohammadbilal@somaiya.edu, [3]asif.shah@somaiya.edu

DOI: 10.1201/9781003596707-50

and is after AI. A detailed process to look at five important aspects of healthcare shortlisted by this study were - accuracy in diagnosis, planning treatments, using resources, involving patients, and overall effectiveness. Our study compares these aspects before and after AI to see the changes. We explore what it means for healthcare workers, policymakers, and everyone involved. The research shows the good things AI brings to healthcare, acknowledges the challenges, and suggests areas for more research to make AI and healthcare even better.

Keywords: Healthcare, Artificial intelligence, AI, Machine learning

1. Introduction

Our journey into understanding the impact of Artificial Intelligence (AI) in healthcare begins by recognizing the increasing importance of AI in this field. We aim to grasp what this means for healthcare as a whole. The introduction serves as a guide, shedding light on the primary objectives of our study and underscoring the transformative nature of AI technologies. Healthcare is undergoing significant changes, and AI is at the forefront of this transformation. The introduction underscores the growing relevance of AI in healthcare settings, emphasizing the need to comprehend its implications. As we delve into the study, our focus is on unraveling the consequences of integrating AI into healthcare practices. The primary objectives of our research are unveiled in this introduction, outlining the key areas where AI is making a substantial impact on healthcare. It is crucial to understand the profound changes AI brings to the healthcare landscape. Our study strives to provide insights that contribute to a broader understanding of how AI transforms healthcare practices. As we navigate the landscape of healthcare transformation, the spotlight is on AI and its burgeoning role. The introduction highlights the importance of comprehending the implications of AI in healthcare, acting as a compass for our research. The primary objectives guide our exploration, stressing the transformative nature inherent in the integration of AI technologies. Our study is propelled by the recognition that AI is reshaping healthcare in fundamental ways. The introduction emphasizes the growing significance of AI in this domain and acts as a gateway to the main objectives of our research. It places a spotlight on the transformative nature intrinsic to AI technologies in the healthcare sector. In addition to understanding the transformative impact, our research also aims to explore the challenges and benefits brought about by AI in healthcare. By systematically analyzing existing literature and conducting a detailed examination of healthcare scenarios before and after AI implementation, we seek to provide a comprehensive understanding of the dynamic intersection between healthcare and AI. Our focus is on simplifying the complex changes happening in healthcare due to AI, making it accessible for everyone to comprehend. Through our study, we hope to contribute valuable insights that will help individuals, healthcare professionals, and policymakers better navigate the evolving landscape of AI in healthcare.

2. Problem Definition

This study aims to analyze and grasp how AI influences healthcare, exploring its impact on the way healthcare operates and delivers services.

3. Literature Review

3.1 Deep Learning for Healthcare: Opportunities and Challenges by Andre Esteva et al. (2019)

In this research study the authors have explored applications of deep learning in medical diagnosis, drug discovery, patient monitoring, and more. demonstrates superior performance of deep learning algorithms compared to traditional methods in specific tasks. Data privacy concerns and potential for bias in algorithms require careful consideration. Interpretability of deep learning models can be challenging, hindering trust and acceptance by healthcare professionals. Technical expertise and computational resources needed for successful implementation pose barriers. In conclusion: deep learning holds immense promise for revolutionizing healthcare by improving accuracy, efficiency, and affordability. However, challenges like data availability, bias, and interpretability need to be addressed for responsible and successful implementation.

3.2 Artificial Intelligence in Medicine: Current Trends and Future Possibilities by Xiaobai Li et al. (2020)

Provides a comprehensive overview of AI applications in medical diagnosis, drug discovery, and personalized medicine. Discusses successful examples of AI-powered systems for early disease detection, treatment optimization, and genomic analysis. Emphasizes the need for collaboration between AI developers and healthcare professionals. Limitation and drawbacks we found while studying this research paper are ethical considerations regarding data ownership, transparency, and accountability need to be addressed. Lack of standardized methods for validation and clinical integration of AI systems. To conclude this paper we would like to address AI has the potential to significantly improve clinical decision-making, personalize healthcare, and advance medical research. Continuous research and development are crucial to address ethical concerns, data privacy issues, and regulatory challenges.

3.3 The Potential for Artificial Intelligence in Healthcare by Eric Topol (2019)

Explores the transformative potential of AI in healthcare across diagnosis, treatment, and prevention of diseases. Discusses AI-powered tools for early detection of cancers, personalized medication plans, and virtual patient assistance. Emphasizes the need for human-centered design and responsible development of AI in healthcare. Potential for job displacement in the healthcare workforce needs to be addressed with training and adaptation. Accessibility and affordability of AI-powered technologies might exacerbate existing healthcare disparities. Over Reliance on AI could lead to de-skilling of healthcare professionals and loss of critical human judgment. AI can significantly improve healthcare outcomes, but its successful implementation requires careful consideration of ethical, social, and economic implications.

3.4 A Review of the Role of Artificial Intelligence in Healthcare by M. Arif et al. (2020)

Provides a comprehensive review of AI applications in medical imaging, virtual patient care, drug discovery, and patient engagement. Discusses successful examples of AI-powered systems for image analysis, remote patient monitoring, and chatbots for health education. Emphasizes the importance of data security, privacy, and transparency in AI-driven healthcare. Lack of interoperability between different AI systems can hinder. AI has the potential to improve access to care, optimize resource allocation, and empower patients, but ethical considerations and regulatory frameworks are essential for responsible implementation.

3.5 Research Methodology

In order to describe or analyze current events or situations, the study uses a descriptive research strategy and a variety of statistical metrics and tools, including frequency, graphs, comparison charts, and t-statistics. The analysis of the advantages for patients in having their health issues resolved by AI and other similar technology in healthcare services is the main topic. Participants in the current study include anyone who has undergone surgery or dealt with health concerns of any kind. People in Mumbai, Maharashtra, provide the primary data source. 160 completed responses to the structured questionnaire were gathered after it was distributed at random using Google Forms.

3.6 Sample Design

The study's population consisted of people from Maharashtra state, including a sample of people living in the Mumbai region.

3.7 Research Technique

The study employed a simple random sampling approach.

4. Hypothesis

H0: Majority of people are not satisfied with healthcare services after digitization mainly using AI and related technologies.

H1: Majority of people are highly satisfied with healthcare services after digitization mainly using AI and related technologies.

5. Healthcare Before Integration of AI

Prior to the AI revolution in healthcare, the sector struggled with time-consuming manual diagnostic procedures, finite analytical power, difficult distribution of resources, a reactive healthcare structure, and limited patient involvement.

1. **Manual Diagnostic Processes**

   Before the AI revolution, healthcare heavily relied on manual diagnostic processes, which were time-consuming and prone to human errors. Professionals faced challenges in efficiently managing patient data, leading to delays in treatment.

2. **Limited Analytical Capabilities**

   The absence of advanced analytics tools hindered healthcare professionals' ability to derive meaningful insights from patient data. This limitation affected the customization of treatment plans and personalized care.

3. **Resource Allocation Challenges**

   Healthcare facilities struggled with suboptimal resource allocation and scheduling. Decisions were often based on historical data and subjective assessments, leading to inefficiencies in the use of both facilities and personnel.

4. **Reactive rather than Proactive Healthcare**

   Without AI, healthcare systems operated in a reactive manner, lacking the ability to proactively address emerging health trends or potential outbreaks. Preventive measures and early interventions were constrained.

5. **Limited Patient Engagement and Information Exchange**

   Patient engagement was primarily face-to-face, and access to medical information was restricted. The traditional model struggled to meet the increasing demands for more efficient, precise, and patient-centric services.

6. Healthcare After Integration of AI

After the AI revolution, healthcare witnessed transformative advances, with enhanced precision diagnostics, predictive analytics enabling personalized medicine, optimized resource allocation, a proactive healthcare model, and improved patient engagement through AI-driven technologies. This revolution has reshaped the healthcare landscape, offering more efficient, accurate, and patient-centric services.

1. **Precision Diagnostics**

   With the AI revolution, healthcare experiences a significant advancement in diagnostics. AI-driven technologies, such as machine learning algorithms, enhance the accuracy and speed of diagnostics, aiding healthcare professionals in detecting diseases and abnormalities more efficiently.

2. **Predictive Analytics and Personalized Medicine**

   AI enables predictive analytics, allowing healthcare providers to anticipate health trends and personalize treatment plans based on individual patient data. This shift towards personalized medicine ensures tailored and more effective healthcare interventions.

3. **Optimized Resource Allocation**

   AI contributes to optimized resource allocation by streamlining administrative tasks, improving workflow efficiency, and enhancing predictive models for patient admission rates. This results in better utilization of healthcare facilities and personnel, ultimately improving overall healthcare management.

4. **Proactive Healthcare Management**

   The integration of AI fosters a proactive approach to healthcare. Predictive modeling and data analytics help in identifying potential health risks and intervening early, contributing to preventive care strategies and reducing the burden on the healthcare system.

5. **Enhanced Patient Engagement**

   AI technologies facilitate improved patient engagement through personalized health monitoring, virtual health assistants, and telemedicine solutions. Patients have increased access to health information and can actively participate in managing their well-being, fostering a more collaborative and patient-centric healthcare approach.

7. Data Analysis

Primary data were gathered, a systematic questionnaire was created, and 160 people from Mumbai, Maharashtra, between the ages of 20 and 65, were questioned (81 females and 79 males), in order to validate the secondary data of healthcare systems both before and after AI integration. Subsequent to the initial binary questions, queries concerning the type of healthcare facilities (public or private), the degree of familiarity with healthcare digitization through artificial intelligence (AI) and related technologies, health concerns, surgeries performed prior to and following AI integration, and the degree of satisfaction with healthcare facilities prior to and following digitization were posed.

The demographic information presented in Table 50.1 includes age group, education level, preferred healthcare facilities, and awareness of healthcare digitization through artificial intelligence and other technologies. It was discovered that the majority of respondents visit private healthcare facilities, and nearly half of them are aware of the digital health sector.

Table 50.1 Demographic data

Parameter	Category	Frequency
Age group in years	Below 25	80
	25-35	23
	35-45	24
	45-55	22
	55-65	7
	Above 65	4
Education	Primary	0
	High School	5
	Undergraduate	62
	Graduate	22
	Post graduate	49
	Ph.D.	19
	Any other	3
Visiting health sector	Private	144
	Public	16
Familiar about healthcare digitization using AI and other technologies	Yes	74
	No	86

The responses regarding respondents' awareness of different digital health care services are shown in Fig. 50.1. Of the 160 respondents, 38 knew about telemedicine, 63 about electronic health records, 123 knew about mobile health applications, and 43 knew about AI-based healthcare.

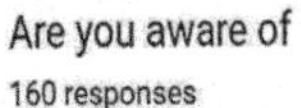

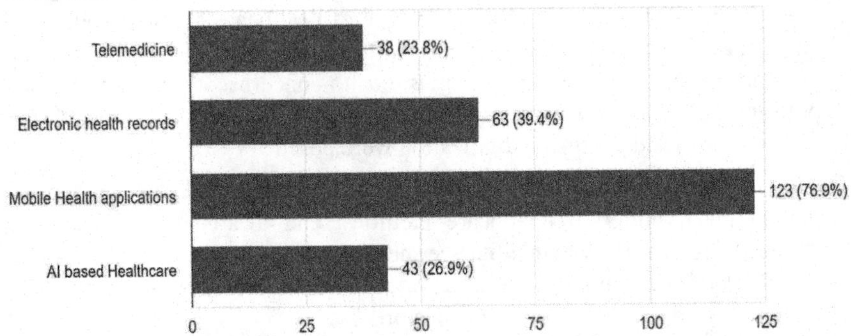

Fig. 50.1 Awareness of type of digitized healthcare services

Further various problems faced for any kind of health issues before digitization and were resolved after digitization were asked, Figure 50.2 compares the health issues respondents faced prior to digitization with the issues that were resolved following the process. Based on the data submitted by respondents, we can conclude that issues such as limited access to healthcare facilities, inadequate road connectivity and transportation, high risk of communicable diseases, lengthy wait times in lines, and delays in receiving reports such as CBC, X-ray, CT scan, etc. were resolved following the digitization of healthcare services.

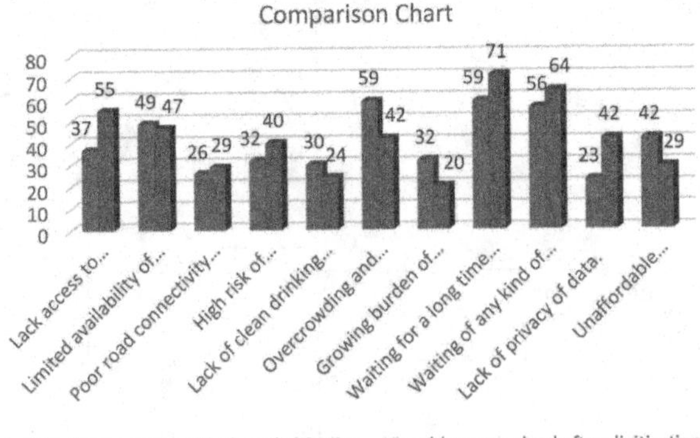

Fig. 50.2 Comparison chart of health issues faced before and resolved after digitization of healthcare services

After knowing about the health issues, the next questions were about surgery issues faced before digitization and resolved after digitization.

Figure 50.3 presents a comparison between the surgical-related problems that respondents encountered before digitization and the problems that were fixed after the procedure. We can infer from the data provided by the respondents that the problems including longer hospital stays, cuts on body parts, more stitches, greater pain, more blood loss, and longer periods of unconsciousness were fixed when healthcare services were digitized.

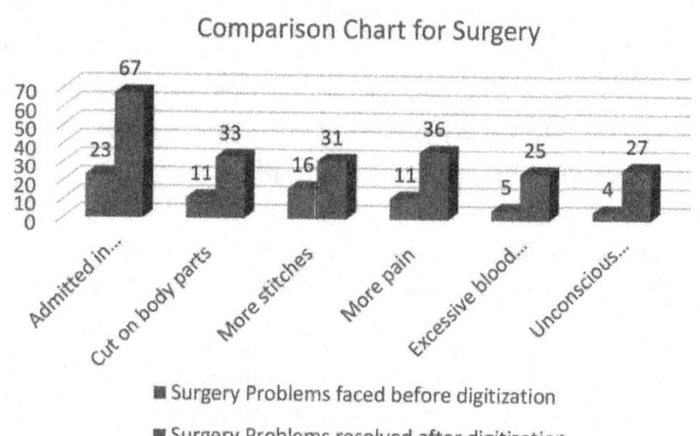

Fig. 50.3 Comparison chart of surgery issues faced before and resolved after digitization of healthcare services

Based on the data presented in Figures 50.2 and 50.3, it can be inferred that, in comparison to health difficulties, surgical issues were more successfully resolved following the digitalization of healthcare services through the use of artificial intelligence and similar technologies.

At the conclusion, a t-test was utilized to assess the hypothesis further. Of the 160 respondents, 93 expressed satisfaction with the healthcare services following digitization.

$$H0: p0<75$$

$$H1: p0>=75$$

$$p0= 0.75$$

$$q0=1-p0= 1-0.75=0.25$$

$$p= 93/160 = 0.58$$

$$t = p - p0 / \mathrm{sqrt}\ (p0q0/n)$$

$$= 0.58 - 0.75 / \mathrm{sqrt}\ ((0.75x0.25)/160)$$

$$= - 0.17 / 0.034$$

$$= -5$$

Absolute value of t= 5 > 2.58

Hence at 1% level of significance H0 is rejected.

This proves the hypothesis that Majority of people are highly satisfied with healthcare services after digitization mainly using AI and related technologies.

8. Conclusion

According to the research, apps for digital healthcare were appropriate for all kinds of patients. Based on the responses provided, it can be inferred that while there are still some unresolved healthcare issues, surgical issues have been resolved to a great extent, and as a result, people are happier with the healthcare services provided by AI and related technologies. However, they would prefer more advanced technologies in the future, particularly those that improve personal productivity, reduce healthcare costs, speed up diagnosis, and enhance healthcare efficiency.

Based on the data, healthcare would be the most impacted e-service in digital India in terms of higher accountability, consistent interpretation, and better information access. Using the best-designed technologies, digitizing e-health information is an aggressive goal to cogently target large numbers of patients across vast geographic areas in India. The acceptance and strategy of digital health efforts hinge on providing the appropriate information to the appropriate individual at the appropriate time and location.

Digital healthcare encompasses more than just technology; it also involves innovative approaches to problem-solving, personalized patient care, and the rapid expansion of healthcare providers. Digital healthcare will demonstrate the value for patients, physicians, and healthcare systems by assisting system-wide procedures and institutions in providing more effective services for a longer amount of time. The enormous field of digital health will only expand, grow, and change further, elevating the significance of healthcare globally.

References

1. Ayden kesarili et al (2022), "Evaluation of digital healthcares services and satisfaction level of outpatients at the city training and research hospital in Turkey during Covid-19 pandemic", journal of international health sciences and management, 8(15), 47–57.
2. Avinash kumar et al (2020), "Aspects of digitization of healthcare in India", International journal of research and analytical reviews, 7(1), 404–407.
3. Craig J and Patterson V. (2005), "Introduction to the Practice of Telemedicine." Journal of Telemedicine and Telecare.11(1): 3–9.
4. Lupton D. (2013), " The digitally engaged patient: Self-monitoring and self-care in the digital health era." , Social Theory & Health 11(3): 256–270.

5. Marc Mitchell and Lena Ken (2019), "Digital technology and the future of health systems", Health system and reform journal, Taylor & Francis,
6. Patrick K et al (2008),." Health and the Mobile Phone." American Journal of Preventive Medicine 35(2): 177–181
7. Dörfler, Viktor. "AI in Medical Diagnosis: AI Prediction & Human Judgement." Artificial Intelligence in Medicine, 149: 102769 (2024): n. pag. Web.
8. Coghlan, Simon. "Readying Medical Students for Medical AI: The Need to Embed AI Ethics Education." arXiv (Cornell University) (2021): n. pag. Print.
9. Hatherley J. (2020). Limits of trust in medical AI. Journal of Medical Ethics. https://doi.org/10.1136/medethics-2019-105935
10. SPARROW R., and HATHERLEY J. (2019). "The promise and perils of AI in Medicine.", International Journal of Chinese and Comparative Philosophy of Medicine, 17(2), 79–109. https://doi.org/10.24112/IJCCPM.171678

Note: All the figures and table in this chapter were made by the authors.

Technologies for Energy, Agriculture, and Healthcare – Shailesh Nikam et al. (eds)
© 2024 Taylor & Francis Group, London, ISBN 978-1-032-98028-7

51

Design of Efficient CNN Architecture for Rice Disease Classification with Comparative Study for Impact Analysis of Batch Size

Sanam Kazi[1]

Research Scholar,
K. J. Somaiya College of Engineering,
Somaiya Vidyavihar University

Bhakti Palkar[2]

Associate Professor,
K. J. Somaiya College of Engineering,
Somaiya Vidyavihar University

Abstract: Deep learning architectures have drawn attention of researchers towards the challenge for reducing the training time. But still huge training time needed for deep learning models. The deep learning architectures tend to overfit. An attempt has been made to overcome overfitting by proposing novel Convolutional Neural Network (CNN) design with three convolutional layers and one dense layer with 500 neurons for rice disease classification. The proposed architecture contains the max pooling layer and the batch normalization layer added after each convolution block and two dropout layers to achieve good accuracy and prevent overfitting. To explore the impact of batch size on training time and accuracy, proposed novel CNN architecture is trained with batch size of 32, 64 and 128. The CNN model trained with the batch size of 64 achieves the highest training accuracy of 98.11% with minimum training time of 563.48 seconds as compared to the models trained with the batch size of 32 and 128. It is realized through experimentation that proper batch size improves the model

[1]sanam.kazi19@gmail.com, [2]bhaktiraul@somaiya.edu

DOI: 10.1201/9781003596707-51

accuracy and training time. The novel architecture proposed with its optimal hyperparameters prevents Overfitting.

Keywords: Batch size, Transfer learning, Training time, Rice disease identification

1. Introduction

Asia contributes for 90% of world rice production as reported by Muthayya et al. (2014). India is the developing country where the agriculture plays very important role in life of a large section of people. Many times, it becomes tedious for farmer to understand the manifestation of the disease on the rice plant. The yield loss due to bacteria, fungus is causing economic loss to the farmer and the nation. Approximately 19% of domestic economy is contributed by agriculture as per study by Krishnamoorthy N. et al. (2021). The application of technological innovations at the rice field can aid and assist framers. Deep learning in research has shown tremendous results and potential (LeCun et al., 2015). CNN has made vital contribution in classification task as it has eliminated the need for explicitly identifying features. CNN extracts the features efficiently (Tunio et al., 2021; Singh N et al., 2023). Bhattacharya A. (2021) has reported problem of overfitting by CNN models in his study.

This study proposes the novel architectural design for CNN for classifying of rice diseases Bacterial blight, Blast, Brown Spot, Tungro and healthy leaf. The challenges of overfitting and training time are overcome through use of batch normalization and dropout layers with optimal hyperparameters in the proposed architecture. The batch size is explored by training the proposed model with different values and then performing comparative analysis on accuracy and training time of proposed CNN architecture. The literature review highlights the technique used by researchers for rice disease classification along with their results. The methodology is discussed in next section along with dataset and implementation details. Further section discusses the comparative result analysis for CNN models implemented using different batch sizes. Lastly the study is concluded.

2. Literature Review

2.1 Review of CNN Models in the Literature

Krishnamoorthy N. et al. (2021) in their study implemented inceptionResNetv2 based on CNN for rice disease detection with accuracy of 95.67%. They collected three diseases and one healthy class from Kaggle containing 5200 images. Vimal

K et al. (2019) have performed classification of three diseases using pretrained CNN and machine learning classifier on 619 images only. They observed accuracy of 91.37% with split of 80, 20 as compared to 60, 40 and 70,30 split of dataset. Muhammad Hanif Tunio et al. (2021) used only 119 images and implemented a hybrid CNN classification model. They predicted for three diseases with accuracy of 90.8%. Singh S. et al. (2021) conducted study to classify four rice diseases by proposing CNN model. 7332 images were collected from the field of Orissa and Manipur for experimentation. They conducted tests with different kernel size of convolution layer but didn't witness any significant change in performance. The highest accuracy 99.83 was obtained by model with optimizer Adam but with the total time of 4 hour 53 minutes which is very large. Prabira Kumar et al. (2020) investigated the 11 CNN models and found ResNet50 plus SVM as higher rank performing classification model. It attained the f1 score of 0.9838. Bhattacharya A. (2021) used transfer learning for classification of five rice disease classes. The author used DenseNet 201 as pretrained model followed by customized architecture. The experiments were run for 15 epochs and noted training accuracy of 97.44 %. The models Inceptionv3, Resnet 50 and DenseNet121 were overfitted after 15th epoch in his study. Chen J. (2022) proposed light weight model for disease detection. The inception block was modified to achieve accuracy of 99.21%. The depth wise separation was employed. Rakesh M. et al. (2022) have reviewed the papers from 2007 to 2017 comparing CNN transfer learning models. The preprocessing, augmentation techniques are reviewed and the highest accuracy of 97% is reported by the VGG19 model in his comparative study. Prajapati et al. (2017) employed image segmentation and extract the diseased portion using K means. SVM was used for classification of three diseases obtaining the training accuracy of 93.33%. they could raise the testing accuracy to 88.57% with cross fold validation. Yakkundimath et al. (2022) used VGG16 and GoogleNet for rice disease classification achieving accuracy of 92.24% and 91.28% respectively.

It was been observed through studies that CNN though used in classification of rice diseases but its requirement of huge data, accuracy and training time is still the challenge. The accuracy reported in some studies is after few epochs but it may degrade with increasing epochs. The impact of batch size is not discussed in any of the above studies. Hence this work is novel as it explores impact of training CNN architecture with different batch sizes on the accuracy and training time.

3. Methodology

3.1 Dataset and Training Details

The images with rice disease symptoms for this study is downloaded from the online dataset of four rice diseases Rice Blast, Bacterial Blight, Tungro and Brown

Spot from Mendeley named as "Rice Leaf Disease Image Samples" created by Sethy et al. (2020). The healthy images for rice are taken from Mendeley. The total images in the dataset for five classes are 6313. There are 4456 images for training and 1857 for test. The validation ratio is 20%. The images are 224 $*$ 224 and are resized to 100 $*$ 100. The dataset prepared for this study is balanced dataset. Further images are augmented to generate 20 images per batch size.

3.2 Proposed CNN Design

The study proposed simple improved novel CNN design for rice disease classification as seen in the figure below. As compared to transfer learning models this design is not very deep but number of filters used is more. It consists of three convolution layers each having 100 filters with size 5 $*$ 5. The max pooling helps to find information and reduces the size. In this design the filter size of 2 $*$ 2 is convolved with image size of 100 $*$ 100. Similarly, batch normalization is use to prevent overfitting but, in this design, placing it in each convolutional block after max pooling is chosen to make proposed CNN model efficient and accurate. The hyperparameters applied to the model are based on experimentation. The dropout layer eliminates 50% of the neurons eliminating chances of overfitting as preventive design measure. Softmax function and 0.001 learning rate is applied.

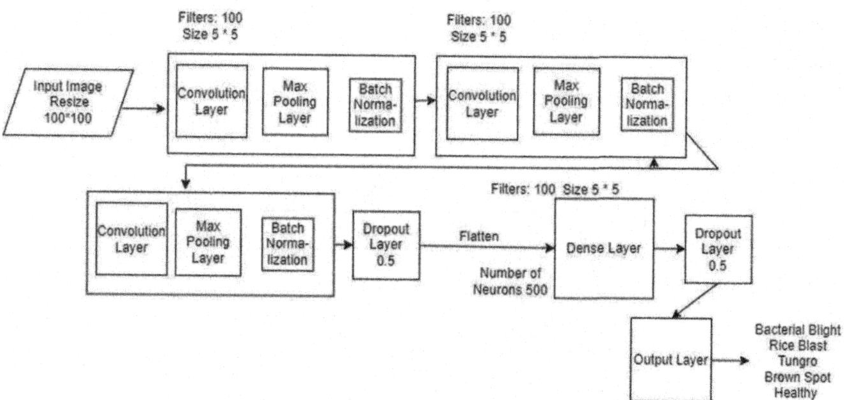

Fig. 51.1 Design of proposed novel CNN model

3.3 Implementation Details

Google Collaboratory (COLAB) is used as platform to create and execute the code for implementation. The GPU is selected for runtime environment as this is the basic requirement of CNN model. The code for rice disease classification using proposed CNN design is implemented using python. The libraries of Tensor flow, Sklearn are used. The high-speed internet of 100 mbps is needed to connect and execute the code.

3.4 Performance Evaluation with Different Batch Size

This study through empirical research carefully observes the effect of batch size on CNN architecture. The batch size of 32, 64 and 128 are selected for training the proposed CNN model. It can be seen that the training accuracy is higher with batch size of 64 from the beginning of the epoch and maintained till 100th epoch in Fig. 51.2. The comparative analysis below shows that CNN model trained with batch size 64 showed the greatest training and testing accuracy from 20th epoch as compared to models trained with batch size 32 and 128. No fitting is observed in the figure below.

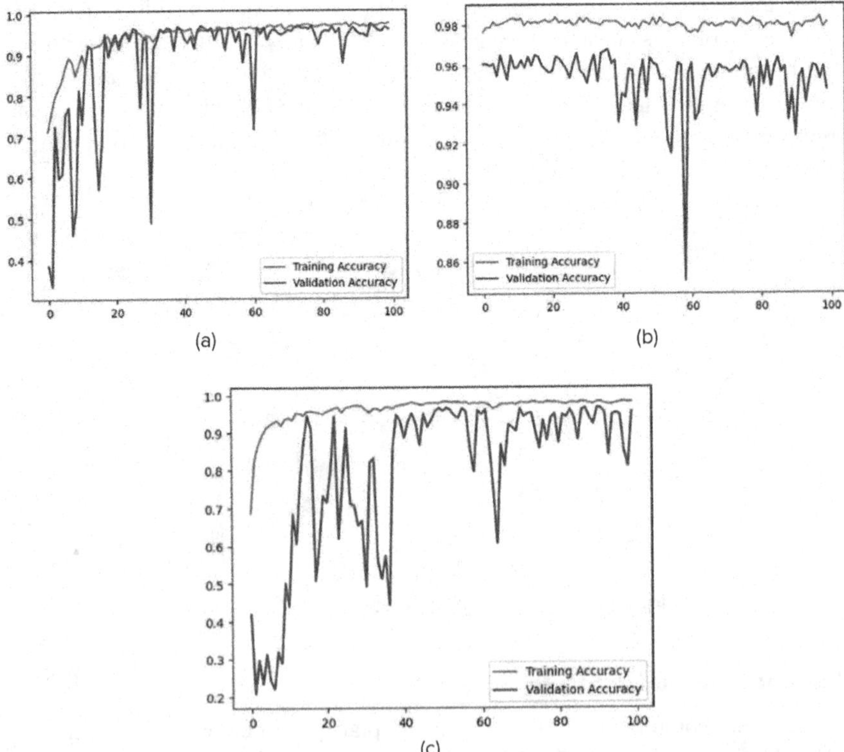

Fig. 51.2 Training accuracy with batch size of (a) 32 (b) 64 (c) 128

4. Empirical Results

4.1 Results with K-Fold Cross Validation

To evaluate the result and performance of the proposed novel CNN architecture with fine-tuned hyperparameters 10-fold cross validation is applied. The model is trained for 100 epochs for 10 folds. Following is the result obtained in Fig. 51.3 below for each fold. The mean accuracy after 10 folds is 92.28%. It indicates that the model is performing well with the proposed architecture.

```
array([0.91704036, 0.94843049, 0.93049327, 0.93946188, 0.83408072,
       0.9103139 , 0.95505618, 0.91011236, 0.94831461, 0.94157303])
```

Fig. 51.3 Accuracies using 10 fold cross validation

4.2 Comparative Result Analysis with Batch Sizes 32, 64 and 128

Table 51.1 shows the performance metrics calculated, compared and evaluated for the proposed CNN model when trained with batch size 32, 64 and 128.

Table 51.1 Result summary

Batch Size	Training Accuracy %	Testing Accuracy %	Precision	Recall	F1 Score	Training Time (s)
32	97.49	95.34	0.94557	0.94453	0.9449	625.376
64	98.11	95.53	0.95593	0.95530	0.95491	563.486
128	98.03	98.00	0.95201	0.95207	0.95191	565.871

It can be seen that when the batch size is changed from 32 to 64 for training it improves all the performance metrics. Also, it reduces the training time to 563.486 seconds which is minimum as compared to other batch sizes under consideration. When the batch size is further increased to 128 it hardly improves the model performance. It increases the training time. Hence it can be concluded that batch size of 64 is a good choice for training the proposed CNN architecture. Also loss is highly reduced as seen in the Fig. 51.4a.

Out of total 1857 testing images the correct predictions given by three models with batch size 32, 64, 128 are 1754, 1774, 1768 respectively as seen in confusion matrix. Labels 0, 1, 2, 3, 4 represent Bacterial Blight, Rice Blast, Brownspot, Tungro, Healthy respectively. Batch size 64 is outperforming other batch sizes under study.

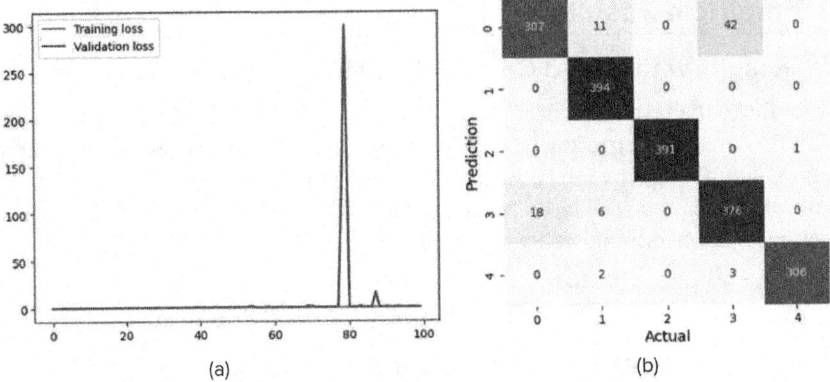

(a) (b)

Fig. 51.4 Best performance using batch size 64 (a) Loss graph (b) Confusion matrix

4.3 Comparison of Result with Existing Models

After comparing the proposed CNN model with the existing models following is the Table 51.2 for comparative analysis. It can be seen that the proposed CNN model is achieving highest accuracy of 98.11% when compared with available models. Hence the proposed CNN architecture and the hyperparameters along with batch size 64 are achieving encouraging results.

Table 51.2 Comparison with existing models

Author	Year of Publication	Technique used	Accuracy %
Yakkundimath, R. et al	2022	CNN	91.28
Debnath O. et al.	2022	CNN	97.7
Rahman, C. R et. al	2020	CNN	93.3
Hossain, S. M. M. et al.	2020	CNN	97.82
Hasan M. J et al.	2019	CNN	97.5
Mique Jr, E. L., & Palaoag, T. D	2018	CNN	90.9
Lu, Y., Yi, S., Zeng, N., Liu, Y., & Zhang, Y.	2017	CNN	95.48
		Proposed CNN	98.11

5. Conclusion

CNN model for rice disease classification needs huge training time and may overfit as reported in some studies. The batch size is not explored deeply in the studies in the literature. This study has proposed a novel CNN model for rice disease classification with improved accuracy and reduced training time. The proposed enhanced design for CNN having max pooling and batch normalization in each block of convolution layer helps to prevent overfitting. The dropout layer placed before and after dense layer further eliminates unwanted features making the designed improved architecture. The impact of batch size on accuracy and training time is analysed by training the proposed CNN models with batch sizes 32, 64, 128. The hyperparameters applied optimizes the proposed CNN model's performance. The empirical analysis shows that the training time reduces when model is trained with the batch size 64 as compared to 32 and 128. Batch size 64 has given maximum correct predictions achieving highest training accuracy of 98.11% and minimum training time of 563.486. The precision, recall and f1 score are highest with batch size 64 making it the most suited hyperparameter for creating optimized CNN model. The correct design decisions for batch normalization and dropout in the proposed architecture has enhanced the CNN architecture preventing overfitting.

References

1. Muthayya, S., Sugimoto, J. D., Montgomery, S., & Maberly, G. F. (2014). An overview of global rice production, supply, trade, and consumption. Annals of the new york Academy of Sciences, 1324(1), 7–14.
2. Krishnamoorthy, N., Prasad, L. N., Kumar, C. P., Subedi, B., Abraha, H. B., & Sathishkumar, V. E. (2021). Rice leaf diseases prediction using deep neural networks with transfer learning. Environmental Research, 198, 111275.
3. LeCun, Y., Bengio, Y., & Hinton, G. (2015). Deep learning. nature, 521(7553), 436–444.
4. Shrivastava, V. K., Pradhan, M. K., Minz, S., & Thakur, M. P. (2019). Rice plant disease classification using transfer learning of deep convolution neural network. The International Archives of the Photogrammetry, Remote Sensing and Spatial Information Sciences, 42, 631–635.
5. Tunio, M. H., Jianping, L., Butt, M. H. F., & Memon, I. (2021, December). Identification and classification of rice plant disease using hybrid transfer learning. In 2021 18th International computer conference on wavelet active media technology and information processing (ICCWAMTIP) (pp. 525–529). IEEE.
6. Singh, N., Tewari, V. K., Biswas, P. K., & Dhruw, L. K. (2023). Lightweight convolutional neural network models for semantic segmentation of in-field cotton bolls. Artificial Intelligence in Agriculture, 8, 1–19.
7. Singh, S. P., Pritamdas, K., Devi, K. J., & Devi, S. D. (2023). Custom Convolutional Neural Network for Detection and Classification of Rice Plant Diseases. Procedia Computer Science, 218, 2026–2040.
8. Sethy, P. K., Barpanda, N. K., Rath, A. K., & Behera, S. K. (2020). Deep feature based rice leaf disease identification using support vector machine. Computers and Electronics in Agriculture, 175, 105527.

9. Bhattacharya, A. (2021, December). A Novel Deep Learning Based Model for Classification of Rice Leaf Diseases. In 2021 Swedish Workshop on Data Science (SweDS) (pp. 1–6). IEEE.
10. Chen, J., Chen, W., Zeb, A., Yang, S., & Zhang, D. (2022). Lightweight inception networks for the recognition and detection of rice plant diseases. IEEE Sensors Journal, 22(14), 14628–14638.
11. Meena, R., Joshi, S., & Raghuwanshi, S. (2022, December). VGG19-based Transfer learning for Rice Plant Illness Detection. In 2022 6th International Conference on Electronics, Communication and Aerospace Technology (pp. 1009–1016). IEEE.
12. Prajapati, H. B., Shah, J. P., & Dabhi, V. K. (2017). Detection and classification of rice plant diseases. Intelligent Decision Technologies, 11(3), 357–373.
13. Yakkundimath, R., Saunshi, G., Anami, B., & Palaiah, S. (2022). Classification of rice diseases using convolutional neural network models. Journal of The Institution of Engineers (India): Series B, 103(4), 1047–1059.
14. Sethy, Prabira Kumar (2020), "Rice Leaf Disease Image Samples", Mendeley Data, V1, doi: 10.17632/fwcj7stb8r.1
15. Hossain, S. M. M., Tanjil, M. M. M., Ali, M. A. B., Islam, M. Z., Islam, M. S., Mobassirin, S., ... & Islam, S. R. (2020). Rice leaf diseases recognition using convolutional neural networks. In Advanced Data Mining and Applications: 16th International Conference, ADMA 2020, Foshan, China, November 12–14, 2020, Proceedings 16 (pp. 299–314). Springer International Publishing.
16. Debnath, O., & Saha, H. N. (2022). An IoT-based intelligent farming using CNN for early disease detection in rice paddy. Microprocessors and Microsystems, 94, 104631.
17. Rahman, C. R., Arko, P. S., Ali, M. E., Khan, M. A. I., Apon, S. H., Nowrin, F., & Wasif, A. (2020). Identification and recognition of rice diseases and pests using convolutional neural networks. Biosystems Engineering, 194, 112–120.
18. Mique Jr, E. L., & Palaoag, T. D. (2018, April). Rice pest and disease detection using convolutional neural network. In Proceedings of the 1st international conference on information science and systems (pp. 147–151).
19. Lu, Y., Yi, S., Zeng, N., Liu, Y., & Zhang, Y. (2017). Identification of rice diseases using deep convolutional neural networks. Neurocomputing, 267, 378–384.
20. Hasan, M. J., Mahbub, S., Alom, M. S., & Nasim, M. A. (2019, May). Rice disease identification and classification by integrating support vector machine with deep convolutional neural network. In 2019 1st international conference on advances in science, engineering and robotics technology (ICASERT) (pp. 1–6). IEEE.

Note: All the figures and tables in this chapter were made by the authors.

Technologies for Energy, Agriculture, and Healthcare – Shailesh Nikam et al. (eds)
© 2024 Taylor & Francis Group, London, ISBN 978-1-032-98028-7

52

Advancing Sustainable Manufacturing: Designing a Supervisory and Control Interface for Sorting Stations

Shaina Rathod[1], Shivam Verma[2], Midhya Mathew[3]

Department of Electronics Engineering,
K J Somaiya College of Engineering,
Mumbai, Maharashtra, India

Abstract: The evolution of industrial production systems is undergoing a significant shift driven by heightened digitalization, fostering a landscape of intelligent, interconnected, and decentralized production. This marks the advent of the fourth industrial revolution, often termed Industry 4.0, which harmonizes advanced manufacturing techniques with cutting-edge digital technologies, yielding dynamic and smart manufacturing ecosystems. Critical to this paradigm is the continual empowerment of underlying components like smart factories, intertwined with the imperative of energy sustainability. Recognizing the profound interplay between Industry 4.0 and energy sustainability, this research employs enhancing energy efficiency but also ensuring sustainability in smart manufacturing. Addressing the escalating demand for effective and environmentally friendly logistics solutions, this research focuses on the development and application of an advanced, energy-efficient sorting station. To reduce overall energy usage, the suggested sorting station makes use of intelligent control systems, smart automation, and energy-efficient components. Because of its scalability and modularity, the system can adjust to changing workloads without consuming extra energy when demand is low. The experiment study's conclusion highlights the significance of sustainability in modern supply chain

[1]shaina.r@somaiya.edu, [2]shivam.verma@somaiya.edu, [3]midhya@somaiya.edu

DOI: 10.1201/9781003596707-52

management by going over the wider ramifications of using such energy-efficient sorting methods in the logistics sector. Such sorting stations help save energy and contribute making a green environment with the possibility of large long-term cost savings. Energy is saved by this smart automation by operating only when it is necessary. They require energy-efficient automatic components which consume less energy but maintain high performance. Modular designs of the sorting stations can be operated based on the current workload, preventing the unnecessary operation of equipment and reducing energy consumption during periods of lower demand.

Keywords: Sustainable energy, Smart manufacturing, SCADA, Sorting station, Automation

1. Introduction

Energy conservation in industrial automation is a critical aspect of sustainable and efficient manufacturing processes. In an era marked by rapid technological advancements and an ever-increasing demand for efficiency in industrial processes, the paradigm of sorting operations has undergone a transformative shift. Traditional manual sorting methods, characterized by labor-intensive and time-consuming processes, are gradually making way for more sophisticated and automated solutions. The emergence of automated sorting stations represents a pivotal milestone in the evolution of material handling and distribution systems across various industries[1]. With the smart automation sorting system, we delve into the world of Industry 4.0, which is an imminent evolution in industrial practices. A primary goal of Industry 4.0 is sustainability and energy efficiency. With almost no manpower requirements, lack of errors, efficient actuation, and high output rate, the sorting system is a very optimized system, which is the core of Industry 4.0[7].

SCADA controlled manufacturing stations provide the tools and capabilities necessary to optimize energy usage, reduce environmental impact, and enhance energy sustainability in industrial operations. SCADA systems continuously monitor energy consumption and production processes in manufacturing stations. By providing real-time insights into energy usage patterns, operators can identify inefficiencies and implement adjustments to optimize energy consumption. SCADA systems enable remote control and automation of manufacturing processes, allowing for precise adjustments to equipment and machinery settings to maximize energy efficiency. This reduces the need for manual intervention, minimizing energy wastage associated with human error or inefficiencies. SCADA systems can analyze equipment performance data to predict potential failures or malfunctions. By detecting issues early, proactive maintenance can be scheduled, preventing costly breakdowns and reducing energy consumption associated with inefficient or malfunctioning equipment. SCADA systems can implement load management strategies to optimize energy usage during peak demand periods. By adjusting production schedules or shifting energy-intensive processes to off-peak hours, manufacturing stations can reduce overall energy costs and minimize strain on the electrical grid. collect vast amounts of data from manufacturing processes, which can be analyzed to identify opportunities for energy efficiency improvements. By leveraging data analytics and optimization algorithms, manufacturing stations can implement targeted measures to minimize energy waste and maximize productivity [8][9].

Sorting station is a facility or area where items or materials are organized and separated based on specific criteria such as size, shape, type, or destination. Sorting stations are commonly used in various industries and contexts. The recycling centers or waste processing facilities, sorting stations separate recyclable materials (such as plastics, glass, paper, and metals) from non-recyclables to facilitate their processing and proper disposal. Sorting stations are utilized in warehouses, distribution centers, and shipping hubs to categorize and route packages and goods efficiently for delivery to their respective destinations. Sorting stations play a crucial role in postal facilities where mail and parcels are sorted based on their destination addresses or postal codes before being dispatched for delivery. In manufacturing plants, sorting stations may be used to organize and segregate components or parts as they move along the production line, ensuring accurate assembly and quality control. Sorting stations are employed in agricultural settings for grading and sorting harvested crops based on factors such as size, ripeness, or quality before packaging and distribution. The fundamental purpose of a sorting station is to categorize and organize diverse items swiftly and accurately based on predetermined criteria. Automated sorting stations leverage cutting-edge technologies such as sensor systems, robotics, artificial intelligence, and machine learning to achieve unprecedented levels of precision and speed in the sorting process. This evolution is driven by the relentless pursuit of operational efficiency, reduced error rates, and the ability to handle a wide array of items with varying characteristics. As we delve deeper into the various facets of this technology, it becomes apparent that automated sorting stations not only streamline operations but also offer a versatile and adaptive solution capable of meeting the complex demands of modern industries.

For the remainder of the paper:

- Section II consists of other related work.
- Section III presents the methodology, algorithm, and work flow of the program.
- Section IV includes the outcome of this program, and images to support it.
- Section V presents the conclusion and a few references

2. Literature Review

This study analyzes and explores the intersection of sustainable manufacturing and Industry 4.0. Using a conceptual framework based on Industry 4.0 principles and sustainable manufacturing, the study identifies research agendas, gaps, and opportunities [1]. Industry 4.0, the fourth industrial revolution, integrates advanced manufacturing with digital technologies to create smart ecosystems. This research, using interpretive structural modeling, identifies and analyzes energy sustainability functions of Industry 4.0, revealing significant relationships and emphasizing opportunities for players to leverage Industry 4.0 for energy sustainability [2]. In recent years, the industrial and IT fields have undergone significant changes, entering the "Industry 4.0" era characterized by the evolution from embedded systems to Cyber Physical Systems (CPS), enabling intelligent manufacturing through Internet-based integration, as exemplified by the Wise Information Technology of 120 (WIT120), with a focus on the transition from digital to intelligent factories [3].

Industry 4.0, originating in Germany in 2011, has become a widely discussed and transformative concept in manufacturing. The existing research on Industry 4.0, emphasizing its integration of information technology, Cyber Physical Systems, and the Internet of Things to bring about automation and presents a comprehensive literature review, highlighting the characteristics and opportunities associated with this paradigm shift in the manufacturing industry [4]. Significance of Industry 4.0, emphasizing its impact on organizations and society through Internet-connected techniques, focusing on the integration of cyber-physical systems and the Internet of Things to enhance efficiency, with a case study on digitizing manufacturing processes is studied in [5]. The study explores how Industry 4.0 technologies, like AI and IoT, reshape performance management by enabling remote monitoring, predictive analytics, and personalized feedback, fostering agility and efficiency but also raising concerns about data privacy and ethics, highlighting their transformative potential in empowering and engaging the workforce for enhanced productivity[6]. The paper presents a distributed control method for industrial automation aligned with Industry 4.0, offering advantages like flexible process reconfiguration, scalability, and wireless data exchange between production equipment and intelligent transport. Information on equipment loading, properties, throughput, and process states is stored in a public cloud, facilitating simultaneous production of multiple parts for various technological processes [7].

3. Methodology and Model Specifications

CODESYS and CIROS software have been used to simulate the Sorting station and develop the logic for the automatic control of the Sorting station. CODESYS (Controlled Development System) is a development environment used for programming and configuring virtual PLC. Figure 52.1 shows the front and top view of the sorting station in the CIROS software respectively. CIROS (Computer Integrated Robotic Off-line Simulation) is a software platform used for the simulation and offline programming of industrial robotic station. This is done to test the project on the programming front. The ladder logic is programmed on CODESYS and then run CIROS to check if it's running smoothly without any errors. Easy Port is used to connect the CODESYS program to the CIROS simulation.

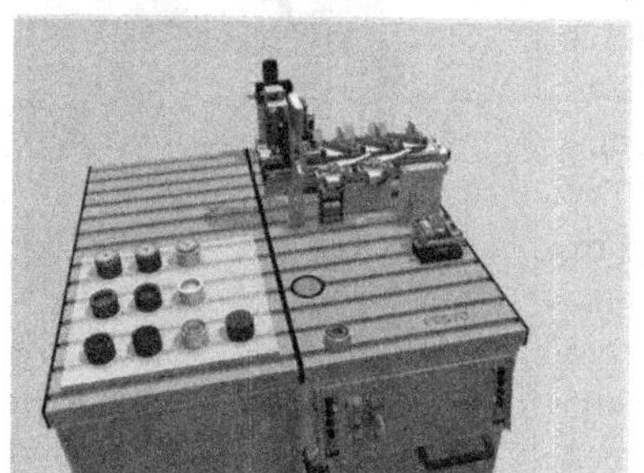

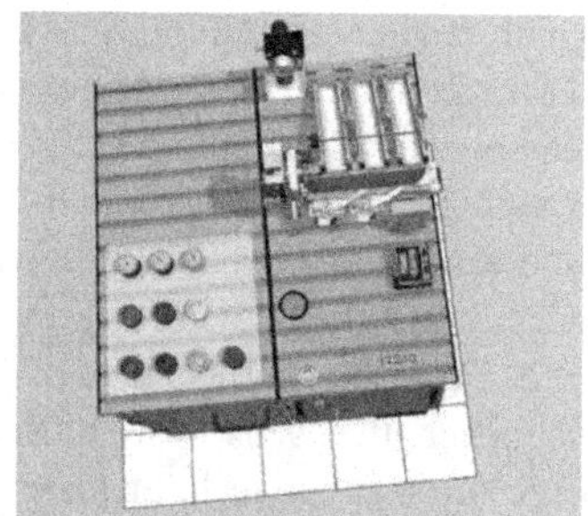

Fig. 52.1 CIROS simulation model of sorting station

Source: Author

Table 52.1 shows the hardware and software components used for the experiment setup. For the experimental setup up the ladder logic is developed on the TIA portal and use the S7 1200 PLC to control the Sorting station. Then established a connection between PLC and Reliance SCADA for the real time monitoring and control of the station. The hardware station consists of two types of sensors, the photo sensor and the inductive sensor for the sorting. The inductive sensor is used to check whether the workpiece is metallic or not, whereas the photo sensor is used to check whether the non-metallic workpiece is of light or dark colour.

Figure 52.2 shows the FESTO Modular Production Station. It consists of three sensors, a stopper and two deflectors. The two deflectors when extended play a crucial role in separating the workpieces.

Table 52.1 Hardware/software components used

Components	Specification
SCADA	Reliance SCADA
Programmable Logic Controller	Siemens S7-1200, CPU 1215C DC/DC/DC
Sensors	Photo sensor, Inductive sensor
Sorting Station	FESTO MPS sorting station
Simulator	CIROS Educational
Programming software	CODESYS, Siemens TIAPortal
Interfacing type(software)	Easy Port
Interfacing type(hardware)	PROFINET

Source: Author

Figure 52.3 shows the block diagram representation of the hardware setup. The sensors and actuators consist of the mini factory sorting station. The sensors obtain input signals which are received by the PLC which is later sent to the SCADA for the real time monitoring. SCADA acts as an interface between the user and the program and provide the supervisory control. The command given to SCADA is received by the PLC and then the actuators perform the command. From the output, it is observed that whenever the sensor identifies the workpiece the respective light turns green in the SCADA.

Fig. 52.2 FESTO modular production sorting station

Source: Adapted image of station

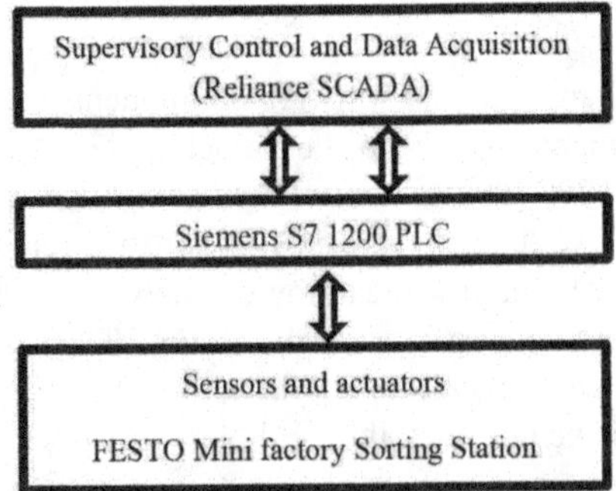

Fig. 52.3 Block diagram representation of the experiment

Source: Author

Figure 52.4 shows the flow chart for the program downloaded to the controller. If a workpiece is detected in the beginning, the conveyer moves forward and the stopper is extended. The inductive sensor present in the station detects whether the workpiece is metallic or non-metallic. If the workpiece is metallic, the stopper is retracted, the conveyer moves forward and the second deflector is extended

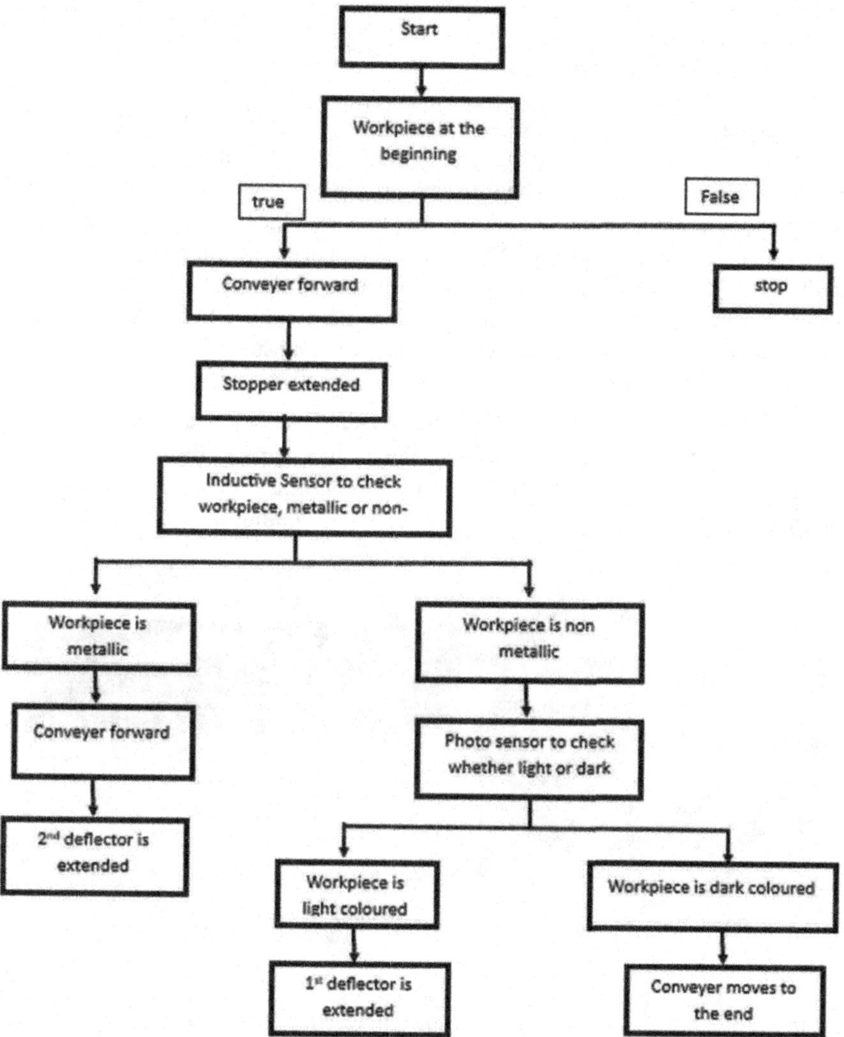

Fig. 52.4 Flow chart of the algorithm used for the sorting station

Source: Author

which forces the workpiece down the second chute. Whereas if the workpiece is non-metallic, the photo sensor present identifies whether it is a dark or light-colored workpiece. If it is light colored, the stopper is retracted, the conveyer moves forward, the first deflector is extended and the workpiece goes down the first chute. Lastly, if the workpiece is dark-colored when the stopper is retracted, the conveyer moves forward and the workpiece goes down the last chute.

4. Experiment Results

Figure 52.4 shows the initial state of the Sorting station in SCADA view. The start button in the SCADA to start the mini factory. The red circle will turn green when the stopper is retracted, the conveyor forward rectangle turns green when the conveyor is moving forward. The three rectangles below will turn green when the workpiece is metallic, light, or dark-colored non-metallic workpiece respectively. Each rectangle box in the SCADA mimic is associated with an individual counter in the PLC program which counts the number of workpieces coming to it. The black screen is the container property option in the Reliance SCADA software to show if there are any errors or if the program is running smoothly. This option can be used to visualize the current events in the automation system.

In Fig. 52.5 it is observed that when the sensor detects the workpiece is light colored because the pink workpiece rectangle turns green, the stopper is also retracted and hence turns green. The counter count turns one since one workpiece is detected. The current event shows the program is running smoothly without any errors.

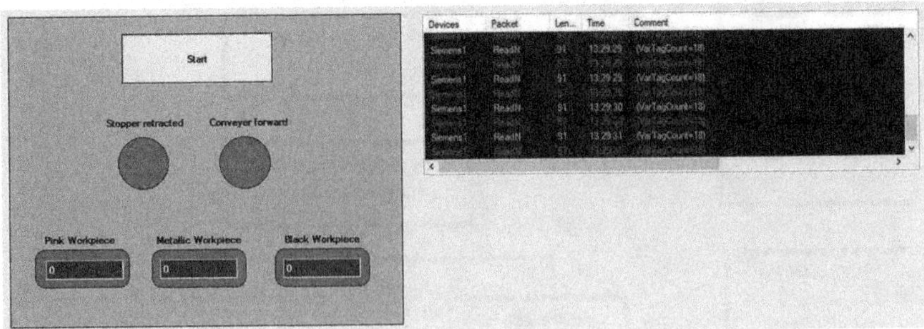

Fig. 52.5 SCADA view for the initial state of the Sorting station

Source: Made by author using reliance SCADA

From Fig. 52.6, we can observe that the stopper is retracted, the conveyer moves forward since it is a metallic workpiece and the second deflector gets extended. Here the counter shows that it has detected four metallic workpieces.

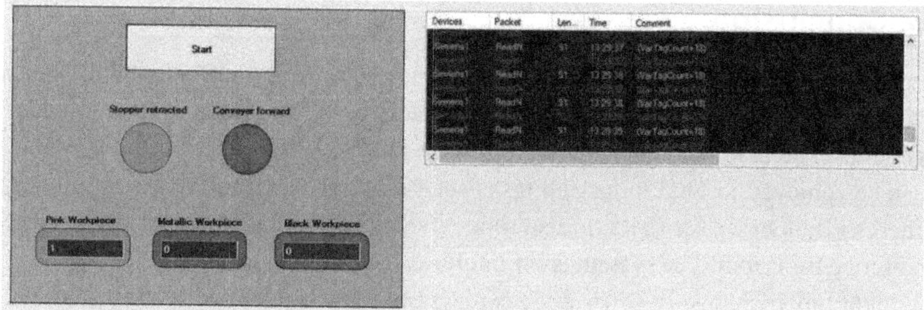

Fig. 52.6 SCADA view when light-coloured workpiece is detected

Source: Made by author using reliance SCADA

Lastly, in Fig. 52.7, once the stopper gets retracted, the conveyor moves forward towards the end since the workpiece is dark coloured/black. Here the counter detected two black workpieces.

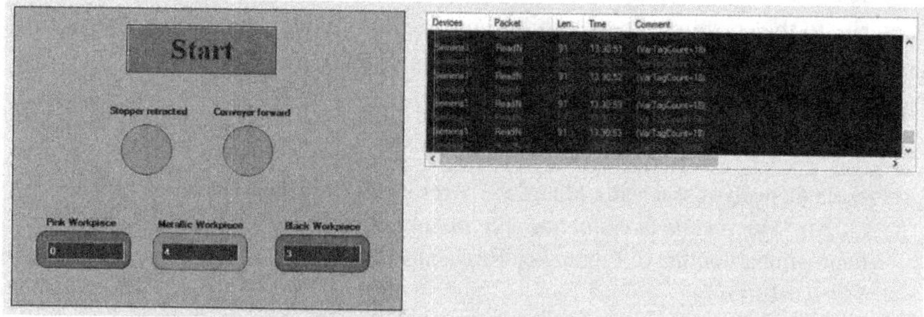

Fig. 52.7 SCADA view when metallic workpiece is detected

Source: Made by author using reliance SCADA

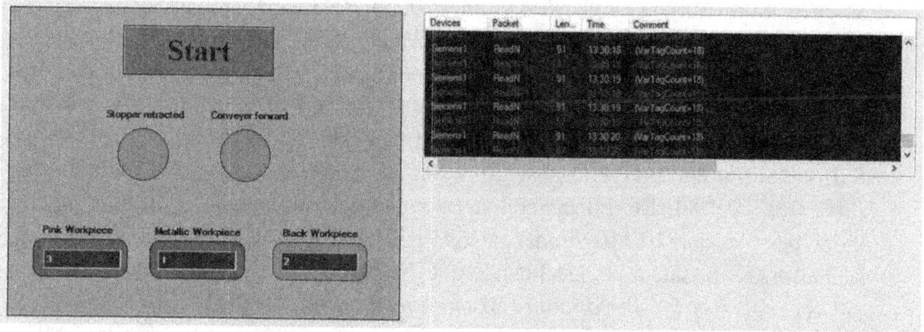

Fig. 52.8 SCADA view when a dark coloured workpiece is detected

Source: Made by author using reliance SCADA

5. Conclusion

In conclusion, an era of unmatched efficiency, precision, and adaptability has been ushered in by the transition of sorting activities from manual to automated systems. The automated sorting station shows the level of energy efficiency that can be achieved in vital industrial tasks such as object sorting. Apart from being energy efficient we see that it is also time efficient as compared to humans and has no scope for errors. The system is infinitely scalable in size and operating capacity without much change in software specifications.With automated sorting stations, enterprises can now handle a wide range of materials with greater speed and accuracy thanks to technical advancement in sorting. These stations have become leaders in modern industrial solutions because of the thorough integration of cutting-edge technology such as robots, artificial intelligence, machine learning, and sensor systems. The essential parts of automated sorting stations collaborate with one another to provide previously unheard-of levels of sophistication in real-time detection, categorization, and decision-making. Robotics adds flexibility and dexterity to the sorting process, while sensor technologies—from photo sensors to vision systems—ensure accurate and dependable item identification.

References

1. Carla Gonçalves Machado, Mats Peter Winroth & Elias Hans Dener Ribeiro da Silva (2020) "*Sustainable manufacturing in Industry 4.0: an emerging research agenda*", International Journal of Production Research, 58:5, 1462–1484, DOI:10.1080/00207543.2019.1652777
2. Tan Ching Ng and Morteza Ghobakhloo, 2020"*Energy sustainability and industry 4.0*", IOP Conf. Ser.: Earth Environ. Sci.463 012090
3. Guo-jian CHENG1, Li-ting LIU2, Xin-jian QIANG3," *Industry 4.0 Development and Application of Intelligent Manufacturing*", Ye LIU4 School of Computer Science, Xi'an Shiyou University Xi'an, 710065, P.R.China
4. S. J. Joshi, S. Mamaniya and R. Shah, 2022 "*Integration of Intelligent Manufacturing in Smart Factories as part of Industry 4.0 - A Review*" 2022 Sardar Patel International Conference on Industry 4.0 - Nascent Technologies and Sustainability for 'Make in India' Initiative, Mumbai, India, pp. 1–5, doi: 10.1109/SPICON56577.2022.10180471.
5. R. Pisal, S. Razdan and P. Kalaskar, 2019 "*Influence of Industry 4.0 on Manufacturing Processes*" 2019 IEEE Pune Section International Conference (PuneCon), Pune, India, pp. 1–5, doi: 10.1109/PuneCon46936.2019.9105836.
6. K. Pawar, K. Kasat, A. P. Deshpande and N. Shaikh, 2023 "*Adoption of Industry 4.0 Technologies for Performance Management*" 2023 1st DMIHER International Conference on Artificial Intelligence in Education and Industry 4.0 (IDICAIEI), Wardha, India, pp. 1–5, doi: 10.1109/IDICAIEI58380.2023.10406888.

7. S. Novoselov and O. Sychova, 2019 "*Using Wireless Technology for Managing Distributed Industrial Automation Objects within the Concept of Industry 4.0*" IEEE International Scientific-Practical Conference Problems of Info communications, Science and Technology (PIC S&T), Kyiv, Ukraine, 2019, pp. 580–584, doi: 10.1109/PICST47496.2019.9061333.
8. Enes Talha Tükez, 2022 "*SCADA System for Next-Generation Smart Factory Environments*" ICONTECH INTERNATIONAL JOURNAL 6(1):48–52 DOI:10.46291/ICONTECHvol6iss1pp48–52.
9. Ubaidullah Mohammad, Cheng Yee Low, Roman Dumitrescu, 2019 "*Smart Factory Reference Model for Training on Industry 4.0*" Journal of Mechanical Engineering, Vol 16(2), 129–144.

Technologies for Energy, Agriculture, and Healthcare – Shailesh Nikam et al. (eds)
© 2024 Taylor & Francis Group, London, ISBN 978-1-032-98028-7

53

Understanding Illicit Opioid Drug References on the Dark Web: A Text Mining Approach to Public Health Analysis

Ashwini Dalvi[1],
Riya Thapar[2], Siddharth Singh[3]
Department of Information Technology,
K. J. Somaiya College of Engineering,
Somaiya Vidyavihar University,
Mumbai, India

Abstract: The escalating opioid epidemic, as seen by an increase in overdose deaths and rates of addiction, calls for novel strategies to comprehend and counteract the distribution of opioids, particularly through covert channels such as the dark web. By examining the discourse and frequency of terms connected to opioids, this study introduces a novel text mining tool to examine the role of the dark web in the opioid epidemic. Leveraging advanced data crawling techniques, this study encompasses the collection of a rich dataset from the dark web, revealing a predominance of discussions around opioids such as Oxycodone, Morphine, and Fentanyl. The results of our analysis show that Oxycodone is the most often referenced opioid, indicating that it is in high demand in the illegal drug trade. Furthermore, the surprising mention of non-steroidal anti-inflammatory medicines (NSAIDs) with opioids suggests a complicated terrain of self-medicating behaviors among users. This paper is notable for its methodological innovation, which uses text mining to provide insights into the dark web's impact on the opioid crisis. The findings not only add to our comprehension of illicit opioid trade dynamics, but also provide vital information for public health policies and policy design targeted at reducing the opioid epidemic. Our study's originality

[1]ashwinidalvi@somaiya.edu, [2]riya.thapar@somaiya.edu, [3]siddharth18@somaiya.edu

DOI: 10.1201/9781003596707-53

lies in its application of text mining to dark web data, a significant step forward in digital public health analytics.

Keywords: Dark web, Opioid, Text analytics, TF-IDF

1. Introduction

The effectiveness of opioids in pain management is well-established, integral to treating various medical conditions, including post-traumatic stress and cancer. Derived from poppy seeds or synthesised, opioids target opiate receptors in the brain and body, diminishing pain perception and altering emotional responses, offering relief for chronic and severe pain sufferers. Despite their benefits, opioids carry significant drawbacks, notably the risk of addiction. The euphoria they induce can lead to both mental and physical dependence, thereby escalating the addiction risk.

The transition from legitimate use to opioid addiction marks a dangerous phase in treatment, with opioid use disorder significantly affecting society and individual well-being [1]. A study linking legal cannabis access to reduced opioid prescriptions suggests potential benefits of legalisation [2]. The 2014 US ruling that tightened opioid access marked a critical step in addressing the epidemic, impacting healthcare, patients, and the illicit drug market alike [3].

Research mentions the rising availability of fentanyl and other illicit opioids on crypto markets, contributing to increased accidental overdoses [4]. The opioid crisis has intensified with more individuals turning to the dark web for illegal drug purchases as the dark web's anonymity facilitates the opioid trade. Researchers highlight the need for advanced internet technologies, effective health policies, and stakeholder collaboration to address the online opioid trade [5]. Text mining is key to filtering relevant information from the unstructured data of the dark web .[6], acknowledging its influence on opioid distribution and the exacerbation of the crisis, guiding efforts towards solutions that enhance public health and safety.

The proposed work used text mining techniques to analyse opioid information on the dark web. This approach brings a new perspective to technology, public health, and drug addiction. Our research directly aligns with the theme of "Healthcare Technologies", encompassing areas such as Digital Health, Artificial Intelligence for Health, and Web-Based Medical Information Systems. Our work bridges the gap between digital technology and medical care by using advanced text mining techniques and data analytics in the context of the dark web.

The necessity of researching illegal opiate references on the dark web is emphasised in the introduction. The work is organised logically, beginning with a review of the literature that looks at the body of knowledge regarding opioid use. After that, it goes over the dataset that was utilised, the methodology (which includes preprocessing, cleaning, and applying the TF-IDF algorithm), and the visualisation techniques. Important information about opioid phrase frequencies and discourse on the dark web is provided in the findings section. The conclusion then explores the ramifications of the findings and offers potential directions for future study.

2. Literature Review

The opioid crisis, marked by a surge in drug-related deaths, demands immediate intervention, largely stemming from the over prescription of painkillers and subsequent illicit opioid abuse, like heroin and fentanyl. To tackle this epidemic, researchers are leveraging diverse data sources, including social media [7]. Specifically, studies have used Twitter's API to identify illegal opioid marketing and sales by targeting relevant keywords, revealing significant insights from Twitter's extensive data [8].

Researchers conducted geospatial and temporal analyses of opioid-related discussions on Twitter, shedding light on the distribution and trends of such conversations over various locations and times [9]. Using the Drug Abuse Ontology, they analysed social media and dark web posts to understand public perceptions and experiences with drugs [10]. The study focused on dark web listings and social media mentions of fentanyl and other synthetic opioids, employing computational methods to identify patterns and instances of substance abuse.

The emergence of dark web markets has brought about significant changes in the drug market [11]. The dark web offers opportunities for users to seek a broader selection and potentially more potent substances. Cryptomarkets can attract new buyers who may have hesitated or could not participate in traditional drug markets [12].

Dark web markets provide an easy avenue for people to access illicit drugs, with opioids being a common type of drug sold within these underground platforms. The challenging task of locating and taking down the top opioid dealers on DNMs falls to law enforcement agencies. Researchers used machine learning and text analysis to predict high-impact opioid products within the Darknet Markets (DNMs) [13]. Researchers studied the reliability and trustworthiness of fentanyl sellers operating on dark web markets [14].

The scope of existing research spans from identifying opioid buyers and sellers to assessing the impact of opioid sales on the dark web, with our proposed work focusing on identifying terms commonly associated with opioids to serve as an indicator of public awareness of their dangers. Highlighting limitations such as correlating dark web sales to geographic impacts, analysing the evolving varieties of opioids available online, and establishing causal links between online sales and local overdose trends, this study aims to improve data collection and analysis strategies. By doing so, it seeks to mitigate the adverse effects of opioid mentions on public health, addressing critical gaps in current research.

3. Dataset Description

This study makes use of an extensive dataset that was scraped from the dark web and contains 1033 onion connections to websites selling illegal drugs. These links are enhanced with thorough HTML excerpts and succinct headlines that highlight important details about each website. The study mainly focuses on opioid markets that are part of the dark web's drug trade. Furthermore, the opioid dataset—which comes from an open-source platform—contributes to the transparency and accessibility of the research by covering the names and brands of different opioids, including their generic equivalents. This dataset is useful for detecting opioids in dark web data, providing insight into the distribution patterns of opioids in the dark web.

4. Methodology

4.1 Data Collection and Processing

To focus on the text content, HTML tags and site formatting were removed during text preprocessing. The raw data was then tokenised into individual words or phrases. To ensure that the focus was on pertinent text, filtering was used to eliminate non-English phrases, numerical values, and words with fewer than three characters. Stop words—common English terms that add little to the intended opioid-related meaning—were removed for data cleaning and normalisation in order to decrease noise. Lemmatization, on the other hand, reduced words to their base forms in order to consider diverse versions of a word as identical and preserve analytical integrity.

To increase the analysis's robustness, words with comparable meanings were also grouped, avoiding terminological variations from diluting the results and guaranteeing thorough coverage of keywords commonly used in conversations concerning opioids.

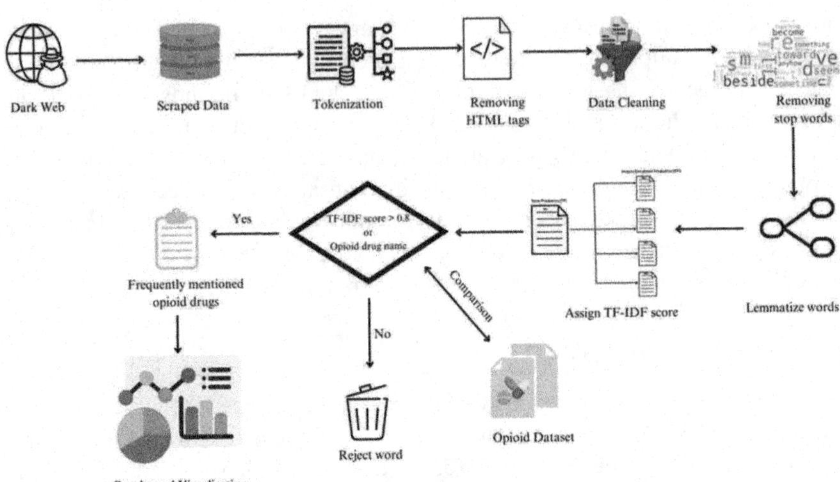

Fig. 53.1 Conceptual diagram of methodology for analysing opioid drug trends in dark web marketplaces

4.2 Implementation of the TF -IDF Algorithm

The Time Inverse Document Frequency (TF-IDF) algorithm measures the importance of a word in a set of documents relative to its frequency in the larger text. The formula is:

$$TF(t,d) = \frac{f_d(t)}{max_{w \in d} f_d(w)}, \ where \ \ f_d(t) := frequency \ of \ term \ t \ in \ document \ d$$

$$IDF(t,D) = \log\left(\frac{|D|}{|\{d \in D : t \in d\}|}\right), where \ D := corpus \ of \ documents$$

$$TF - IDF(t, d, D) = TF(t, d) \cdot IDF(t, D)$$

Where: *TF(t,d)* represents the word frequency *d* of the input middle.

IDF(t, D) represents the inverse frequency data at time *t* in the entire corpus *D*.

After this, each word is assigned a TF-IDF score, indicating its relevance to information in a larger corpus. Articles with higher TF-IDF scores are considered more suitable for analysis by highlighting essential issues in the opioid industry.

4.3 Filter and Compare Opioid Data

In the proposed study, communications pertaining to opioid trafficking on the dark web were particularly targeted to focus on the most pertinent data for tackling this pressing issue. A threshold for the TF-IDF score was set at 0.8 to identify significant content for additional analysis. To identify themes and generate frequency lists, the process entails cross-referencing keywords from opioid databases and the dark web datasets. To emphasie their importance, keywords that match the opioid dataset and have high TF-IDF scores are chosen for these lists; those that don't fit these requirements are eliminated, guaranteeing a focused effort on important content that clarifies the intricacies of the opioid crisis in relation to the dark web environment.

5. Results

The bar chart in Fig. 53.2 shows a clear hierarchy in the frequency of opioid mentions in the darknet market; Oxycodone ranks first with 136 mentions, followed by Morphine and Fentanyl with 34 and 28 mentions. This data suggests that Oxycodone may be the most popular opioid on the hidden market, possibly due to its potency or user demand. The number of comments made about Morphine and Fentanyl also shows that they are active in the illegal drug trade over the dark web. The importance of these opioids may reflect relative rates of abuse and may be related to factors in the broader context of the opioid epidemic.

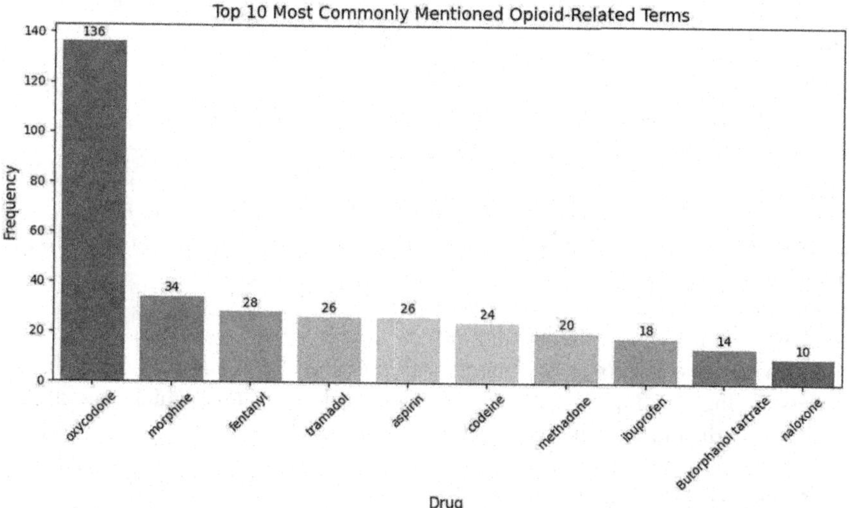

Fig. 53.2 Top 10 most commonly mentioned opioid-related terms

It is particularly surprising to note that the dataset includes references to non-steroidal anti-inflammatory medicines (NSAIDs) in these dark web marketplaces, including Ibuprofen and aspirin. Given how frequently these comments occur—26 for aspirin and 18 for ibuprofen—it is possible that persons who buy opioids may also buy NSAIDs to lessen the negative effects of using them or to supplement with other pain relief techniques. The combined intake of NSAIDs and opioids shown by the data may point to a pattern in users' self-medication strategies, most likely to alleviate issues like opioid tolerance or improve the analgesic effects of opioids.

Table 53.1 Opioid-related terms usage frequency on the dark web

Opioid-Related Terms	Frequency
Oxycodone	136
Morphine	34
Fentanyl	28
Tramadol	26
Aspirin	26
Codeine	24
Methadone	20
Ibuprofen	18
Butorphanol tartrate	14
Naloxone	10
Hydromorphone	8
Opium	6
Buprenorphine	4

Table 53.1 summarises the opioid-related phrases identified in the data set and their all-time frequency. The detailed data in the table can be used to track the availability of individual drugs, provide public health alerts, and guide authorities to target products at most businesses.

Figure 53.3 illustrates the word cloud created from the data, which is a useful visual tool containing the relative words of all opioid-related words. In the word cloud, frequently mentioned drugs appear prominently, immediately drawing attention to the big players in the darknet opioid business. This view not only helps quickly communicate critical findings in the data but also improves the interpretation of the study, allowing for a visual understanding of complex data and helping identify the rate at which opioids are available in the most illicit places online.

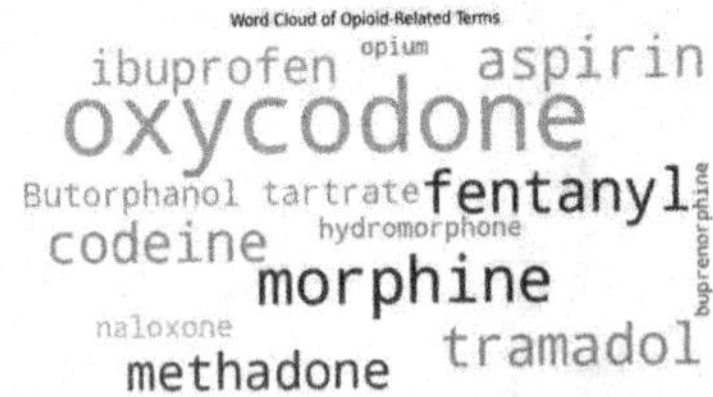

Fig. 53.3 Word cloud of opioid-related terms

6. Conclusion

The study provides a comprehensive examination of the dark web market and reveals trends in the use of opioids and NSAIDs such as Aspirin and Ibuprofen. Statistical results showing the frequency of opioid-related terms illustrate the prevalence of illegal opioid drugs as well as their role in the illegal drug trade.

The addition of NSAIDs to these regimens has the potential to significantly alter user behavior in terms of self-medication for pain management or prevention of opioid side effects. These findings highlight the critical need for public health and education programs that address the risks of opioid use and the implementation of prevention measures.

Detailed information and context, including a comprehensive table of drug references and word cloud, make it an excellent resource. Understanding the opiate black market allows stakeholders to create more effective solutions. This could mean stricter prescribing practices, increased monitoring of drug sales, and targeted disruption of illicit online drug networks. Looking ahead, the scope for future research is vast and necessary for continued progress in the fight against the opioid crisis. It is imperative that future research be based on opioid classification, on evolving trends in the dark web drug market, and integrate interdisciplinary expertise to develop a more comprehensive response. This can significantly contribute to the development of informed strategies aimed at minimising the impact of the opioid epidemic. Finally, integrating technical expertise and policies can have the potential to improve public health and lessen the harm that opioids do to individuals and communities.

References

1. Strang, J., Volkow, N. D., Degenhardt, et al., (2020). Opioid use disorder. Nature reviews Disease primers, 6(1), 3.
2. McMichael, B. J., Van Horn, R. L., & Viscusi, W. K. (2020). The impact of cannabis access laws on opioid prescribing. Journal of health economics, 69, 102273.
3. John, Carson. (2018) "Opioids on the dark web", Nature Human Behaviour, 2(8), 536–536. doi: 10.1038/S41562-018-0386-4
4. Lokala, U., Lamy, F. R., Daniulaityte, R., Sheth, A., Nahhas, R. W., Roden, J. I., ... & Carlson, R. G. (2019). Global trends, local harms: availability of fentanyl-type drugs on the dark web and accidental overdoses in Ohio. Computational and mathematical organization theory, 25, 48–59.
5. Mackey, T. K. (2018). Opioids and the internet: Convergence of technology and policy to address the illicit online sales of opioids. Health services insights, 11, 1178632918800995.
6. Li, Z., Du, X., Liao, X., Jiang, X., & Champagne-Langabeer, T. (2021). Demystifying the dark web opioid trade: content analysis on anonymous market listings and forum posts. Journal of Medical Internet Research, 23(2), e24486.
7. Balsamo, D., Bajardi, P., Salomone, A., & Schifanella, R. (2021). Patterns of routes of administration and drug tampering for nonmedical opioid consumption: data mining and content analysis of Reddit discussions. Journal of Medical Internet Research, 23(1), e21212.

8. Mackey, T., Kalyanam, J., Klugman, J., Kuzmenko, E., & Gupta, R. (2018). Solution to detect, classify, and report illicit online marketing and sales of controlled substances via twitter: using machine learning and web forensics to combat digital opioid access. Journal of medical Internet research, 20(4), e10029.
9. Sarker, A., Gonzalez-Hernandez, G., Ruan, Y., & Perrone, J. (2019). Machine learning and natural language processing for geolocation-centric monitoring and characterization of opioid-related social media chatter. JAMA network open, 2(11), e1914672-e1914672.
10. Lokala, U., Phukan, O. C., Dastidar, T. G., Lamy, F., Daniulaityte, R., & Sheth, A. (2023). " Can We Detect Substance Use Disorder?": Knowledge and Time Aware Classification on Social Media from Darkweb. arXiv preprint arXiv:2304.10512.
11. Angus, Bancroft. (2022). Potential Influences of the Darknet on Illicit Drug Diffusion. Current Addiction Reports, 9(4), 671–676. doi: 10.1007/s40429-022-00439-2
12. Bancroft, A. (2023). The darknet, bitcoins and the role of the internet in drug supply. In Understanding Drug Dealing and Illicit Drug Markets (pp. 311–324). Routledge.
13. Du, P. Y., Ebrahimi, M., Zhang, N., Chen, H., Brown, R. A., & Samtani, S. (2019, July). Identifying high-impact opioid products and key sellers in dark net marketplaces: An interpretable text analytics approach. In 2019 IEEE international conference on intelligence and security informatics (ISI) (pp. 110–115). IEEE.
14. Maras, M. H., Arsovska, J., Wandt, A. S., & Logie, K. (2023). Keeping pace with the evolution of illicit darknet fentanyl markets: Using a mixed methods approach to identify trust signals and develop a vendor trustworthiness index. Journal of contemporary criminal justice, 39(2), 276–297.

Note: All the figures and tables in this chapter were made by the authors.

Technologies for Energy, Agriculture, and Healthcare – Shailesh Nikam et al. (eds)
© 2024 Taylor & Francis Group, London, ISBN 978-1-032-98028-7

54

Named Entity Recognition in Healthcare: An Evaluation and Implementation of Deep Learning Models for Biomedical Data

Mohammed Danish Ansari[1]
Computer Engineering, KJSCE,
Mumbai, India

Soham Patil[2]
Computer Engineering, KJSCE,
Mumbai, India

Grishma Sharma[3]
Computer Engineering, KJSCE,
Mumbai, India

Abstract: The increasing volume of biomedical publications demands for efficient knowledge extraction through Natural Language Processing (NLP). Technologies like BioFlair, BioBERT, and Deep Learning play pivotal roles in developing robust biomedical text mining models. This paper conducts a comprehensive comparison of CRF, BERT and other deep learning-based technologies for Biomedical Named Entity Recognition (BioNER), elucidating their strengths and weaknesses. BioBERT, tailored for biomedical text, outperforms general language models in entity recognition. The study evaluates metrics such as training time, memory consumption, and loss per epoch using CRF, Bi-LSTM-CNN, and BERT models. Implementation involves curating a custom dataset from various sources and utilizing SparkNLP library. Evaluation reveals much about performance, with CRF demonstrating efficiency in training time and Bi-LSTM-CNN showcasing superior precision and recall.

Keywords: Bioinformatics, Text mining, Entity classification, Natural language processing, BioNER, BERT

[1]mohammed.da@somaiya.edu, [2]soham14@somaiya.edu, [3]neelammotwani@somaiya.edu

DOI: 10.1201/9781003596707-54

1. Introduction

With the exponential growth of biomedical publications annually, the field of Natural Language Processing (NLP) has become indispensable for extracting knowledge efficiently from vast scientific literature. Technologies such as BioFlair, BioBERT, and Deep Learning have emerged as essential tools in this endeavor, facilitating the development of effective biomedical text mining models. These technologies enable machines to comprehend and analyze complex biomedical texts, aiding researchers in extracting valuable insights and accelerating scientific discoveries.[1]

Among these technologies, BioBERT stands out as a pre-trained language model specifically designed for biomedical text. It surpasses general-purpose language models in tasks such as Biomedical Named Entity Recognition (BioNER) and biomedical question answering by capturing domain-specific context and terminology.[2]

As the volume of biomedical data continues to surge at a rapid pace, the demand for efficient and accurate BioNER technology becomes increasingly imperative. This research paper aims to address this need by providing a comprehensive comparison of various technologies, including CRF, BERT and deep learning-based models, used in biomedical named entity recognition. By examining their strengths, weaknesses, and potential for improvement, this study aims to contribute to the advancement of biomedical text mining and enhance our understanding of these critical NLP techniques in the context of biomedical research.[3]

2. Literature Survey

BioBERT[4] is a pre-trained language model based on Bidirectional Encoder Representations from Transformers (BERT)[7] specifically designed for biomedical text. It outperforms general-purpose language models in tasks like biomedical named entity recognition and biomedical question answering. BioBERT's architecture captures domain-specific context and terminology, making it a valuable tool for various biomedical natural language processing (NLP) applications. Bioflair[6][13] NER[5] is a named entity recognition tool specifically designed for the biomedical domain, using pre-trained contextualized embeddings to identify and classify named entities like genes, proteins, and chemicals in biomedical text. It supports various NER tasks, including detection, normalization, and species tagging. A Few-shot learning[9] approach was used to enhance Named Entity Recognition (NER) in medical text with ten annotated examples. Five improvements were evaluated, including layer-wise initialization with pre-trained weights, hyperparameter tuning, combining pre-training data, custom word embeddings, and optimizing out-of-vocabulary words. Bidirectional Long Short-Term Memory (BiLSTM)[8][10] and Convolutional Neural Networks (CNN)[11][12] play distinct roles in NER. BiLSTM captures global context, while CNN extracts local features, allowing the NER model to effectively identify and classify named entities in textual data.

3. Methodology

The BioNER models, including CRF[14], BERT, and deep learning-based architectures like Bi-LSTM-CNN, were compared for their effectiveness in biomedical text processing. A custom dataset was curated from openly available corpora and datasets from platforms like HuggingFace, Kaggle, and GitHub. Preprocessing steps were taken to ensure dataset compatibility and consistency across models. Data formats like IOB, IOBES, and CoNLL were standardized for uniformity. The sources include BC2GM, BC4CHEMD, BC5CDR, JNLPBA, NCBI, and s800 among others. The SparkNLP library was used for implementation, and models were instantiated and trained on the curated dataset. GloVe 100d embeddings were used for CRF and Bi-LSTM-CNN models, while BERT embedding small model was used for BERT. Models were trained using default configurations on Google Colab's CPUs with 12GB RAM. Evaluation metrics like precision, recall, and F1-score were computed to assess the performance of the models on the testing dataset.

4. Implementation

This section describes the implementation of the CRF, Bi-LSTM-CNN, BERT models. Implementation was done on Google Colab free version. The program was executed on the freely available CPUs of Google Colab with default configuration of RAM of 12Gib (Gigabytes).

In our paper, a custom dataset was curated by combining openly available datasets from platforms like HuggingFace, Kaggle, GitHub, and others. Key contributors to the final dataset include BC2GM, BC4CHEMD, BC5CDR, JNLPBA, NCBI, and s800, among others. The dataset was created using tokens from mentioned datasets which have a total of 16091 training tokens and 5139 testing tokens. The combined created dataset provided more tokens but because high RAM cost of CRF model (more in analysis section) the tokens were reduced and to keep the comparison unbiased the same number of tokens were used for other models. The datasets were not available in same format and were in different standards like IOB (Inside-Outside-Beginning), IOBES (IOB-End-Start) and CoNLL format. It was a challenging task to make every dataset compatible to be consumed by SparkNLP library models.

The program was implemented using SparkNLP library which has built in modules like TensorFlow and PyTorch to implement NLP tasks. The architecture of the model was created, and datasets were converted to CoNLL format from JSON, IOB and IOBES format which were available on HuggingFace. For CRF and Bi-LSTM-CNN, GloVe (Global Vectors for Word Representation)[15] embeddings 100d were used and for BERT the BERT embedding small model was used. The implementation was kept as raw as possible to make it a fair comparison of all models. Next section describes analysis of the models based on previously specified metrics.

5. Comparative Analysis

5.1 Training Analysis

Training analysis is divided into comparison of RAM consumed, time taken to train and loss per epoch step.

5.2 RAM Consumed

During model training, RAM consumption was monitored and recorded at peak levels, presented in Table 54.1. The CRF model showed the highest RAM usage at 11.4 GiB, possibly due to its older architecture lacking advanced memory management features. The other models exhibited lower RAM consumption, aligning with expectations.

Table 54.1 Maximum memory/RAM Consumed by BERT, Bi-LSTM-CNN, CRF during training

	Maximum memory/ RAM consumed
CRF	11.4 GiB
Bi-LSTM-CNN	3.3 GiB
BERT	5.7 GiB

5.3 Time Taken

Among the models CRF was the fastest one to train followed by Bi-LSTM-CNN and then BERT (except the 100-epoch step). CRF, Bi-LSTM-CNN, BERT took on average 0.283 minutes, 8.332 minutes, 12.878 minutes to train from 10, 20, 30, 50, 100 epochs respectively. The Fig. 54.1 shows a graphical comparison of the same.

There were no surprises in here, everything was according to eh specifications of the models. Only the 100 epoch step of Bi-LSTM-CNN can be classified as an anomaly, the dynamic resources of Google Colab might have hands behind it or some outlier randomness initialization of models can also be the cause of it.

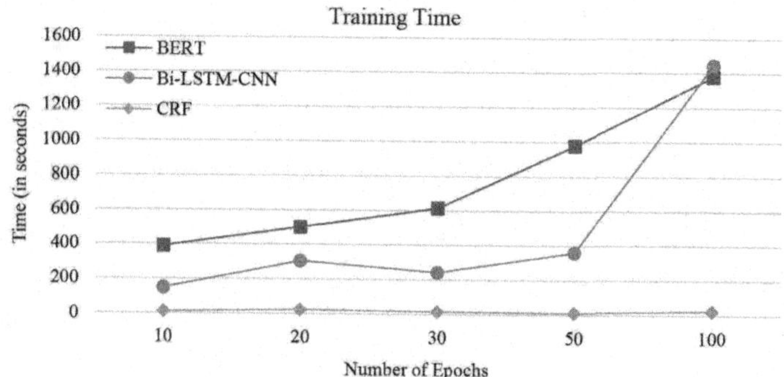

Fig. 54.1 Comparison of training time for CRF, Bi-LSTM-CNN, BERT

5.4 Loss Per Epoch

The CRF model was not included in this comparison as it would seem unjust to this older machine learning model from class of statistical modelling methods which differs vastly in architecture and learning process than BERT and Bi-LSTM-CNN.

In Fig. 54.2, the Bi-LSTM-CNN exhibited its lowest point during the 50th epoch, transitioning from two-digit to a single-digit loss. The anomaly at the 100th epoch, previously noted, is reaffirmed as the loss deviates from the expected trend. Figure 54.3 illustrates the loss pattern of the BERT model, following a consistent trend from its prior epochs.

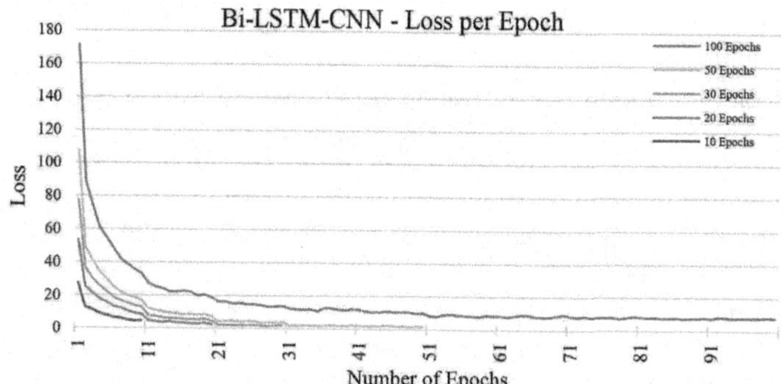

Fig. 54.2 Loss per epoch for Bi-LSTM-CNN

5.5 Testing Analysis

Testing analysis involves metrics like precision, recall, and F1-score for each epoch step. The CRF model has the potential to further improve accuracy with

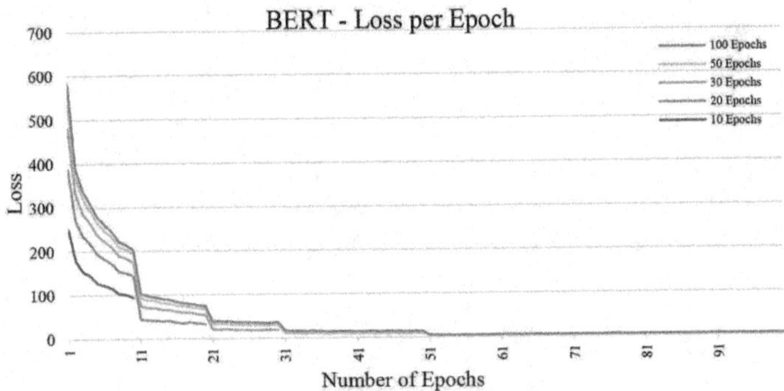

Fig. 54.3 Loss per epoch for BERT

more epochs, and it's a time-efficient training option, though RAM consumption should be monitored. The Bi-LSTM-CNN achieves peak accuracy between 30-50 epochs, but the anomaly at 100 epochs lacks clear reasoning. In contrast, the BERT model maintains consistent accuracy, attributed to the strength of BERT embeddings. Specific precision, recall, and F1-score numbers are detailed in Table 54.2.

Table 54.2 Comparison of CRF, Bi-LSTM-CNN, BERT on precision, recall and f1-score for each epoch step of 10, 20, 30, 50, 100

Epochs	CRF					Bi-LSTM-CNN					BERT				
	10	20	30	50	100	10	20	30	50	100	10	20	30	50	100
Precision	**0.92**	0.84	0.86	0.85	0.89	0.92	0.91	**0.95**	0.93	0.72	**0.74**	0.71	0.67	0.71	0.70
Recall	0.46	0.56	0.60	0.66	**0.72**	0.86	0.86	0.87	**0.88**	0.65	0.63	**0.65**	0.63	0.63	0.63
F1 score	0.52	0.65	0.69	0.73	**0.79**	0.88	0.88	**0.90**	**0.90**	0.68	0.66	**0.67**	0.65	**0.67**	0.66

6. Results and Discussions

The BERT model shows a non-linear relationship between training time and epoch number, with a significant increase in training time as epochs increase. This suggests that the model may overfit to the training data, resulting in minimal improvements in performance. The BiLSTM model's training time pattern is irregular, with a significant increase from 50 epochs to 100 epochs suggesting overfitting. Unlike deep learning models, CRF models have minimal and consistent training times across different epochs, suggesting efficient convergence and quicker experimentation. CRF models are less prone to overfitting due to fewer parameters and simpler assumptions about the data, making them less prone to overfitting compared to deep learning models.

Conditional Random Field (CRF) model demonstrates competitive precision values across epochs, indicating its ability to correctly identify named entities in the biomedical text. However, its recall and F1 score values, while showing improvement with more epochs, remain relatively lower compared to other models. This suggests that the CRF model may struggle with capturing a comprehensive range of named entities, potentially due to its reliance on local context and linear-chain structured predictions. On the other hand, the Bi-LSTM-CNN model showcases superior precision, recall, and F1 score metrics, particularly in earlier epochs. This model benefits from its ability to capture both short and long-range dependencies in the text through its combination of bidirectional Long Short-Term Memory (Bi-LSTM) and Convolutional Neural Network (CNN) layers. The higher precision values suggest fewer false positive predictions, while the robust recall and F1 scores indicate the model's effectiveness in capturing a wide range of biomedical named entities. Interestingly, the BERT model, a state-of-the-art transformer-based architecture pretrained on large corpora, presents mixed performance across epochs. While exhibiting relatively higher precision values compared to recall, its F1 score remains moderate. This suggests that while BERT can effectively capture the nuances of biomedical named entities, it may struggle with generalization or recall, potentially due to overfitting or domain-specific challenges.

7. Applications

Biomedical research is being accelerated by automated tools for extracting relevant information from scientific literature, enabling the identification of novel biomarkers, genes, proteins, and drug targets. This aids in the discovery of potential therapeutic interventions and enhances drug discovery and development. Integrating BioNER technologies into clinical information systems improves the efficiency and accuracy of literature searches for healthcare professionals. Developers guide tool development and optimization, providing insights into strengths, weaknesses, and performance metrics. Bioinformatics is contributing to the bioinformatics community, fostering a deeper understanding of BioNER methodologies and their applications. BioNER technologies are applied in clinical document understanding, disease and drug entity recognition, biomedical literature mining, health information management, public health surveillance, medication information extraction, entity linking for knowledge graphs, and clinical trial information extraction.

8. Conclusions

This research paper evaluates and implements Named Entity Recognition (NER) models in the context of biomedical data in the healthcare domain. The study examines CRF, a Bidirectional Long Short-Term Memory Convolutional Neural Network (Bi-LSTM-CNN), and BERT models. The comparative analysis reveals insights into their performance on biomedical datasets, revealing strengths and weaknesses in handling healthcare-specific information extraction tasks. Challenges such as data heterogeneity and domain-specific terminology are explored. Each model has unique advantages and limitations, making them suitable for specific aspects of biomedical named entity recognition. The implementation and analysis phases consider factors such as training time, memory consumption, and loss per epoch, providing a holistic perspective on the efficiency and effectiveness of the models. The research has potential implications for healthcare practitioners, researchers, and policymakers, contributing to advancements in healthcare analytics, aiding tasks like drug discovery, clinical decision support, and literature mining.

Acknowledgment

This paper acknowledges the invaluable support and advice provided by our mentor, Grishma Sharma, who significantly enhanced our understanding of the topic. Her dedication to academic and personal development inspired Soham Patil and Mohammed Danish Ansari, demonstrating the significant impact of Grishma Sharma on our research and academic activities. We are deeply appreciative of her guidance and contribution to our intellectual growth.

References

1. Lample, Guillaume, et al. (2016). "Neural architectures for named entity recognition." Proceedings of NAACL-HLT. Association for Computational Linguistics, 260–270.
2. Ma, Xuezhe, and Eduard Hovy. (2016). "End-to-end Sequence Labeling via Bi-directional LSTM-CNNs-CRF." Proceedings of the 54th Annual Meeting of the Association for Computational Linguistics, Volume 1, Pages 1064–1074. doi: 10.18653/v1/P16-1101
3. Alshaikhdeeb, B., Ahmad, K.: Biomedical Named Entity Recognition: A Review. Artificial Intelligence Review, 6(6), pp. 889, (2016)
4. Lee, Jinhyuk, Wonjin Yoon, Sungdong Kim, Donghyeon Kim, Sunkyu Kim, Chan Ho So, and Jaewoo Kang. "BioBERT: a pre-trained biomedical language representation model for biomedical text mining." *Bioinformatics 36, no. 4 (2020): 1234–1240.*
5. Sharma, Shreyas, and Ron Daniel Jr. "BioFLAIR: Pretrained pooled contextualized embeddings for biomedical sequence labeling tasks." *arXiv preprint arXiv:1908.05760 (2019).*

6. Akbik, Alan, Tanja Bergmann, Duncan Blythe, Kashif Rasul, Stefan Schweter, and Roland Vollgraf. "FLAIR: An easy-to-use framework for state-of-the-art NLP." *In Proceedings of the 2019 conference of the North American chapter of the association for computational linguistics (demonstrations), pp. 54–59. 2019.*
7. Devlin, Jacob, Ming-Wei Chang, Kenton Lee, and Kristina Toutanova. "Bert: Pre-training of deep bidirectional transformers for language understanding." *arXiv preprint arXiv:1810.04805 (2018).*
8. Panchendrarajan, Rrubaa, and Aravindh Amaresan. "Bidirectional LSTM-CRF for named entity recognition." *In Proceedings of the 32nd Pacific Asia conference on language, information and computation. 2018.*
9. Hofer, Maximilian, Andrey Kormilitzin, Paul Goldberg, and Alejo Nevado-Holgado. "Few-shot learning for named entity recognition in medical text." *arXiv preprint arXiv:1811.05468 (2018).*
10. Zhao, Sendong, Ting Liu, Sicheng Zhao, and Fei Wang. "A neural multi-task learning framework to jointly model medical named entity recognition and normalization." *In Proceedings of the AAAI Conference on Artificial Intelligence, vol. 33, no. 01, pp. 817–824. 2019.*
11. Zhou, Baohang, Xiangrui Cai, Ying Zhang, and Xiaojie Yuan. "An end-to-end progressive multi-task learning framework for medical named entity recognition and normalization." *In Proceedings of the 59th Annual Meeting of the Association for Computational Linguistics and the 11th International Joint Conference on Natural Language Processing (Volume 1: Long Papers), pp. 6214–6224. 2021.*
12. Chiu, Jason PC, and Eric Nichols. "Named entity recognition with bidirectional LSTM-CNNs." *Transactions of the association for computational linguistics 4 (2016): 357–370.*
13. Peters, Matthew E., Mark Neumann, Mohit Iyyer, Matt Gardner, Christopher Clark, Kenton Lee, and Luke Zettlemoyer. "Deep contextualized word representations. CoRR abs/1802.05365 (2018)." *arXiv preprint arXiv:1802.05365 (1802).*
14. Patil, Nita, Ajay Patil, and B. V. Pawar. "Named entity recognition using conditional random fields." *Procedia Computer Science 167 (2020): 1181–1188.*
15. Pennington, Jeffrey, Richard Socher, and Christopher D. Manning. "Glove: Global vectors for word representation." *In Proceedings of the 2014 conference on empirical methods in natural language processing (EMNLP), pp. 1532–1543. 2014.*

Note: All the figures and tables in this chapter were compiled by the authors.

Technologies for Energy, Agriculture, and Healthcare – Shailesh Nikam et al. (eds)
© 2024 Taylor & Francis Group, London, ISBN 978-1-032-98028-7

55

ResNet Models in Real World Action: A Comparative Study on Tomato Disease Recognition

Manori Sangani[1]
Student, K.J. Somaiya College of Engineering, Mumbai

Jheel Gala[2]
Student, K.J. Somaiya College of Engineering, Mumbai

Chaitanya Patil[3]
Student, K.J. Somaiya College of Engineering, Mumbai

Smita Sankhe[4]
Assistant Professor, K.J. Somaiya College of Engineering, Mumbai

Abstract: In recent years, deep learning architectures have revolutionised the field of computer vision, offering powerful tools for image classification and object detection tasks. In this study, we investigate the performance of three ResNet (Residual Network) models - ResNet-50, ResNet-101, and ResNet-152 - in the context of detecting tomato leaf diseases using real-world field datasets. The ResNet models, known for their deep architectures and skip connections, are trained and tested on a dataset comprising images of tomato leaves affected by various diseases. The evaluation metrics include accuracy, precision, recall, f1-score, and confusion matrix, providing a comprehensive assessment of the models' performance. Our results reveal significant variations with ResNet-152 demonstrating superior results in terms of disease detection accuracy. Overall, this

[1]manoribsangani@gmail.com,manori.s@somaiya.edu;[2]jheel0802@gmail.com,jheel.g@somaiya.edu; [3]chaitanyapatil12@rediffmail.com, chaitanya.hp@somaiya.edu; [4]smitasankhe@somaiya.edu

DOI: 10.1201/9781003596707-55

study contributes to the understanding of deep learning models' effectiveness in agricultural applications and provides insights into selecting optimal architectures for tomato leaf disease detection tasks.

Keywords: Tomato disease recognition, Deep learning, Transfer learning, Computer vision, Image classification, Convolutional neural networks (CNNs), Pre-trained models, Comparative analysis, Real-world dataset, Performance evaluation, Data augmentation, Training efficiency, ResNet-50, ResNet-101, ResNet-152, Agricultural applications, Model comparison, Model architecture

1. Introduction

Tomato plants (Solanum lycopersicum) are crucial in global agriculture, providing nutrition and income, yet they face constant threats from diseases, risking significant yield and economic losses for farmers [7] if not promptly addressed. Early disease detection is essential for timely interventions and mitigating the impact on crop productivity. [1].

Image classification, a cornerstone of computer vision, uses convolutional neural networks (CNNs) to label images based on their content, benefiting from accessible large-scale datasets and advances in deep learning. The advancements in computer vision and machine learning represent the forefront in automatically extracting crucial and distinguishing features of plant disease detection [8] [5]. Transfer learning adapts pre-trained models to specific tasks, enhancing proficiency by transferring knowledge from interconnected source domains, yielding effectiveness across various domains. [3]. This research focuses on leveraging transfer learning using ResNet architectures, specifically ResNet50, ResNet101, and ResNet152, for the detection of tomato plant diseases [4]. We use the tomato disease images from the plant village dataset [9], a widely-used comprehensive collection of plant disease images. However, to enhance the practical applicability of our findings, we introduced real field images for testing. This approach simulates the challenges encountered in actual agricultural settings [6], where disease symptoms may vary, and the conditions may differ from controlled environments.

In a comparative analysis of ResNet50, ResNet101, and ResNet152 models, we seek to determine the optimal architecture for tomato plant disease detection, assessing metrics like accuracy, precision, recall, and f1 score. The findings promise to advance automated disease detection, offering valuable insights for farmers to enhance crop management decisions.

Our paper on tomato plant disease detection using transfer learning with ResNet architectures aligns with the conference theme of Agriculture Technologies. By employing advanced deep learning methods and data mining techniques, we contribute to automating disease detection in agriculture. Additionally, our integration of real-world images alongside curated datasets addresses segmentation challenges in agricultural imagery, advancing technology for precise farming practices. Overall, our research offers practical solutions for early disease detection in tomato plants, supporting the conference's focus on innovative tools and techniques for smart agriculture.

The rest of the paper is structured as follows. Section 2 reviews the related work of Transfer Learning in Image Classification with deep learning models and the advantages of pre-trained models, minimising time loss and improving accuracy. Section 3 describes the dataset used in the study and the preprocessing techniques employed. Section 4 explains the research methodology. Section 5 discusses the empirical findings. Section 6 summarises the paper.

2. Literature Review

In contemporary image classification, deep learning models are often integrated with transfer learning to minimise time loss and improve object classification. This survey examines transfer learning's importance in image classification, highlighting its impact on model accuracy and computational efficiency.

The study conducted by [1] classifies 18 types of fasteners using TL with three scenarios, showing efficacy on smaller datasets. They achieved 0.97 accuracy with ResNet50 using feature extraction and 0.93 with fine-tuning. [4] initially obtained 0.87 accuracy with a self-trained CNN on the Plant Village Dataset. Incorporating TL with ResNet50 boosted training accuracy to 0.9840 and validation accuracy to 0.93, mitigating time loss by utilising pre-trained models for efficient object classification. Due to the high parameter count and computational expense associated with CNN models, researchers [8] replaced standard convolution with depth separable convolution to reduce parameters and computational costs. ResNets are preferred under computational constraints due to their balanced accuracy and computational efficiency compared to other DNN models. [2] demonstrates that ResNet usually stays on the higher end of the graph for accuracy compared to other DNN models considering computational costs for ImageNet1k competition data. [3] explored 39 different models on the EuroSAT dataset under the land use domain that are mainly different versions of ResNet50, EfficientNetV2B0, and ResNet152 where ResNet50 wins. [5] utilised ResNet34, leveraging its automatic feature extraction capabilities, eliminating the need for separate feature extraction methods. They employed a pretrained ResNet34 model on small datasets, addressing degradation issues with skip connections, achieving impressive test accuracy of 0.994 and precision of 0.9651. Their study, conducted on an open dataset derived from the Plant Village dataset, showcased ResNet's high accuracy, reaching 0.994 on a test set compared to traditional machine learning methods. [6] proposed research achieving an accuracy of 0.984 for grapes and 0.9571 for tomatoes trained on plant village dataset and tested on real field captured images on CNN models. [7] uses plant village dataset where ResNet emerges as the close winner with 0.9968 accuracy with other models like DenseNet, AlexNet, VGG16, and GoogleNet close behind.

In conclusion, the literature underscores TL's crucial role in deep learning-based image classification, highlighting the efficiency gains of using pre-trained models like ResNet50, ResNet101, and ResNet152 to boost accuracy and reduce computational overhead. TL remains a promising approach for optimising image classification tasks, enhancing model performance and resource efficiency across diverse domains as research advances.

3. Data and Variables

The initial step involved the ingestion and preparation of the plant village dataset. This dataset consists of tomato leaf images classified into 10 distinct classes as shown in Table 55.1. The dataset is split into two sections: a training set with 18,345 images and a validation set with 4,585 images.

Table 55.1 Dataset summary

Classes	Bacterial Spot	Early Blight	Late Blight	Leaf Mold	Septoria Leaf Spot	Two-spotted spider mite	Target spot	Yellow Leaf Curl virus	Mosaic Virus	Healthy	Total
Number of Images	1702	1920	1851	1882	1745	1741	1827	1961	1790	1926	**18345**
Mapping	0	1	2	3	4	5	6	7	8	9	

The image data undergoes preprocessing, including resizing and data augmentation techniques such as random transformations, flipping, and colour jittering, creating variations to mimic real-world scenarios. This augmentation helps the model learn robust features and generalise better to unseen data. Finally, all images are normalised to standardise pixel intensities across all channels (red, green, blue) and centre the data distribution, improving training efficiency and convergence by reducing sensitivity to illumination and colour balance variations.

4. Methodology and Model Specifications

ResNet-50, 101, and 152 are deep convolutional neural networks (CNNs) known for their residual connections. These connections help overcome vanishing gradients, allowing effective training of deep models. All utilise convolutional layers, batch normalisation, and ReLU activations. Bottleneck layers with convolutions improve efficiency. With increasing depth (50, 101, 152 layers), these models excel at capturing complex features and hierarchical representations in images, making them valuable for computer vision tasks like image classification and object detection.

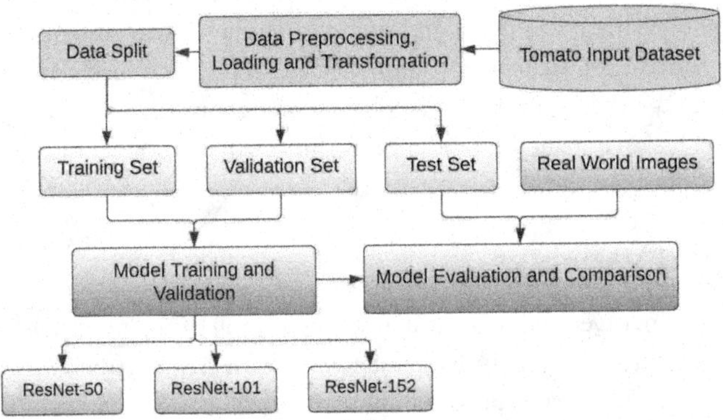

Fig. 55.1 Proposed architecture for tomato disease classification

5. Empirical Results and Discussion

Table 55.2 Precision, Recall, F1-Score of ResNet - 50, 101, 152

Classification report											
Classes	**Model**	**0**	**1**	**2**	**3**	**4**	**5**	**6**	**7**	**8**	**9**
Precision	ResNet - 50	0.92	0.90	0.84	0.89	0.85	0.85	0.97	0.94	0.91	0.88
	ResNet - 101	0.94	0.92	0.86	0.91	0.86	0.88	0.96	0.91	0.83	0.91
	ResNet - 152	0.94	0.90	0.91	0.86	0.86	0.85	0.96	0.88	0.91	0.93
Recall	ResNet - 50	0.90	0.80	0.92	0.82	0.92	0.81	0.97	0.94	0.96	0.90
	ResNet - 101	0.94	0.78	0.92	0.85	0.92	0.77	0.97	0.96	0.99	0.88
	ResNet - 152	0.92	0.81	0.88	0.88	0.89	0.76	0.97	0.98	0.99	0.93
F1 - Score	ResNet - 50	0.91	0.85	0.88	0.86	0.88	0.83	0.97	0.94	0.93	0.89
	ResNet - 101	0.94	0.84	0.89	0.88	0.89	0.82	0.97	0.94	0.90	0.89
	ResNet - 152	0.93	0.85	0.89	0.87	0.88	0.80	0.97	0.93	0.95	0.93

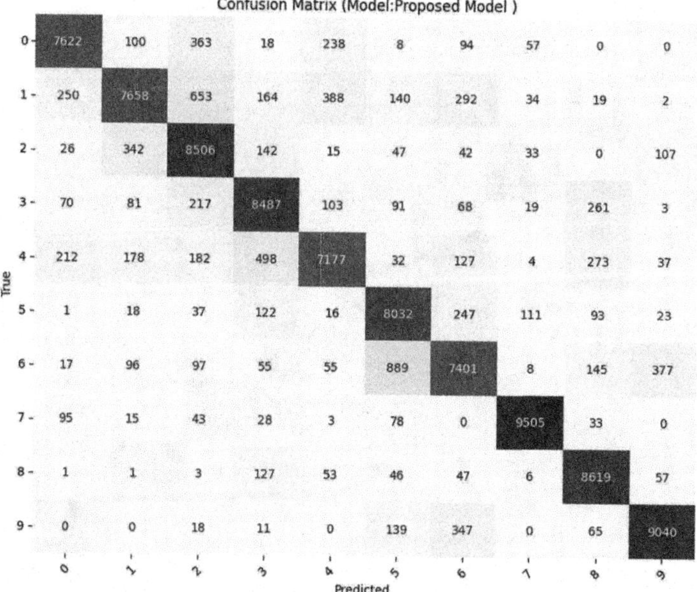

Fig. 55.2 ResNet50 confusion matrix

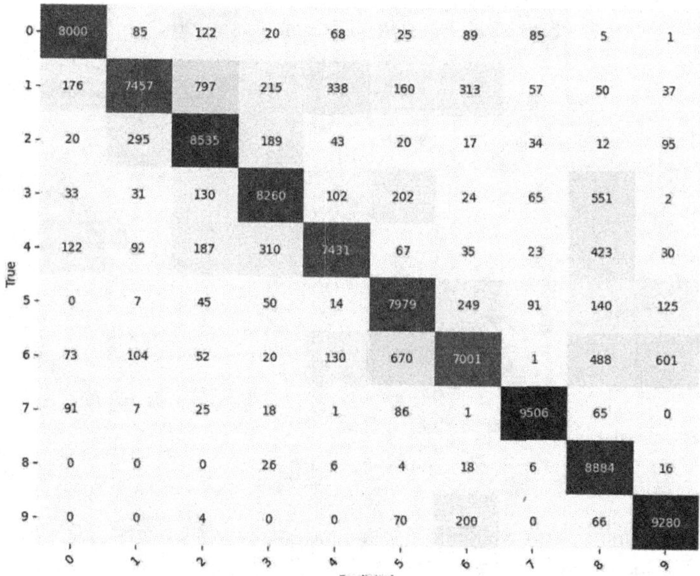

Fig. 55.3 ResNet101 confusion matrix

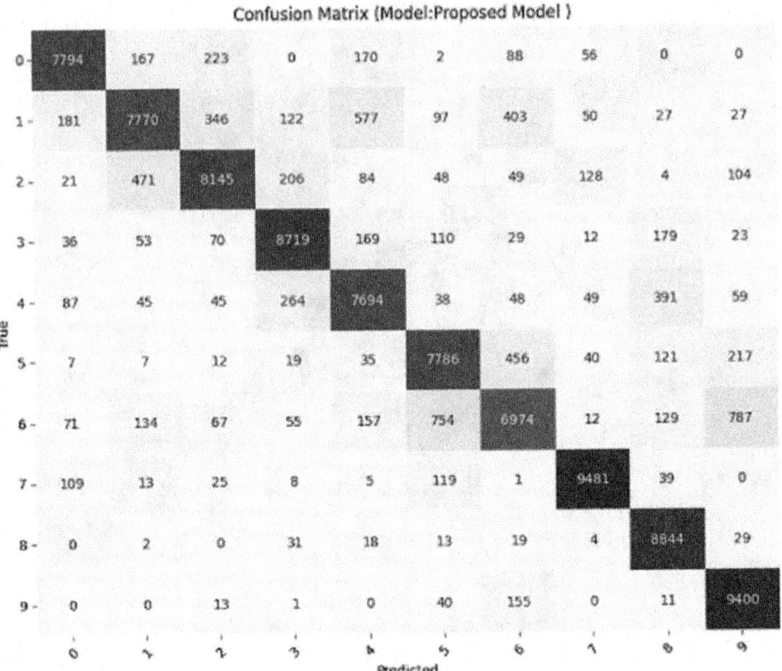

Fig. 55.4 ResNet152 confusion matrix

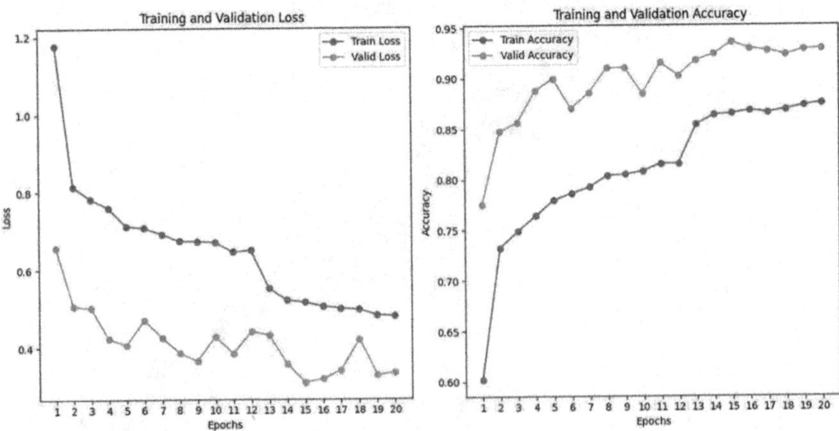

Fig. 55.5 Training and validation loss and accuracy against 20 epochs for ResNet-50

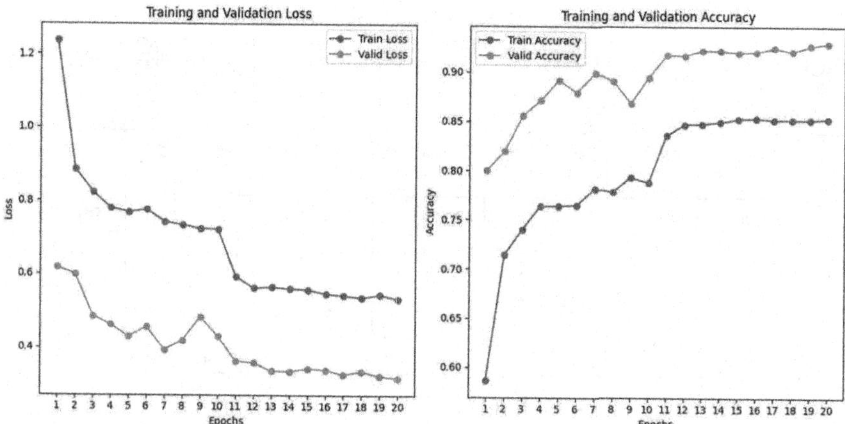

Fig. 55.6 Training and validation loss and accuracy against 20 epochs for ResNet-101

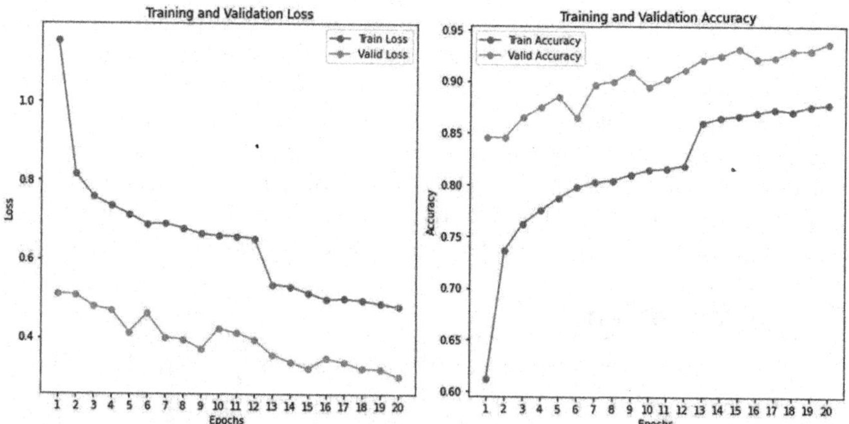

Fig. 55.7 Training and validation loss and accuracy against 20 epochs for ResNet-152

As seen in Table 55.2, all three models achieve relatively high performance, with ResNet-152 having the highest (93%). However, the difference in test accuracy between the three models is small as seen in Table 55.3. The models perform well for most classes, with high values on the diagonal of the confusion matrix as shown in Fig. 55.2, Figure 55.3, and Fig. 55.4. This indicates a good ability to correctly classify most images.

Table 55.3 Accuracy for test and real field images

Plant Village Test Set Accuracy	ResNet – 50		ResNet – 101		ResNet – 152	
	92%		94%		96%	
Actual	Predicted	Confidence %	Predicted	Confidence %	Predicted	Confidence %
Leaf Mold	Leaf Mold	99.88%	Leaf Mold	99.88%	Leaf Mold	99.68%
Early Blight	Early Blight	99.89%	Early Blight	81.19%	Early Blight	98.16%
Late Blight	Late Blight	86.54%	Late Blight	99.61%	Late Blight	96.31%
Tobacco Mosaic Virus	Leaf Mold	65%	Late Blight	89.27%	Tobacco Mosaic Virus	49.79%

6. Conclusion

In conclusion, this research underscores the efficiency of transfer learning using ResNet50, ResNet101, and ResNet152 architectures for tomato plant disease detection. The comparative analysis reveals a distinction in performance, with ResNet152 exhibiting a superior accuracy on the benchmark test dataset as well as confidence percent on Real Field images compared to ResNet101 and ResNet50. The future scope of this study lies in delving deeper into the optimization of hyperparameters, aiming to unlock further potential in model accuracy. By systematically fine-tuning these key settings, future research endeavours could contribute to a more nuanced understanding of how hyperparameter adjustments impact the performance of deep learning models in real-world agricultural scenarios.

Acknowledgement

Plant Village Dataset: https://data.mendeley.com/datasets/tywbtsjrjv/1

References

1. TAŞTİMUR, Canan, and Erhan AKIN. "A Hybrid Classification Approach for Fasteners Based on Transfer Learning with Fine-Tuning and Deep Features." Turkish Journal of Science and Technology 18, no. 2 (September 1, 2023): 461–75. https://doi.org/10.55525/tjst.1317713.
2. Bianco, Simone, Remi Cadene, Luigi Celona, and Paolo Napoletano. "Benchmark Analysis of Representative Deep Neural Network Architectures." IEEE Access 6 (2018): 64270–77. https://doi.org/10.1109/access.2018.2877890.
3. Dastour, Hatef, and Quazi K. Hassan. "A Comparison of Deep Transfer Learning Methods for Land Use and Land Cover Classification." Sustainability 15, no. 10 (May 11, 2023): 7854. https://doi.org/10.3390/su15107854.

4. Chellapandi, Bincy, M. Vijayalakshmi, and Shalu Chopra. "Comparison of Pre-Trained Models Using Transfer Learning for Detecting Plant Disease." 2021 International Conference on Computing, Communication, and Intelligent Systems (ICCCIS), February 19, 2021. https://doi.org/10.1109/icccis51004.2021.9397098.98
5. Kumar, Vinod, Hritik Arora, Harsh, and Jatin Sisodia. "ResNet-Based Approach for Detection and Classification of Plant Leaf Diseases." 2020 International Conference on Electronics and Sustainable Communication Systems (ICESC), July 2020. https://doi.org/10.1109/icesc48915.2020.9155585.
6. Paymode, Ananda S., and Vandana B. Malode. "Transfer Learning for Multi-Crop Leaf Disease Image Classification Using Convolutional Neural Network VGG." Artificial Intelligence in Agriculture 6 (2022): 23–33. https://doi.org/10.1016/j.aiia.2021.12.002.
7. Gehlot, Mamta, and Madan Lal Saini. "Analysis of Different CNN Architectures for Tomato Leaf Disease Classification." 2020 5th IEEE International Conference on Recent Advances and Innovations in Engineering (ICRAIE), December 1, 2020. https://doi.org/10.1109/icraie51050.2020.9358279.
8. Hassan, Sk Mahmudul, Arnab Kumar Maji, Michał Jasiński, Zbigniew Leonowicz, and Elżbieta Jasińska. "Identification of Plant-Leaf Diseases Using CNN and Transfer-Learning Approach." Electronics 10, no. 12 (June 9, 2021): 1388. https://doi.org/10.3390/electronics10121388.
9. J, ARUN PANDIAN; GOPAL, GEETHARAMANI (2019), "Data for: Identification of Plant Leaf Diseases Using a 9-layer Deep Convolutional Neural Network", Mendeley Data, V1, doi: 10.17632/tywbtsjrjv.1
10. Alkaff, Ardiansyah Kamal, and Budi Prasetiyo. "Hyperparameter Optimization on CNN Using Hyperband on Tomato Leaf Disease Classification." 2022 IEEE International Conference on Cybernetics and Computational Intelligence (CyberneticsCom), June 16, 2022. https://doi.org/10.1109/cyberneticscom55287.2022.9865317.
11. Kapucuoglu, Koksal, and Murvet Kirci. "Tomato Leaf Disease Detection Using Hyperparameter Optimization in CNN." 2021 13th International Conference on Electrical and Electronics Engineering (ELECO), November 25, 2021. https://doi.org/10.23919/eleco54474.2021.9677637.
12. Trivedi, Naresh K., Vinay Gautam, Abhineet Anand, Hani Moaiteq Aljahdali, Santos Gracia Villar, Divya Anand, Nitin Goyal, and Seifedine Kadry. "Early Detection and Classification of Tomato Leaf Disease Using High-Performance Deep Neural Network." Sensors 21, no. 23 (November 30, 2021): 7987. https://doi.org/10.3390/s21237987.

Note: All the figures and tables in this chapter were made by the authors.

Technologies for Energy, Agriculture, and Healthcare – Shailesh Nikam et al. (eds)
© 2024 Taylor & Francis Group, London, ISBN 978-1-032-98028-7

56

Disease Severity Estimation of Sunflower Leaf using Sequential Model

Rupali Sarode[1]
Assistant Professor,
Computer Engineering Department
Thadomal Shahani Engineering College, Bandra west,
Mumbai, University of Mumbai, India

Arti Deshpande[2]
Associate Professor,
Computer Engineering Department
Thadomal Shahani Engineering College, Bandra west,
Mumbai, University of Mumbai India

Abstract: Worldwide, sunflower is one of the most important oilseed crops that is grown. Sunflower oilseed crops have several uses, such as food (as a high-protein source), medicine, and decoration. It is a significant industrial oilseed crop as well. The growth of oilseed plants is influenced annually by biotic and abiotic factors. Numerous illnesses have an impact on oilseed production and yield. By stopping infections from spreading throughout the entire farm, early detection of infections in sunflowers lowers financial losses for farmers. This work offers a foundation for automating the assessment of illness severity. This work includes information on Alternaria leaf blight and powdery mildew diseases. Images of Indian sunflower leaves are included in the dataset. Classes of disease stages are generated for the combined images of the two diseases based on the disease stage. This study classifies the disease stage and determines the disease severity using a sequential model based on convolution neural networks.

Keywords: Alternaria leaf blight, Convolution neural network, Powdery mildew, Sequential model

[1]rupali.patil@thadomal.org, [2]arti.deshpande@thadomal.org

DOI: 10.1201/9781003596707-56

1. Introduction

Amongst all the oilseed crops, sunflower is the most widely grown oilseed crop. Both biotic and abiotic factors annually impact oilseed plants. As a result, sunflower diseases proliferate and have an effect on production and yield. Sunflowers can be infected by at least 30 different fungus, bacteria, and viruses, however, only a limited number of these have a substantial economic impact on production [1]. Bacterial leaf spot, Crown gall, apical chlorosis, Bacterial wilt, etc. are a few examples of bacterial diseases. On the other hand, fungi that cause fungal diseases include Downy mildew, Powdery mildew, Sclerotinia basal stalk rot, head rot, Alternaria leaf blight, Charcoal rot, Phytophthora stem rot, stem spot, and rust, among others [2]. Different sunflower diseases are present in various regions of India based on the season and climatic conditions [3][4]. Three diseases, however, have been observed in the Marathwada region: Alternaria leaf spot[5], Powdery mildew[5] and Downy mildew[6]. The extent of a plant disease's damage is estimated based on disease severity. The traditional method of examining the severity of plant diseases is by naked eye vision. Hence this kind of assessment is very time-consuming. However, as modern agriculture advanced quickly, the cost and effectiveness of disease stage evaluation increased, enabling it to reach new heights.

2. Literature Review

Numerous studies indicate that, as compared to human visual assessments, image-based assessment techniques produce more accurate and consistent outcomes. Using the image analysis method, the disease symptoms of wheat leaves caused by Zymoseptoria tritici are investigated. When compared to human visual estimations of virulence, this approach gave improved accuracy and precision since it allowed for the quantification of pycnidia size and density as well as other characteristics and their relationships [7]. This researcher [8] presented a method for detecting plant leaf infection using Convolution Neural Network (CNN), which distinguishes between images of normal and abnormal leaves. The Raspberry Pi module is then used to identify the affected area from the abnormal leaves. Also, after detecting the affected area, the name of the pesticide to handle the leaf sickness is recommended.

Researchers suggested in this work [9] that hyperspectral imaging could be used in conjunction with machine learning classifiers and variable selection approach, and other methods to detect tobacco disease before symptoms appear. Images of leaves free of disease and those infected with the tobacco mosaic virus (TMV) are taken at intervals of two, four, and six days using efficient wavelengths and spectral ranges. Four texture features are retrieved using the grey-level co-occurrence matrix. Along with the grey-level co-occurrence matrix, machine learning algorithms are employed to identify and categorize disease stages using efficient wavelengths, textural features, and data fusion. To improve the accuracy of disease severity estimation, the classification accuracy of three models—the back propagation neural network, the extreme learning machine, and the least squares support vector machine—is examined. The author [10] analyzed the historical viewpoint of visual severity evaluation, taking into account concepts, instruments, shifting paradigms, and techniques. Since 1892, people have been aware of the value of accurate visual severity evaluations. To aid in determining the extent of rust in wheat, Author Cobb produced a set of illustrations known as Standard Area Diagrams (SAD). Despite the fact that sensor technology exists to assess the severity of diseases using visible or other spectral range imaging, visual perception, and vision sense still rule, especially in the field of research. The advantages and disadvantages of sixteen research works based on CNN-based plant disease severity assessment using traditional CNN frameworks, improved CNN architectures and CNN-based segmentation networks were thoroughly comparison presented in the paper [11]. Also, the authors [12] discussed practical issues of disease severity estimation. Using CNN-based models to estimate disease severity supercomputing power is required such as PaaS as cloud computing service model. The CNN model employed to evaluate tea diseases was deployed on the cloud service model Platform as a Service, and a smart mobile device can access the deployed model's hyperlink. A smart mobile device camera can snap an image of the tea and post it to the cloud. Using the deployed CNN model, the cloud system automatically predicts the disease, which is then displayed on a smart mobile phone display. To deploy the CNN model on the cloud, the model should be small, but a small model doesn't give good feature extraction. Larger models, such as AlexNet, with DCNN models, such as VGG and GoogleNet, exhibit stronger effects on feature extraction. Model structure is an important factor to take into account when balancing model size and performance.

3. Image Dataset and Disease Information

The study is conducted using infected sunflower leaf images from 378 samples of Alternaria leaf blight and Powdery mildew, along with their corresponding disease stages, and 105 healthy images. The dataset is then augmented to include 7,000 images. Table 56.1 shows how this dataset is organized. Every image from dataset is scaled successively to a consistent width and height of 512 × 512. Under the supervision of a botanist, stage classes are determined based on the affected leaf area and are labeled as stage0, stage1, stage2, and stage3. Stages 1, 2, and 3 for

Table 56.1 Dataset distribution with their classes

Sr. No	Disease Name	Disease Grade & Stage	Count of Original images
1.	Powdery Mildew	1-Stage1	64
		2-Stage2	84
		3-Stage3	38
2.	Alternaria Leaf Blight	1-Stage1	34
		2-Stage2	33
		3-Stage3	20
3	Disease Free	Healthy-Stage0	105
	Total		378

Powdery Mildew are depicted in Figs. 56.1(a), (b), and (c), while for Alternaria leaf light, they are illustrated in Figs. 56.2(a), (b), and (c). Stage 0 refers to leaves free of disease, Stage 1 denotes the initial stage with 0 to 30% affected leaf area, Stage 2 denotes the middle stage with 31% to 60% affected leaf area, and Stage 3 denotes the final stage with 61% to 100% affected leaf area.

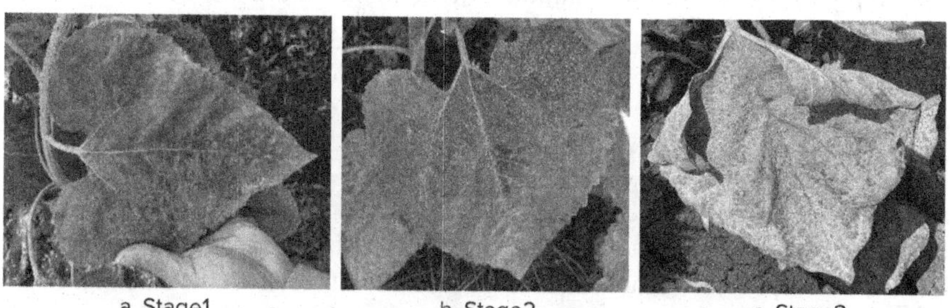

a. Stage1 b. Stage2 c. Stage3

Fig. 56.1 Powdery mildew stage-wise images

a. Stage1 b. Stage2 c. Stage3

Fig. 56.2 Alternaria leaf blight stage-wise images

3.1 Disease Information

Powdery Mildew

White to grey mildew develops on the leaf surface. The same mildew occurs on the stem and petiole as shown in Fig. 56.1[13].

Alternaria Leaf Blight

Little dark angular brown leaf patches are caused by this bacterial disease. Leaf spots frequently develop at the tips of the leaves, along the leaf margins, and in the spaces between the main veins. As illustrated in Fig. 56.2, a considerable degree of yellowing (chlorosis) is followed by leaf death and total leaf browning [13].

Image Augmentation and Pre-processing

Organizing and improving the images is necessary because the obtained images were captured by chance under normal sunlight. This will increase the number of samples that deep learning models can use to improve classification accuracy. To increase the sample size, data augmentation is applied [14]. This process includes:

- Rotating the image ten times by Nighty degrees to the left and right
- Applying a zoom factor of minimum 1.1 and maximum 1.6.

Following the process of augmenting the image dataset, which yields seven thousand images, all target images are resized to an identical size of 512 × 512.

4. Proposed Methodology

The classification of images based on their content is one of the deep learning domains that has been studied [16]. The aim of this project is to develop an automated deep learning model based on convolutional neural networks (CNNs) that can classify disease stages based on the severity of the disease. CNN is capable of extracting unique features from image input. Numerous individual neurons make up each layer of the CNN, which alternates between convolutional and subsampling layers. TensorFlow and Kera's Sequential API are used in this work to construct a CNN model [17][18][19]. For image classification, the sequential model takes the input image size as 512 × 512. Images of sunflower leaves have been grouped into classes based on disease stages. Grouped image samples get split into training and testing sets with a ratio of 80:20. The model receive images in batch size of 32.

A convolutional neural network comprising four convolutional layers and four dense layers has been implemented. A model has the Conv2D, batch normalization, ReLU, MaxPool2D layers. The image size of the input layer is 512 × 512. The subsequent layer consists of 32 filters with a 3 × 3 kernel size, MaxPool2D, batch normalization, and a dropout layer at 0.2. Similar to the first layer, there are the next three, Conv2D layers with sixty-four, one hundred and twenty-eight, two hundred and fifty-six filters followed by MaxPool2D, batch normalization, and a dropout layer at 0.2. After that, there are four fully connected layers where each fully connected layer starts with the flatten layer and ends with a dropout layer at 0.2. As seen in Fig. 56.3, there is an output layer with a SoftMax activation function at the end to determine the probability of each of the four output classes. The ReLu activation function [19] is applied as follows at each layer:

$$\text{ReLU, } f(x) = \max(0, x) \tag{1}$$

Where x is the input value to the function. Selecting the maximum value intends to put all negative values of x to zero and only allow positive values of x to flow through the network. In an attempt to underrate the loss function, the network

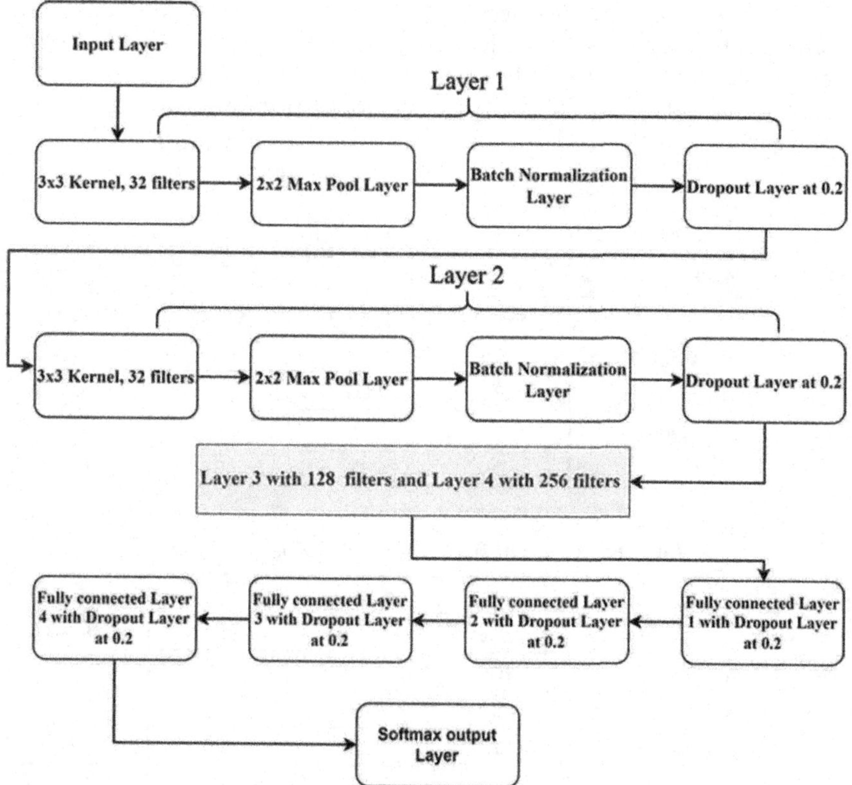

Fig. 56.3 Sequential model architecture

weights were iteratively changed using the Adam optimizer (with learning rate 0.001) throughout the compilation of the model based on training data [20]. Given that the problem involves multiclass image classification, a sparse categorical cross-entropy function with metric accuracy is employed as the loss function. The difference between the expected result and the input label, which is equal to the total cross entropy, is computed by the loss function [21].

5. Results and Observations

The disease stage reveals the threshold at which the disease worsens. Powdery mildew and Alternaria leaf blight images are processed using a CNN-based sequential model. The accuracy of the tests is evaluated at different epochs as the model is trained over time.

According to Table 56.2, the disease accuracies for Alternaria Leaf Blight are as follows: 98.4% and 100% for stages 1 and 3, and 98.5 and 100% for stages 2 and 0 respectively. The accuracy rates for stage 0 and stage 3 powdery mildew disease are 100% and 99.5%, respectively, as shown in Table 56.3. Either of the other two classifications (Stage 1 or Stage 3) are applicable to stage 2 powdery mildew. The sequential model's validation accuracy for Alternaria leaf blight and Powdery mildew disease is 99.25 percent and 98.88 percent, respectively.

Table 56.2 Confusion matrix for alternaria leaf blight

		Predicted			
		Stage0	**Stage1**	**Stage2**	**Stage3**
Ground Truth	Stage0	**204**	0	0	0
	Stage1	3	**187**	0	0
	Stage2	0	3	**202**	0
	Stage3	0	0	0	**201**

Table 56.3 Confusion matrix for powdery mildew

		Predicted			
		Stage0	**Stage1**	**Stage2**	**Stage3**
Ground Truth	Stage0	**204**	0	0	1
	Stage1	0	**190**	0	0
	Stage2	7	2	**196**	0
	Stage3	0	0	0	**201**

The results shown in Tables 56.2 and 56.3 indicate that the sequential model performs better in classifying Alternaria leaf blight disease stages, but stage 2 of Powdery mildew disease is more likely to be identified incorrectly. At epoch 50, training loss and validation loss for Alternaria leaf blight disease are recorded as 0.0132 and 0.0330, respectively. The observed training loss and validation loss for powdery mildew at epoch 70 are 0.0016 and 0.0103, respectively. The training loss obtained is almost identical for both diseases, and the validation loss is nearly close as well.

6. Conclusion

In this work, images collected during the Rabi season for two diseases, which are used to estimate disease severity using a sequential model. After training the sequential model over 50 epochs, it is able to classify Alternaria leaf blight disease into stable stages. Except for stage 2, the model is providing a reliable stage classification of powdery mildew in the 70th epoch. Future research can expand the size of the dataset by collecting more photos at various stages and by redefining stage classifications in accordance with the severity of the condition. The sequential model's findings can also be enhanced by employing other image classification models, such as Visual Geometry Group16, Visual Geometry Group19, etc. This will help plant pathologists and growers make more informed decisions. After submitting an infected leaf image, the system will estimate the disease stage, which in turn helpful for assisting growers to detect the disease stage early and take preventative action based on the disease severity.

References

1. Kandel, H., Endres, G. and Buetow, R. 1984. Diseases- National Sunflower Association Home | https://www.sunflowernsa.com/. Accessed on 29th May 2023.
2. Andre Klapper. (2022). List of sunflower diseases https://en.wikipedia.org/wiki/List_of_sunflower_diseases. Accessed on 29th May 2023.
3. A. Bheemaraya, M.M. Jamadar. and Shalini Huilgol. (2018). Correlation between Weather Parameters and Sunflower Powdery Mildew Caused by Erysiphe cichoracearum DC. International Journal of Current Microbiology and Applied Sciences. ISSN: 2319–7706. Volume 7. Number 09.
4. V. B. Akashe, J. D. Jadhav, V. R. Bavadekar, P. B. Pawar and V. M. Amrutsagar. (2016). Forewarning model for sunflower thrips (Thrips palmi Karny) in western Maharashtra scarcity zone Journal of Agrometeorology 18(1):68–70.
5. Shivani R. Biological Science | Diseases of Sunflower: Necrosis, Leaf Blight, Mildew and Other Diseases. https://www.biologydiscussion.com/plants/plant-diseases/diseases-of-sunflower-necrosis-leaf-blight-mildew-and-other-diseases/43079. Accessed on 27th July 2023.

6. M. K. Ghodke, S. P. Shirshikar and M.Y. Dudhe. (June 2016). (6303) Sunflower Breeding Strategy for Resistance to Downy Mildew Disease in India. 19th International Sunflower Conference, Edirne, Turkey.
7. E. L. Stewart and B. A. McDonald. (2014). Measuring quantitative virulence in the wheat pathogen zymoseptoria tritici using high-throughput automated image analysis. Phytopathology vol. 104. no. 9. 985–992.
8. B. Srikanth, Srinivasa Rao, Chunchu, Naveen Mukkapati, N. Sridevi and Konduru Kranthi Kumar. (2021). Design and Development of Image Based Plant Leaf Disease Monitoring System Using Deep Learning Algorithms". Plant Cell Biotechnology and Molecular Biology. 22(33&34):516–526. ISSN: 0972-2025.
9. Hongyan Zhu, Bingquan Chu, Chu Zhang, Fei Liu, Linjun Jiang & Yong He. 23rd June 2017. Hyperspectral Imaging for Presymptomatic Detection of Tobacco Disease with Successive Projections Algorithm and Machine-learning Classifiers. Scientific Reports. 7: 4125 I DOI:10.1038/s41598-017-04501-2.
10. Clive H. Bock & Kuo-Szu Chiang & Emerson M. Del Ponte. (2022) Plant disease severity estimated visually: a century of research, best practices, and opportunities for improving methods and practices to maximize accuracy. Tropical Plant Pathology. https://doi.org/10.1007/s40858-021-00439-z 47:25–42.
11. Tingting Shi, Yongmin Liu, Xinying Zheng, Kui Hu, Hao Huang, Hanlin Liu & Hongxu Huang. Recent advances in plant disease severity assessment using convolutional neural networks. Scientific Reports. (2023). https://doi.org/10.1038/s41598-023-29230-7, 13:2336.
12. Lanjewar, M. G., Panchbhai, K.G. (2022). Convolutional neural network-based tea leaf disease prediction system on smartphone using paas cloud. Neural Computing Application. 1–17.
13. Samuel Markell, Extension Plant Pathologist, North Dakota State University Robert Harveson, Extension Plant Pathologist, University of Nebraska, Charles Block, Plant Pathologist, USDA, Ames, IA Thomas Gulya, USDA Sunflower Pathologist, (Retired), Fargo, N.D. Febina Mathew, Broadleaf/Oilseed Crops Pathologist, North Dakota State University, Sunflower Disease Diagnostic Series. PP172. Reviewed (Jan. 2023). https://www.ndsu.edu/agriculture/ag-hub/publications/sunflower-disease-diagnostic-series. Accessed on Feb 2023.
14. Marcus D. Augmentor, Docs I Augmentor I User Guide. (2016). https://augmentor.readthedocs.io/en/stable/. Accessed on 24th March 2023.
15. Keiron O'Shea and Ryan Nash. An Introduction to Convolutional Neural Networks, (2 Dec 2015). Department of Computer Science, Aberystwyth University, Ceredigion, SY23 3DB keo7@aber.ac.uk 2 School of Computing and Communications, Lancaster University, Lancashire, LA1 4YW nashrd@live.lancs.ac.uk arXiv:1511.08458v2 [cs.NE].
16. Kishan S.Athrey. (2018). Tutorial on Keras -Centre for Research in Computer Vision. https://www.crcv.ucf.edu/wpcontent/uploads/2019/03/CAP6412_Spring2018_KerasTutorial.pdf. Accessed on Jan 2023.
17. Mohd Azlan Abu, Nurul Hazirah Indra, Abdul Halim Abd Rahman, Nor Amalia Sapiee and Izanoordina Ahmad. (2019). A study on Image Classification based on

Deep Learning and TensorFlow. International Journal of Engineering Research and Technology. Volume 12. Number 4. ISSN 0974–3154. 563–569.

18. Argha Chatterjee, Bidesh Roy, Koushik Majumder, Roni Mondal, Shamim Ahmed. (2022). Image Classification using Keras. International Journal for Research Trends and Innovation IJRTI2210021. Volume 7. Issue 10. ISSN: 2456–3315.
19. Jatin Kayasth. November (2022). Image Classification using CNN. Yeshiva University - Katz School of Science and Health. jkayasth@mail.yu.edu.
20. D. P. Kingma and J. Ba. Adam. (2015). A method for stochastic optimization. 3rd International Conference for Learning Representations. San Diego. arXiv preprint arXiv:1412.6980.
21. Guan Wang, Yu Sun, and Jianxin Wang School of Information Science and Technology, Beijing Forestry University, Beijing 100083. China 5th July 2017. Automatic Image-Based Plant Disease Severity Estimation Using Deep Learning. Computational Intelligence and Neuroscience. Hindawi. Volume 2017. Article ID 2917536. https://doi.org/10.1155/2017/2917536.

Note: All the figures and tables in this chapter were made by the authors.

Technologies for Energy, Agriculture, and Healthcare – Shailesh Nikam et al. (eds)
© 2024 Taylor & Francis Group, London, ISBN 978-1-032-98028-7

57

Numerical Simulation of Effect of Injection Angle on Flow Mixing in A Constant Area Duct

Adit V. Patel[1]
Student,
B.E. Department of Aeronautical Engineering,
SVIT Vasad

Nandisha J. Shah[2]
Student,
B.E. Department of Aeronautical Engineering,
SVIT Vasad

Jigar B.Sura[3]
Assistant Professor,
Department of Aeronautical Engineering,
SVIT Vasad

Abstract: The scramjet combustor is a fundamental part of scramjet engines, which are important for uses beyond supersonic travel. The design and development of the scramjet combustor are made extremely difficult by the extremely complicated flow field inside the engine. A crucial step occurring in the combustor is the mixing of the flow. The goal of this work was to understand the mixing features of flow for non-reacting flow situations by attempting to make complete numerical predictions for the mixing of flow characteristics with injection of a subsonic jet of air in a supersonic flow at various angles. A 2D stable, density-based SST turbulence in a constant area duct of length 650 mm and width 60 mm was studied numerically using an implicit numerical scheme. For the study, a total of four instances were collected at various angles of 80, 90, 100, and 110 degrees. The findings demonstrate flow regime mixing, shockwave

[1]aditpatel.v@gmail.com, [2]nandisha.jshah@gmail.com, [3]jigarsura.aero@svitvasad.ac.in

DOI: 10.1201/9781003596707-57

formation, and shocktrain. This research aids in our understanding of flow mixing at various injection angles and in identifying the most effective scenario.

Keywords: Hypersonic, Scramjet, Combustor, Mixing of flow, Shockwave

1. Introduction and Literature Review

Scramjet engines are a promising technology for hypersonic propulsion, but their effective functioning depends on a full understanding of the flow dynamics inside combustors. This work explores the complex relationship between injection angle and flow mixing in a constant area duct using numerical modelling using Computational Fluid Dynamics (CFD). The study examines the effects of angled jet injection on flow mixing and shockwave production utilizing a 2-D duct arrangement. This work aims to disentangle the fundamental principles controlling flow behaviour and evaluate their consequences for scramjet combustor performance by experimenting with different injection angles. Detailed analysis of flow events at microscopic sizes is made possible by the application of CFD, which provides essential insights into the optimization of combustion processes in hypersonic conditions.

A comprehensive study was conducted to study the effect of the duct length on mixing. The result in the form of shockwaves and boundary layer formation provided an insight of the prevention of backflow. As an advancement of the same has been portrayed here where the effect of angle of incidence is being studied of the mixing properties.

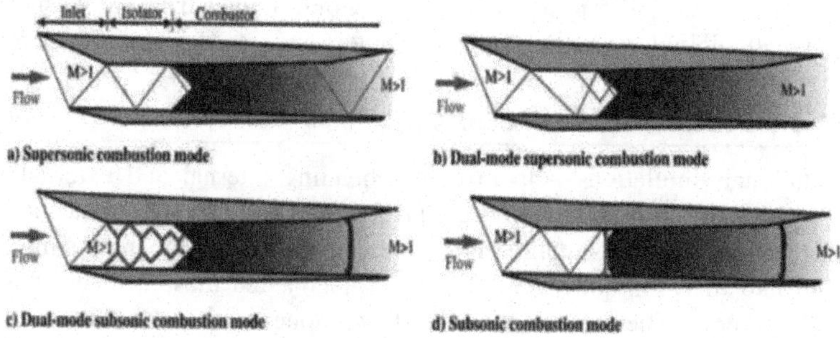

Fig. 57.1 Supersonic combustor schematic diagram

Source: Flame propagation and stabilization in dual-mode scramjet combustors: A survey, Progress in Aerospace Sciences, Vol 101

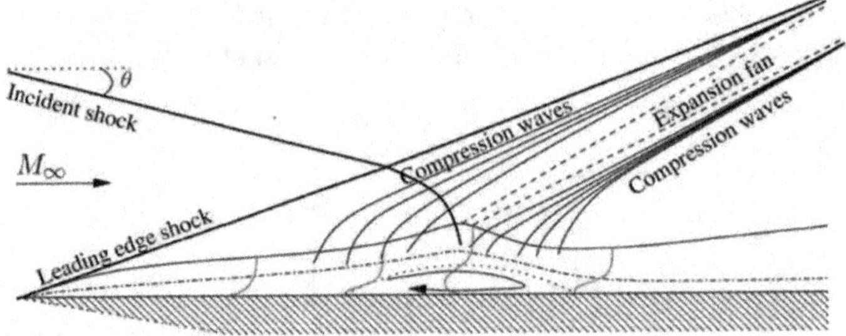

Fig. 57.2 Formation of shockwaves and boundary layer

Source: 3D Steady & Unsteady bifurcations in shockwave/laminar boundary layer interaction: A numerical study

Many studies on the subject have been conducted by various scientists and researchers. Studies from (Anderson 1989) provide data on the relations of hypersonic flow and shockwave, formation of shockwaves and principles of hypersonic flow, hypersonic-expansion wave, shock-boundary layer formation, Mach No. Independence and effects of inviscid Newtonian fluid on flow which helps in understanding flow characteristics, interaction of flow and relations with shockwaves. Studies from (Segal 2009) provide data on working of scramjet engine cycle, shock-boundary layer formation, and mixing of angled flows which helps in understanding the characteristics of scramjet engine and combustor. Studies from (Sura 2020) shows the formation of oblique shock and characteristic of shock travelling downstream in the duct and the change in shock formation with change in Mach No. Studies from (Patel, Shah and Sura 2023) shows data on shockwave formation at the interaction of jet and grid independency of the simulations. Studies from (Ahuja and Hartfield 2009) shows how strongly accelerated fluid in the field affects jet trajectory and mixing behaviour. Studies from (Mahesh 2013) shows a methodology for setting flow parameters and introduction of jet angles.

2. Computational Methodology

Computational simulations help in comprehending internal and external flow fields without developing any physical prototype resulting in reduced time and cost of prototype experimentation. For this kind of work, commercial simulation software like Ansys is quite helpful. The geometry and meshing were done in Ansys Workbench. The analysis of the mesh was done in the FLUENT package of Ansys Workbench for a steady 2D density-based system with an implicit solution method with a k-ω SST turbulence fluid dynamic model.

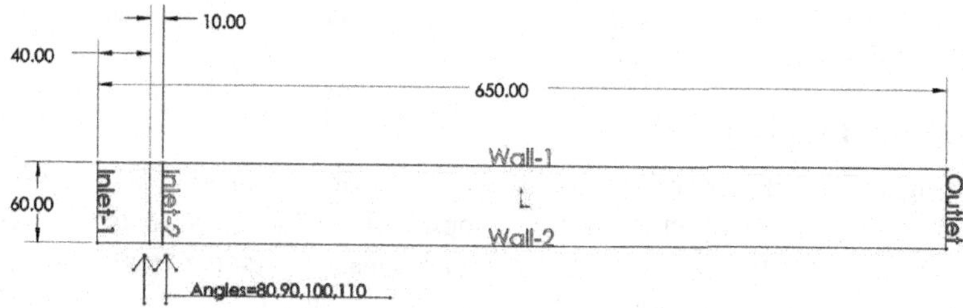

Fig. 57.3 Geometry of constant area duct

Source: Author's compilation

Table 57.1 Geometrical data of combustor duct

Model	Data
[L/D] ratio	10.83
Length of duct	650mm
Width of duct	60mm
Inlet-2 width	10mm
Position of inlet-2	40mm from the origin
Length of duct in focus	600mm

Source: Author's Compilation

Table 57.2 Boundary conditions & parameters of fluid

Fluid-1	Air
Fluid-2	Air
Density-type	Ideal-gas
Sp. Heat [J/(kg K)]	1006.43 (constant)
Viscosity [kg/(m s)]	1.7894e-05 (constant)
Molecular weight[kg/kmol]	28.966 (constant)

Source: Author's Compilation

2.1 Duct Geometry

A model of duct in the present numerical analysis is generated in Ansys Workbench Geometry as shown in the figure.

The Dimensions were 60mm depth and 650mm length which included inlet section of 40mm and a second inlet section of 10mm with the duct of 600mm which gave an L/D ratio of 10 for it respectively. All the dimensional aspects are given in the table.

Since the nozzle design is a 2D constant area duct, the Ansys Workbench Grid Generator could be used to create the structured mesh. Before accepting the simulation's results, the grid independency test is required to be conducted. For any analysis, simulation results are required to be acceptable irrespective of mesh/grid size. The mesh was created with 2 divisions per mm axially along the inlet_1 and 1.5 divisions per mm radially along the walls. The inlet section had 1 division per mm radially. The inlet_2 has been provided with a thicker mesh of 3 divisions per mm radially to capture contact of flow and formation of shockwaves and start of the boundary layer more accurately.

2.2 Boundary Conditions

Utilizing a 2D implicit energy equation with shear stress transport k-ω turbulence model, no-slip boundary conditions are applied in parametric studies, neglecting heat transmission and boundary layer interactions. Supersonic flow is achieved through the duct under specific boundary conditions. Inlet air conditions are provided in a table. The solution converges to IE-6 by adjusting convergence conditions for various parameters. Simulations employ the Implicit Roe-Flux Difference Splitting Method (Roe-FDS) for effective contact discontinuity capture and robust boundary layer resolution. The converged solution serves as a basis for further analysis.

3. Results and Discussion

A comprehensive numerical simulation was performed to understand the influence of scramjet combustor's flow characteristics with injectors placed at different angles in the duct. Computations were performed until residuals converged. Spectrums for static-pressure and velocity-magnitude were captured for the models. Contours of static pressure gradients benefit in visualizing the pressure variation along the duct and also the formation of shock train. Contours of Velocity-Magnitude gradients benefit in visualizing the variation in velocity in the duct along the centre line and the walls. It also helps in understanding the formation of boundary layer along the walls.

Table 57.3 Boundary conditions & parameters of constant area duct

Zone	Boundary type	p (atm)	V (Mach No)	T (K)	ρ (kg/m^3)	M (kg/m-s)
Air-Inlet 1	Pressure Far-field	1	2	300	1.225	1.789401e-05
Air-Inlet 2	Pressure Far-field	5	0.5	300	1.225	1.789401e-05
Mixture outlet	Pressure Outlet	0	-	300	1.225	1.789401e-05
\Wall_1	Wall	-	-	300	1.225	-
Wall_2	Wall	-	-	300	1.225	-

Source: Author's Compilation

3.1 Contours of Static-Pressure Velocity-Magnitude Gradient

The inlet pressure experiences a minimum until the introduction of a high-pressure jet in the flow, leading to the formation of lambda shockwaves with expansion fan regions in the duct. Pressure fluctuations occur around these shockwave regions, and the shocktrain diminishes towards the outlet. In Fig. 57.4(i), where the jet is introduced at an 80-degree angle, the pressure change is minimal, gradually increasing with angles of 90, 100, and 110 degrees in Fig. 57.4(ii)-(iv). Such angles were taken in order to study the pattern of change occurring with an increase in every 10 degrees. These angles provide flow direction in both forward and reverse order, providing significant cases to understand the effects of angles to improve mixing. The highest pressure change is observed in Fig. 57.4(iv), while the lowest is in Fig. 57.4(i). The diminishing of shockwaves toward the outlet is highest in Fig. 57.4(iv) and lowest in Fig. 57.4(i). Shocktrain formation leads to boundary layer separation on the walls.

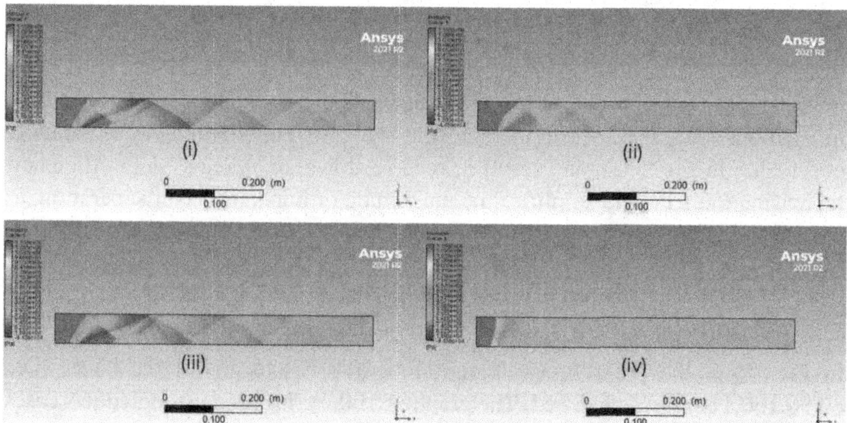

Fig. 57.4 Static-pressure contour (i) 80-degree, (ii) 90-degree, (iii) 100-degree, (iv) 110-dgeree

Source: Author's Compilation

Maintaining a supersonic flow velocity at the inlet (Mach 2), the subsonic-speed jet is introduced at Mach 0.5, resulting in the interaction of subsonic and supersonic flows, causing the formation of shockwaves. In Fig. 57.5(i)-(iv), with jet angles of 80, 90, 100, and 110 degrees, changes in velocity and the formation of lambda shockwaves impeaching on walls are evident, forming a shocktrain toward the outlet. Constant velocity changes occur when jets are introduced at

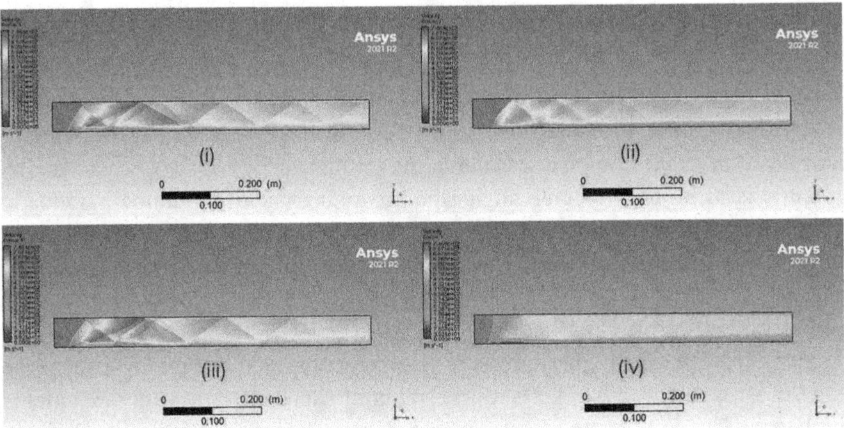

Fig. 57.5 Velocity magnitude contour (i) 80-degree, (ii) 90-degree, (iii) 100-degree, (iv) 110-degree

Source: Author's Compilation

lower angles, transitioning to instant changes as the jet angle increases. Jet-flow interactions and shockwave impact contribute to boundary layer separation, observable in Fig. 57.5, and result in reverse flows, reducing system efficiency. Increasing the jet angle amplifies the magnitude of boundary layer separation, as seen in Fig. 57.5(i)-(iv).

3.2 Graphs of Pressure Variation along X-axis Velocity Magnitude along Y-axis

In Fig. 57.6, the pressure variations along the X-axis of all the cases (i.e., 80,90,100,110) were observed. The graph of 100- & 110-degree angled jets cases have similar variation but also has a considerable change in the magnitude of pressure. While in the case of jets angled at 80- & 90-degree angles, the graphical representation shows similar variations in both the cases and a similar magnitude also. The pressure magnitude observed along the inlet is low and consistent in 80- & 90-degree angled cases till a high-pressure jet is introduced. In the case of 100- &110-degree angles, due to opposing angles, the flow gets disrupted before reaching the jet-inlet. There is a sudden pressure drop at the outlet is observed as no specific boundary conditions were provided there due to the primary focus on understanding mixing inside the duct.

In Fig. 57.7, (i) & (ii), the variations in velocity magnitude along the Y-axis for 80- & 110-degree angled cases respectively, were observed. There were considerable

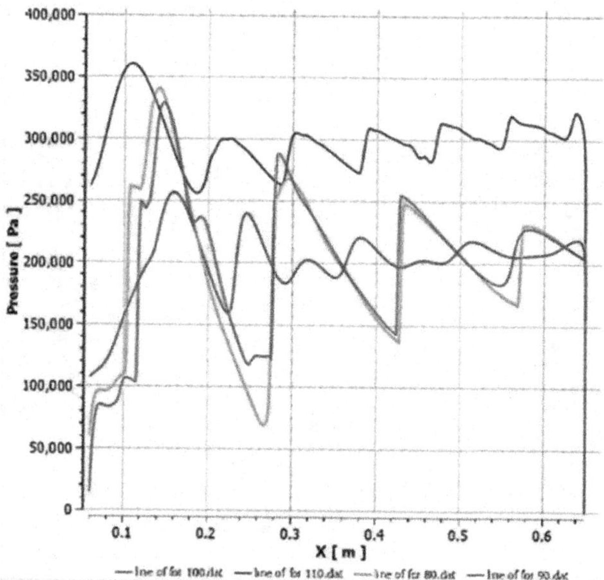

Fig. 57.6 Graph of pressure variation along x-axis in 80,90,100,110-degree angled cases

Source: Author's Compilation

variations observed in the case of an 80-degree angled jet due to the shockwave formed in the flow. While in the jet introduced at 110-degree angle, the variations are smooth along the axis compared to that of 80-degree case observed due to less impact of shockwaves on the flow. A magnitude drop is observed along the walls of the duct due to the boundary layer separation caused due to the interaction and impeachment of shockwaves on the walls. The aspect of Boundary layer separation is not kept in focus in this work as a means of study for the same exists for reference. The creation of boundary layer is a byproduct of mixing. Therefore, this work mainly focuses on the mixing of flow inside the duct.

It can be inferred from the simulation that the data obtained shows the effects on injection angle in a constant area duct. To verify the data achieved, literature (Murty and Chakraborty 2011) is taken into consideration as a reference. On comparing the graphical data of velocity-magnitude along the y-axis of both studies given in Fig. 57.7 (i)-(ii) and Fig. 57.8, on a qualitative basis, it can be observed to attain similar data in both the studies which verify that the data obtained is accurate.

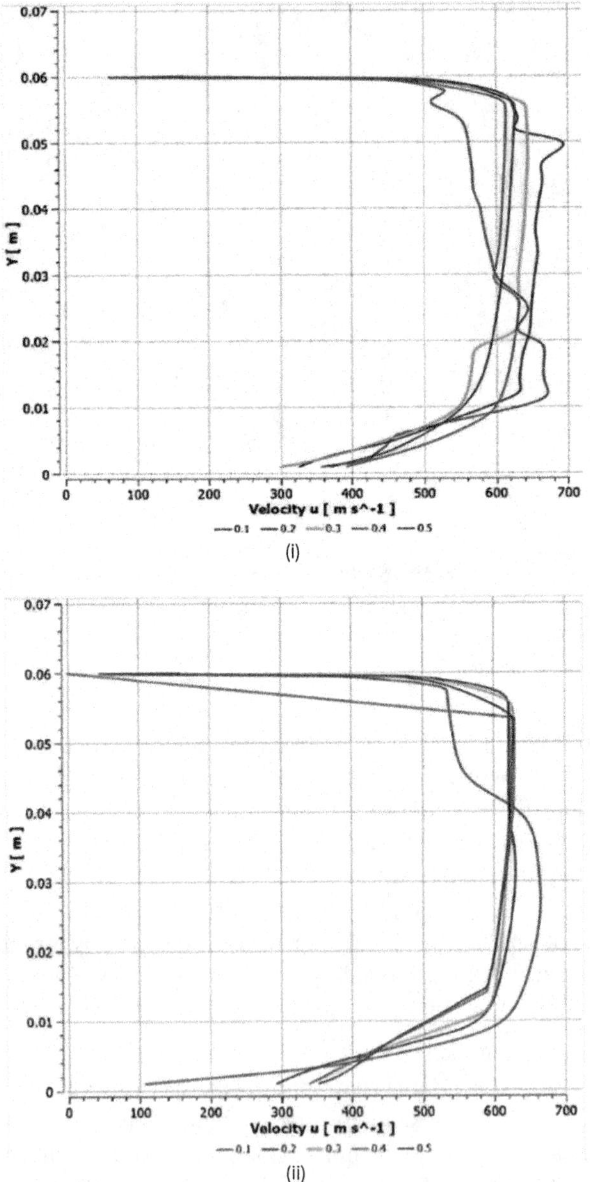

Fig. 57.7 Graph of velocity-magnitude variation along the y-axis in (i) 80-degree angled case, (ii) 100-degree angled cases

Source: Author's Compilation

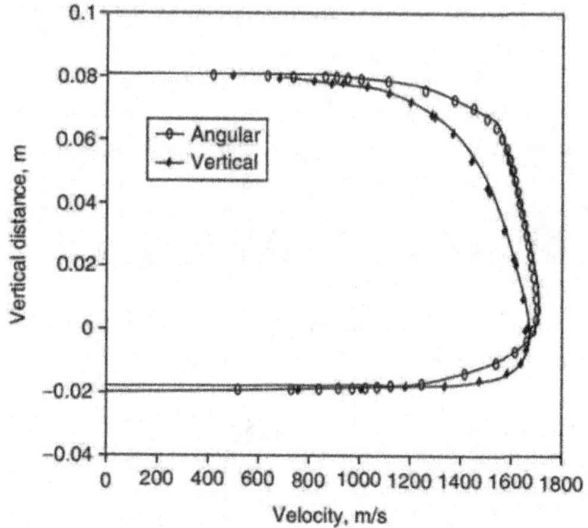

Fig. 57.8 Graph of velocity-magnitude variation along y-axis

Source: Effect of Injection Angle in Mixing and Combustion Characteristics of Scramjet Combustor

4. Conclusion

The Computational Fluid Dynamics (CFD) study concludes by showing the critical influence of injection angle on shockwave dynamics and flow mixing in a constant area duct, which is essential for maximizing the performance of scramjet combustors in hypersonic circumstances. By carefully analysing angled jet injection in a 2-D duct arrangement, the authors found that mixing occurs at different lengths of the duct with respect to the injection angle. This information is important as it helps in improving stability and combustion efficiency. These discoveries have a significant impact on the development of hypersonic propulsion technology and move us one step closer to the development of dependable and effective scramjet engines for use in upcoming aerospace projects.

Acknowledgement

The authors are thankful to the aeronautical department of SVIT Vasad, Gujarat for giving an opportunity to carry out the research work.

References

1. F Bazdidi-Tehrani. 2004. "Numerical analysis of a single row of coolant jets injected into a heated crossflow." *Journal of Computational and Applied Mathematics; Vol 168 Issues1-2* 53-63.

2. "3D Steady & Unsteady bifurcations in shockwave/laminar boundary layer interaction: A numerical study."
3. Ahuja, Vivek, and Roy Hartfield. 2009. "Optimization of Fuel-Air mixing for a Swept Ramp Scramjet Combustor Geometry using CFD and a Genetic Algorithm." *45th AIAA/ASME/SAE/ASEE Joint Propulsion Conference & Exhibit.* Denver.
4. Anderson, J D. 1989. *Hypersonic and high-temperature gas dynamics.* New York: McGraw-Hill.
5. "Flame propagation and stabilization in dual-mode scramjet combustors." *Progress in Aerospace Sciences, Vol 101.*
6. Mahesh, Krishnan. 2013. "The Interaction of Jets in Crossflow." *Annual Review of Fluid Mechanics Volume 45.*
7. Murty, MSR Chandra, and Debasis Chakraborty. 2011. "Effect of Injection Angle in Mixing and Combustion Characteristics of Scramjet Combustor." *International Journal of Hypersonics.*
8. Patel, A V, N J Shah, and J B Sura. 2023. "Numerical Simulation of a Jet in Cross Flow in a Constant Area Duct." *Reimagining Tomorrow: Shaping the Future through Disruptive and Interdisciplinary Technologies.* Jabalpur. 45-47.
9. Segal, Corin. 2009. *The Scramjet Engine; Processes & Characteristics.* Cambridge: Cambridge University Press.
10. Sura, J B. 2020. "CFD Simulation of Hypersonic Shock Tunnel Nozzle." *In Recent Advances in Theoretical, Applied, Computational and Experimental Mechanics: Proceedings of ICTACEM 2017.* Kharagpur: Springer, Singapore. 381-386.
11. Yao, Y., J. Yao, D. Petty, P. Barrigton, and P. Mason. 2006. "Direct numerical simulation of jets in cross-flow." *International Journal of Computational Fluid Dynamics* 279-285.

Technologies for Energy, Agriculture, and Healthcare – Shailesh Nikam et al. (eds)
© 2024 Taylor & Francis Group, London, ISBN 978-1-032-98028-7

58

A Low-cost Acetone Sensor using Amorphous InGaZnO Transistor for Potential Diagnosis of Diabetes by Breath Analysis

Maruti Zalte[1]
Associate Professor,
K J Somaiya College of Engineering,
Somaiya Vidyavihar University,
Mumbai, India

Maryam Shojaei Baghini[2]
Professor,
Department of Electrical Engineering,
Indian Institute of Technology Bombay,
Mumbai, India

Abstract: Diabetes is among the fastest growing chronic health condition globally. Currently, diagnosis and monitoring of diabetes involves blood tests to measure the glucose level. The test requires pricking a finger to obtain a drop of blood, which is then applied to the sensitive area of the glucometer strip. Although the test is simple, it can be painful, invasive, costly, and potentially unsafe if not handled correctly. Analysing human breath provides a non-invasive, cost-effective, and rapid technique for identifying volatile organic compounds that indicate various diseases. In individuals with diabetes mellitus, the body generates excessive amounts of ketones like acetoacetate, beta-hydroxybutyrate, and acetone. The estimation of the acetone concentration in the exhaled human-breath can provide a convenient way to screen diabetes at the early stage and daily monitoring of the health condition of the diabetics. In this study, amorphous Indium Gallium Zinc Oxide (a-IGZO) Thin-film Transistors (TFTs) were fabricated and demonstrated for acetone sensing through in-situ electrical characterization. The a-IGZO thin

[1]marutizalte@somaiya.edu, [2]mshojaei@ee.iitb.ac.in

DOI: 10.1201/9781003596707-58

films were produced using a straightforward, cost-effective solution processing method. These a-IGZO TFTs were employed to detect acetone concentrations at ppm levels under UV irradiation at 30 mW/cm². The change in the performance of a-IGZO TFTs due to acetone adsorption under UV light was utilized to achieve high sensitivity. The corresponding sensitivity obtained with acetone exposure is 15 μA/ppm and 28 mV/ppm in terms of saturation drain current (I_{dsSAT}) and threshold voltage shift (ΔV_{TH}), respectively. The a-IGZO TFT showed identical response for 20 repeated cycles of the acetone exposure.

Keywords: Diabetes, Acetone, a-IGZO, TFT, Sensitivity, Threshold voltage shift, Saturation drain current, ppm

1. Introduction

There are several volatile organic compounds (VOC) found in human breath with specific amount of concentration in healthy person [1]. The abnormal concentration VOCs in breath are useful biomarkers for disease detection. Acetone is generated through fatty acid oxidation in conditions such as diabetes due deficit of insulin. Excess acetone in the bloodstream is expelled through the lungs. An abnormal acetone concentration (1.7-3.7 ppm) can be detected in the breath of diabetic patients. The acetone concentration in the breath of a healthy person is below 0.8 ppm [2]. The detection of acetone at sub-ppm level can play an important role to develop a non-invasive, low-cost and faster diagnostic tool for early detection of diabetes. The conventional and very accurate equipment are used for analysis of very low ppm traces of VOCs. However, such equipment are table-top, expensive and are not suitable to be used in portable devices for clinical or home applications for regular monitoring. The alternative approaches summarized in Table 58.1 are being proposed to detect the sub-ppm traces of acetone VOC so as develop portable and economical breath analyser.

Table 58.1 Reported acetone VOC sensors

Ref	Transducer	Material	Operating Temp	Limit of Detection
[1]	Chemo-resistor	InN	150°C	10 ppm
[3]	Chemo-resistor	Si:WO_3	300-500°C	4 ppb
[4]	Chemo-resistor	SnO_2-MWCNT	250°C	1 ppm
[5]	Chemo-resistor	Fe_2O_3	150°C	1 ppm
[6]	Chemo-resistor	TiO_2	200°C	1.5 ppm
[this work]	Amorphous SMO- TFT	IGZO	Room temp + UV	3-4 ppm

The high-temperature operation (150–500°C) for achieving better sensitivity to target analytes limits their application in portable and body-wearable devices. Using ultraviolet light activation during sensor operation is a superb alternative to high-temperature methods. [8]. In this study, a-IGZO TFTs are fabricated using a low-cost solution processing approach without need of clean room facility. TFTs are explored for sensing acetone at room temperature with UV activation (photoelectrocatalysis (PEC)) to trigger the surface reactions.

2. Experimental

2.1 Synthesis of IGZO Solution

The SMO used in the sensing TFT devices is a-IGZO. The IGZO solution was used to prepare thin films of a-IGZO semiconductor. The 0.1-M solution was synthesized as reported in literature [9]. The 99.999 purity chemicals were received from Sigma Aldrich. The metal precursors namely, Indium, Gallium and Zinc with quantity (225.6 gm, 21.6 gm and 31.5 gm respectively) were dissolved 10 mL 2-ethoxyethanol solvent. The atomic ratio of indium gallium and zinc was 9:1:2. The mixture was vigorously stirred at 900 rpm for 20 hours at room temperature, followed by filtration through a 0.25 μm Polytetrafluoroethylene syringe filter. The resulted light yellow coloured stock solution was stored in the amber coloured chemical glass bottles.

2.2 Sensor Fabrication

The sensor TFT structure is shown in Fig. 58.1(a). The highly doped silicon wafer <100> is used as a substrate and the gate terminal of TFT. The thermally grown silicon dioxide (SiO_2) of thickness 100 nm acts as a gate dielectric of the TFT. The source and drain electrodes are realized using the interdigitated Electrode

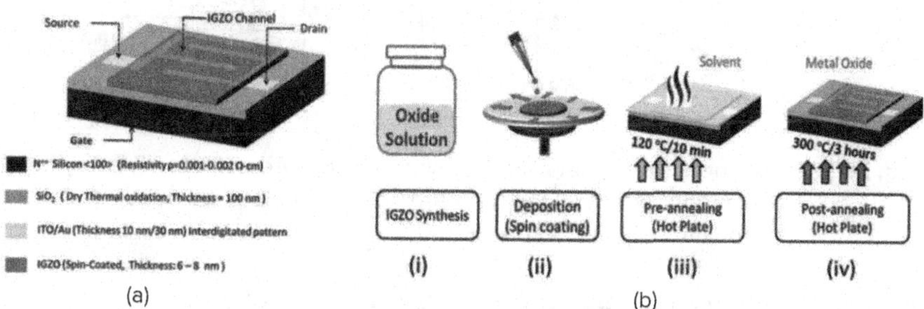

Fig. 58.1 (a) Schematic of a-IGZO TFT device structure (b) Solution process for fabrication of a-IGZO TFT (i) Synthesized IGZO solution (ii) spin-coating IGZO film (iii) Pre-annealing (iv) Post annealing

(IDE) patterns. IDEs are realized using Indium-Tin-Oxide (ITO) and gold (Au). The ITO of 10 nm was deposited on SiO_2 and underneath the gold to ensure the proper adhesion of gold with SiO_2. The a-IGZO thin film forms the semiconductor channel of the TFT. The a-IGZO film was deposited using a solution process. The steps (i-vi) for IGZO film deposition are indicated in Fig. 58.1(b). The details of the material characterization of a-IGZO film and electrical characterization of the sensor TFT are reported in our research studies [10-11].

2.3 In-Situ Acetone Sensing Setup

The gas sensing set-up with UV source as indicated Fig. 58.2 was used to perform acetone sensing experiments. In order to ensure high sensitivity to acetone vapours at room temperature, the photoelectrocatalytic (PEC) reactions were initiated using UV activation. The UV-LED of wavelength (390 - 395 nm) was positioned 1 cm above the sensor TFT. The UV light intensity was controlled using 2.2 kΩ potentiometer connected in series with UV LED across a fixed 5V DC supply. The in-situ measurement of I-V characteristics of a-IGZO TFTs were performed using two Keithley SMUs. SMUs were operated remotely from a personal computer using 'Keithley Kick-Start Software' installed on the PC. It automated data collection from two SMUs and enabled quick test setup and data visualization.

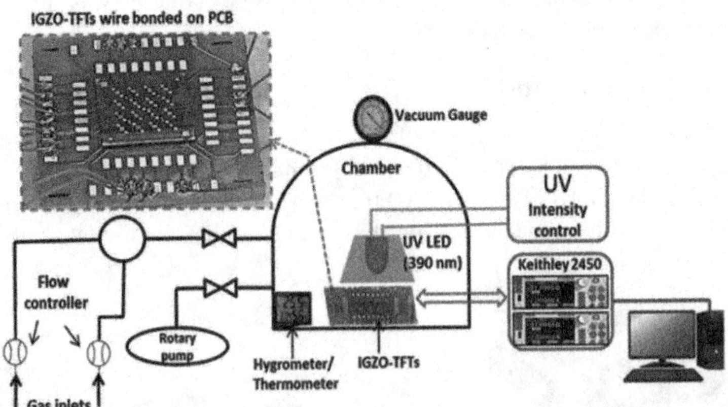

Fig. 58.2 Acetone sensing set-up for in-situ I-V measurement of TFTs exposed to acetone at room temperature. UV activation provided to initiate photoelctrocatalytic reactions

3. Results and Discussions

The fabricated a-IGZO TFT device's performance was evaluated in an acetone gas environment at various concentrations. The current-voltage characteristics of the devices were measured in-situ for acetone gas concentrations ranging from 0 to 30 ppm.

3.1 Response of a-IGZO TFTs to Acetone Gas in Dark (without UV-Activation)

The performance of a-IGZO TFT exposed to acetone in the dark was investigated. I-V characteristics of TFTs were measured in dark (UV OFF) and the evacuated gas chamber (temperature 20°C and relative humidity < 5%) before exposure to acetone. Further, acetone vapours were flown into the chamber. The I_{ds}-V_{ds} and I_{ds}-V_{gs} characteristics of a sample TFT exposed to acetone are shown in Fig. 58.3 (a)-(b), respectively. There was insignificant increase (< 5 μA) in the drain current after exposure of 30 ppm of acetone at V_{gs} = 8 V bias, as shown in Fig. 58.3 (a). A minimal threshold voltage shift (ΔVTH ≈ -0.2 V) was detected in the Ids-Vgs characteristics, as depicted in Fig. 58.3(b). Therefore, in a dark environment and at room temperature, the a-IGZO TFT exhibited no significant response to acetone gas due to the low electrical conductivity of the film and the higher activation energy of the adsorbed oxygen species. Additionally, UV activation was employed to enhance the sensitivity of a-IGZO TFT to acetone.

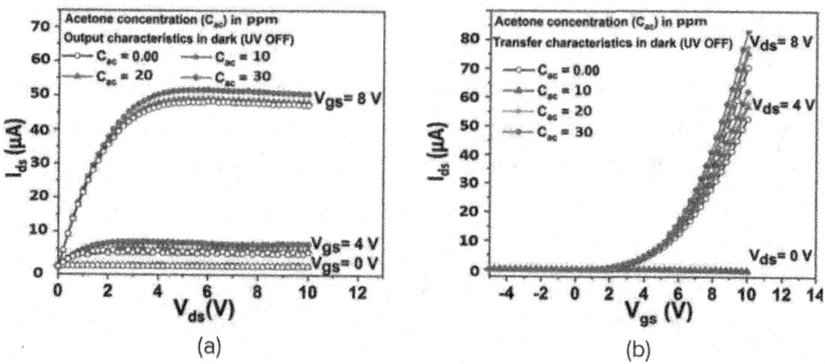

Fig. 58.3 I-V characteristic of a-IGZO TFT (a) output (I_{ds}-V_{ds}) and (b) transfer (I_{ds}-V_{gs}) exposed to acetone in dark at room temperature (UV OFF)

3.2 Acetone Sensing using a-IGZO TFT with UV-Activation

Prior to acetone exposure, I-V characteristics were measured in dark and under UV irradiation to study impact of UV activation on TFTs' electrical performance. The drain current (I_{ds}) was initially 50 μA (at V_{gs} = 8 V) in the dark, which increased to 140 μA after turning the UV source ON. The negative V_{TH} shift (ΔV_{TH} = - 2 V) was observed. This is due to photo-generated carriers in the sensing film exposed to UV light. Further, acetone vapours were introduced into the gas chamber and I-V characteristics of TFTs were measured. As seen from Fig. 58.4 (a), I_{ds}

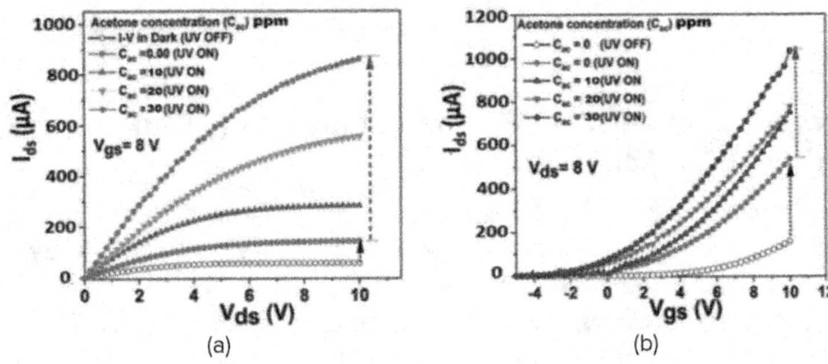

Fig. 58.4 I-V characteristic (a) output (I_{ds}-V_{ds}) and (b) transfer (I_{ds}-V_{gs}), of a-IGZO TFT under UV irradiation (30 mW/cm^2) and exposed to different acetone concentrations

increased from 140 μA to 800 μA for 30 ppm acetone concentration. The negative V_{TH} shift ($\Delta V_{TH} = -2.3$ V) was observed as shown in Fig. 58.4 (b). After acetone gas exposure, numerous oxygen molecules desorb from the surface of a-IGZO, diminishing the space charge layer and augmenting the carrier concentration and conductivity of an IGZO film [179].The intensity of UV has a noticeable impact on the response of the sensor. I-V characteristics of a-IGZO TFTs were measured under 15 mW/cm^2 UV irradiation showed reduced Ids and ΔV_{TH} (180 μA and –1.1 V, respectively) for 30 ppm acetone concentration. The maximum saturation drain current I_{ds} and ΔV_{TH} (extracted from $\sqrt{I_{ds}}$-V_{gs}) of a sample TFT exposed to acetone in dark, under 15 mW/cm^2 and 30 mW/cm^2 UV illumination are indicated in indicated in Fig. 58.5 (a-b).

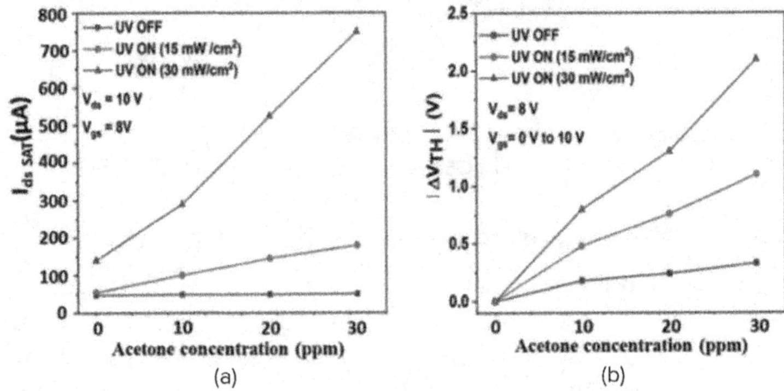

Fig. 58.5 Variation of (a) saturation drain current (I_{dsSAT}) (b) threshold Voltage shift (ΔV_{TH}) with acetone concentration (0-30ppm) in dark and under UV irradiation (15-30 mW/cm^2)

As it can be seen the response of the sensor increased with UV intensity. The sensitivity is defined as $\Delta I_{dsSAT}/\Delta Cac$ (μA /ppm) and $\Delta V_{TH}/\Delta Cac$ (mV /ppm) where Cac is acetone concentration. The sensitivity was extracted from the slope of above plots as 15μA/ppm and 28 mV/ppm in terms of I_{dsSAT} and ΔV_{TH}, respectively.

3.3 Sensing and Recovery Mechanism of Acetone Exposed a-IGZO TFTs

The UV activation was used to realize room-temperature sensing of acetone. The process of UV activation used to trigger the surface reactions is called photoelectrocatalysis (PEC). Free charge carriers are generated when a semiconductor is exposed to UV light having energy more than band gap of the film. The generation of these excess charge carriers can influence the rate of adsorption of a gas on the surface of semiconducting material. The PEC process possible on the surface of a-IGZO is depicted in Fig. 58.8. The adsorbed oxygen in turn generates hydroxyl radical (OH*) or O- centres, which initiate the PEC reaction [12]. Under UV irradiation on a-IGZO TFT, numerous electron-hole pairs are created within the film, as outlined in equation (1), thereby enhancing its conductivity. Furthermore, UV exposure may lead the photocatalytic adsorption of oxygen molecules at the film surface, as depicted in equation (2) [12].

$$h\upsilon \rightarrow h^+ + e^- \tag{1}$$

$$O_{2(gas)} + e^- \rightarrow O^-_{2(h\upsilon)} \tag{2}$$

The chemisorbed oxygen species exhibit strong bonding to the film surface, whereas the photo-catalytically adsorbed species have weaker bonds. This results in a more active surface for targeting gas molecules. Upon introduction of acetone gas into the chamber, the acetone molecules can interact with the surface-adsorbed oxygen species, releasing electrons to the film, as described in equation (3) [13].

$$CH_3COCH_3 + 4O^-_{2(ads)} \rightarrow 3CO_2 + 3H_2O + 4e^- \tag{3}$$

Under constant UV light, the film exhibits heightened reactivity to acetone gas, resulting in higher conductivity compared to dark conditions. Consequently, exposure to acetone prompts the desorption of numerous oxygen molecules from the film , leading to a reduction in the space charge layer and a subsequent rise in the conductivity of the film. After completion of I-V characterization at various concentrations of acetone under UV irradiation, the UV was switched OFF and the gas chamber was purged with dry nitrogen. I-V characteristics of acetone-exposed TFTs were measured. It was observed that the negative V_{TH} shift and the increased drain current did not recover to its pristine I-V characteristics. TFTs were exposed to UV irradiation in the ambient air for 10 minutes and then I-V characteristics of TFTs measured in an evacuated chamber in dark. As seen from Fig. 58.6 (a-b), TFT showed excellent recovery after acetone sensing. The acetone sensing experiment was repeated using the same TFTs to ensure the repeatability of the sensor.

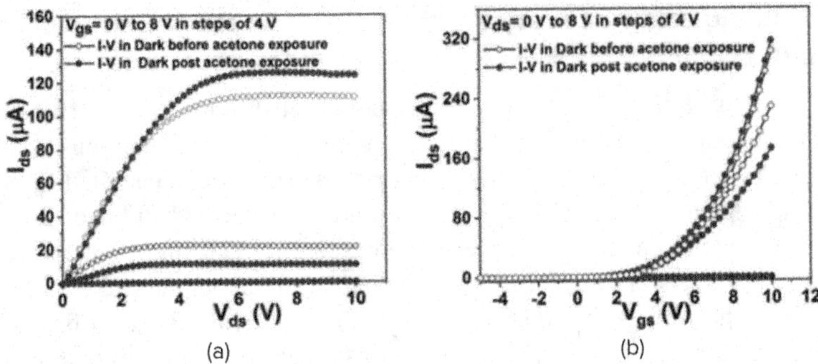

Fig. 58.6 I-V characteristic (a) output (I_{ds}-V_{ds}) and (b) transfer (I_{ds}-V_{gs}) of recovered a-IGZO TFT after acetone sensing. TFT exposed to UV irradiation of 15 mW/cm^2 for 15 minutes

4. Conclusion

The a-IGZO TFT is used for sensing acetone at room temperature. The acetone sensing was performed at room temperature using UV activation. The a-IGZO TFT-based acetone sensor showed sensitivities of 9 µA/ppm and 28 mV/ppm expressed in terms of saturation drain current and threshold voltage shift, respectively. The acetone exposed TFTs showed complete recovery after exposure to UV irradiation in the ambient air for 10 minutes. The amalgamation of cost-effective production, exceptional sensitivity, and efficient recovery of solution-processed a-IGZO-TFT makes it a promising candidate for deployment in breath analysis systems intended for early diabetes detection.

References

1. K.Kao, M. Hsu, Y. Chnag, S. Gwo, and J. Yeh, "A Sub-ppm Acetone Gas Sensor for Diabetes Detection Using 10 nm Thick Ultrathin InN FETs," *Sensors* Vol. 12, No.6 (2012), pp. 7157–7168,
2. C. Deng, J. Zhang, X.Yu, W.Zhang, and X. Zhang. "Determination of acetone in human breath by gas chromatography–mass spectrometry and solid-phase microextraction with on-fiber derivatization," Journal of Chromatography B 810, no. 2 ,2004: pp. 269–275.

3. M. Righettoni, A. Tricoli, and S. Pratsinis, "Si:WO 3 Sensors for Highly Selective Detection of Acetone for Easy Diagnosis of Diabetes by Breath Analysis," Analytical Chemistry, Vol. 82, No. 9, 2010: pp. 3581–3587
4. M. Narjinary, P. Rana, A. Sen, M. Pal, "Enhanced and selective acetone sensing properties of SnO2-MWCNT nanocomposites: Promising materials for diabetes sensor," Materials & Design, Vol.115, 2017: pp. 158–164.
5. S. Chakraborty, D. Banerjee, I. Ray and A. Sen, "Detection of biomarker in breath: a step towards non-invasive diabetes monitoring, Current Science. Vol. 94 2008:pp. 237–242.
6. L. Deng, C. Zhao, Y. Ma, S. Chen, and G. Xu, "Low cost acetone sensors with selectivity over water vapour based on screen printed TiO2 nanoparticles," Analytical Methods, Vol. 5 , 2013 :pp. 3709–3713.
7. F. Vajhadin, M. M.Ardakani and A.Amini, "Metal oxide-based gas sensors for the detection of exhaled breath markers," Medical Devices Sensors. Wiley periodicals LLC, 4:e10161 2021.
8. R.Jaisutti, J.Kim, S. K. Park, and Y.H.Kim, "Low-Temperature Photochemically Activated Amorphous Indium-Gallium-Zinc Oxide for Highly Stable Room Temperature Gas Sensors," ACS Applied Material Interfaces, 8, 20192–99, 2016.
9. Y. Rim, H. Chen, Y. Liu, S. Bae, H. Kim, and Y. Yang, "Direct light pattern Integration of low-temperature solution-processed all oxide flexible electronics,"ACS Nano, vol. 8, no. 9, 2014: pp. 9680–9686.
10. M. Zalte, V. Kumar, S. Surya and M. Baghini, "A Solution Processed Amorphous InGaZnO Thin-Film Transistor-Based Dosimeter for Gamma-Ray Detection and Its Reliability," IEEE Sensor Journal, Vol. 21, 9, 2021:pp 10667–74
11. M.. Zalte , T. Naik , A. Alka, M. Ravikanth, V. Rao and M. Baghini, "Passivation of Solution-Processed a-IGZO Thin-Film Transistor by Solution Processable Zinc Porphyrin Self-Assembled Monolayer," IEEE Transaction On Electron Devices, VOL. 68, NO. 11, 2021: pp 5920–5924.
12. R.Jaisutti, J.Kim, S. K. Park, and Y.H.Kim, "Low-Temperature Photochemically Activated Amorphous Indium-Gallium-Zinc Oxide for Highly Stable Room Temperature Gas Sensors," ACS Applied Material Interfaces, 8, 20192–20199, 2016.
13. F. Shaobin, F. Farha, Q. Li, Y. Wan, Yang Xu, Tao Zhang, and H. Ning, "Review on Smart Gas Sensing Technology" Sensors 19, no. 17: 3760. 2019

Note: All the figures and table in this chapter were made by the authors.

Technologies for Energy, Agriculture, and Healthcare – Shailesh Nikam et al. (eds)
© 2024 Taylor & Francis Group, London, ISBN 978-1-032-98028-7

59

Torrefied Briquettes: A Sustainable and Renewable Energy Source

M. Nikam*, A. Bhongade,
V. Bhirud, R. Mahajan, D. Gode, R. Wankhede
Department of Mechanical Engineering,
Bharati Vidyapeeth College of Engineering,
Navi Mumbai, Maharashtra, India

S. Jadhav
Bharati Vidyapeeth College of Engineering,
Navi Mumbai, Maharashtra, India

Abstract: This review examines the status of research, production, and commercialization of torrefied biomass briquettes. The torrefaction process and its impacts on biomass feedstocks are analysed. Torrefied briquettes are compared to wood, charcoal, and coal across physical, chemical, and combustion characteristics to evaluate their efficacy as fuels. Their performance in industrial and small-scale applications is reviewed to assess suitability. The economics and life-cycle greenhouse gas emissions of torrefied briquette production is also evaluated to determine market viability and environmental sustainability. This review synthesizes current knowledge to provide a comprehensive overview of the opportunities and challenges for torrefied biomass briquettes. With the continued depletion of fossil fuel reserves and rising energy costs, there is an urgent need to develop renewable, sustainable solid fuel alternatives. One promising option is torrefied biomass briquettes. Torrefaction is a thermal pretreatment process that partially pyrolyzed biomass at 200-300°C to remove moisture, volatiles, and oxygen. Results showed torrefied briquettes achieved higher density (1.2 g/cm3), durability (97%), grindability, hydrophobicity, calorific value (22 MJ/kg), and combustion efficiency compared to conventional briquettes from raw

*Corresponding author: manoj.nikam133@gmail.com, vedantgb@gmail.com

DOI: 10.1201/9781003596707-59

biomass. Optimal production conditions were found to be 12% molasses binder, 6% moisture, and 120 MPa pressure. Comprehensive models based on response surface methodology were developed to predict the effects of production variables on briquette properties. Torrefied biomass can then be densified into briquettes that have comparable handling, transport, and combustion properties to coal, but a reduced environmental impact.

Keywords: Torrefied biomass, Briquettes, Renewable fuel, Green future, Sustainable fuel

1. Introduction

Energy has always played an important role in life, survival, and the development of mankind. Even though it has been superseded by other more potent fossil energy sources during the last 200 years, biomass has played a major role in supplying energy since the beginning of civilization and still plays an important role in economies of developing countries. Biomass has recently received renewed attention worldwide, mainly because of high and volatile oil prices and global climate change caused by increased fossil fuel consumption. Moreover, rapid economic growth in developing countries, high dependence on global and local transportation, pollution, depletion of sources, and endangered national security of energy importing countries have raised the awareness of the need for non-fossil based renewable energy sources. Among renewable energy sources, such as wind, solar, geothermal, wave, tidal and ocean thermal, biomass is the most likely short term energy source with mature and readily applicable conversion technologies for the production of transportation compatible liquid fuels. With increasing energy consumption worldwide, finding renewable fuel sources is critical to reduce reliance on fossil fuels. Agricultural residues like coconut shells have potential to be converted into sustainable solid biofuels. However, raw biomass has disadvantages like low energy density, high moisture content, and poor grindability compared to coal (Van der Stelt et al., 2011). Torrefaction can upgrade biomass to improve its fuel properties. Biomass briquettes are gaining popularity as renewable energy sources, but have limitations in India and globally. Torrefaction, first developed in the 1980s, can upgrade biomass to improve fuel properties.

Torrefaction is a thermal process conducted at 200-300°C under inert atmosphere that partially decomposes biomass (Chen et al., 2015). It was initially researched by the French (Arias et al., 2008). Torrefaction degrades hemicellulose and makes the biomass hydrophobic (Dudyński et al., 2015). This increases energy density, grindability, and stability of biomass. Studies show torrefaction improves fuel properties of coconut shells. Medic et al. (2012) reported a 25-30% increase in energy density. Arias et al. (2008) found 80% reduction in grinding energy for torrefied coconut shells. The briquettes meet coal standards (Phanphanich & Mani, 2011).

The limitations of torrefied biomass from coconut shells encompass technical, economic, and environmental factors. The torrefaction process itself requires significant energy input and infrastructure investment, potentially increasing operational costs and limiting scalability. Ensuring a reliable and sustainable supply chain for coconut shell biomass may also be challenging due to competition with other uses and logistical constraints. Environmental considerations include energy consumption, greenhouse gas emissions, and proper management of torrefaction by-products.

But after all these limitations studies show torrefaction improves fuel properties of coconut shells. Medic et al. (2012) reported a 25-30% increase in energy density. The briquettes meet coal standards (Phanphanich & Mani, 2011). While torrefied briquettes can cost 20-30% more than normal briquettes, the fuel property improvements justify the price premium by reducing the ash content, increasing the energy density, and reduction of volatile matter.

2. Methodology and Model Specifications

The aim of this study is to produce torrefied coconut shell briquettes using starch binder and evaluate their combustion performance compared to raw briquettes burned in indigenous conditions. Methodology of work is shown in Fig. 59.1.

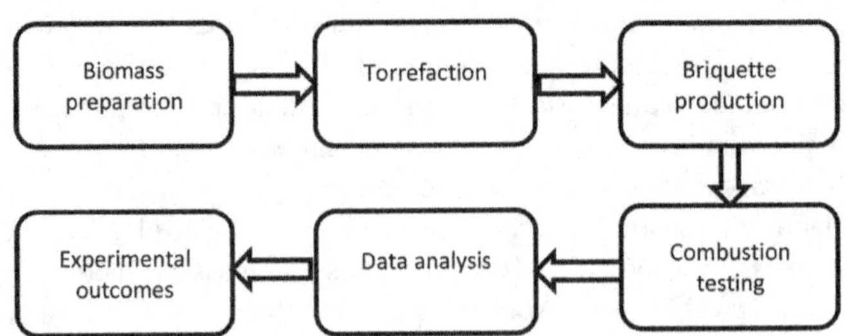

Fig. 59.1 Methodology flowchart for torrefied briquettes

2.1 Biomass Preparation J

Coconut shells will be procured from local street markets in the region. They will be washed to remove dirt and debris and then sun-dried for 5 days as shown in the Fig. 59.2. The shells will be crushed into 10–20 mm sized pieces using manual labour and indigenous mortar and pestles.

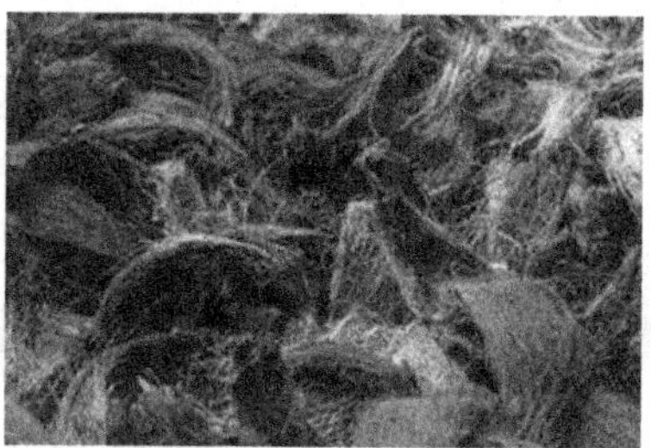

Fig. 59.2 Coconut shell

2.2 Torrefaction

Torrefaction is a thermal process converting biomass into a more energy-dense and stable material. Studies also explore torrefaction's environmental benefits, like reduced greenhouse gas emissions and improved handling characteristics. Terrified biomass can be seen in Fig. 59.3 (a). Current research focuses on optimizing torrefaction parameters for various feedstocks, enhancing energy yield, and integrating torrefied biomass into existing energy systems. Typically conducted at 200-300°C in low oxygen environments, it reduces moisture content by 70-80% and increases energy density by 20-30% this comparison can be seen in Fig. 59.3 (b).

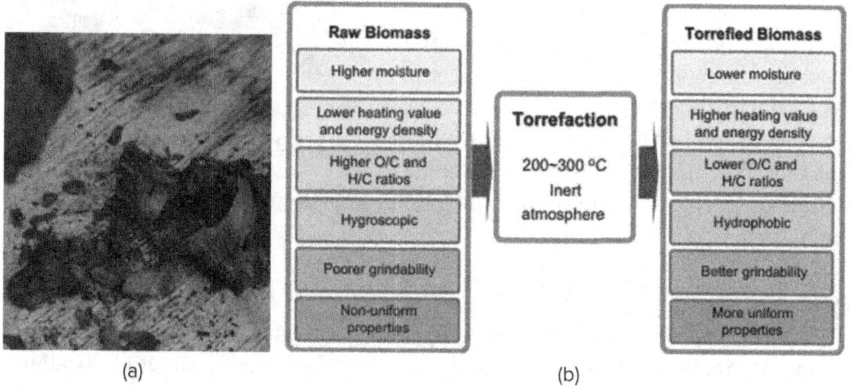

Fig. 59.3 (a) Torrefied coconut shell, (b) Difference between raw and torrefied biomass [1]

2.3 Briquette Production

The crushed coconut shell pieces will be mixed with starch binder solution in 1:1 ratio by weight. The starch binder will be prepared by gelatinizing rice starch in water. The coconut shell-starch mixture as shown in Fig. 59.4(a) will be compressed into ≈50 mm diameter briquettes using a manually operated hydraulic press as shown in Fig. 59.4(b)

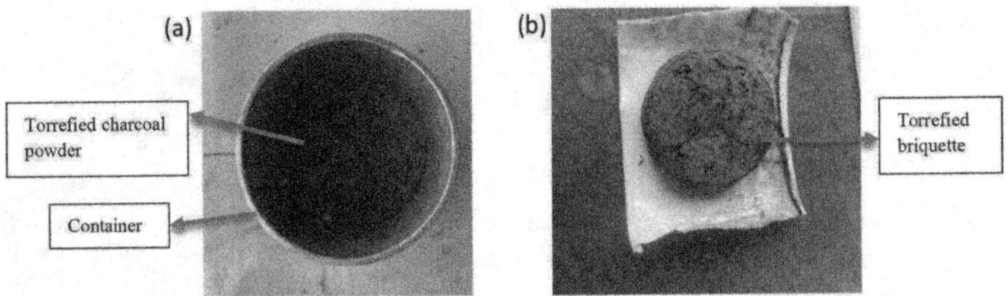

Fig. 59.4 (a) Powdered torrefied coconut shells, (b) Compressed briquette

2.4 Combustion Testing

Experimental combustion trials were conducted as shown in Fig. 59.5, comparing the burning characteristics of both raw and torrefied briquettes. Utilizing a rudimentary steel barrel stove, parameters such as burning rate, flame intensity, and residue production were meticulously observed and juxtaposed between the two types of briquettes. These trials aim to discern any notable differences in combustion behaviour, offering insights into the potential advantages of torrefied biomass for stove applications, including improved efficiency and reduced emissions.

Fig. 59.5 Burning of briquette

2.5 Experimental Outcomes

It is observed that the torrefied briquettes will have improved combustion properties including faster ignition, higher calorific value of about 6000cal/gm, and reduced residual char i.e about 6-7%. The results will provide insights into the effects of simple torrefaction on upgrading coconut shell briquettes for cooking applications using local methods and materials.

3. Comparative Analysis of Conventional and Torrefied Briquette

To evaluate the advantages of torrefied biomass briquettes, their properties were compared against normal type of briquettes produced from raw, untreated biomass feedstocks.

3.1 Conventional Briquettes

The exact calorific value, moisture and ash contents can vary depending on factors like the source of coconut shells, pre-treatment methods, and briquetting techniques used which can be seen in Table 59.1. But the below ranges represent reported values for densified coconut shell briquettes based on research literature.

Table 59.1 Fuel properties of coconut shell and torrefied coconut shell briquettes

Parameter	Coconut shell	Torrefied Coconut Shell
Gross calorific value	4000 Cal/gm	6085 Cal/gm
Moisture content	8%	6.90%
Ash content	7%	6.80%

Well-produced coconut shell briquettes meet typical quality, efficiency, and low emissions criteria for solid biofuels through their standardised calorific value, moisture content, and ash content specifications. The parameters indicated in Table 59.1 serve as reasonable reference values for further coconut shell briquette production and applications.

3.2 Comparison of Burning Characteristics for Raw vs. Torrefied Briquettes

Table 59.2 compares how raw and torrefied briquettes burn. Torrefied briquettes are drier, burn slower, and release less smoke. They also have higher energy content due to less moisture and more fixed carbon. While needing slightly higher ignition temperatures, torrefied briquettes offer a cleaner and more efficient burn.

Table 59.2 Comparison of burning characteristics for Raw vs. torrefied briquettes

Characteristic	Raw Briquettes	Torrefied Briquettes
Moisture Content (%)	15	6.90
Volatile Matter (%)	75	55
Fixed Carbon (%)	17	23
Heating Value (MJ/kg)	16	20
Ignition Temperature (°C)	Lower	Higher (by 50-100°C)
Burning Rate (kg/hr)	Higher	Slower, more controlled burn
Smoke Emissions	Higher	Lower

3.3 Torrefied Briquettes

The fuel property data presented in this work was obtained through experimental testing of torrefied coconut shell briquette samples at Italab Private Ltd facilities available at Mumbai as show in Table 59.1. The lab analysis aims to evaluate key fuel metrics to determine combustion suitability as an energy-dense solid biofuel.

The fuel properties of torrefied coconut shell briquettes were investigated through proximate analysis. Torrefaction was utilised as a thermochemical pre-treatment to transform the raw coconut shells into an upgraded briquette fuel. As shown in Table 59.1 the torrefied briquettes demonstrated a high gross calorific value of 6085 cal/g, indicating substantial energy density for fuel applications. This is 35% higher than raw, untreated coconut shell briquettes typically ranging from 4000-4500 cal/g. A relatively low moisture content of 6.9% as illustrated in Fig. 59.6. was achieved for the torrefied briquettes, likely a result of volatiles expelled during the >=250°C torrefaction temperature exposure. This moisture level meets general briquette fuel quality standards (<10%) for efficient combustion. The ash content at 6.8% by weight is reasonably within norms for densified biomass briquettes, avoiding slagging/fouling issues during burning. This confirms the torrefaction process did not excessively concentrate problematic inorganic compounds like alkali metals or chlorine from the original biomass. The mild conditions decomposed organics while maintaining a manageable ash composition, avoiding issues like slagging, corrosion, and emissions that could occur if troublesome inorganics were highly concentrated. The torrefied biomass and resulting briquettes thus have an ash profile comparable to conventional biomass fuel, without disproportionate levels of detrimental inorganic components.

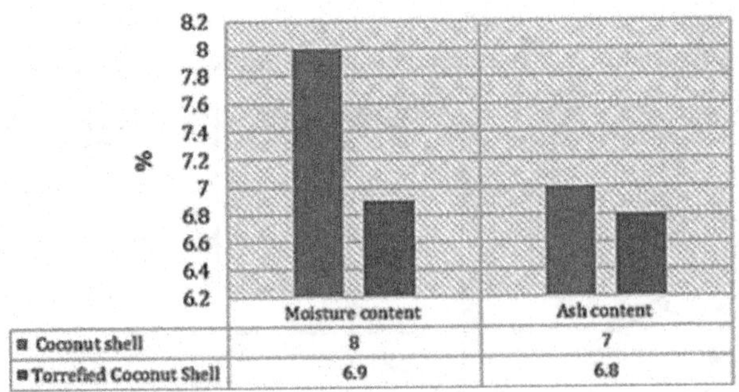

Fig. 59.6 Graph for comparison of ash & moisture content

Overall, the torrefaction modification generated superior fuel properties compared to regular coconut shell briquettes. Specifically enhanced energy value (<6000cal/

gm) as illustrated in Fig. 59.7. kept moisture low (<8%) and maintained ash within acceptable levels i.e 6-9% for practical combustion performance as a biofuel.

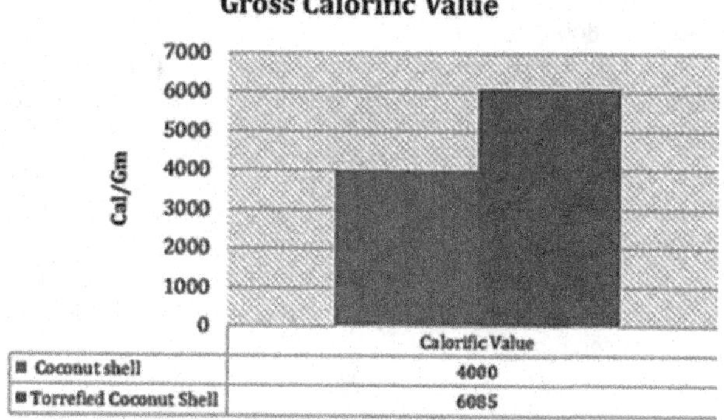

Fig. 59.7 Graph for comparison of calorific value

4. Conclusion

This study successfully optimized the production of torrefied briquettes from coconut shell biomass feedstock. Coconut shells were torrefied at optimal conditions of 275°C for 45 minutes before briquetting. The ideal briquette production parameters were determined to be 15% molasses binder, 8% moisture content, and 130 MPa compression pressure. The grindability was improved over conventional coconut shell briquettes. Hydrophobicity was enhanced by the torrefaction process, leading to better resistance to moisture absorption and improved storage stability.

Most notably, the calorific value of the torrefied coconut shell briquettes was 6085 Cal/gm - over 40% higher than the 4000 Cal/gm for conventional coconut shell briquettes. The ash content remained reasonably low at 6.8%. Combustion performance was also significantly improved, with more even and efficient burning compared to untreated coconut shell briquettes.

The response surface models developed accurately predicted the effects of production factors on all critical quality metrics. This optimization framework provides valuable insights for maximizing fuel properties of torrefied coconut shell and other similar biomass briquettes.

In conclusion, torrefaction briquetting can unlock the full potential of coconut shells and other agricultural residues as renewable, high-performance solid biofuels. The enhanced energy density, durability, hydrophobicity, and combustion behaviour of optimized torrefied coconut shell briquettes make them an ideal sustainable fuel source for industrial and domestic energy applications.

References

1. *Chen, W.H., J. Peng, and X.T. Bi. 2015.* A state-of-the-art review of biomass torrefaction, densification and applications. Renewable and Sustainable Energy Reviews 44: 847–866.
2. *Li, J., A. Brzdekiewicz, W. Yang, and W. Blasiak. 2012.* Co-firing based on biomass torrefaction in a pulverised coal boiler with aim of 100% fuel switching. Applied Energy 99: 344–354.
3. *Abnisa, F., W.M.A. Wan Daud, W.N.W. Husin, and J.N. Sahu. 2013.* Utilisation possibilities of palm shell as a source of biomass energy in Malaysia by producing bio-oil in pyrolysis process. Biomass and Bioenergy 56: 72–82.
4. *Dudyński, M., K. Kwiatkowski, and M. Sosnowska. 2015.* Biomass gasification: influence of torrefaction on syngas production and tar formation. Fuel Processing Technology 131: 203–212.
5. *Van der Stelt, M.J.C., H. Gerhauser, J.H.A. Kiel, and K.J. Ptasinski. 2011.* Biomass upgrading by torrefaction for the production of biofuels: A review. Biomass and Bioenergy 35(9): 3748–3762.
6. *Medic, D., M. Darr, A. Shah, B. Potter, and J. Zimmerman. 2012.* Effects of torrefaction process parameters on biomass feedstock upgrading. Fuel 91(1): 147–154.
7. *Lu, K.M., W.J. Lee, W.H. Chen, S.H. Liu, and T.C. Lin. 2012.* Torrefaction and low temperature carbonization of oil palm fibre and eucalyptus in nitrogen and air atmospheres. Bioresource Technology 123: 98–105.
8. *Arias, B., C. Pevida, J. Fermoso, M.G. Plaza, F. Rubiera, and J.J. Pis. 2008.* Influence of torrefaction on the grindability and reactivity of woody biomass. Fuel Processing Technology 89(2): 169–175.
9. *Phanphanich, M., and S. Mani. 2011.* Impact of torrefaction on the grindability and fuel characteristics of forest biomass. Bioresource Technology 102(2): 1246–1253.
10. *Bach, Q.V., K.Q. Tran, Ø. Skreiberg, E. Trømborg, and M.A. Gjennestad. 2016.* Effects of wet torrefaction on reactivity and kinetics of wood under air combustion conditions. Fuel 177: 342–350.

Note: All the figures and tables in this chapter were made by the authors.

Technologies for Energy, Agriculture, and Healthcare – Shailesh Nikam et al. (eds)
© 2024 Taylor & Francis Group, London, ISBN 978-1-032-98028-7

60

Energy Optimization of Highly Nonlocal Soliton via Diffraction Management

Soumendu Jana
SPMS, Thapar Institute of Engineering and Technology,
Patiala, Punjab, India

Mohit Sharma, Brajraj Singh
Mody University of Science and Technology,
Lakshmangarh, Sikar, Raj., India

Manoj Mishra*
SciSER, SKSC, Somaiya Vidyavihar University,
Vidyavihar, Mumbai, India

Abstract: The Optical solitons are self-localized beams or pulses originating under nonlinear confinement and can be used in numerous applications, namely, all-optical communication, data processing, encryption and all-optical devices. In this study, we present the energy optimization of optical solitons via a diffraction-managed microstructure system made by alternating positive and negative diffracting media. The system, modelled by a nonlocal nonlinear Schrodinger equation, is solved through analytical and numerical methods. The initial beam energy needed for diffraction-managed soliton formation has been determined under varying local and average diffraction values. Results indicate reduced energy requirements for soliton formation at higher diffraction values compared to a constant diffraction system, while slightly more energy is needed at lower diffraction values. The closed contours in phase trajectories under perturbation ensures stable soliton generation within the system.

Keywords: Diffraction management, Nonlocal soliton, Lorentz variational method, and SSFM

*Corresponding author: manoj2712@gmail.com

DOI: 10.1201/9781003596707-60

1. Introduction

Energy optimization is an essential task for any tool or system, starting from a gigantic ocean liner to a microscopic quantum device and so for optical devices. Optical solitons have wide potential applications and essentially energy optimization is relevant for them. Optical solitons are self-localized beam or pulses that arises due to the balance of the self-diffraction of the beam by the nonlinearity-induced self-focusing or due to the balance of dispersion-induced pulse broadening by the nonlinearity induced self-phase modulation of the pulse. The spatial soliton found in highly nonlocal media is called accessible soliton (AS). Following their identification by Snyder and Mitchell [1], numerous theories and experiments on ASs have been documented [2–9]. The nonlocality characteristics have three classifications: weakly nonlocal, generally nonlocal, and highly nonlocal [2, 3, 10, 11]. Nonlinearity is considered weakly nonlocal when $\sigma \approx w$, strongly nonlocal when $\sigma >> w$, and the intermediate case is termed general nonlocality, here w is beam width and σ is response function's characteristic length [6]. Theoretical studies typically employ two types of response functions for nonlocal media: Gaussian-type [2] and exponential-decay type [10]. ASs have been observed to display large phase shifts [6], attraction dynamics between out-of-phase solitons [10], attraction dynamics between dark solitons [12], and other noteworthy phenomena.

During the late 1990s, a pioneering approach for temporal optical solitons emerged, termed as *dispersion managed soliton.* This method presents numerous advantages in comparison to conventional soliton-based systems [13, 14]. Conversely, the diffraction managed (DM) soliton is formed using an arrangement of waveguides featuring alternating normal and anomalous diffraction segments, akin to the dispersion managed soliton [15].

This article introduces a theoretical model for the highly nonlocal soliton via diffraction management (DM) using a nonlocal nonlinear Schrodinger equation (NNLSE). The soliton dynamics are explored employing both the Lagrangian variational method (LVM) and the split-step-Fourier method (SSFM). Section 2 outlines the mathematical formulation, while the analysis of results is presented in section 3. The article concludes in section 4.

2. Mathematical Modelling

The dynamics of an optical soliton in a highly nonlocal nonlinear media alternating between positive and negative diffraction can be modelled using the NNLSE [4]:

$$i\frac{\partial Y}{\partial z}+\frac{d(z)}{2}\frac{\partial^2 Y}{\partial x^2}+\rho Y\frac{P_0}{\sqrt{\pi}\sigma}\left(1-\frac{x^2}{\sigma^2}\right)=0 \tag{1}$$

The equation is defined with $Y(x,z)$ representing the beam profile, where z signifies the propagation distance along the z-axis. The local diffraction parameter is denoted by $d(z)$, and $\rho = k\eta$ is defined in terms of the wavenumber k and material constant η. The characteristics length of the response function is represented by σ. The symmetric diffraction management map is adopted, with the diffraction coefficient $d(z)$ varying along the solid blue line, as depicted in Fig. 60.1 two-unit cells.

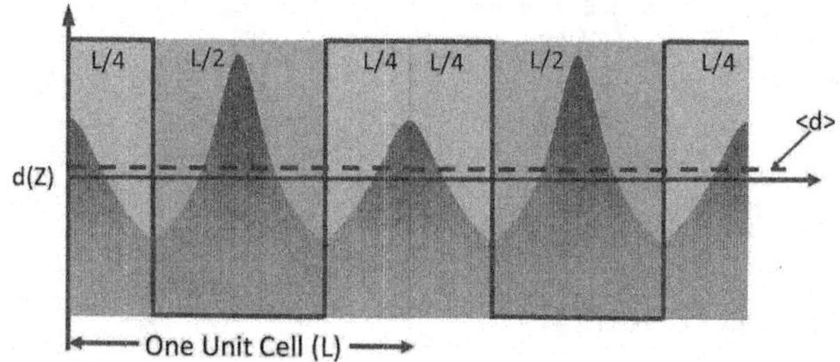

Fig. 60.1 A DM map featuring a periodic pattern of alternating positive and negative diffraction over two-unit cells

The LVM [16], an approximate analytical approach, is employed due to the absence of a direct solution for the NNLSE(1). We adhered to the standard procedure to derive the Lagrangian density, as outlined in [16].

$$L=\frac{1}{2}\left(Y*\frac{\partial Y}{\partial z}-Y\frac{\partial Y^*}{\partial z}\right)-\frac{d(z)}{2}\left|\frac{\partial Y}{\partial x}\right|^2+\frac{\rho P_0|Y|^2}{2\sqrt{\pi}\sigma}\left(1-\frac{x^2}{\sigma^2}\right) \tag{2}$$

We use a Gaussian profile to solve (1) as:

$$Y(x,z)=A(z)\exp\left(-\frac{x^2}{2w^2(z)}\right)\exp\left(ic(z)x^2+i\theta(z)\right) \tag{3}$$

The beam's amplitude, width, phase front curvature, and phase are represented by $A(z)$, $w(z)$, $c(z)$, and $\theta(z)$, respectively. By substituting the expression (3) into (2) and integrating, we obtain the average Lagrangian $\langle L\rangle$.

$$\langle L\rangle=\frac{\sqrt{\pi}\rho w^2A^4}{2\sqrt{2w^2+\sigma^2}}-\sqrt{\pi}A^2\left(w\frac{\partial\theta}{\partial z}+\frac{w^3}{2}\frac{\partial c}{\partial z}+\frac{d(z)}{2}\frac{(1+4c^2w^4)}{2w}\right) \tag{4}$$

The Euler-Lagrange equation,

$$\frac{\partial\langle L\rangle}{\partial r_j}-\frac{\partial}{\partial Z}\left(\frac{\partial\langle L\rangle}{\partial\left(\frac{\partial r_j}{\partial Z}\right)}\right)=0,$$

is applied to get following differential equations,

$$\frac{\partial c}{\partial z}=\frac{d(z)}{2}\left(-4c^2+\frac{1}{w^4}\right)-\frac{\rho E_0}{\sqrt{\pi}\,(2w^2+\sigma^2)^{3/2}}, \tag{5}$$

$$\frac{\partial\theta}{\partial z}=\frac{\rho E_0(5w^2+2\sigma^2)}{2\sqrt{\pi}(2w^2+\sigma^2)^{3/2}}-\frac{d(z)}{2}\frac{1}{w^2}, \text{ and} \tag{6}$$

$$\frac{\partial w}{\partial z}=\frac{d(z)}{2}4cw. \tag{7}$$

Also, $E_0=\sqrt{\pi}wA^2$, where E_0 is the initial beam power.

3. Results and Discussion

The governing NNLSE, represented by Eq. (1), has been numerically solved using the SSFM. Figure 60.2 illustrates the presence of DM solitons, characterized by periodic oscillations in width and intensity, within a single DM cell length for both the average positive DM map and the average negative DM map.

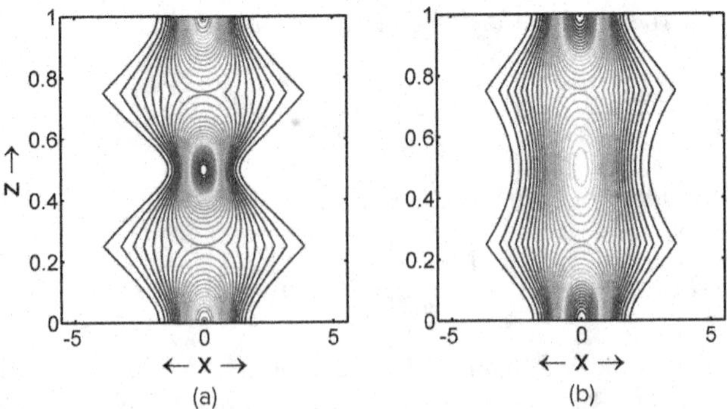

Fig. 60.2 Evolution pattern of a DM breather soliton over a single cell length is depicted for both (a) a +ve ⟨*d*⟩ and (b) a -ve ⟨*d*⟩

To elucidate the beam evolution characteristics during the propagation of DM solitons, we employ the Range-Kutta method to solve the coupled ordinary differential equations (ODEs) from Eq.(5) to Eq.(7) under periodic boundary conditions $w(0) = w(L)$ and $c(0) = c(L)$. Throughout this investigation, the initial values for parameters w, c, ρ, and σ are consistently set at 1,0, 1/6, and 10, respectively, unless specified otherwise. Notably, these solitons exhibit breather-like behaviour during propagation.

As an example, when we set the map parameters to $d_1 = 10$ and $d_2 = -9$, the average diffraction becomes positive (·*d*Ò = +0.5), leading to a closed loop in the $w - c$ phase diagram (Fig. 60.3a). This closed-loop indicates periodic oscillations in the width (w) of the DM breather soliton, ensuring stability in a highly nonlocal nonlinear medium. Additionally, the pulse width w exhibits periodic variations (Fig. 60.3b), while the chirp c does not show regular changes along the propagation length z (Fig. 60.3c). Similarly, when the map parameters are set to $d_1 = -10$ and $d_2 = 9$, resulting in a negative average diffraction (·*d*Ò = −0.5), the dynamics exhibit similarities, as depicted in the second row of Fig. 60.3.

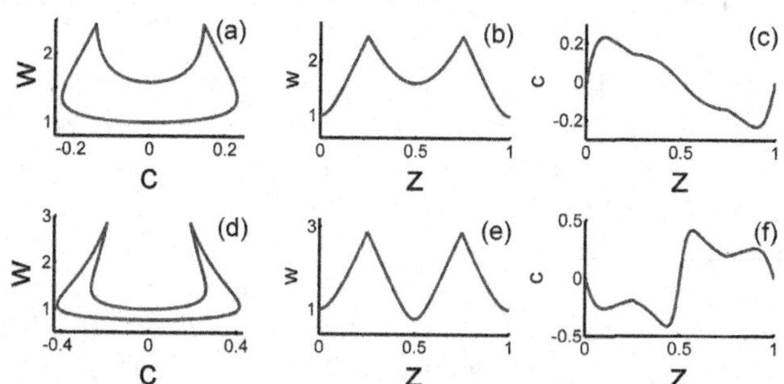

Fig. 60.3 (a & d) Trajectory in the beam width (w) - chirp (c) phase space, (b & e) Chirp (c) variation with propagation distance z, and (c & f) Beam width (w) variation with z. The first-row data corresponds to one DM cell length with positive average diffraction, using parameters $\langle d \rangle = +0.5$, $d_1 = 10$, $d_2 = -9$, and $E_0 = 4082$. The second-row data is for one DM cell length with negative average diffraction, with parameters $\langle d \rangle = -0.5$, $d_1 = -10$, $d_2 = 9$, and $E_0 = 2649$

Moreover, a w-c phase diagram is graphed across single or multiple DM unit cells, illustrating the real, imaginary, and complex Eigenvalues (see Fig. 60.4). Notably, all the phase trajectories form simply connected closed contours, indicating the reliable generation of stable DM breather spatial solitons. Here λ_1 and λ_2 are the Eigenvalues obtained in linear stability analysis of Eq. 7 and Eq. 5.

We calculated the impact of variations in the local diffraction parameter on the required input energy E_0 by setting $\langle d \rangle$ as negative (i.e., $\langle d \rangle = -0.5$) and adjusting

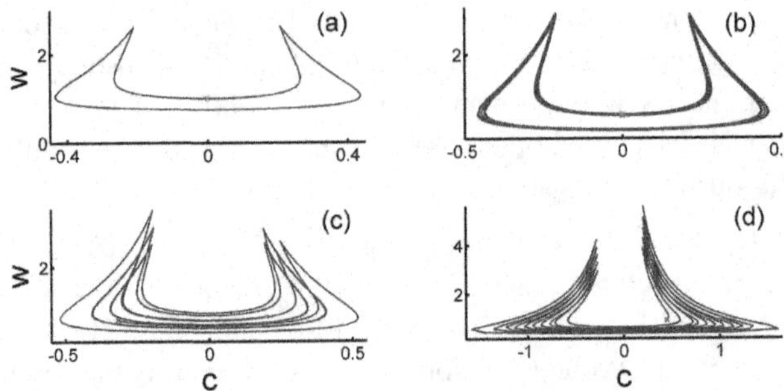

Fig. 60.4 Phase trajectories for (w-c) are depicted for different cases: (a) when both Eigenvalues are purely imaginary, (b) when both Eigenvalues are purely imaginary, (c) when λ_1 and λ_2 are complex conjugates, and (d) when both λ_1 and λ_2 are real. Only case (a) is illustrated for a single DM cell length, while the remaining cases are presented for five DM cell lengths

the combination of d_1 and d_2 values. For higher values of d_1, the value of E_0 exhibits a rapid decrease (Fig. 60.5a). A similar dataset was computed for positive $\langle d \rangle$, and the trend of E_0 with d_1 is nearly the opposite of its negative $\langle d \rangle$ counterpart (Fig. 60.5b). Notably, although the range of d_1 and d_2 variations is consistent for both cases, the negative $\langle d \rangle$ demonstrates a much broader range of E_0 (38,803 units) compared to the positive $\langle d \rangle$ (1,393 units). Furthermore, the negative $\langle d \rangle$ yields lower values of E_0 for larger d_1 or d_2, while the positive $\langle d \rangle$ results in higher values of E_0. Consequently, as depicted in Fig. 60.5(a & b), one can intelligently select

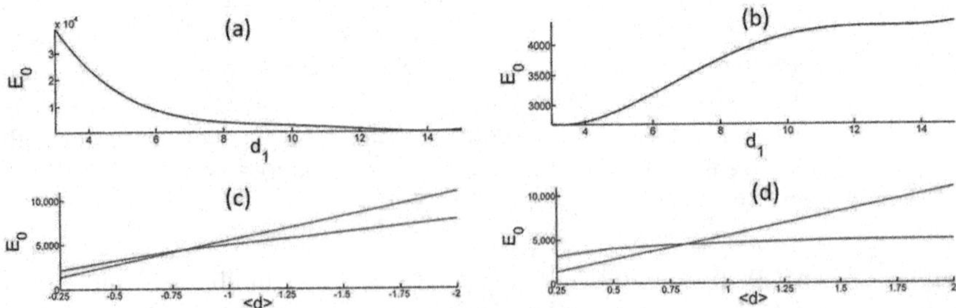

Fig. 60.5 Variation of initial soliton energy (E_0) is depicted against (a) d_1 for $\langle d \rangle = -0.5$, (b) d_1 for $\langle d \rangle = +0.5$, (c) -ve $\langle d \rangle$, with a constant local diffraction coefficient $d_2 = +8.0$, and (d) +ve $\langle d \rangle$, with $d_2 = -8.0$. The variation of E_0 is presented for both cases with a constant diffraction (CD) system (in the second row)). The green line represents the soliton in the CD system, while the magenta line corresponds to the soliton in the DM system

the parametric region for low or high-beam energy operations. The reason behind the above-mentioned observation lies in the dependency of the system behaviour on the coupled action of the nonlinearity, nonlocality and nature of dispersion. It may be noted that the E_0 and other terms are in normalized units. So, the results can be used for different scales. For the present case, the energy will be generally in milli Joule order.

The superiority of the DM system over a CD system regarding the energy needed for stable spatial soliton formation is evident in Fig. 60.5(c & d). The DM system exhibits a notably slower rate of energy increase with dispersion compared to the CD system. Consequently, the DM system demands significantly less energy for stable soliton formation, providing flexibility in selecting the operating energy level based on specific requirements.

4. Conclusion

The article explores the energy optimization of optical solitons via diffraction management in highly nonlocal nonlinear media. In comparison to conventional CD systems, these solitons demand less energy for their formation, rendering them suitable for low-energy applications such as beam switching and sensing. The breather characteristics of DM solitons have been validated through both numerical and analytical methods, providing valuable insights for designing experiments related to all-optical devices.

References

1. Allan W Snyder and D John Mitchell. Accessible solitons. *Science*, 276(5318): 1538–1541, 1997.
2. Wieslaw Krolikowski, Ole Bang, Jens Juul Rasmussen, and John Wyller. Modulational instability in nonlocal nonlinear kerr media. *Physical Review E*, 64(1):016612, 2001.
3. Wies law Krolikowski and Ole Bang. Solitons in nonlocal nonlinear media: Exact solutions. *Physical Review E*, 63(1):016610, 2000.
4. Manoj Mishra, Sandeep Kumar Kajala, Mohit Sharma, Swapan Konar, and Soumendu Jana. Energy optimization of diffraction managed accessible solitons. *JOSA B*, 39(10):2804–2812, 2022.
5. Manoj Mishra, Sandeep Kumar Kajala, Mohit Sharma, Swapan Konar, and Soumendu Jana. Generation, dynamics and bifurcation of high power soliton beams in cubic-quintic nonlocal nonlinear media. *Journal of Optics*, 24(5):055504, 2022.
6. Qi Guo, Boren Luo, Fahuai Yi, Sien Chi, and Yiqun Xie. Large phase shift of nonlocal optical spatial solitons. *Physical Review E*, 69(1):016602, 2004.
7. Marco Peccianti, Katarzyna A Brzdkakiewicz, and Gaetano Assanto. Nonlocal spatial soliton interactions in nematic liquid crystals. *Optics letters*, 27(16):1460–1462, 2002.

8. M. Mishra, S. K. Kajala, M. Sharma, B. Singh, and S. Jana. Stabilizing the optical beam in higher-order nonlocal nonlinear media. In *Frontiers in Optics + Laser Science 2022 (FIO, LS)*, page JTu5A.42, 2022.
9. S. Jana, M. Sharma, B. Singh, S. K. Kajala, and M. Mishra. Interaction dynamics of accessible solitons in highly nonlocal cubic-quintic nonlinear media. In *Frontiers in Optics + Laser Science 2022 (FIO, LS)*, page JTu5A.21, 2022.
10. Per Dalgaard Rasmussen, Ole Bang, and Wieslaw Krolikowski. Theory of nonlocal soliton interaction in nematic liquid crystals. *Physical Review E*, 72(6):066611, 2005.
11. Daniel Buccoliero, Anton S Desyatnikov, Wieslaw Krolikowski, and Yuri S Kivshar. Laguerre and hermite soliton clusters in nonlocal nonlinear media. *Physical review letters*, 98(5):053901, 2007.
12. Alexander Dreischuh, Dragomir N Neshev, Dan E Petersen, Ole Bang, and Wieslaw Krolikowski. Observation of attraction between dark solitons. *Physical review letters*, 96(4):043901, 2006.
13. S Konar, Manoj Mishra, and S Jana. Dispersion-managed optical solitons with higher-order nonlinearity. *Fiber and integrated optics*, 24(6):537–548, 2005.
14. Manoj Mishra and WP Hong. The role of the asymmetry of a dispersion map in a dispersion managed optical communication system possessing quintic nonlinearity. *J. Korean Phys. Soc.*, 58:1614, 2011.
15. HS Eisenberg, Yaron Silberberg, R Morandotti, and JS Aitchison. Diffraction management. *Physical Review Letters*, 85(9):1863, 2000.
16. Dan Anderson. Variational approach to nonlinear pulse propagation in optical fibers. *Physical review A*, 27(6):3135, 1983.

Note: All the figures in this chapter were made by the authors.

Technologies for Energy, Agriculture, and Healthcare – Shailesh Nikam et al. (eds)
© 2024 Taylor & Francis Group, London, ISBN 978-1-032-98028-7

61

A Comparative Study on Spatial Diversity Techniques in Multiantenna Systems over Various Fading Channels

Aditi Upadhyay[1]
Student, K J Somaiya College of Engineering,
Somaiya Vidyavihar University,
Mumbai

Vaidehi Bhiwapurkar[2]
Student, K J Somaiya College of Engineering,
Somaiya Vidyavihar University,
Mumbai

Kartik Patel[3]
Student, K J Somaiya College of Engineering,
Somaiya Vidyavihar University,
Mumbai

Abstract: This paper presents an investigation into the advantages of Multiple Input Multiple Output (MIMO) systems in augmenting wireless network capacity and data rates. MIMO, characterized by the utilization of multiple antennas in both transmission and reception, surpasses alternative systems by offering superior data rates and capacity. With a specific emphasis on Binary Phase Shift Keying (BPSK), the study delves into MIMO's lower Bit Error Rate (BER) compared to other modulation techniques, indicative of enhanced signal transmission quality and minimized impact on transmitted bits. The paper confronts challenges associated with multipath fading in communication channels, particularly in Single Input Single Output (SISO) systems, and undertakes an analysis of Rayleigh fading effects. It evaluates the design and performance of MIMO antennas for wireless communication systems, with a particular focus on the 2x2 MIMO system

[1]aditi.u@somaiya.edu, [2]v.bhiwapurkar@somaiya.edu, [3]kartik@somaiya.edu

DOI: 10.1201/9781003596707-61

employing adaptive measures to mitigate wireless channel impairments, including multipath fading. Utilizing SIMULINK in MATLAB, the study undertakes the modeling of the SISO communication system and assesses the BER performance of various spatial diversity techniques. It evaluates the Alamouti Space-Time Block Code (STBC) 2x2 and Maximum Ratio Combining (MRC) code under both Line of Sight and Multipath propagation scenarios. Among the modulation techniques considered, BPSK and QPSK exhibit the least Probability of Error (POE) of -5.91 for an Eb/No value of 10 dB. Regarding system configurations, MIMO offers the lowest BER of 0.022% for an Eb/No value of 2 dB. Additionally, among SISO, SIMO, MISO, and MIMO, MIMO demonstrates the highest capacity of 28.4599 b/s/Hz for an SNR value of 25 dB.

Keywords: MIMO, BPSK, STBC, MRC, Diversity, Multiple antennas, Multipath fading, Rayleigh fading

1. Introduction

5G, the latest wireless standard, connects various entities with features like ultra-low latency, high data speeds, reliability, availability, network capacity, and user experience. With projected rates of up to 20 Gbps, 5G emphasizes spectrum efficiency for effective data delivery without compromising quality. [5]

Spectral efficiency, vital in cellular networks, sets the maximum data transmission rate for a user count without compromising service quality. The utilization of Multiple Input Multiple Output (MIMO)[1], a high-efficiency transmission technology with multiple antennas, enhances system performance and enables higher data rates. Smart devices adhering to the 802.11n wireless standard can effectively leverage MIMO capabilities.

As mobile networks advance, optimizing spectral efficiency becomes crucial, requiring focused efforts to address wireless communication challenges. This research employs MATLAB and Simulink to scrutinize diverse spatial diversity strategies to enhance spectral efficiency across varying fading channels.[9]

2. Literature Review

2.1 Study the Performance of Capacity for SISO, SIMO, MISO, and MIMO in Wireless Communication

The expanding realm of wireless communication requires a reliable system to cater to customers' increasing needs for channel capacity and data transfer rates. Multiple

Input Multiple Output (MIMO) systems, featuring multiple antennas at both the transmitter and receiver, provide spatial diversity and enable multiplexing. This study offers a comparative analysis of capacity performance in SISO, SIMO, MISO, and MIMO systems, emphasizing MIMO systems for their enhanced capacity and faster data transmission speeds. The abundant antennas in the MIMO system substantially improve its capacity, positioning it as an optimal choice for contemporary communication technologies such as 3G, 4G, WLAN, LTE, and others.

The MIMO system is superior to modern communication technology due to its heightened capacity and accelerated data transmission speeds facilitated by multiple antennas on both the transmitter and receiver sides.

2.2 BER Analysis of SIMO and MIMO Systems with Rayleigh Fading Using SIMULINK

This paper investigates Rayleigh fading in communication channels employing SIMO and MIMO models under AWGN Channel conditions. It analyzes BER performance across different SNR values. As the number of receivers increases, BER approaches that of an ideal SISO system. Moreover, the Alamouti STBC 2×2 MIMO system with QAM exhibits superior performance compared to PSK, FSK, and PAM within an SNR range of 0–15 dB.

QAM signals achieve higher data rates and better BER performance than PSK, PAM, and FSK within an SNR range of 0–15 dB. However, increasing M in M-ary modulation can raise BER despite higher data rates.

2.3 Performance of MIMO Systems over Rayleigh Channels

Explored MIMO systems, MIMO technologies, space-time coding, spatial multiplexing, and OFDM to tackle temporal and spatial challenges. Stressed the importance of spatial variation in reducing error rates for enhanced signal reception. The use of transmitters and receivers allows for unlimited capacity expansion. The integration of MIMO-OFDM eliminates channel selectivity and reduces symbol interference, resulting in improved signal reception. Higher diversity order is associated with increased reception rates. In optimal receiving conditions, errors decrease, especially in high signal-to-noise ratio (SNR) scenarios.

MIMO and OFDM integration enhances capacity, reduces errors, and mitigates interference, especially in high SNR conditions.

3. Theory and Concept

3.1 Modulation Techniques

In this study, modulation transforms information into radio waves by adding data to an electrical or optical carrier signal. This involves modifying amplitude, frequency, phase, polarization, and even quantum-level phenomena like spin for optical signals. With a focus on digital signals, the research employs digital modulation methods. Industry-prevalent techniques include amplitude, frequency, and phase modulation.

1. **Binary Phase Shift Keying**

   Binary Phase-Shift Keying (BPSK) is the simplest PSK form, using two-phase offsets for logic high and low. A significant 180° phase difference between these states is employed to ensure system robustness. In response to one logic state, BPSK inverts the carrier, while for the other logic state, the carrier remains unchanged, contributing to system resilience.

2. **Quadrature Phase Shift Keying**

   Quadrature Phase-Shift Keying (QPSK), derived from BPSK, is a Double Side Band Suppressed Carrier (DSBSC) modulation scheme that simultaneously transmits two digital information digits. By converting digital bits into bit pairs, QPSK halves the data transmission rate, creating space for additional users.

3. **8 Phase Shift Keying**

   In 8PSK modulation, the carrier suppresses the upper sideband signals of the two first-order sidebands, dividing them into eight phases on average before synthesis.

4. **16 – Quadrature Amplitude Modulation**

   In 16-QAM, a type of Quadrature Amplitude Modulation (QAM), a fixed-frequency carrier wave is utilized, offering sixteen unique states. Each state is represented by a symbol within a constellation comprising 16 levels of amplitude and phase. The plot's in-phase (X-axis) and quadrature (Y-axis) axes are perpendicular, exhibiting a 90-degree phase disparity.

3.2 Diversity Techniques

Employing multiple channels to capture signal replicas enhances wireless communication system performance over fading channels by leveraging path characteristics and increasing receiver sensitivity. The core principle is information replication. Time diversity minimizes codeword loss during deep fading by retransmitting at intervals, introducing temporal variety without increased power. Frequency diversity includes techniques like Frequency Hop and Direct Sequence Spread Spectrum. Spatial diversity, using multiple antennas, generates diverse signal replicas at the transmitter and receiver, requiring no extra bandwidth. Implementing spatial diversity may involve adjusting transmit antennas for a sufficient signal-to-noise ratio, although this is uncommon due to closed-loop control complexity.

3.3 Classification of Multi-Antenna System

1. SISO

$$C = Blog_2\left(1 + \frac{S}{N}\right) \quad (1)$$

In this expression, C denotes the capacity, B represents the signal's bandwidth, and S/N signifies the signal-to-noise ratio.

2. SIMO

$$C = Blog_2\left(1 + M_R \frac{S}{N}\right) \quad (2)$$

In this context, C denotes capacity, M_R specifies receiver-side antennas, B stands for signal bandwidth, and S/N represents signal-to-noise ratio.

3. MISO

$$C = Blog_2\left(1 + M_T \frac{S}{N}\right) \quad (3)$$

In this context, C denotes capacity, M_T specifies transmitter-side antennas, B stands for signal bandwidth, and S/N represents signal-to-noise ratio.

4. MIMO

$$C = Blog_2\left(1 + M_R \cdot M_T \frac{S}{N}\right) \quad (4)$$

In this context, C denotes MIMO system capacity, M_T represents transmitting antennas, M_R represents receiving antennas, and S/N signifies signal-to-noise ratio.

3.4 Fading Channels

1. **Rayleigh Fading**

   Rayleigh fading, a statistical model depicting radio signal variations from multipath reception, is applicable in diverse conditions like the troposphere, ionosphere, and urban landscapes, where no direct line of sight exists. Mobile antennas encounter reflected waves, creating a random variable in received power due to wave cancellation effects.

2. **Rician Fading Channel**

   Like Rayleigh fading, the Rician fading model includes a dominant component, usually the line-of-sight wave. Advanced models treat this dominant wave as a phasor sum, representing a deterministic process, and address shadow attenuation in satellite channel modeling.

3. **Nakagami Fading Channel**
Utilized to characterize signal variations, the Nakagami-m fading model posits that the received signal's magnitude follows a Nakagami distribution, with 'm' as the shape parameter and 'ω' as the scale parameter.

4. Implementation and Execution

Modulation techniques were evaluated based on Energy per bit to noise power spectral density (Eb/N_0), a normalized signal-to-noise ratio (SNR) in decibels (dB). Higher E_b/N_0 values indicate better signal quality. The considered modulation schemes include BPSK, QPSK, 8-PSK, 16-PSK, 16-QAM, and 64-QAM. The probability of bit error (POE) is inversely proportional to E_b/N_0, with lower POE indicating better performance. MATLAB simulations generated a graph depicting POE against E_b/N_0 values for the different modulation schemes.

1. BPSK

$$P_e = \frac{1}{2} erfc\left(\sqrt{\frac{E_b}{N_0}}\right) \tag{5}$$

2. M-PSK

$$P_e = \frac{1}{M} * erfc\left(\sqrt{\left(\frac{E_b}{N_0} * M\right)} * \sin\left(\frac{\pi}{i}\right)\right) \tag{6}$$

3. M-QAM

$$\frac{2}{M} * \left(1 - \frac{1}{\sqrt{M}}\right) * erfc\left(\sqrt{\frac{\left(3 * \frac{E_b}{N_0} * M\right)}{(2 * (M-1))}}\right) \tag{7}$$

After choosing a modulation technique, performance was evaluated across various multi-antenna systems, considering parameters like Bit Error Rate (BER) and Channel Capacity. Systems assessed include Single Input Single Output (SISO), Single Input Multiple Output (SIMO), Multiple Input Single Output (MISO), and Multiple Input Multiple Output (MIMO).

1. **Bit Error Rate (BER):** BER represents the number of bit errors per unit of time. It is calculated as the ratio of bit errors to the total transferred bits during a specified time interval, expressed as a percentage.

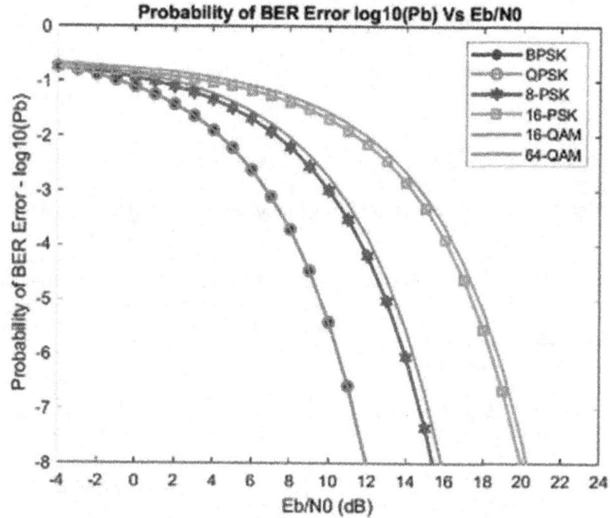

Fig. 61.1 Comparison of modulation techniques

2. **Channel Capacity:** Channel capacity (C) is the maximum rate at which information can be transmitted through a channel, expressed in bits per second. Higher capacity corresponds to greater data rates.
3. **Line-of-sight propagation (LoS)** occurs when stations can only exchange data signals if they are in direct view of each other. LoS propagation depends on the transmitter's circular zone and increases with transmitter height.
4. **Multipath Propagation** involves radio signals reaching the receiving antenna through multiple paths, leading to time dispersion and inter-symbol interference. Causes include atmospheric ducting, ionospheric reflection, and reflection from various surfaces.

The evaluation considers both Line-of-Sight and Multipath Propagation to determine which multiantenna system minimizes BER.

4.1 Simulation of Rayleigh Fading Channel

To explore the impact of various fading channels on MIMO performance, a simulation using MATLAB focuses on the commonly encountered Rayleigh Fading channel. The simulation employs the SISO Fading Channel block filter, representing the input signal through a SISO multipath fading channel that mimics Rayleigh fading.

The input data signal is structured as an NS-by-1 vector, with NS denoting the number of samples in the input signal. Similarly, the output data signal from

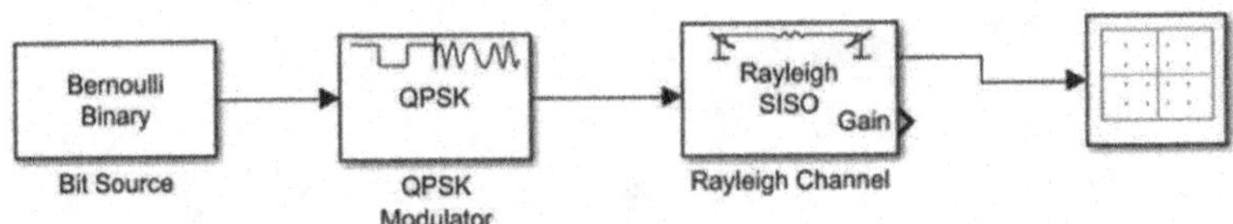

Fig. 61.2 Block diagram of Rayleigh Fading channel

Table 61.1 Input parameters of Rayleigh Fading channel

Sr no.	Input Parameters	
1.	Bit Rate	1 Mbits/second
2.	Bits per frame	2000
3.	Maximum Doppler shift	200 Hz which corresponds to a mobile speed of 65 mph (30 m/s) and a carrier frequency of 2 GHz
4.	For 4 paths, path delay	
	Delay Vector	[0, 200 ns, 400 ns, 800 ns]
	Gain Vector	[0 -3 -6 -9]
5.	For 6 paths, path delay	
	Delay Vector	[0, 0.2ms 0.4ms 0.8ms 1.6ms 3.2ms]
	Gain Vector	[0 -3 -6 -9 -12 -15]
6.	For 2 paths, path delay	
	Delay Vector	[0, 0.2ms]
	Gain Vector	[0 -3]

Source: Author's compilation

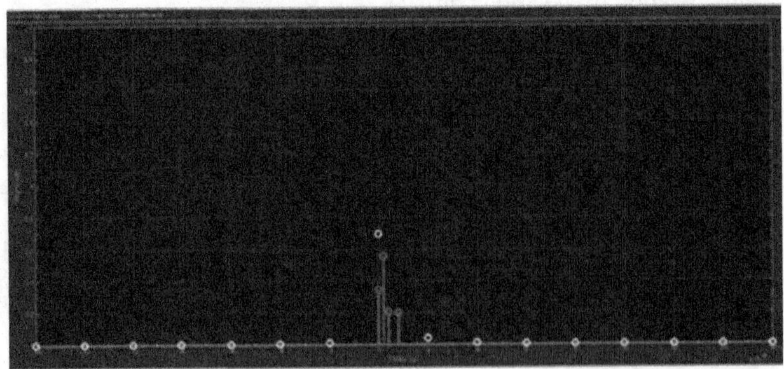

Fig. 61.3 Four path impulse response

the fading channel is obtained as an NS-by-1 vector. The discrete path gains of the underlying fading process are captured in an NS-by-NP matrix, where NP signifies the number of channel paths.

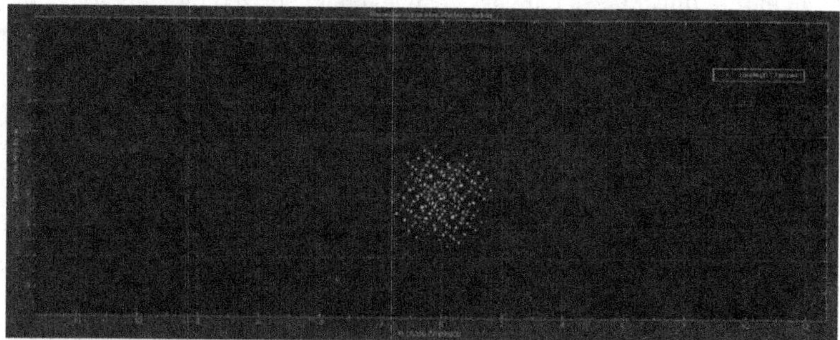

Fig. 61.4 Four path constellation response

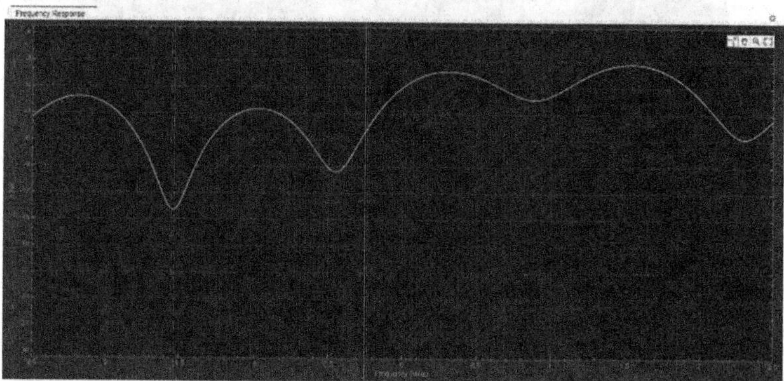

Fig. 61.5 Four path wideband response

Fig. 61.6 Four path narrowband response

Due to limited time dispersion and very small ISI from the impulse response, the frequency response is roughly flat when the bandwidth is too small to discern the individual components.

Rayleigh fading illustrates a wireless channel characterized by numerous random scatterers inducing phase shifts and amplitude variations. Typically, Rayleigh fading manifests in urban areas abundant with obstacles.

4.2 Rayleigh VS Rician

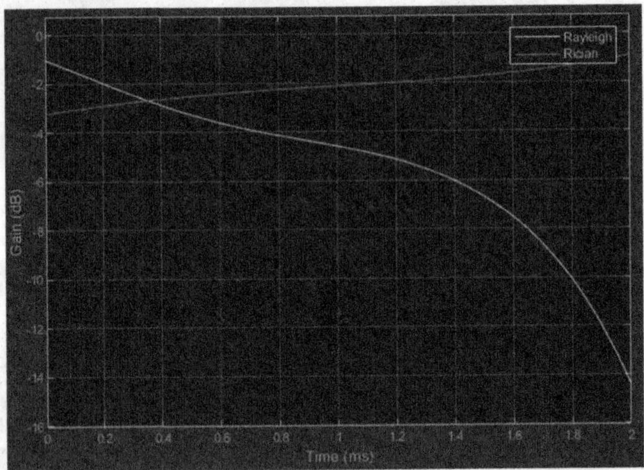

Fig. 61.7 Four path gain of rayleigh and rician

Rayleigh fading characterizes a wireless channel populated by numerous random scatterers, introducing fluctuations in phase and amplitude. This phenomenon is prevalent in urban environments characterized by a high density of obstacles.

On the other hand, Rician fading signifies the influence of both line-of-sight and random scatterers in the wireless channel, leading to unpredictable phase shifts and amplitude variations. This type of fading is commonly observed in rural areas where there is a direct line-of-sight to the receiver, accompanied by substantial scattering.

4.3 Nakagami VS Rayleigh

For a given value of Eb/No, the BER decreases with an increase in the value of m. The greater the value of m, the least is the BER and so the better the signal. In the Nakagami fading channel, for the given value of Eb/No as 12dB, the value of shape factor as 5 provided the least BER of 25.51%, implying that an increase in m leads to a decrease in BER

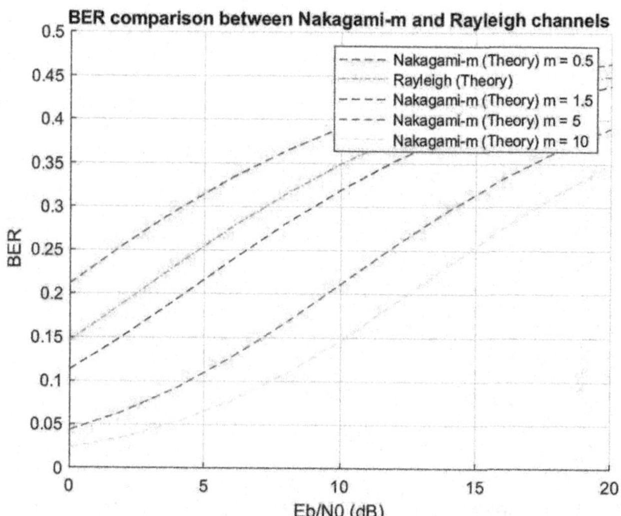

Fig. 61.8 Comparison between rayleigh and nakagami m=5.0

5. Results and Discussion

Out of all modulation techniques considered, BPSK and QPSK had the least POE of -5.91 for the value of Eb/No given as 10 dB. Among SISO, SIMO, MISO, and MIMO, MIMO provided the least BER of 0.022% for the value of Eb/No given as 2dB. Out of all the 4 techniques, MIMO provided the highest capacity of 28.4599 b/s/Hz for the given value of SNR as 25dB. The path gains of Rayleigh and Rician Fading Channels show that Rayleigh is more prevalent in urban settings(more obstacles) while Rician is prevalent in rural areas. (less obstacles).

6. Conclusion

In conclusion, our investigation effectively categorizes multi-antenna systems, emphasizing MIMO's exceptional performance in mitigating signal degradation across diverse scenarios. With its impressive capacity and high data transmission rates, MIMO emerges as a promising option for modern wireless technologies. Our comparative analysis highlights the superior performance of the Alamouti scheme, aligned with a theoretical second-order transit diversity scheme, compared to MRC.

Our examination of fading channels, encompassing Rayleigh, Rician, and Nakagami models, provides valuable insights for optimizing wireless communication. Rayleigh fading is suited to urban environments with numerous reflections, while

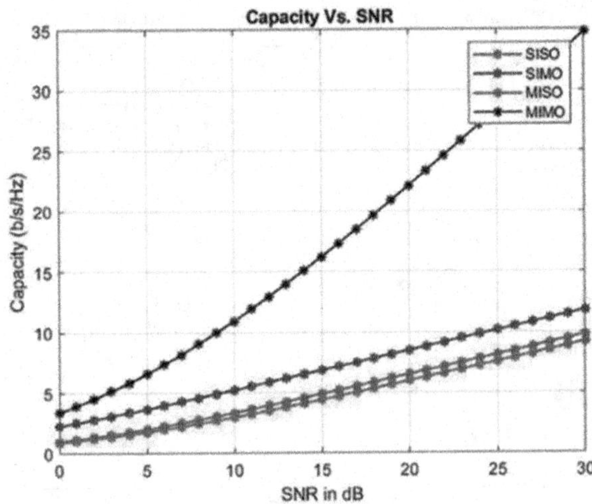

Fig. 61.9 Comparison of SISO, SIMO, MISO and MIMO based on capacity

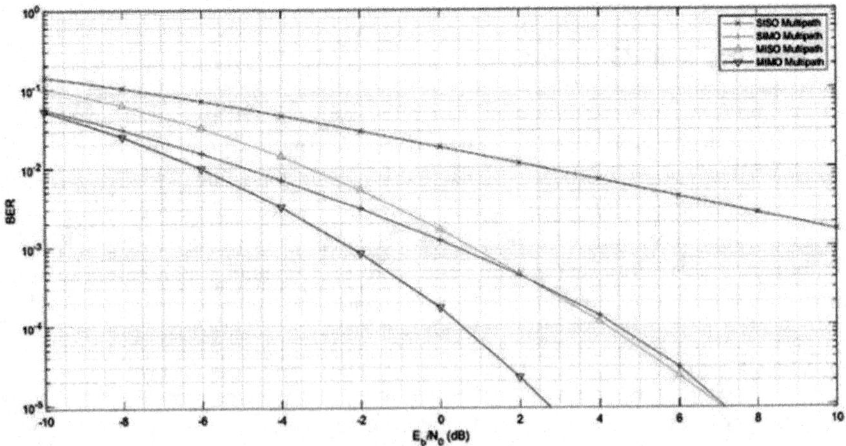

Fig. 61.10 Comparison of SISO, SIMO, MISO, and MIMO for BER

Rician fading is more appropriate for rural areas with less obstruction. Notably, our observations regarding the Nakagami fading model underscore the correlation between higher 'm' values and improved signal transmission quality.

In essence, our research offers a comprehensive understanding of multi-antenna systems and diverse fading channel models, providing insights to significantly enhance wireless communication across various deployment scenarios.

References

1. Belattar, Mounir, and Mohamed Lashab. (2018). "Performance of MIMO Systems over Rayleigh Channels." networks (4G networks and more) 3: 4.
2. Alkaf, Omar, Raed M. Shubair, and Khaled Mubarak. (IEEE, 2012.) "Improved performance of MIMO antenna systems for various fading channels." 2012 International Conference on Innovations in Information Technology (IIT).
3. Liu, Jun.(IEEE, 2016.) "Wireless multipath fading channels modeling and simulation based on Sum-ofSinusoids." 2016 First IEEE International Conference on Computer Communication and the Internet (ICCCI).
4. Mandalay, M. I. I. T. (2018): "BER performance analysis of MIMO-OFDM over the wireless channel." International Journal of Pure and Applied Mathematics 118.5: 195–206.
5. Shah CR. (2017) Performance and comparative analysis of SISO, SIMO, MISO, and MIMO. Int. J. Wirel. Commun. Simul ;9(1):1–4
6. Munshi, Ami, and Srija Unnikrishnan.(2017) "Design Simulation and Evaluation of SISO/MISO/MIMO OFDM Systems." IJLTEMAS 6.VIIIS : 63–66.
7. Varshney, Neeraj, and Aditya K. Jagannatham. (IEEE, 2016.) "Performance analysis of MIMO-OSTBC based selective DF cooperative wireless system with node mobility and channel estimation errors." 2016 twenty second national conference on communication (NCC).
8. Kaushik, D., Rajkhowa, A., Gogoi, P., & Saikia, B. Performance Analysis of MIMO-LTE for MQAM over Fading Channels. IOSR Journal of Electronics and Communication Engineering (IOSRJECE), e-ISSN, 2278–2834.
9. Khan, Zuhaib Ashfaq, Imran Khan, and Nandana Rajatheva. (IEEE, 2010.) "Performance analysis of MIMO based multi-user cooperation diversity system using hybrid FDMA-TDMA technique." ECTI-CON2010: The 2010 ECTI International Conference on Electrical Engineering/Electronics, Computer, Telecommunications and Information Technology.
10. Munshi, A., & Unnikrishnan, S. (2017, December). Modeling and simulation of MIMO-OFDM systems with classical and Bayesian channel estimation. In 2017 International Conference on Advances in Computing, Communication and Control (ICAC3) (pp. 1–4). IEEE.
11. Bala, D., Waliullah, G. M., Hena, M. A., Abdullah, M. I., & Hossain, M. A. (2020). Study the Performance of Capacity for SISO, SIMO, MISO, and MIMO in Wireless Communication. Journal of Network and Information Security, 8(1&2), 01–06.
12. H. -G. Yeh and V. R. Ramirez.(2007)" Implementation and Performance of a M-ary PSK and QAMOFDM System in a TMS320VC5416 Digital Signal Processor," 2007 Second International Conference on Digital Telecommunications (ICDT'07), San Jose, CA, USA, 2007, pp. 19–19, doi: 10.1109/ICDT.2007.20.
13. Daljeet Singh, Hem Dutt Joshi, (2018.) "Generalized MGF based analysis of line-of-sight plus scatter fading model and its applications to MIMO-OFDM systems", AEU - International Journal of Electronics and Communications, Volume 91, 2018.

Note: All the figures and table in this chapter were compiled by the authors.

Printed in the United States
by Baker & Taylor Publisher Services